Cell to Cell Signalling:
From Experiments to Theoretical Models

These papers were first presented at the NATO Advanced Research Workshop 'Theoretical Models for Cell to Cell Signalling' held in Knokke-Zoute, Belgium, during September 1988. The Workshop was further supported by the Commission of the European Communities.

Cell to Cell Signalling: From Experiments to Theoretical Models

Edited by

A. GOLDBETER

Faculté des Sciences, Université Libre de Bruxelles, Brussels, Belgium

ACADEMIC PRESS
Harcourt Brace Jovanovich, Publishers
London   San Diego   New York   Berkeley   Boston
Sydney   Tokyo   Toronto

This book is printed on acid-free paper. ∞

ACADEMIC PRESS LIMITED
24–28 Oval Road
London NW1 7DX

United States Edition published by
ACADEMIC PRESS, INC.
San Diego, CA 92101

Copyright © 1989 by
ACADEMIC PRESS LIMITED

British Library Cataloguing in Publication Data
Cell to cell signalling.
1. Organisms. Cells. Interactions
574.87'6

ISBN 0–12–287960–0

Typeset by Latimer Trend & Company Ltd, Plymouth.
Printed in Great Britain by University Press, Cambridge.

Contents

List of contributors*

*Where a paper is by more than two authors, only the first is listed below.

ADAMS, W.B.: Department of Pharmacology, Biozentrum der Universität Basel, Klingelbergstrasse 70, CH-4056 Basel, Switzerland.

ALLESSIE, M.A.: Department of Physiology, Rijksuniversiteit Limburg, Biomedical Center, P.O. Box 616, 6200 MD Maastricht, The Netherlands.

ANDERSON, R.M.: Department of Pure and Applied Biology, Imperial College, London SW7 2BB, UK.

BENSON, J.A.: CIBA-GEIGY Ltd., Entomology Basic Research, R-1093.P.47, CH-4002 Basel, Switzerland.

BERRIDGE, M.J.: Department of Zoology, Cambridge University, Cambridge CB2 3EJ, UK.

BONNER, J.T.: Department of Biology, Princeton University, Princeton, NJ 08544, USA.

COOKE, J.: National Institute for Medical Research, MRC, Mill Hill, London NW7 1AA, UK.

CUTHBERTSON, K.S.R.: Dept. of Human Anatomy and Cell Biology, University of Liverpool, Liverpool L69 3BX, UK.

DE BOER, R.J.: Bioinformatics Group, University of Utrecht, Padualaan 8, 3584 Utrecht, The Netherlands.

DEMONGEOT, J.: Service d'Informatique Médicale, Faculté de Médecine, Université de Grenoble, Domaine de la Merci, F-38700 La Tronche, France.

DE REFFYE, Ph.: CIRAD, Laboratoire de Biomodélisation, Av. du Val de Montferrand, BP 5035, F-34032 Montpellier Cedex, France.

DEVREOTES, P.N.: Department of Biological Chemistry, Johns Hopkins University, 725 N. Wolfe Street, Baltimore, MD 21205, USA.

DUDAI, Y.: Department of Neurosciences, The Weizmann Institute of Science, 76100 Rehovot, Israel.

DUPONT, G.: Service de Chimie Physique, Université Libre de Bruxelles, Campus Plaine, C.P. 231, B-1050 Brussels, Belgium.

ERMENTROUT, G.B.: Department of Mathematics, University of Pittsburgh, Pittsburgh, PA 15260, USA.

FILICORI, M.: Department of Reproductive Medicine, University of Bologna, Via Masserenti 13, 40138 Bologna, Italy.

GOLDBERGER, A.L.: Harvard Medical School, Beth Israel Hospital, 330 Brookline Ave, Boston, Mass. 02215, USA.

GOLDBETER, A.: Service de Chimie Physique, Faculté des Sciences, Campus Plaine, CP 231, Université Libre de Bruxelles, B-1050 Brussels, Belgium.

GOODWIN, B.C.: Department of Biology, The Open University, Walton Hall, MK7 6AA Milton Keynes, UK.

GRILLNER, S.: The Nobel Institute for Neurophysiology, Karolinska Institutet, Box 60400, S-104 01 Stockholm, Sweden.

GUEVARA, M.R.: Department of Physiology, McGill University, 3655 Drummond Street, Montreal, Quebec H3G 146, Canada.

GUNDERSEN, R.E.: Department of Biological Chemistry, Johns Hopkins University, 725 N. Wolfe Street, Baltimore, MD 21205, USA.

HESS, B.: Max-Planck-Institut für Ernährungsphysiologie, 201 Rheinlanddamm, 46 Dortmund, FRG.

HINDMARSH, J.: Department of Applied Mathematics, University College, Cardiff CF1 1XL, UK.

HUNDING, A.: Kemisk Laboratorium III, H.C. Ørsted Institutet, Universitetsparken 5, 2100 Kobenhavn Ø, Denmark.

KAUFFMAN, S.A.: School of Medicine, Dept. of Biochemistry and Biophysics. University of Pennsylvania, Philadelphia, Pa. 19104-6059, USA.

KAUFMAN, M.: Service de Chimie Physique, Université Libre de Bruxelles, Campus Plaine, C.P. 231, B-1050 Brussels, Belgium.

KEENER, J.P.: Department of Mathematics, University of Utah, Salt Lake City, Utah 84112, USA.

KNOBIL, E.: Laboratory for Neuroendocrinology, The University of Texas Health Science Center at Houston, Houston, Texas 77225, USA.

KOPELL, N.: Boston University, Department of Mathematics, Boston, Mass. 02215, USA.

LACKER, H.M.:   Department of Biomathematical Sciences, Mount Sinai Medical School, Box 1023, Gustave Levy Place, New York, NY 10029, USA.

LEFEVER, R.:   Service de Chimie Physique, Faculté des Sciences, Campus Plaine, CP 231, Université Libre de Bruxelles, B-1050 Brussels, Belgium.

LI, Y.X.:   Service de Chimie Physique, Faculté des Sciences, Université Libre de Bruxelles, Campus Plaine, C.P. 231, B-1050 Brussels, Belgium.

LLINAS, R.:   Dept. of Physiology and Biophysics, New York University Medical Center, 550 First Avenue, New York, NY 10016, USA.

MAINI, P.K.:   Department of Mathematics, University of Utah, Salt Lake City, Utah 84112, USA.

MAY, R.M.:   Department of Zoology, University of Oxford, Oxford OX1 3PS, UK.

MEINHARDT, H.:   Max-Planck-Institut Für Entwicklungsbiologie, Spemanstrasse 35/IV, D-7400 Tübingen I, FRG.

MOORE, H.-P.H.:   Department of Biology, University of California, Berkeley, Ca. 94720, USA.

MÜLLER, S.C.:   Max-Planck-Institut für Ernährungsphysiologie, 201 Rheinlanddamm, 46 Dortmund, FRG.

MURRAY, J.D.:   Department of Applied Mathematics FS-20, University of Washington, Seattle, Wa. 98195, USA.

NANJUNDIAH, V.:   Departments of Microbiology and Cell Biology, Centre for Theoretical Studies, Indian Institute of Science, Bangalore 560012, India.

OSTER, G.F.:   Departments of Entomology and Biophysics, Wellman Hall, University of California, Berkeley, CA 94720, USA.

PARNAS, H.:   Department of Neurobiology, The Hebrew University, Jerusalem, Israel.

PARNAS, I.:   Department of Neurobiology, The Hebrew University, Jerusalem, Israel.

PERELSON, A.S.:   T-10 Division, Mail Stop 710, Los Alamos Natl. Labs., Los Alamos, NM 87545, USA.

RIGNEY, D.R.:   Harvard Medical School, Beth Israel Hospital, 330 Brookline Ave, Boston, Mass. 02215, USA.

RINZEL, J.:   Mathematical Research Branch, NIADDK, National Institutes of Health, Building 31, Room 4B-54, Bethesda, MD 20892, USA.

ROSE, R.M.:   Department of Physiology, University College, Cardiff CF1 1XL, UK.

SEGEL, L.A.:   Department of Applied Mathematics, The Weizmann Institute of Science, 76100 Rehovot, Israel.

SHERMAN, A.:   Mathematical Research Branch, NIADDK, National Institutes of Health, Building 31, Bethesda, MD 20892, USA.

THALABARD, J.-C.:   Unité 292, INSERM, Paris, France, presently at Laboratory for Neuroendocrinology, The University of Texas Health Science Center at Houston, Houston, Texas 77225, USA.

TYSON, J.J.:   Department of Biology, Virginia Polytechnic Institute and State University, Blacksburg, Virginia 24061, USA.

URBAIN, J.:   Service d'Immunologie, Département de Biologie Moléculaire, Université Libre de Bruxelles, Rhode St-Genèse, Belgium.

VAN CAPELLE, F.J.L.:   Academisch Ziekenhuis, Universiteit Amsterdam, Department of Cardiology, Meibergdreef 9, 1105 AZ Amsterdam, The Netherlands.

WAGNER, T.O.F.:   Medizinische Hochschule Hannover, Abt. Klinische Endokrinologie, Konstanty-Gutschow-Str. 8, D-3000 Hannover 61, FRG.

WINFREE, A.T.:   Department of Ecology and Evolutionary Biology, University of Arizona, Tucson, AZ 85721, USA.

WOLPERT, L.:   Anatomy and Developmental Biology, University College and Middlesex School of Medicine, Windeyer Building, Cleveland Street, London W1P 6DB, UK.

WURSTER, B.:   Fak. Biologie, Universität Konstanz, Postf. 5560, D-7750 Konstanz, FRG.

Preface

Cell to cell signals govern the development of multicellular organisms and most of their functions. Given the complexity of dynamic phenomena involved in intercellular communication, it is often advantageous to complement their experimental study by a theoretical approach. The aim of this book is to present an up-to-date account of the contribution and usefulness of theoretical models in the main fields of cell to cell signalling. The peculiarity of this volume, reflected in its title, is to link experimental and theoretical work. This link is emphasized in all 44 contributions written by leading experts in the fields. Moreover, the models considered are primarily discussed with respect to their biological rather than mathematical interest. Although far from being exhaustive, the volume is divided into seven parts which cover most domains where cell to cell signals play a prominent role. The scope of the project explains the unusual size of this book.

It is only fitting that the first part be devoted to nerve cells and neural networks. After all, the fertility of theoretical models in neurobiology has long been recognized since the work of Hodgkin and Huxley on the nerve impulse: the series of papers that they published in 1952 dealt with experimental aspects as well as with the mathematical modelling of the action potential. This seminal work still represents a prototypic combination between the experimental and theoretical approaches in biology.

As a follow-up of these classical studies devoted to excitability of the squid axon, papers in Part 1 start at the cellular level by addressing more complex modes of behaviour such as bursting or multiple modes of neuronal oscillations. Thalamic neurons and the *R15* burster cell of *Aplysia* are dealt with in detail. Another topic is that of neurotransmitter release. At the multicellular level, several aspects are considered. The first is how the coupling between secretory cells modifies their dynamic behaviour. The second is how neural networks behave as central pattern generators that control, for example, locomotion in invertebrates; the latter topic is covered from an experimental and theoretical point of view. Finally, the process of learning is discussed at the molecular level and in terms of neural networks.

Part 2 is devoted to morphogenesis and development. Here also, a classical paper published in 1952 by A.M. Turing set the theoretical foundations for understanding the physicochemical bases of pattern formation. Current theories of morphogenesis are presented in relation to experimental observations. Included is a comparative discussion of reaction–diffusion and mechanochemical models, together with applications of the two types of structure-generating mechanisms. Particular attention is given to pattern formation in the amphibian embryo and in *Drosophila*. The closing paper deals with the modelling of plant growth.

Immunology is one of the fastest growing fields for the application of theoretical ideas in biology. In Part 3, some of the most salient examples of immunological modelling are presented. These pertain to the dynamics of immune networks based on clonal or idiotypic regulation, and to the immune control of tumour growth. The last paper investigates the complex immune dynamics that follows infection with HIV.

Hormonal signalling is one of the primary modes of intercellular communication. Rather than covering all types of hormonal control, Part 4 focuses on the signals which govern the reproductive function in mammals. Experimental, clinical and theoretical papers address two main aspects, namely, selection of the dominant follicle in the ovarian cycle, and the effectiveness of pulsatile hormone secretion which allows for encoding of periodic signals in terms of their frequency. The case of the gonadotropin-releasing hormone secreted by the hypothalamus represents a prototype for the other hormones or growth factors which exert their physiological effect only when delivered in a pulsatory manner.

An increasing number of hormones and neurotransmitters are found to act on target cells by triggering intracellular calcium oscillations. The experimental evidence and the theoretical bases for this phenomenon are discussed in Part 5.

Another system in which intercellular communication occurs through pulsatile signals is the slime mould *Dictyostelium discoideum* to which Part 6 is devoted. The temporal aspects of signalling are considered in papers dealing with the molecular basis of cyclic AMP oscillations which control aggregation of the amoebae after starvation. The mechanism of this wavelike aggregation is also discussed and related to experimental examples of spatiotemporal order in chemical systems.

The last part of the volume deals with signal propagation in the heart. Here also, experimental and theoretical aspects are closely intertwined. Among the topics discussed are the relationship between chaos and cardiac dynamics, the response of periodically stimulated cardiac cells, and the role of anisotropic impulse propagation in ventricular tachycardia and in the process of re-entry.

Is there a common theme that emerges from the presentation of experimental and theoretical aspects of intercellular communication in such different fields of cell biology? Looking for general, unifying principles can be misleading and somewhat ludicrous, since the usefulness of models is directly related to their predictive power; the latter, in turn, requires them to be highly specific rather than vague. However, despite the diversity of the phenomena considered, some common themes do appear: excitable and oscillatory behaviour occurs for nerve as well as cardiac cells; pulsatile signalling is a neuronal property which is also found for an increasing number of hormones and for *Dictyostelium* amoebae. Encoding of an external signal in terms of its frequency is observed in all these instances of intercellular communication and might also underlie signal transduction based on intracellular calcium oscillations. On the other hand, wavelike phenomena are observed in *Dictyostelium* as in the myocardium. Wave propagation in excitable media has become a subject of choice for modelling studies in biology. It is in view of this long-standing fascination exerted by waves that the original drawing by Royer (which owes its spiral jolt to a suggestion from Art Winfree) is included at the end of the book.

Being closely related to experimental data, the models discussed in this volume retain the specificities of their respective fields. Most studies nevertheless deal with dynamic events which often lead to instabilities and nonequilibrium self-organization. Therefore, bifurcations associated with oscillations or with transitions between multiple steady states are recurrent themes, from neurobiology to immunology and hormonal regulation, while spatial pattern formation arises in morphogenesis as in cardiac physiology or slime mould aggregation.

This volume originates from a NATO Workshop that took place in September 1988 in Knokke-Zoute (Belgium). The suggestion to organize the Workshop came from the panel of the special NATO programme 'Cell to Cell Signals in Plants and Animals'. I would like to express my gratitude to the members of this panel, particularly Professor Kees Libbenga, for their invitation to organize this meeting, and to Dr Alain Jubier, in charge of this programme at the NATO Division for Scientific Affairs, for his constant help. Thanks are also due to Professor Paolo Fasella and the Commission of the European Communities for their complementary support. Further help came from the Belgian National Fund for Scientific Research (FNRS) and from the Belgian Ministry for Education, as well as from the Science Policy Programming Unit of the Prime Minister's Office (SPPS).

Finally, I am deeply grateful to the two co-organizers of this Workshop, John Rinzel, from the Mathematical Research Branch of the NIH (Bethesda, USA), and Lee A. Segel, from the Weizmann Institute of Science (Rehovot, Israel) for their most precious support and help in selecting themes and contributors.

In a lithograph entitled 'Ophtalmologie', the Belgian painter Pierre Alechinsky represented a human face probing the world with two pairs of eyes. This figure could serve as metaphor to illustrate the purport of this book, which is to demonstrate the added insight that the combined approaches of experiment and theory provide into the dynamics of living systems and, in particular, into the various modes of intercellular communication.

Albert Goldbeter
Brussels

Ophtalmologie, lithograph (1972) by P. Alechinsky.

The role of the intrinsic electrophysiological properties of central neurones in oscillation and resonance

RODOLFO R. LLINÁS

Department of Physiology and Biophysics, New York University School of Medicine, New York, USA

INTRODUCTION

In attempting to assess the importance of the intrinsic electrophysiological properties of central neurones, one should perhaps begin by reviewing the 'neurone doctrine' as these two concepts are intimately related. The idea that the central nervous system is, like other organs, a collection of individual and separable cellular elements was proposed, amongst others, by Waldeyer as the 'neurone doctrine' at the end of the nineteenth century (1891). However, the establishment of this hypothesis on firm footing, as well as the realization of its momentous significance, really belongs to Ramon y Cajal. He pointed out that the nervous system is fundamentally an organized set of individual elements separated physically from each other and having as the mechanism for their interaction specified contacts between cells (1888, 1934). These contacts were named 'synapses' by Sir Charles Sherrington. Opposing this view were scientists such as Max Schultze (1861) and Camilo Golgi (1898) who considered that the nervous system was composed of a complex continuous network of protoplasmic bridges between cells. This randomly organized protoplasmic network in which nuclei were imbedded, was viewed by him as forming an enormous structure, referred to as the 'reticulum'. It is quite clear that attempting to understand the nervous system in the absence of the neurone doctrine would have been impossible as it represents the single most fundamental concept in modern neuroscience. Indeed, the ionic basis of electrophysiology, neuronal integration, synaptic transmission and the modulation of genome expression by hormones and transmitters are but corollaries to the existence of the nerve cell.

Cell to Cell Signalling: From
Experiments to Theoretical Models
ISBN 0–12–287960–0

3

On the other hand, while the study of single cell physiology has emphasized that neurones are independent anatomical entities, it has not always been obvious that neurones are to a certain extent *independent* functional entities. At present, many neuroscientists still believe that central neurones are brought to electrical activity or to quiescence by synaptic input exclusively. Indeed, central neurones are thought to serve as mere *relay elements* in a process which allows the conductance of impulses along the different pathways in a rapid race to some portion of the brain that 'puts it all together'. This view of the organization of the nervous system is, at best, incomplete. Rather, modern electrophysiology suggests that central neurones are endowed with voltage-dependent electrophysiological properties that allow them to have truly intrinsic electrical properties. Examples of such interesting electrical properties will be given below when considering the activity in the inferior olivary or thalamic neurones as studied *in vitro*.

The recognition of the intrinsic electrophysiological activity of neurones *de facto* implies that the overall activity in the nervous system is most probably governed by the interplay of synaptic input and intrinsic membrane properties. This being the case, we come to the conclusion that intrinsic oscillation, and resonance (the ability of neurones to respond preferentially to given frequencies of stimulation), must play an important role in the organization of nervous system function. This is in contrast to the view that most activity originates from the periphery via sensory systems, or from the corollary discharge of motor output.

NEURONAL OSCILLATION IN MAMMALIAN CNS

One of the truly remarkable findings relating to the electrical activity of the brain was the discovery by Hans Berger (1929) of rhythmic electrical activities on the surface of the cranium. Equally remarkable was the discovery that different rhythms related to states of consciousness. Today, 60 years later, we continue to be amazed by the fact that a structure as complex and as massive, in the sense of the number of neuronal elements (4×10^{10}), can recruit sufficiently coherent neuronal activity to generate such macroscopic oscillatory states. In fact, the actual mechanism for such temporal cohesiveness over such large cellular populations continues to escape us.

For the most part, in the 1960s and 1970s, the accepted site of 'origin' for the most prominent oscillatory rhythms displayed by the brain, the alpha rhythm, was the thalamus. The oscillatory behaviour was considered to be a property of the neuronal circuits, where the system (the thalamo-cortical network) produced an early excitatory postsynaptic potential (EPSP) followed by synaptic inhibition (IPSP). These EPSP–IPSP sequences were assumed to occur in a cyclic manner so as to generate the previously mentioned rhythm. The minimal neuronal machinery required was thought to be a two-neuronal chain organized in a loop. Thus, a thalamic afferent to the cortex would excite, by way of axon collaterals, an inhibitory interneurone which would return axons to the cell of origin in the thalamus. This primitive circuit could in principle generate the oscillatory behaviour of the cortex by way of negative feedback implemented by the interneurone (Anderson and Sears, 1964).

The *in vitro* study of the electrical activity in brain slices, in particular, the electrophysiology of the inferior olive (IO) and thalamus, has provided us with further information regarding possible mechanisms for neuronal oscillations. The most significant mechanism in the generation of the oscillatory properties of the CNS is the intrinsic ionic properties of individual neurones (Llinás and Yarom, 1981b, 1986; Jahnsen and Llinás, 1984b; Llinás, 1988; Llinás and Mühlethaler, 1988; Steriade and Llinás, 1988). This represents a clear shift of emphasis from the properties of circuits to the properties of single neurones. Undoubtedly feedback circuits will always play a role in the generation and support of neuronal oscillations, if only by aiding the synchronizations of single oscillators into sets of coupled oscillators, which in turn give the macroscopic field potentials observable at the surface of the cranium and, in the special case of the IO, electrotonic junctions between these cells help to entrain large numbers of cells into rhythmic firing.

REBOUND EXCITATION AND OSCILLATION OF INFERIOR OLIVARY CELLS

In vitro experiments using brainstem slices (Llinás and Yarom, 1981a,b; 1986) have demonstrated that IO neurones have a set of ionic conductances that are activated in such a way as to give these cells intrinsic oscillatory properties. The firing of IO cells is characterized by an initial fast-rising action potential (a somatic sodium spike), which is prolonged to 10–15 ms by an after-depolarization (Fig. 1A). The abrupt, long-lasting after-hyperpolarization following the plateau after-depolarization totally silences the spike-generating activity. This hyperpolarization is typically terminated by a sharp, active rebound response, which arises when the membrane potential is negative to the resting level. When this rebound reaches threshold for an action potential, the cell is again activated. In this way the cell will fire at a frequency determined largely by the characteristics of the after-hyperpolarization. The oscillatory tendency of the IO neurone shown in Fig. 1 was enhanced by addition of harmaline to the bath (see later discussion).

Experiments in which a calcium-free Ringer solution or calcium channel blockers were used, or in which barium was substituted for calcium, have demonstrated that the after-depolarization–hyperpolarization sequence following the initial spike is due to the sequential activation of a calcium and a potassium conductance (Llinás and Yarom, 1981a). Analysis of extracellular

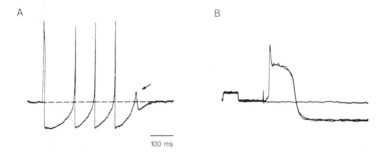

Fig. 1. Oscillatory properties of IO neurones *in vitro*. (A) Initial spike is followed by a large after-hyperpolarization itself followed by three full action potentials and a small rebound response indicated by an arrow. Note that the action potentials arise from a level negative to the resting potential. (B) Detail of the activation of the IO neurones. The initial component is a sodium spike. The plateau that follows is a dendritic calcium spike which activates the long after-hyperpolarization due to the activation of a calcium-dependent potassium conductance. (Modified from Llinás and Yarom, 1981a.)

field potentials showed that the after-depolarization is produced by activation of a voltage-dependent calcium channel located in the dendrite (Llinás and Yarom, 1981b). This conductance is very similar to that first demonstrated in Purkinje cells (Llinás and Hess, 1976; Llinás and Sugimori, 1980a,b) and is present in the dendrites of other central neurones (Llinás, 1988).

As shown in Fig. 1B, the fast sodium-dependent action potential is followed by a prolonged after-depolarization (adp) generated by a dendritic calcium spike. This adp is itself followed by a powerful after-hyperpolarization as seen at a faster sweep speed in Fig. 1A. The membrane conductance during the hyperpolarization is large enough to shunt even powerful synaptic input. During this period the cell is virtually 'clamped' at the potassium equilibrium potential. This potassium conductance is calcium-dependent and similar to that initially described in invertebrate neurones (Meech and Standen, 1975). Indeed, substituting barium for calcium completely abolished the after-hyperpolarization (Llinás and Yarom, 1981b; Eckert and Lux, 1976).

The amount of calcium entering the dendrites during the after-depolarization modulates the duration of the after-hyperpolarization. Thus, if the dendritic calcium action potential is smaller, the duration of the after-hyperpolarization is decreased, while prolonged dendritic spikes may generate after-hyperpolarizations as long as 250 ms. This may be seen in Fig. 1B where the first spike has the broadest after-depolarization (generated by a calcium-dependent dendritic spike) followed by a prolonged after-hyperpolarization. This point is central to the oscillatory properties of IO cells, as it indicates that calcium entry determines the cycle time of the oscillator.

Perhaps the most important new finding of the *in vitro* studies with respect to neuronal oscillation, was the abruptness with whicn the membrane potential returned to the baseline at the end of the after-hyperpolarization, even overshooting the initial resting potential. This rebound response (shown in Fig. 1A, arrows) is due to the activation of a somatic calcium-dependent action potential and results from a second voltage-dependent calcium conductance, which is unusual in that it is inactive at resting membrane potential (-65 mV). Membrane hyperpolarization deinactivates this conductance, and thus, as the membrane potential returns to baseline, a 'low-threshold' calcium spike is generated (Llinás and Yarom, 1981a,b). The rebound potential can be modulated by small changes in the resting membrane potential such that a full sodium spike, which in turn can set forth the whole sequence of events once again, is activated. Calcium-dependent spikes may be generated if the cell is hyperpolarized.

PHARMACOLOGICAL ASPECTS OF THE REBOUND EXCITATION

Administration of harmaline, an alkaloid of *Pegamus harmala*, elicits a very regular tremor in higher vertebrates. This tremor, which has been known since the turn of the century (Neuner and Tappeiner, 1894), is now known to result from the effect of harmaline on the IO as described above. It produces a 10–12 Hz tremor in intact and decerebrated cats (Villablanca and Riobo, 1970; Lamarre *et al.*, 1971; de Montigny and Lamarre, 1973), as well as in other mammals. Indeed, intracellular recording from Purkinje cells demonstrated that harmaline tremor was accompanied by all-or-none Purkinje cell EPSPs (de Montigny and Lamarre, 1973; Llinás and Volkind, 1973). The reversal of these large synaptic potentials (Llinás and Volkind, 1973) demonstrated that this activation of Purkinje cells was due to the activation of the inferior olivary cell, which terminates on the Purkinje cell as climbing fibres (Eccles *et al.*, 1966).

In vitro experiments using brainstem slices have shown that the application of harmaline hyperpolarizes IO neurones and produces an exaggerated rebound response (Yarom and Llinás, 1981; Llinás, 1988). Figure 2 illustrates the action of harmaline on the firing properties of an inferior olivary cell recorded *in vitro* from a brainstem slice (Llinás and Yarom, 1986). A train of depolarizing current pulses were delivered before and during addition of 10^{-5} M harmaline to the bath. Note the slow membrane hyperpolarization. Figure 2B shows selected traces from the series in Fig. 2A at an expanded time base. Before harmaline addition, the current pulse elicited only a subthreshold response (Fig. 2B, trace 1). When the membrane was hyperpolarized by a few millivolts (trace 2) a slowly rising response is seen, with further hyperpolarization this increases in amplitude (trace 3) until

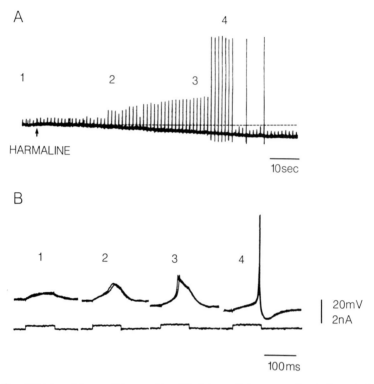

Fig. 2. Effects of harmaline on the IO. Intracellular recordings from inferior olivary neurones *in vitro*. Harmaline is added to the bath at the arrow. Note that the membrane becomes hyperpolarized and subthreshold current injections produce low-threshold calcium spikes of increasing amplitude until full action potentials are generated. Examples of the different levels of calcium activation are shown in B. In 1–4, points in A are illustrated. 1 is prior to harmaline, 2 is after-hyperpolarized, 3 is a full calcium spike, 4 is a sodium spike riding on a calcium spike. (Modified from Llinás and Yarom, 1986.)

it is large enough to bring the membrane to threshold for a fast action potential (trace 4). The cell continues to fire rhythmically until the pulse once again becomes subthreshold. The increase in the Ca spike in the presence of harmaline is due in part to the ability of this drug to hyperpolarize the cell, in addition there is the direct action of harmaline on the rebound Ca conductance.

OSCILLATORY NEURONAL INTERACTIONS WITHOUT ACTION POTENTIALS

In the slice preparation treated with harmaline, and occasionally in the absence of harmaline, the membrane potential of IO neurones tends to

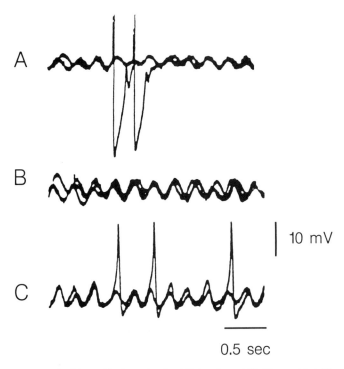

Fig. 3. Subthreshold oscillations in the IO *in vitro*. (A) Sinusoidal-like oscillations producing full action potentials when the cell is artificially depolarized from rest. (B) Membrane oscillations seen at rest. (C) Low-threshold spikes observed riding on the rising phase of the oscillations when the membrane is artificially hyperpolarized. (Modified from Llinás and Yarom, 1986.)

oscillate, proceeding in an almost perfect sinusoidal waveform. An example of such spontaneous oscillation is shown in Fig. 3. Spontaneous rhythmic membrane oscillations may be generated by inferior olivary neurones *in vitro* as shown in Fig. 3. The oscillations in Fig. 3A were recorded at the resting potential, several sweeps are superimposed to demonstrate the regularity of the oscillations. Two dendritic calcium action potentials (truncated) were evoked. As the membrane was hyperpolarized (by 15 mV in Fig. 3B, and 24 mV in Fig. 3C), somatic action potentials were generated (note that the frequency of the oscillation was not affected by the change in membrane potential). Both types of action potentials occurred during the depolarizing phase of the oscillation. This sinusoidal modulation of membrane potential, which was not related to the generation of action potentials in IO neurones, was observed throughout the IO nucleus. The frequency (5 cycles per second) was the same for neighbouring cells, and there is total coherence in the oscillation of different neurones. Activation of a stimulating electrode placed in the centre of the inferior olivary nucleus affected a large number of cells. As shown in Fig. 4A, extracellular stimulation elicited an action potential for each stimulus (delivered at 5 Hz in this experiment). At the end of the stimuli the membrane was too hyperpolarized to support the spontaneous sub-threshold oscillations; this is shown at an expanded time base in Fig. 4B. The oscillations returned as the membrane reached the resting potential. If the impaled cell was directly depolarized at the same frequency used for the extracellular stimulation, the cell fired (Fig. 4C), but the oscillatory rhythm of the ensemble was not modified.

That harmaline acts to enhance the spontaneous oscillations is shown in Fig. 5. Before addition of harmaline to the bathing solution the membrane potential of the cell oscillated rhythmically (Fig. 5B, trace 1); after addition of the drug the membrane potential increased and the somatic calcium spikes

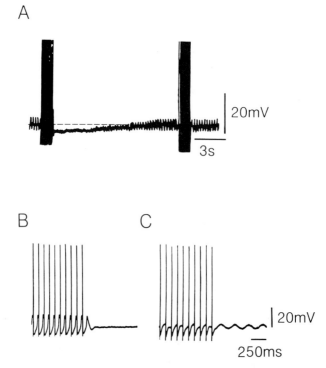

Fig. 4.   Blockage of subthreshold IO oscillations by electrical stimulation. (A) Intracellular recording from an IO neurone demonstrating subthreshold oscilla-tions: 200 ms after the beginning of the trace, the slice was electrically stimulated using an extracellular macroscopic bipolar electrode. The traces in A are shown at a faster sweep speed in B. The stimulus elicited firing of the recorded cell and silenced the oscillations for more than 3 s (A). After this silence the oscillatory rhythm organized itself to a full amplitude. A second electrical stimulus, through the intracellular electrode (i.e. the stimulus was restricted to the impaled cell) fired the cell but did not reset the membrane oscillations. (Modified from Llinás and Yarom, 1986.)

A

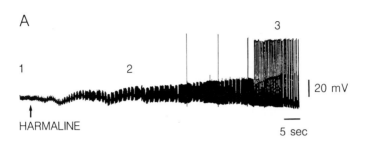

1   2   3

20 mV

↑
HARMALINE   ——
                                            5 sec

B

1   2   3

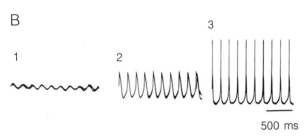

                              ——
                            500 ms

Fig. 5.   Effect of harmaline on spontaneous IO oscillation *in vitro*. Addition of harmaline to the bathing solution is observed intracellularly in an inferior olivary cell. Harmaline produces an increased oscillation and finally, continuous firing. Details are illustrated in B. The numerals 1, 2, 3 refer to the points in A where the records were obtained. (Modified from Llinás and Yarom, 1986.)

were seen with each oscillation (Fig. 5B, trace 2). These increased in amplitude and reached threshold for spike generation (Fig. 5B, trace 3). In each record in Fig. 5B three traces have been superimposed to show the regularity of the response as well as the progressive effect of harmaline.

The frequency of oscillations has been shown to vary in different preparations and under different pharmacological conditions. This different frequency in the oscillatory rhythm probably reflects the metabolic state of the given preparation. Regardless of this variability, the synchronicity of the ensemble emphasizes the importance of the electronic coupling between cells and suggests the presence of an underlying chemical oscillatory mechanism (Neu, 1980) which modulates the membrane conductance.

Since this oscillatory firing reflects intrinsic conductance changes in each inferior olivary neurone, its basic frequency cannot be easily modified. That is, individual IO cells are limit cycle oscillators. Indeed, normally, IO cells cannot be made to discharge with frequencies much higher than 15 cycles per second but their axons generate a short burst of repetitive firing for the peak of each cycle. The interval between these bursts is determined by the powerful after-hyperpolarizations separating the calcium plateaus.

THE THALAMUS: ITS OSCILLATORY PROPERTIES

A second example of a neurone with intrinsic properties that enables it to fire rhythmically has been described *in vitro* in the guinea-pig thalamus (Llinás and Jahnsen, 1982; Jahnsen and Llinás, 1984a,b). This oscillatory behaviour was found in all parts of the thalamus, including the medial and lateral geniculate nuclei. It differs from that in the inferior olive in that thalamic cells fire at one of two preferred frequencies, near 6 Hz or 10 Hz. The ionic mechanisms underlying thalamic neurone oscillatory behaviour are similar to those encountered in IO cells and also display an early potassium

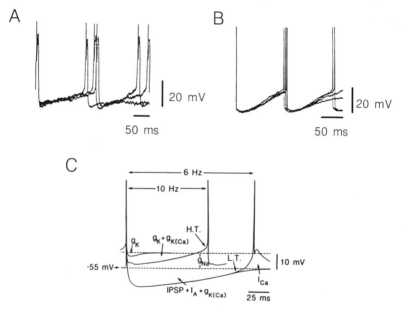

Fig. 6. Oscillatory properties of thalamic cells. (A) 6 Hz oscillation produced as a rebound following a short hyperpolarizing pulse (not shown). The oscillation was enhanced by the presence of 4-AP in the bath. (B) 9–10 Hz oscillation elicited by membrane depolarization. (C) Diagram showing the oscillatory mechanisms in the thalamus for the 6 Hz and 9–10 Hz oscillations. In the generation of the 10 Hz oscillation a fast Na$^+$ spike is followed by an after-hyperpolarization generated by voltage-sensitive (g_K) and Ca^{2+}-sensitive ($g_{K(Ca)}$) K$^+$ conductances. The membrane potential is brought back to the threshold for the fast spike by activation of the slow N$^+$ conductance (g_{Na}). The 6 Hz oscillations can occur by facilitating rebound excitation or by repeated hyperpolarizing potentials simulating IPSPs. The hyperpolarization deinactivates the transient K$^+$ current, I_A, which increases the duration of the after-hyperpolarization. This after-hyperpolarization deinactivates the low-threshold Ca^{2+} conductance (g_{Ca}) generating a rebound low-threshold spike which triggers the process once again. L.T. and H.T. are the thresholds for low-threshold and fast spikes, respectively. (Modified from Jahnsen and Llinás, 1984b.)

conductance (A current) similar to that described in invertebrate neurones (Connor and Stevens, 1971) and a non-inactivating sodium conductance similar to that seen in Purkinje cells (Llinás and Sugimori, 1980a).

A strong TTX-insensitive rebound calcium spike can be generated at membrane potential levels negative to -60 mV when combined with a hyperpolarizing potassium conductance and an A current, thalamic cells generate oscillatory responses at frequencies near 6 Hz (Fig. 6A). When they are depolarized, the fast sodium-dependent spike is followed by an after-hyperpolarization, due to the increase in voltage- and calcium-activated potassium conductance. Activation of a slow sodium conductance returns the membrane to the resting potential (Fig. 6B). Tonically depolarized thalamic cells can also fire at high frequencies. This is due to the fact that the dendritic calcium conductance is not as powerful as that in the IO, and somatic firing may not activate the after-hyperpolarization–depolarization sequence. Thus their intrinsic properties allow thalamic neurones to display a versatility whereby they switch between tonic and phasic responses as shown in Fig. 6C. The point to be emphasized here is not the difference between these two groups of cells but rather that they both have intrinsic properties that give them a distinct 'point of view' and preferred firing characteristics.

ROLE OF THE CYCLIC ACTIVITY IN THE CNS

For the IO to exert an influence on motoneuronal pools along the neuroaxis, the oscillatory behaviour of single neurones must be synchronized to yield oscillations in an ensemble of cells. Such a mechanism may be subserved by the electrical coupling between IO neurones (Llinás et al., 1974; Llinás and Yarom, 1981a). This coupling is most probably related to the presence of gap junctions (Bennett and Goodenough, 1978) at the olivary glomeruli as well as directly between dendrites (Sotelo et al., 1974; King, 1976; Gwyn et al., 1977). The oscillatory behaviour of single cells would become synchronized through coupling such that the IO would generate a phasic modulation of the motoneurones in brainstem and spinal cord by way of vestibulo- and reticulospinal pathways (de Montigny and Lamarre, 1973; Llinás and Volkind, 1973). The main function of this oscillatory input would be to synchronize the activation of muscle groups throughout the body to generate organized motor responses. The ability of brain regions to recruit groups of motoneuronal pools in this way is essential to the generation of even the simplest coordinated movement, since sets of muscles must be activated in a specific temporal sequence. With respect to the thalamic oscillation, it seems evident that oscillatory activity at that nucleus is closely related to changes in states of awareness. Indeed, oscillatory spindling is an early sign of incipient sleep. Perhaps some of the most interesting avenues for further study relate to the mathematical analysis of oscillatory behaviour. In the case of thalamic cells two sets of mathematical approaches have been developed (Goldbeter and Moran, 1988; Rose and Hindmarsh, 1985) and are being refined in this volume.

SIGNIFICANCE OF INTRINSIC PROPERTIES OF NEURONES

The question to be considered next is that of distribution of oscillatory neurones. It seems evident that at least two cell types in the CNS are capable of generating intrinsic oscillatory activity with frequencies very close to those observed clinically, and experimentally in some motor tremors. Further, a link may exist between the IO specific nuclei and specific forms of tremor. In fact, the IO and associated nuclei may be directly related to the 8–10 Hz physiological tremor observed in higher vertebrates (Llinás, 1984), whereas phenomena such as the alpha rhythm seem clearly to be related to the thalamus and to the state of consciousness. Both cases have the common characteristic of cell activation and reactivation by way of a rebound excitatory phenomenon.

It is important to note that anodal break firing (postanodal exaltation) is also a type of rebound response and a general property of excitable tissues observable to varying degrees in most excitable elements from axons to central dendrites. At this juncture, then, an important point must be made: rebound excitation is a special example of a general phenomenon usually produced by voltage deinactivation of sodium channels following membrane hyperpolarization. At the normal resting membrane potential, a certain percentage of sodium channels are in the inactivated state, and hyperpolarizing the membrane can reincorporate them into the active channel pool. Inferior olive and thalamic cells are examples of a special case where a new ionic mechanism, deinactivation of a g_{Ca}, is preceded by a hyperpolarization that insures a maximal level of sodium-dependent electroresponsive 'readiness' during the rebound. Hyperpolarization of the membrane may be viewed metaphorically as the stretching of a bow, the rebound as the release of the arrow. Synchronization is then attained by the simultaneous release of arrows, and the interval is the time necessary to stretch the bow once again.

More fundamental, however, is the possibility that intrinsic rhythmicity may generate the necessary functional states required to represent, in intrinsic coordinates, external reality (cf. Pellonisz and Llinás, 1985). The fact that the massive oscillatory changes that are observed in brain function correspond well with such global states as being awake and attentive, being asleep or dreaming or hallucinating with open eyes, indicate that the oscillation and resonance are most probably the scaffolding system that allows the computational state that we know as consciousness.

REFERENCES

Anderson, P. and Sears, T. (1964) The role of inhibition in the phasing of spontaneous thalamo-cortical discharge. *J. Physiol.* **173**, 459–80.

Bennett, M. V. L. and Goodenough, D. A. (1978) Gap junctions, electrotonic coupling and intercellular communication. *Neurosci. Res. Prog. Bull.* **16** (3), 377–463.

Berger, H. (1929) Uber das elektrenkephalogramm des menschen. *Arch. Psychiat.* **87**, 527.

Connor, J. A. and Stevens, C. F. (1971) Inward and delayed outward membrane currents in isolated neural somata under voltage clamp. *J. Physiol., Lond.* **213**, 1–20.

Eccles, J. C., Llinás, R. and Sasaki, K. (1966) The excitatory synaptic action of climbing fibers on Purkinje cells of the cerebellum. *J. Physiol., Lond.* **182**, 268–96.

Eckert, R. and Lux, H. D. (1976) A voltage-sensitive persistent calcium conductance in neuronal somata of *Helix*. *J. Physiol., Lond.* **254**, 129–51.

Goldbeter, A. and Moran, F. (1988) Dynamics of a biochemical system with multiple oscillatory domains as a clue for multiple modes of neuronal oscillations. *Eur. Biophys. J.* **15**, 277–87.

Golgi, C. (1898) Sur la structure des cellules nerveuses. *Arch Ital. Biol.* **30**, 60–71.

Gwyn, D. G., Nicholson, G. P. and Flumerfelt, B. A. (1977) The inferior olivary nucleus in the rat: a light and electron microscopic study. *J. Comp. Neurol.* **174**, 489–520.

Jahnsen, H. and Llinás, R. (1984a) Electrophysiological properties of guinea pig thalamic neurones: An *in vitro* study. *J. Physiol., Lond.* **349**, 205–26.

Jahnsen, H. and Llinás, R. (1984b) Ionic basis for the electroresponsiveness and oscillatory properties of guinea-pig thalamic neurones *J. Physiol., Lond.* **349**, 227–47.

King, J. S. (1976) The synaptic cluster (glomerulus) in the inferior olivary nucleus. *J. Comp. Neurol.* **165**, 387–400.

Lamarre, Y., DeMontigny, C., Dumont, M. and Weiss, M. (1971) Harmaline-induced rhythmic activity of cerebellar and lower brain stem neurons. *Brain Res.* **32**, 246–50.

Llinás, R. R. (1984) Rebound excitation as the physiological basis for tremor: A biophysical study of the oscillatory properties of mammalian central neurons *in vitro*. In *Movement Disorders: Tremor* (ed. L. J. Findley and R. Capildeo), pp. 165–82. Macmillan, London.

Llinás, R. (1988) The intrinsic electrophysiological properties of mammalian neurons: A new insight into CNS function. *Science, N.Y.* **242**, 1654–64.

Llinás, R. and Hess, R. (1976) Tetrodotoxin-resistant dendritic spikes in avian Purkinje cells. *Proc. Natn. Acad. Sci. U.S.A.* **73**, 2520–3.

Llinás, R. and Jahnsen, H. (1982) Electrophysiology of mammalian thalamic neurons *in vitro*. *Nature* **297**, 406–408.

Llinás, R. and Mühlethaler, M. (1988a) An electrophysiological study of the *in vitro*, perfused brainstem-cerebellum of adult guinea pig. *J. Physiol., Lond.* **404**, 215–40.

Llinás, R. and Mühlethaler, M. (1988b) Electrophysiology of guinea-pig cerebellar nuclear cells in the *in vitro* brainstem-cerebellar preparation. *J. Physiol., Lond.* **404**, 241–58.

Llinás, R. and Sugimori, M. (1980a) Electrophysiological properties of *in vitro* Purkinje cells somata in mammalian cerebellar slices. *J. Physiol., Lond.* **305**, 171–195.

Llinás, R. and Sugimori, M. (1980b) Electrophysiological properties of *in vitro* Purkinje cells dendrites in mammalian cerebellar slices. *J. Physiol., Lond.* **305**, 197–213.

Llinás, R. and Volkind, R. A. (1973) The olivo-cerebellar system: functional properties as revealed by harmaline-induced tremor. *Expl Brain Res.* **18**, 69–87.

Llinás, R. and Yarom, Y. (1981a) Electrophysiology of mammalian inferior olivary neurones *in vitro*. Different types of voltage-dependent ionic conductances. *J. Physiol., Lond.* **315**, 549–67.

Llinás, R. and Yarom, Y. (1981b) Properties and distribution of ionic conductances generating electroresponsiveness of mammalian inferior olivary neurones *in vitro*. *J. Physiol., Lond.* **315**, 569–84.

Llinás, R. and Yarom, Y. (1986) Oscillatory properties of guinea-pig inferior olivary neurones and their pharmacological modulation: An *in vitro* study. *J. Physiol., Lond.* **376**, 163–82.

Llinás, R., Baker, R. and Sotelo, C. (1974) Electrotonic coupling between neurons in cat inferior olive. *J. Neurophysiol.* **37**, 560–71.

Meech, R. W. and Standen, N. B. (1975) Potassium activation in *Helix aspersa* neurones under voltage clamp: A component mediated by calcium influx. *J. Physiol.* **249**, 211–39.

Montigny, C. de and Lamarre, Y. (1973) Rhythmic activity induced by harmaline in the olivo-cerebellar-bulbar system of the cat. *Brain Res.* **53**, 81–95.

Neu, J. C. (1980) Large populations of coupled chemical oscillators. *SIAM J. Appl. Math.* **38**(2), 305–16.

Neuner, A. and Tappeiner, H. (1894) Uber bei wirkungen der alkaloide von *Peganum harmala*, insbesonders des Harmalins. *Arch. exp. Pathol. Pharmak.* **36**(I), 69.

Pellonisz, A. and Llinás, R. (1985) Tensor network theory of the metaorganization of functional geometries in the central nervous system. *Neuroscience* **16**, 245–73.

Ramon y Cajal, S. (1888) La estructura de los centros nerviosos de las aves. *Rev. Trimestr. Histol.* **1**, 1–10.

Ramon y Cajal, S. (1934) Les preuves objectives de l'unité anatomique des cellules nerveuses. *Trab. Lab. Invest. Biol. Univ. Madr.* **29**, 1–137.

Rose, R. M. and Hindmarsh, J. L. (1985) A model of a thalamic neuron. *Proc. R. Soc. Lond.* **225**, 161–93.

Schultze, M. (1861) Ueber Muskelkorperchen und das was man eine Zelle zu nennen habe. *Mueller's Arch.* pp. 1–27.

Sotelo, C., Llinás, R. and Baker, R. (1974) Structural study of the inferior olivary nucleus of the cat: morphological correlates of electrotonic coupling. *J. Neurophysiol.* **37**, 541–59.

Steriade, M. and Llinás, R. (1988) The functional states of the thalamus and the associated neuronal interplay. *Physiol. Rev.* **68**, 649–742.

Villablanca, J. and Riobo, F. (1970) Electroencephalographic and behavioral effect of harmaline in intact cats and in cats with chronic mesencephalic transection. *Psychopharmacology* **17**, 302–13.

Waldeyer, W. (1891) Ueber einige neuere forschungen im gebiete der anatomie des central-nervensystems. *Dtsch Med. Wochenschr.* **17**, 1213–356.

Yarom, Y. and Llinás, R. R. (1981) Oscillatory properties of inferior olive cells: A study of guinea-pig brain stem slices *in vitro*. *Soc. Neurosci. Abst.* **7**, 864.

A three-dimensional model of a thalamic neurone

J. L. HINDMARSH[a] AND R. M. ROSE[b]

[a]School of Mathematics and [b]Department of Physiology,
University of Wales College of Cardiff, Cardiff CF1 1SS, UK

INTRODUCTION

The simplest non-linear oscillator used to model the firing cycle of a nerve cell is that devised by Fitzhugh (1961). A typical example of his equations is given by:

$$\dot{x} = y - x^3 + 3x^2 + I$$
$$\dot{y} = 1 - 5x - y \tag{1}$$

where x is a variable analogous to membrane potential, y is a generalized recovery variable, and I is the externally applied current. The nullclines (curves where $\dot{x} = 0$ or $\dot{y} = 0$) are given by:

$$y = x^3 - 3x^2 - I \quad \text{for } \dot{x} = 0 \text{ and}$$
$$y = 1 - 5x \quad \text{for } \dot{y} = 0$$

In the x,y phase plane the cubic x-nullcline intersects the y-nullcline at a single equilibrium point (EP). For a suitable choice of I this EP is unstable and is surrounded by a stable limit cycle. For almost any choice of initial conditions the system will approach this limit cycle. When the oscillation is plotted against time it can be seen to resemble a nerve action potential (Fitzhugh, 1961).

A problem with these equations is that they do not model the fact that in many cells the action potential has a very much shorter duration than the interspike interval. About 7 years ago we considered the problem of how the Fitzhugh model could be modified to produce a long interspike interval. The

Cell to Cell Signalling: From
Experiments to Theoretical Models
ISBN 0–12–287960–0

17

solution of this problem was eventually to lead us to a model of a thalamic neurone.

QUALITATIVE MODELS

Repetitive firing

A simple way to produce a long interspike interval using eqns (1) would be to introduce time constants which were functions of x, but this is not very satisfactory. Our solution (Hindmarsh and Rose, 1982) can be described as follows. Firstly we note that in eqns (1) the rate of movement of the phase point in the x,y phase plane depends on its vertical distance from the nullclines. Since this distance is similar for both suprathreshold and subthreshold values of x, the oscillation is almost symmetrical. Our solution was to

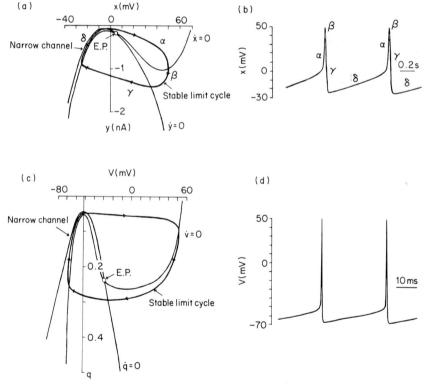

Fig. 1.   Comparison of qualitative and quantitative two-dimensional models of repetitive firing. Qualitative model: (a) x,y phase plane showing stable limit cycle and narrow channel, (b) corresponding time course of x. Quantitative model (eqns (2)): (c) v,q phase plane, (d) corresponding time course of v. (Based on Rose and Hindmarsh, 1989a.)

bend the y-nullcline downwards on the left hand side (i.e. in the subthreshold region) so that it became parabolic in shape. As shown in Fig. 1a, the x- and y-nullclines again intersect at a single unstable EP which is surrounded by a stable limit cycle. The difference between this model and the Fitzhugh model is that because the x- and y-nullclines now lie close together in the subthreshold region, the phase point is forced to move slowly in this region. We refer to this close proximity of the x- and y-nullclines as the 'narrow channel' property. When the phase point enters the narrow channel its vertical distance from the nullclines is much smaller than its vertical distance during the corresponding recovery phase of the action cycle (compare points labelled β and δ in Fig. 1a). This gives the long interspike interval shown in Fig. 1b.

Burst model

Because of the closeness of the x- and y-nullclines in the subthreshold region, it only required a *small* deformation of one of them to give a model with three EPs (Fig. 2a). If a third slow differential equation is now added to represent adaptation, we obtain the following three-dimensional system which gives the bursting solution shown in Fig. 2d:

$$\dot{x} = y - x^3 + 3x^2 - z + I$$
$$\dot{y} = 1 - 5x^2 - y \qquad\qquad (2)$$
$$\dot{z} = r\,[s(x - x_r) - z]$$

In these equations x and y have the same meaning as before. Note that these equations are very similar to eqns (1) except that the $\dot{y}$ equation now has an x^2 term to give a parabolic y-nullcline, and the adaptation (z) variable now contributes to the $\dot{x}$ equation. The equation for the z variable is such that z changes slowly towards the value $s(x - x_r)$. The parameters s and x_r are chosen so that during firing the value of z will increase, and when firing stops the value of z will slowly decrease. Because z is slowly varying we treat it as a parameter in the two-dimensional subsystem obtained from eqns (2) by ignoring the $\dot{z}$ equation. We then sketch the nullcline diagrams for this subsystem for different values of the parameter z. To illustrate how this model works we will now explain the bursting cycle shown in Fig. 2d in terms of the nullcline diagrams shown in Fig. 2a, b and c. For a more complete description see Hindmarsh and Rose (1984).

Beginning at time point α, the z variable has a high value as a result of the preceding firing, and the system is at the stable EP at A in Fig. 2a. During the interburst period the value of z slowly declines, and this displaces the x-nullcline downwards. As a result the stable EP at A and the saddle EP at B move towards each other, coalesce, and then disappear. The phase point moves up along the narrow channel as shown in Fig. 2b, and then enters a

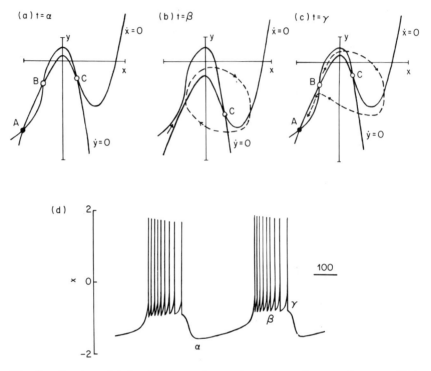

Fig. 2.   Burst model. (a–c) Schematic x,y phase plane diagrams for eqns (2) for different fixed values of z. These values are those at the time points α, β, γ, indicated in (d) which shows the time course of x. Parameter values are r=0.001 and s=4. (Reproduced from figure 6b of Hindmarsh and Rose, 1984.)

temporary limit cycle, which we will refer to as an action cycle, around the remaining unstable EP at C. There follows a succession of action cycles corresponding to firing. As firing progresses the value of z begins to increase, and the x-nullcline is displaced upwards again. This recreates the EPs at A and B, with A moving to the left and downwards and B moving to the right and upwards as z continues to increase. Eventually the saddle point at B moves so far to the right that it falls inside the path of the final action cycle. As shown in Fig. 2c, the phase point then returns to the stable EP at A to complete the burst cycle. For a discussion of this and similar models see Rinzel (1986).

Thalamic model

The model given by eqns (2) has the property that if the external current is chosen so that the system is at rest at the EP at A, then a current pulse can trigger continuous firing. This is eventually terminated by the adaptation variable z whereupon the system returns to the EP at A. An example of such a triggered burst is given in Hindmarsh and Rose (1984, figure 5). We next considered whether it was possible to trigger not just one burst as above, but a succession of bursts. We found that this could be achieved if we made a *second* deformation in the narrow channel region to give a model with five EPs. This property of triggered bursting was investigated because it was interesting and was not an attempt to explain any known phenomenon. At about the same time as we were investigating the properties of this five-EP model Jahnsen and Llinás (1984a,b) published a series of papers on the properties of thalamic neurones in the guinea-pig *in vitro* slice preparation. By altering several of the parameter values in our five-EP model we found a set of parameter values which were such that many of the properties of thalamic neurones could be explained, at least qualitatively, using this model. The most important property of both the cells and the model is shown in Fig. 3a, b and c.

Figure 3 shows a series of three responses to a current step of fixed amplitude at different values of initial membrane potential. At rest (Fig. 3b) the model gives a passive response. When slightly depolarized from rest (Fig. 3c) the response is tonic firing. Neither of these two properties are unusual in themselves. However, if the model is hyperpolarized and the same current

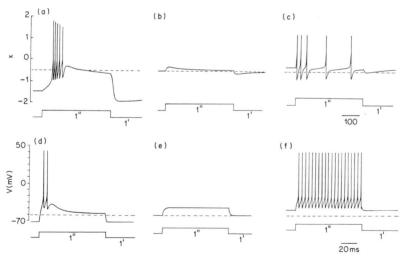

Fig. 3. Comparison of qualitative and quantitative three-dimensional thalamic models. Qualitative model: (a–c) show burst, rest and tonic responses for current steps of fixed amplitude I'' with the applied current I' set at different values. (Reproduced from figure 7 of Rose and Hindmarsh, 1985.) Quantitative model: (d–f) show similar responses to a current step of 20 µA/cm². Parameter values are those of the three-dimensional z-model of appendix C of Rose and Hindmarsh (1989a) with σ=10 mV. In (d) $v(0)=-71.5$ mV, $I=-15$ µA/cm², (e) $v(0)=-60$ mV, $I=0$ µA/cm², (f) $v(0)=-53$ mV, $I=15$ µA/cm².

step is applied (Fig. 3a) it responds with a burst. We refer to these three responses as the burst, rest and tonic responses. Other responses such as post-inhibitory rebound, and an unusual firing pattern when the model is given a slowly rising and falling current ramp are described in Rose and Hindmarsh (1985). A biochemical model having similar qualitative features has been described by Goldbeter and Moran (1988).

QUANTITATIVE MODELS

Although the responses of this five-EP model are qualitatively similar to those of a thalamic neurone, the main difficulty with the model is that the equations have polynomial expressions on the right hand side of the differential equations. In real cells we would expect ionic currents of the Hodgkin–Huxley type (Hodgkin and Huxley, 1952). We therefore looked for an approximation of the Hodgkin–Huxley equations that resembled our qualitative model. With such an approximation we hoped to be able to use the results of our qualitative models to see how the Hodgkin–Huxley equations could be modified to describe the thalamic neurone.

The Hodgkin–Huxley equations, which describe repetitive firing in the squid giant axon, are as follows:

$$\dot{v} = C^{-1}[-g_{Na}m^3h\,(v-v_{Na}) - g_Kn^4(v-v_K) - g_L(v-v_L) + I]$$
$$\dot{m} = \tau_m^{-1}\,(v)\,(m_\infty(v) - m)$$
$$\dot{h} = \tau_h^{-1}\,(v)\,(h_\infty(v) - h)$$
$$\dot{n} = \tau_n^{-1}\,(v)\,(n_\infty(v) - n) \qquad\qquad (3)$$

We will not describe these equations in detail except to note that $m_\infty(v)$, $h_\infty(v)$ and $n_\infty(v)$ are each sigmoidal functions of voltage.

In the 1970s a number of attempts were made to reduce eqns (3) to a second- or a third-order system (Krinskii and Kokoz, 1973; Kokoz and Krinskii, 1973; Plant, 1976; Rinzel, 1978). In the analysis given by Krinskii and Kokoz (1973) the m variable in the $\dot{v}$ equation is set equal to $m\infty(v)$ and the h variable is replaced by $G - n$, where G is a constant. This gives the following two-dimensional system:

$$\dot{v} = C^{-1}[-g_{Na}m_\infty^3(v)(G-n)(v-v_{Na}) - g_Kn^4(v-v_K) - g_L(v-v_L) + I]$$
$$\dot{n} = \tau_n^{-1}(v)\,(n_\infty(v) - n) \qquad\qquad (4)$$

Although these equations are a good approximation to the Hodgkin–Huxley equations, neither they nor the Hodgkin–Huxley equations describe repetitive firing with a comparatively long interspike interval. From the point of view of our model this is not surprising since the nullclines do not exhibit our narrow channel property. This forced us to look for some modification of the Hodgkin–Huxley equations that would give a long interspike interval, or equivalently, low frequency firing.

Repetitive firing

In 1971 Connor and Stevens gave the first quantitative description of a current in a molluscan neurone now referred to as the A-current. This outward current rises to a peak during the after-hyperpolarization which follows the action potential, and then inactivates slowly. The effect of the slow inactivation of this additional outward current is to slow the return of the membrane potential to spike threshold giving low-frequency repetitive firing. There is a possibility therefore that our narrow channel property could be related to the presence of this current in the real cell. A cell having an A-current in addition to the ionic currents of the Hodgkin–Huxley model is described by the following sixth-order system (Connor and Stevens, 1971):

$$\dot{v} = C^{-1}[-g_{Na}m^3h(v-v_{Na}) - g_L(v-v_L) - g_K n^4(v-v_K) - g_A a^3 b(v-v_A) + I]$$
$$\dot{m} = \tau_m^{-1}(v)(m_\infty(v)-m)$$
$$\dot{h} = \tau_h^{-1}(v)(h_\infty(v)-h)$$
$$\dot{n} = \tau_n^{-1}(h)(n_\infty(v)-n) \tag{5}$$
$$\dot{a} = \tau_a^{-1}(v)(a_\infty(v)-a)$$
$$\dot{b} = \tau_b^{-1}(v)(b_\infty(v)-b)$$

where $v_A \simeq v_K$.

To reduce this system to a second-order system we first replaced m^3 by $m_\infty^3(v)$ and eliminated the $\dot{m}$ equation. We then wrote the sum of the $I_K(v)$ and I_A currents as $g_K\, q\, (v-v_K)$, where $q = n^4 + (g_A/g_K)\, a^3 b$, and found that this could be approximated by $q = n^4 + 0.21\, (g_A/g_K)b$. If we differentiate q to obtain the $\dot{q}$ equation we find that this equation is rather complicated. However, if we use the simpler equation $\dot{q} = \tau_q^{-1}(q_\infty(v)-q)$ where $q_\infty(v) = n^4_\infty(v) + 0.21(g_A/g_K)\, b_\infty(v)$ and $\tau_q = 0.5(\tau_b + \tau_n)$ we find that this is a good approximation. Finally we approximated h by $0.85 - 3n^4 = 0.85 - 3(q - 0.21(g_A/g_K)b)$, and b by $b_\infty(v)$. Replacing b by $b_\infty(v)$ eliminates the $\dot{b}$ equation and gives the following second-order system:

$$\dot{v} = C^{-1}\{-g_{Na}m_\infty^3(v)(-3(q - Ab_\infty(v) + 0.85)(v-v_{Na}) - g_L(v-v_L)$$
$$\qquad - g_K q(v-v_K) + I\} \tag{6}$$
$$\dot{q} = \tau_q^{-1}(v)(q_\infty(v)-q) \qquad A = 0.21 g_A/g_K$$

The nullclines and time solution for these equations are shown in Fig. 1c and d (see Rose and Hindmarsh, 1989a for further details). Note the presence of the narrow channel and long interspike interval.

Burst and thalamic models

We next found that we could obtain a model for both a bursting neurone and a thalamic neurone by adding a non-inactivating subthreshold inward current and a slow outward current to the repetitive firing model given by eqns (6). This gives the following three-dimensional system (Rose and Hindmarsh, 1989a):

$$\dot{v} = P(v,q,z) + C^{-1}I \equiv C^{-1}\{-g_{Na}m_\infty^{3}(v)(-3(q-Ab_\infty(v))+0.85\,(v-v_{Na})$$
$$\quad - g_L(v-v_L) - g_Kq(v-v_K) - g_{sa}s_{a\infty}(v)(v-v_{Ca}) - g_{out}z(v-v_K) + I\}$$
$$\dot{q} = Q(v,q) \equiv \tau_q^{-1}(v)(q_\infty(v)-q)$$
$$\dot{z} = R(v,z) \equiv \tau_z^{-1}(z_\infty(v)-z)$$
(7)

where $s_{a\infty}(v)$ is the steady-state activation curve for the inward current and z is the slow outward current variable.

Whether we obtain a bursting solution or responses similar to those of thalamic neurones depends on the choice of parameter values in eqns (7). An example of the burst, rest and tonic responses which can be obtained using eqns (7) is shown in Fig. 3c, d and f. These responses correspond more closely to those obtained experimentally (Jahnsen and Llinás, 1984a,b) than do the corresponding series of responses obtained using our earlier qualitative model.

The analysis of eqns (7) is more complicated than that of the qualitative model. The main difference between eqns (2) and eqns (7) appears when we consider the two-dimensional subsystem obtained by regarding z as a parameter in both cases. In the qualitative model changing z displaces the x-nullcline vertically by an amount that is independent of x. In the quantitative model the v-nullcline is displaced by an amount which depends on v. That is, the v-nullcline does not simply move vertically as z changes but changes shape as well.

In eqns (7) the variable z is slowly changing compared with v and q. We therefore consider the first two of eqns (7) and treat the z variable as a parameter. We call these equations the one-parameter family of systems D_z.

Consider a point $\mathbf{x}_\infty(v) = (v,q_\infty(v))$ (which may or may not be an EP depending on the value of I) in the two-dimensional phase space of D_z. Define the matrix $\mathbf{A}(\mathbf{x}_\infty(v), z)$ by:

$$A(\mathbf{x}_\infty(v), z) = \begin{bmatrix} \dfrac{\partial P}{\partial v}, \dfrac{\partial P}{\partial q} \\[2ex] \dfrac{\partial Q}{\partial v}, \dfrac{\partial Q}{\partial q} \end{bmatrix}_{q=q_\infty(v)}$$

This will not depend on the applied current, I, but only on v and z. It is a stable matrix if both its eigenvalues have strictly negative real parts. Otherwise we call it an unstable matrix. We call a point (v,z) in the v,z plane a stable (unstable) point if the matrix $A(\mathbf{x}_\infty(v),z)$ is a stable (unstable) matrix. In Fig. 4a stable points are in the unshaded regions and unstable points in the shaded regions (saddle points, unstable spirals and nodes). We call this diagram the stability diagram.

The point $\mathbf{x}_\infty(v)$ will be an EP for the system D_z provided that $P(\mathbf{x}_\infty(v),z) + C^{-1}I = 0$. Points satisfying this equation form what we call the nullcurve for external current I. In Fig. 4a this is drawn for $I = 5\ \mu A/cm^2$, which is the value of the applied current during the current step of the burst response shown in Fig. 3d.

Consider a point (v,z) on the nullcurve. If this point lies in a stable (unstable) region of the stability diagram then $\mathbf{x}_\infty(v)$ is a stable (unstable) EP of the system D_z. Segments of the nullcurve corresponding to stable (unstable) EPs are shown by a continuous (broken) line. We call an EP of D_z a temporary EP (TEP) of the three-dimensional system of eqns (7).

We now define the state diagram which is drawn in the v,z plane. It consists of a nullcurve with the stability of the TEPs taken from the stability diagram, the projection of the phase path, and the graph of $z_\infty(v)$ against v. In Fig. 4b and c the nullcurve is labelled $\dot{v} = \dot{q} = 0$, the projection of the phase path is shown with arrows indicating the time direction, and the graph of $z_\infty(v)$ is labelled $\dot{z} = 0$.

We now illustrate with a simple example the use of the state diagram to explain the time course of the potential shown in Fig. 3d. This is the response to a current step. We describe our current steps by giving the magnitude of the step, I'', which is applied to a background current I'. Since we have two values of total current, I' and $I' + I''$, we have two nullcurves. These are shown in Fig. 4c and b respectively. In each case the stability of the TEPs on the nullcurves is shown by continuous lines (stable) or broken lines (unstable).

Before the step is applied the projection of the phase path onto the state diagram is at A (shown in Fig. 4c). After the application of the step the relevant state diagram is Fig. 4b in which the projection of the phase path shows the cell firing twice whilst z increases until the phase path encounters the stable TEPs on the nullcurve. It follows these to the (projection of the) EP at A'. When the step terminates the relevant state diagram is Fig. 4c which

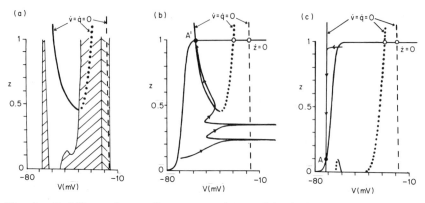

Fig. 4.   Stability and state diagrams used to explain the burst response of Fig. 3d. (a) Stability diagram with v,q nullcurve for $I=5\,\mu A/cm^2$ superimposed. (b) State diagram for $I=5\,\mu A/cm^2$ (during the current step). (c) State diagram for $I=-15\,\mu A/cm^2$ (before and after current step). Parameter values as in Fig. 3d.

shows the (projection of the) phase path following the stable TEPs from A′ to A.

This example shows how the time course of the potential can be explained once the state diagram is known. The more interesting situation occurs when we record an unusual potential time course. We then conjecture what changes in the structure of the state diagram might explain this time course. These changes will probably require changes in both the stability regions and the shape of the nullcurves. From experience we are learning how to bring about these changes by parameter variation and the addition of extra currents activated at various potentials. We found it impossible to make the necessary changes in parameter values and currents to produce a given potential time course without the help of the intermediate step provided by the stability and state diagrams. Elsewhere (Rose and Hindmarsh, 1989b) we have used these diagrams to help us explain such features as post-inhibitory rebound, plateau potentials, rebound followed by a plateau and sub-threshold oscillations.

FURTHER DEVELOPMENTS OF THE MODEL

We have described above a simple model in which the burst response was terminated by the activation of a slow outward current. However experiments by Jahnsen and Llinás (1984a,b) suggest that the burst is terminated by the inactivation of a slow inward current. We therefore modified our model accordingly. This new model (Rose and Hindmarsh, 1989b) was also three-dimensional and allowed a similar explanation using state diagrams. These three-dimensional models used the somewhat artificial q variable. It is therefore reassuring to note that the changes that these simple models taught us to make can be made in the same way to the six-dimensional system that we started with. The resulting seven-dimensional system provides an even better model of a thalamic neurone (Rose and Hindmarsh, 1989c). The stability and state diagrams can also be defined and used directly for these seven-dimensional models.

Acknowledgement: This work was supported by a grant from the Wellcome Trust.

REFERENCES

Connor, J. A. and Stevens, C. F. (1971) Prediction of repetitive firing behaviour from voltage-clamp data on an isolated neurone soma. *J. Physiol., Lond.* **213**, 31–53.

Fitzhugh, R. (1961) Impulses and physiological states in theoretical models of nerve membrane. *Biophys. J.* **1**, 445–66.

Goldbeter, A. and Moran, F. (1988) Dynamics of a biochemical system with multiple oscillatory domains as a clue for multiple modes of neuronal oscillations. *Eur. Biophys. J.* **15**, 277–87.

Hindmarsh, J. L. and Rose, R. M. (1982) A model of the nerve impulse using two first order differential equations. *Nature, Lond.* **296**, 162–4.

Hindmarsh, J. L. and Rose, R. M. (1982) A model of neuronal bursting using three coupled first order differential equations. *Proc. R. Soc. Lond. B* **221**, 87–102.

Hodgkin, A. L. and Huxley, A. F. (1952) A quantitative description of membrane current and its application to conduction and excitation in nerve. *J. Physiol., Lond.* **117**, 500–44.

Jahnsen, H. and Llinás, R. (1984a) Electrophysiological properties of guinea-pig thalamic neurones: an *in vitro* study. *J. Physiol., Lond.* **349**, 205–26.

Jahnsen, H. and Llinás, R. (1984b) Ionic basis for the electroresponsiveness and oscillatory properties of guinea-pig thalamic neurones *in vitro*. *J. Physiol., Lond.* **349**, 227–47.

Krinskii, V. I. and Kokoz, Yu. M. (1973) Analysis of equations of excitable membranes – I. Reduction of the Hodgkin–Huxley equations to a second order system. *Biofizika* **18**, 533–9.

Kokoz, Yu. M. and Krinskii, V. I. (1973) Analysis of equations of excitable membranes – II. Method of analysing the electrophysiological characteristics of the Hodgkin–Huxley membranes from the graphs of the zero-isoclines of a second order system. *Biofizika* **18**, 937–44.

Plant, R. E. (1976) The geometry of the Hodgkin–Huxley model. *Computer Programs in Biomedicine* **6**, 85–91.

Rinzel, J. (1978) Integration and propagation of neuroelectric signals. In *Studies in Mathematical Biology I* (ed. S. A. Levin), pp. 1–66. Math. Assoc. Amer.

Rinzel, J. (1986) A formal classification of bursting mechanisms in excitable systems. In *Proceedings of the International Congress of Mathematicians* (ed. A. M. Gleason), pp. 1578–93.

Rose, R. M. and Hindmarsh, J. L. (1985) A model of a thalamic neuron. *Proc. R. Soc., Lond. B* **225**, 161–93.

Rose, R. M. and Hindmarsh, J. L. (1989a) The assembly of ionic currents in a thalamic neuron. I. The three-dimensional model. *Proc. R. Soc., Lond. B* **237**, 267–88.

Rose, R. M. and Hindmarsh, J. L. (1989b) The assembly of ionic currents in a thalamic neuron. II. The stability and state diagrams. *Proc. R. Soc., Lond. B* **237**, 289–312.

Rose, R. M. and Hindmarsh, J. L. (1989c) The assembly of ionic currents in a thalamic neuron. III. The seven-dimensional model. *Proc. R. Soc., Lond. B* **237**, 313–34.

Rhythmic neuronal burst generation: Experiment and theory

WILLIAM B. ADAMS[a] AND JACK A. BENSON[b]

[a]Department of Pharmacology, Biozentrum, University of Basel,
Klingelbergstrasse 70, CH-4056 Basel, Switzerland
[b]CIBA-GEIGY Ltd, Agricultural Division, R1093.P.47, CH-4002
Basel, Switzerland

INTRODUCTION

In his brief paper 'On the pitfalls of modeling and not-modeling', Dan Hartline made this important point: 'models which lack physiological detail initially are likely to fail, or if they succeed, it will be fortuitous and hence unuseful to the researcher' (Hartline, 1976). Hartline meant 'the experimental researcher', and his main examples were drawn from the bursting activity of the lobster stomatogastric neuronal network. For the electrophysiologist, models of neuronal activity, such as endogenous rhythmic bursting in neurones, should be quantified versions of physiological hypotheses based on data from a single, identifiable neuronal class with uniform properties. They need to be tied as rigorously as possible to the experimental data and their purpose is two-fold:

(a) To provide evidence that the physiological hypothesis can, in fact, account for the data.
(b) To reveal unexpected consequences that can be investigated experimentally. This means not producing what is in effect no more than a glorified mnemonic for the results to date, but rather providing insights for further experimentation that might not otherwise emerge readily from the data.

What follows is a brief outline of the physiology of certain neuronal bursters with comments on the directions modelling has taken and could take in this field. Our comments are to a large extent directed to modellers who are

Cell to Cell Signalling: From
Experiments to Theoretical Models
ISBN 0–12–287960–0

29

not experimentalists but who might be interested in the kinds of data obtained by electrophysiologists, and the problems that they hope modellers might address.

MOLLUSCAN ENDOGENOUS BURSTERS

In the absence of synaptic input, the rhythmic activity of molluscan endogenous bursters is a regular cycle composed of a burst followed by a hyperpolarized interburst interval (Fig. 1A). The burst consists of a train of action potentials with a frequency that increases until mid-burst and then decreases. The positive overshoot of the action potentials increases to mid-burst and then remains fairly constant. The undershoot (negative after-potential) decreases almost throughout the burst. At the end of the burst there is a slow depolarizing after-potential (DAP) followed by the post-burst hyperpolarization. The hyperpolarization decays during the interburst interval at an increasing rate leading to a depolarizing inflection that triggers the first action potential of the next burst.

The following description of the ionic mechanism of bursting probably applies to many but not all molluscan bursters. However, the computer simulation that was used to test this physiological model was based on quantitative data from the *Aplysia* neurone R15 (the BASIC program of the model is given in the Appendix; for further details, see Adams and Benson, 1985). On purely theoretical grounds, one can predict the participation of at least three currents in generating bursting. First, some depolarizing current

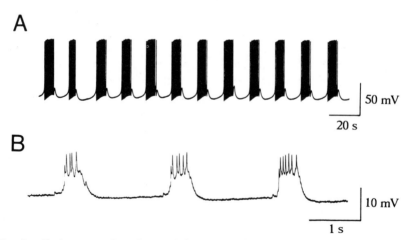

Fig. 1. Endogenous bursting activity recorded by intracellular microelectrode from: (A) the neurone R15 in an isolated abdominal ganglion of the mollusc, *Aplysia californica*, (B) a motor neurone in an isolated cardiac ganglion from the crustacean, *Portunus sanguinolentus*.

must be present to get the burst started (in R15, I_{NSR}, an inward current activated by depolarizing steps at membrane potentials below the action potential threshold). Second, given the presence of a depolarizing current, an activity-activated hyperpolarizing current is needed to bring the burst to an end (I_H, a transient, hyperpolarizing after-current that arises from a calcium-dependent inactivation of a background, steady-state calcium current). Finally, in order that the burst keep going, and not be ended by I_H after the first action potential, a local (in time), regenerative, excitatory current is necessary (I_D, an inward current activated by action potentials and decaying with a time course of hundreds of milliseconds). In fact, it is the presence of I_D that distinguishes bursting cells from autoactive beating cells.

At the point in the burst cycle just before the first action potential in a burst, I_D is completely deactivated, while I_H remains partially activated from the preceding burst. I_{NSR}, which activates with depolarization, is larger than I_H and, thus, the net current is inward, or depolarizing. As the membrane potential reaches threshold, the first action potential is triggered. The action potential, in turn, incrementally activates I_D and I_H. The incremental activation of I_D is larger than that of I_H so that the net inward current is larger than it was before the first action potential. As a result, the depolarization toward the next action potential is faster. With succeeding action potentials this sequence of events is repeated. The incremental activations of I_D and also of I_H are approximately the same for each action potential in the burst and the two currents summate temporally. For the first few action potentials in the burst, the rate of depolarization between action potentials successively increases, as does the rate of action potential production. Later in the burst, the summated I_D reaches a maximum, as its incremental activation with each succeeding action potential only just replaces the amount by which it has deactivated since the preceding action potential. However, this is not the case for I_H, with its much longer time constant of temporal summation. I_H continues to grow with each action potential in the burst. As it does so, the net inward current between action potentials becomes less, and the rates of depolarization and action potential production decrease. Finally, following what will consequently be the last action potential in the burst, the sum of I_{NSR}, I_D, and I_H is no longer inward and the membrane potential cannot reach threshold for generation of another action potential. The membrane hyperpolarizes, beginning the interburst. I_D decays rapidly, with a time constant of a few hundred milliseconds. I_{NSR} deactivates because of its voltage dependence. Since I_H is still large and outward at this time, the net result is a rapid 'swoop' into the interburst phase. It should be emphasized that the rapid hyperpolarization at the beginning of the interburst interval is not the result of a sudden activation of a hyperpolarizing current but, rather, of the rapid decay of I_D and the deactivation of I_{NSR}. What is left behind is the DAP at the end of the burst, actually the remnants of the incremental activation of I_D by the last action potential in the burst.

During the interburst interval, the predominant current is I_H. As I_H decays, the membrane potential begins to depolarize. I_{NSR}, which is activated by depolarization, begins to increase. Toward the end of the interburst interval, the regenerative character of the activation of I_{NSR} becomes obvious, as the membrane depolarizes faster and faster. Finally, the threshold is reached for the generation of the first action potential in the burst, and the cycle begins again. This model differs significantly from many other models for molluscan bursters in not attributing the post-burst hyperpolarization at normal temperatures to an increase in K^+ conductance activated by the influx of Ca^{2+} ions during the burst (discussed by Benson and Adams, 1988). Such a mechanism probably does occur, however, in some vertebrate brain bursters (see below).

A good example of the value of the computer simulation of this model is provided by our efforts to account for the characteristic changes in overshoot, undershoot, shape and frequency of the action potentials during a burst (Fig. 1A). In many other neurones, action potentials are truly 'all-or-none' events, but in R15 they go through a continual change in shape during the burst. Using quantitative voltage-clamp data from the literature and our own experiments, it was not difficult to account for action potential shape. The rising phase is controlled by influxes first of Na^+ and then later of Ca^{2+}. I_{Na}, the classical fast Na current, with its rapid kinetics and more negative voltage range for activation, dominates action potential initiation and rise time. The conductances for these ions increase with depolarization, leading to a regenerative increase in inward current and a continuing increase in the rate of rise of the action potential. As I_{Na} begins to inactivate and I_K, the delayed rectifier, begins to activate, the rate of rise slows until the action potential reaches its peak amplitude. A sizeable inward current is carried by Ca^{2+} and I_{Ca}, the inward Ca current, shows only slow inactivation. As a result of this continued inward current flow, the membrane potential falls rather more slowly than in the squid axon. Only when the voltage falls to approximately 0 mV do the Ca^{2+} channels begin to deactivate as a result of their voltage dependence. Deactivation of I_{Ca} results in an increase in net outward current and the resulting more rapid hyperpolarization leaves a distinct 'shoulder' on the falling phase of the action potential. Repolarization is mediated by I_K and possibly by I_C, the Ca-activated K^+ current, as well. Action potential broadening during the burst is due to inactivation of I_K, and recovery from inactivation must be quite slow, since action potential durations continue to increase during bursts that are several seconds long. Our first simulations reproduced these effects fairly well, although not perfectly, with a minimum of 'fine-tuning' of the experimentally determined parameters (Fig. 2Ai).

However, the changes in overshoot and undershoot could not be accounted for by even the most unlikely alterations in activation and inactivation parameters, as well as in the voltage-dependence and time

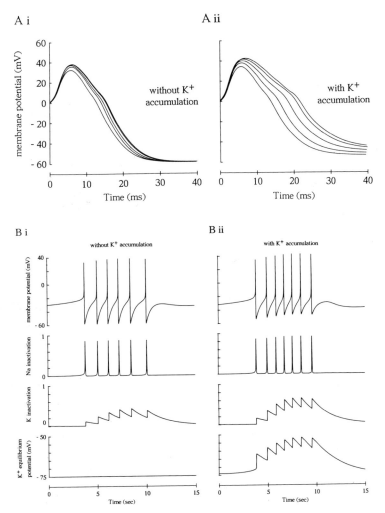

Fig. 2. Computer simulation of bursting in R15 using a mathematical representation of identified currents with experimentally determined parameters. (The
BASIC program for the model is given in the Appendix.) In (Ai) and (Bi), the K^+
equilibrium potential remains constant. In (Aii) and (Bii), the K^+ equilibrium
potential changes due to K^+ accumulation in an extracellular pool. (A) Successive
action potentials from a single burst superimposed to show the changes in shape
that occur during the burst. (B) Membrane potential, Na and K inactivation, and
K^+ equilibrium potential during the course of a single full burst cycle.

constants, of the K^+ currents (Fig. 2Bi). We knew that, in brain, alterations
in the ionic microenvironment that occur *in vivo*, such as increased $[K^+]_0$ and
decreased $[Ca^{2+}]_0$, are of sufficient magnitude to induce depolarizing shifts
and epileptiform field potentials *in vitro* (e.g. Heinemann *et al.*, 1977), and
that, in *Aplysia*, positive voltage-clamp steps produce marked changes in the

K$^+$ equilibrium potential, E_K (Eaton, 1972), probably due to accumulation of K$^+$ within the numerous infoldings of the somal membrane. A localized extracellular K$^+$ pool that increases in proportion to the outward flow of K$^+$ and equilibrates with the bulk medium with a diffusion constant of 2 s, was added to the simulation. This modified model predicted changes in E_K from -74 mV during the interburst to -62 mV following the first action potential to -52 mV following the sixth action potential in a burst. Correspondingly, the predicted negative undershoots decrease from -53 mV after the first action potential to -47 mV after the sixth. This closely resembles the behaviour observed in R15 (Fig. 2Bii). In a similar manner, the increase in positive overshoot of the action potentials was accounted for in part by I_K inactivation and changes in E_K, and in part by changes in I_{Na} inactivation. The change in action potential shape (Fig. 2Aii) also coincided much more closely with the experimental data (cf. figure 19 of Strumwasser, 1967).

The shift in E_K required by the model and confirmed experimentally has an additional significant consequence that became evident during simulation runs on the computer. The response of R15 to serotonin is of considerable scientific interest since it is mediated by cyclic-AMP and protein phosphorylation, and R15 is one of the few neural systems in which such a second messenger system has been elucidated in almost complete detail (reviews by Levitan and Benson, 1981; Benson et al., 1983; Lemos et al., 1986). Serotonin increases an inwardly rectifying, subthreshold K$^+$ current (I_R). Because it is an imperfect inward rectifier, I_R, under steady-state voltage clamp, carries some current over the membrane potential range -60 to -80 mV (Benson and Adams, 1987). However, the latter range is often below the maximum hyperpolarization attained during a normal burst cycle, so that it was difficult to see how an increase in I_R could alter the activity of a bursting neurone. The solution to this apparent paradox is the positive shift in E_K which ensures that the active range for I_R coincides exactly with the membrane potentials reached during the interburst interval.

ENDOGENOUS BURSTERS IN CRUSTACEA

The hearts of crustacea are provided with a cardiac ganglion that controls the heart beat frequency and amplitude. The cardiac ganglia of crabs and lobsters consist of five large and four small electrotonically-coupled neurones that produce periodic bursts of action potentials. The form of the activity recorded intracellularly from one of the large neurones consists of a square-shaped plateau potential surmounted by attenuated action potentials and excitatory synaptic potentials (Fig. 1B). The tight coupling between the neurones means that the intact ganglion can be treated analytically as a relaxation oscillator. From experimentally determined phase response curves, this very simple model allowed the successful quantitative prediction of the details of frequency control (limits of entrainment and phase angle difference between driver cycle and oscillator) during entrainment to periodically applied current pulses (Benson, 1980). The details of the phase response curves, especially at the phases corresponding to interruption of the burst by current injection, were consistent with the ionic model for the system hypothesized by Tazaki and Cooke (see below). However, the relaxation oscillator model *per se* did not contribute a great deal to our knowledge of the underlying mechanism of bursting in these cells. A detailed analysis of the ion currents was a prerequisite for a 'useful' model. These experiments were carried out by Tazaki and Cooke who described a TTX-resistant, slow, regenerative depolarization in the cardiac ganglion motor neurones that underlies the square-shaped plateau potential seen in normal saline. They called this depolarization a 'driver potential' because it appears to be the driving force underlying burst formation. It is also an integral part of the cycle of events underlying rhythmicity in these neurones.

The physiological model of Tazaki and Cooke (1983) proposes that depolarizing input to the soma triggers a driver potential there. The soma is thus electrically excitable although it does not support action potentials. The depolarizing input probably arises from a leak current that manifests itself as the slow ramp pacemaker potential during the interburst interval. The other potential source of depolarizing drive is the input from chemical and electrotonic synaptic potentials from the other bursting neurones in the ganglion. The driver potential is not propagated: it is confined to the soma and proximal axon, and appears to arise from the activation of a very slow calcium current and to be terminated by inactivation of this current combined with activation of voltage- and calcium-dependent K^+ currents. However, by electrotonic spread along the axon, the driver potential evokes action potentials in a more distal part of the axon where Na^+ channels are present and the action potentials are then propagated away as a burst to the heart muscles. This means that the ionic mechanism for burst organization is physically separated from the action potential initiation site. In contrast to the molluscan bursters, action potential activity is not a crucial part of the depolarizing driving force, and driver potentials can be recorded in cardiac somata separated from their axons by ligaturing. The unstructured internal organization of the cardiac burst reflects both the difference in the burst organizing mechanism as well as the fact that the cardiac bursters are part of a neuronal network where intercellular interaction plays an important role.

The integration of the burst-organizing driver potential into the overall mechanism resulting in rhythmic bursts is proposed by Tazaki and Cooke to be as follows. At the end of the driver potential, the somal membrane potential is at its most hyperpolarized, as in the molluscan bursters at the end of a burst. Around this time, three conditions coincide: firstly, in the cardiac motor neurones this hyperpolarization reflects a high K^+ conductance rather than a current analogous to I_H, providing a low-resistance membrane that conducts incoming events poorly; secondly, the fast transient K^+ current, I_A, an outward current activated by depolarizations from hyperpolarized membrane potentials, is in a responsive state; thirdly, the slow calcium current recovers only slowly from inactivation. These three conditions reduce the efficacy of incoming depolarizing influences and hold the membrane potential below driver potential threshold for some time, thus ensuring an interburst interval. As these influences fade, the depolarizing effects of the 'leak' current, in combination with synaptic input from other cardiac neurones, move the membrane potential towards the threshold for the slow calcium current and a new driver potential is generated. It is sustained not only due to the properties of the calcium current but probably also by electrotonic feedback from the axonal action potentials which can be recorded in attenuated form in the soma.

The Tazaki and Cooke model illustrates a still not widely recognized contribution to burst generation and organization, namely, the spatial distribution of the various conductances over the soma and neurites (Thompson and Coombs, 1988; reviews by Crill and Schwindt, 1983, Benson and Cooke, 1984, and Adams and Benson, 1987). This aspect is important in physiological models for both molluscan and crustacean bursting mechanisms, as summarized schematically in Fig. 3, but has yet to be taken into account in a mathematical model of single-cell endogenous rhythmicity. This is despite the considerable literature, both theoretical and physiological, devoted to regional variations in neurones and the effect of neurite size and shape on electrical activity. The crustacean cardiac ganglion is an example of neural systems that are open to several productive forms of analysis, from which a great deal of quantitative information is available, but for which there is as yet no mathematical model or computer simulation.

From the preceding sections on molluscan and crustacean bursters, it can be seen that although the activity recorded from the output axons of the two systems is quite similar, the ionic mechanisms by which the output is generated are fundamentally different processes. This should be reflected in any attempt at mathematical modelling of these systems that pretends to physiological relevance. A model that can account for both forms of activity merely as a result of altering the parameters of constituent 'currents' is not likely to be useful according to our initial definition (e.g. Chay and Cook, 1988; Rinzel and Lee, 1987).

So far, we have been considering the relevance of modelling to experimentation on systems the same as or closely related to the one that was the basis of the model itself. This is in accord with the two criteria of validity for a useful model given at the beginning of this paper. In addition, a well-constructed physiological model can provide hints for experiments on different systems. It serves as a vehicle for conveying information in a coherent and comprehensible form. For various reasons, particularly individual identifiability and large size, the molluscan and crustacean burs-

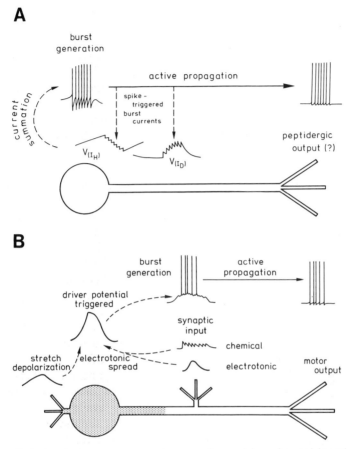

Fig. 3. Relation between neurone topography and ion channel location, and generation and modulation of bursting activity in (A) the molluscan neurone R15 and (B) a crab cardiac neurone. (From Adams and Benson, 1987.)

ters have yielded data on ionic burst mechanisms more readily than the bursters in the vertebrate brain. It is therefore worthwhile to ask whether the models of invertebrate bursters can shed light on those in the brain.

ENDOGENOUS BURSTERS IN THE VERTEBRATE CNS

The vertebrate brain exhibits a wide range of periodic neural activities. The possibilities for mechanisms based on neuronal networks have proved extremely attractive to both physiologists and mathematical modellers, and for many years oscillatory behaviour was hypothesized to arise from processes such as recurrent excitation and inhibition (e.g. Andersen *et al.*,

1964). Indeed, an excellent recent example of a model based on a combination of electrophysiology, anatomy, behaviour and computer simulation attempts to account for saccadic eye movements in terms of a feedback network (Scudder, 1988). However, it is also clear that single-cell properties are an important and perhaps even dominant aspect of burst generation in many brain systems, even where the neurones are part of an oscillatory network (e.g. Traub and Wong, 1983; Jahnsen and Llinás, 1984). We confine ourselves here to single-cell endogenous bursters and, from the vast literature on this subject, we have chosen a very few examples for a brief discussion of the current understanding of bursting mechanisms in the vertebrate brain and the relationship of modelling to this.

Do endogenous bursters exist in the vertebrate brain? For molluscan bursters, continued bursting in neurones isolated from one another, and a frequency change induced by current injection are two important criteria for establishing the endogenous nature of the burst mechanism. These criteria are not easily fulfilled in vertebrate preparations because of the tight synaptic interconnections. Lesion of structures that, when stimulated, evoke bursting in thalamic neurones does not interfere with its appearance (Andersen and Andersson, 1968) suggesting that these various inputs all trigger an endogenous cellular mechanism, and, indeed, the requisite subthreshold slow inward current (see the molluscan model above) has been recorded in these cells (Dêschenes et al., 1982). Cell and explant culture techniques allow more persuasive data to be gathered. Gähwiler and Dreifuss (1979) showed that bursting continued in some but not all cultured rat hypothalamic neurones synaptically decoupled by Co^{2+} treatment, and that the burst frequency changed in proportion to the amount of current injected into the cell body. Also using cultured brain neurones, Legendre et al. (1985) fulfilled the two main criteria as well as the additional one of being able to phase shift the rhythm with a single pulse of injected current.

Conceptually it has been useful to distinguish between the mechanism of burst generation per se and the process by which these bursts are generated periodically. However, as described above, in molluscan bursters the two processes are part of a single, indivisible mechanism, and only in the cardiac ganglia of the crustacea can burst formation take place when the pacemaker system is inoperative. Both combinations appear to occur in the brain.

Taking the former model first, we have already pointed out that a slow, subthreshold depolarizing current has been recorded in some vertebrate bursters. It is probably carried by calcium ions in bursting CA3 pyramidal hippocampal neurones, for example (Johnston et al., 1980). This fulfils the requirement for a depolarizing drive to bring the neurone to action potential threshold, thus initiating a burst. What about termination of the burst? The post-burst hyperpolarization that terminates or at least limits the duration of bursts in many endogenously active vertebrate CNS neurones is accompanied by a change in membrane conductance that has been interpreted as arising from an increase in K^+ conductance (e.g. in frog spinal motor neurones (Barrett and Barrett, 1976)). Although this mechanism is not important in R15 and some other molluscan bursters, there is no reason why it should not play a more significant role in vertebrate bursters, and it has been suggested, often without a great deal of evidence, that the K^+ current is activated by an influx of calcium (reviewed by Prince, 1978). Data in favour of this hypothesis were provided by Alger and Nicoll (1980) for bursting neurones in the CA1 region of the hippocampus. Intracellular injection of EGTA, which binds free calcium, blocked the post-burst hyperpolarization and resulted in prolonged cell discharge, and the hyperpolarization reversed at the K^+ equilibrium potential. Schwartzkroin and Stafstrom (1980) support this conclusion for spontaneous or current-induced bursts in CA3 neurones, but they find that epileptiform bursts are terminated by a non-EGTA-sensitive process. An important point here, as with molluscan neurones, is that differing mechanisms can be involved in similar or even identical neurones under different conditions (see Benson and Adams, 1988).

Do vertebrate bursters exhibit a current to fill the role of I_D? This problem appears to have received comparatively little attention although, by analogy with the molluscan model, a current that sustains the burst should be present. Bursting activity induced by constant depolarizing current in cultured spinal cord neurones seems to reveal a depolarizing after-potential (DAP) in many of the published records (Legendre et al., 1985) and Andrew and Dudek (1983) present evidence for the summation of DAPs that result from somal action potentials in mammalian endocrine bursters. Little information is available regarding the ionic basis and other properties of these putative I_D homologues.

Is there any evidence in the vertebrate CNS for a burst-organizing mechanism resembling the crustacean driver potential? Working with cultures of foetal mouse hypothalamic neurones, Legendre et al. (1982) recorded prolonged plateau potentials with certain similarities to crustacean cardiac plateau potentials. They represent a low resistance state of the neuronal membrane, with Ca^{2+} implicated as the depolarizing current carrier and K^+ acting to repolarize the cell. Inhibition of plateau generation by Co^{2+} or Cd^{2+} but not by TTX is also a common feature. On the other hand, the repolarization process is slow and not accompanied by the hyperpolarizing after-potential that characterizes the crustacean plateau potential. Recent work suggests that the duration of the plateau potentials is controlled by the progressive activation of I_K and I_C, but not by inactivation of calcium currents. The interburst interval seems to be due to activation by depolarization of I_C (Legendre et al., 1988). A high level of synaptic input (not necessarily rhythmic) results in a recognizably rhythmic repetition of plateau potentials and it is very likely that the plateau potential and associated 'slow depolarizing potential' reflect an endogenous property associated with burst organization and possibly rhythmicity. Although not identical, this system is

reminiscent of the crustacean cardiac bursters in many respects, and in fact
the investigation of the vertebrate system was influenced by the crustacean
cardiac model.

It will have been noticed that in the preceding few paragraphs observations
have been drawn from a range of vertebrate cell types, contrary to the
important requirement that the quantitative physiological data for a useful
model be derived from a single, clearly-defined and uniform source. Unfortunately, no neuronal type has so far provided all the data required to model
vertebrate neurone bursting behaviour due to the difficulty of voltage-
clamping vertebrate neurones and manipulating their ionic environment and
synaptic input in a controlled manner. Possibly neuronal culturing methods
provide the best hope in this regard.

CONCLUDING REMARKS

It has already been noted that single models can, with the adjustment of one
or a very few parameters, simulate neuronal activity that is known to
originate from entirely different ionic mechanisms. The analogous case is
succinctly summarized in the concluding statement by Schwindt and Crill
(1982), in their paper reporting voltage-clamp data on currents in non-
bursting but rhythmically firing cat lumbar motor neurones: 'Apparently, a
variety of models can reproduce rhythmic firing even though the models
differ greatly from each other and differ in important ways from the voltage-
clamp data that are now available. It is hoped that the data presented here
and elsewhere will provide a basis for construction of more realistic models,
which will increase our understanding of motoneuron rhythmic firing.'

REFERENCES

Adams, W. B. (1985) Slow depolarizing and hyperpolarizing currents which mediate
  bursting in *Aplysia* neurone R15. *J. Physiol.* **360**, 51–68.
Adams, W. B. and Benson, J. A. (1985) The generation and modulation of
  endogenous rhythmicity in the *Aplysia* bursting pacemaker neurone R15. *Prog.
  Biophys. molec. Biol.* **46**, 1–49.
Adams, W. B. and Benson, J. A. (1987) Ion channel distribution and the generation
  and modulation of endogenous bursting activity in molluscan and crustacean
  neurones. In *Neurobiology, Molluscan Models* (eds H. H. Boer, W. P. M. Geraerts
  and J. Joose), pp. 178–85. North Holland, Amsterdam and New York.
Adams, D. J. and Gage, P. W. (1979) Characteristics of sodium and calcium
  conductance changes produced by membrane depolarization in an *Aplysia* neurone. *J. Physiol.* **289**, 143–61.
Alger, B. E. and Nicoll, R. A. (1980) Epileptiform burst afterhyperpolarization: calcium-
  dependent potassium potential in hippocampal CA1 pyramidal cells. *Science, N.Y.*
  210, 1122–4.

Andersen, P. and Andersson, S. A. (1968) *Physiological Basis of the Alpha Rhythm.* Appleton-Century-Crofts, New York, 235 pp.

Andersen, P., Brooks, C. McC., Eccles, J. C. and Sears, T. A. (1964) The ventro-basal nucleus of the thalamus: potential fields, synaptic transmission and excitability of both presynaptic and postsynaptic components. *J. Physiol.* **174**, 348–69.

Andrew, R. D. and Dudek, F. E. (1983) Burst discharge in mammalian neuroendocrine cells involves an intrinsic regenerative mechanism. *Science, N.Y.* **221**, 1050–2.

Barrett, E. and Barrett, J. N. (1976) Separation of two voltage sensitive potassium currents and demonstration of a tetrodotoxin-resistant calcium current in frog motoneurones. *J. Physiol.* **255**, 337–74.

Benson, J. A. (1980) Burst reset and frequency control of the neuronal oscillators in the cardiac ganglion of the crab, *Portunus sanguinolentus. J. exp. Biol.* **87**, 285–313.

Benson, J. A. and Adams, W. B. (1987) The control of rhythmic neuronal firing. In *Neuromodulation. The Biochemical Control of Neuronal Excitability* (eds L. K. Kaczmarek and I. B. Levitan), pp. 100–18. Oxford University Press, New York.

Benson, J. A. and Adams, W. B. (1988) The ionic mechanisms of endogenous activity in molluscan burster neurons. In *Neuronal and Cellular Oscillators* (ed. J. W. Jacklet), pp. 87–120. Marcel Dekker, New York.

Benson, J. A. and Cooke, I. M. (1984) Driver potentials and the organization of rhythmic bursting in crustacean ganglia. *Trends NeuroSci.* **7**, 85–91.

Benson, J. A., Adams, W. B., Lemos, J. R., Novak-Hofer, I. and Levitan, I. B. (1983) Molecular mechanisms of neuromodulator action on a target neurone. In *Molluscan Neuro-endocrinology* (eds J. Lever and H. H. Boer), pp. 31–7. North-Holland, Amsterdam and New York.

Chay, T. R. and Cook, D. L. (1988) Endogenous bursting patterns in excitable cells. *Math. Biosci.* **90**, 139–53.

Crill, W. E. and Schwindt, P. C. (1983) Active currents in mammalian central neurons. *Trends NeuroSci.* **6**, 236–40.

Deschênes, M., Roy, J. P. and Steriade, M. (1982) Thalamic bursting mechanism: an inward slow current revealed by membrane hyperpolarization. *Brain Res.* **239**, 289–93.

Eaton, D. C. (1972) Potassium ion accumulation near a pacemaking cell of *Aplysia. J. Physiol.* **224**, 421–40.

Gähwiler, B. H. and Dreifuss, J. J. (1979) Phasically firing neurons in long-term cultures of the rat hypothalamic supraoptic area: pacemaker and follower cells. *Brain Res.* **177**, 95–103.

Hartline, D. K. (1976) On the pitfalls of modeling and not-modeling. *Brain Theory Newsletter* **2**, 25–7.

Heinemann, U., Lux, H. D. and Gutnick, M. J. (1977) Extracellular free calcium and potassium during paroxysmal activity in the cerebral cortex of the cat. *Exp Brain Res.* **27**, 237–43.

Jahnsen, H. and Llinás, R. (1984) Ionic basis for the electroresponsiveness and oscillatory properties of guinea-pig thalamic neurones *in vitro. J. Physiol.* **349**, 227–47.

Johnston, D., Hablitz, J. J. and Wilson, W. A. (1980) Voltage clamp discloses slow inward current in hippocampal burst-firing neurones. *Nature* **286**, 391–3.

Legendre, P., Cooke, I. M. and Vincent, J.-D. (1982) Regenerative responses of long duration recorded intracellularly from dispersed cell cultures of fetal mouse hypothalamus. *J. Neurophysiol.* **48**, 1121–41.

Legendre, P., McKenzie, J. S., Dupouy, B. and Vincent, J. D. (1985) Evidence for bursting pacemaker neurones in cultured spinal cord cells. *Neuroscience* **16**, 753–67.

Legendre, P., Poulain, D. A. and Vincent, J. D. (1988) A study of ionic conductances involved in plateau potential activity in putative vasopressinergic neurons in primary cell culture. *Brain Res.* **457**, 386–91.

Lemos, J. R., Adams, W. B., Novak-Hofer I., Benson, J. A. and Levitan, I. B. (1986) Regulation of neuronal activity by protein phosphorylation. In *Neural Mechanisms of Conditioning* (eds D. L. Alkon and C. D. Woody), pp. 397–420. Plenum Press, New York and London.

Levitan, I. B. and Benson, J. A. (1981) Neuronal oscillators in *Aplysia*: modulation by serotonin and cyclic AMP. *Trends NeuroSci.* **4**, 38–41.

Prince, D. A. (1978) Neurophysiology of epilepsy. *Annu. Rev. Neurosci.* **1**, 395–415.

Rinzel, J. and Lee, Y. S. (1987) Dissection of a model for neuronal parabolic bursting. *J. Math. Biol.* **25**, 653–75.

Schwartzkroin, P. A. and Stafstrom, C. E. (1980) Effects of EGTA on the calcium-activated afterhyperpolarization in hippocampal CA3 pyramidal cells. *Science, N.Y.* **210**, 1125–6.

Schwindt, P. C. and Crill, W. E. (1982) Factors influencing motoneuron rhythmic firing: results from a voltage-clamp study. *J. Neurophysiol.* **48**, 875–90.

Scudder, C. A. (1988) A new local feedback model of the saccadic burst generator. *J. Neurophysiol.* **59**, 1455–75.

Strumwasser, F. (1967) Types of information stored in single neurons. In *Invertebrate Nervous Systems* (ed. C. A. G. Wiersma), pp. 291–319. University of Chicago Press, Chicago.

Tazaki, K. and Cooke, I. M. (1983) Neuronal mechanisms underlying rhythmic bursts in crustacean cardiac ganglia. In *Neural Origin of Rhythmic Movements*, Society for Experimental Biology Symposium XXXVII (eds. A. Roberts and B. Roberts), pp. 129–57. Society for Experimental Biology, Cambridge.

Thompson, S. H. (1977) Three pharmacologically distinct potassium channels in molluscan neurones. *J. Physiol.* **265**, 465–88.

Thompson, S. and Coombs, J. (1988) Spatial distribution of Ca currents in molluscan neuron cell bodies and regional differences in the strength of inactivation. *J. Neurosci.* **8**, 1929–39.

Traub, R. D. and Wong, R. K. S. (1983) Synchronized burst discharge in disinhibited hippocampal slice. II. Model of cellular mechanism. *J. Neurophysiol.* **49**, 459–71.

APPENDIX: COMPUTER SIMULATION OF BURSTING

The simulation was written in BASIC. Na^+, Ca^{2+} and K^+ currents were described in Hodgkin–Huxley form (subroutine Ipoly). Parameters for Na^+ and Ca^{2+} currents were estimated from the work of Adams and Gage (1979) on cell R15 in *Aplysia juliana*; those for K^+ currents from Thompson's (1977) work on *Tritonia*. The parameters were then adjusted to fit our own, less extensive, voltage-clamp studies on R15 in *Aplysia californica* (subroutines Param and Pspike). The burst currents I_D and I_H were incremented by spikes (subroutine Iburs) and decayed with the kinetics found by Adams (1985) (subroutine Param). Changes in E_K were determined by $[K^+]_O$ in a local pool (lines 130–140, see text); line 130 was deleted for simulations in which E_K remained constant. The simulation was initialized (subroutine Init) to the most negative point of the midburst ($dV/dt = 0$) by setting the sum of I_H and the steady-state values of I_l (leak current), I_{Na}, I_{Ca} and I_K to zero. Iterative integration (subroutine Integrate) was carried out using a variable δt which kept changes in activation and inactivation functions within prescribed limits (subroutine Deltas). Machine-specific commands for data storage and graphics output have been deleted from the program.

```
10      ! BURST SIMULATION
20      !
30      GOSUB Scalet
40      GOSUB Param
50      GOSUB Init
60      GOSUB Integrate
70      STOP
80      END
90      !
100     !
110 Integrate:  ! ITERATIVE INTEGRATION
120 Tcheck:     IF T>Tmax THEN RETURN
130     Dko=K1*Ik*Delt+(10-Ko)*(1-EXP(-K2*Delt))
140     Ko=Ko+Dko               ! Delete if Vk is to remain constant
150     Vk=58*LGT(Ko)-133
160     !
170     GOSUB Iburs             ! Burst currents
180     GOSUB Ispike            ! Spike currents
190     !
200     Imem=Il+Id+Ih+Ina+Ica+Ik         ! Total membrane current
210     Dv=-Delt*Imem/C
220     V=V+Dv                  ! Increment membrane voltage
230     T=T+Delt                ! Increment time counter
240     GOTO Tcheck
250     !
260 Scalet:  ! TIME LIMITS AND SCALING
270     Tmax=35000
280     T0=-23000
290     Delt_max=100
300     Delt_min=.1
310     Delt=Delt_min
320     RETURN
330     !
340 Param:  ! VALUES FOR Cm, CONDUCTANCES AND EQUILIBRIUM POTENTIALS
350     C=50
360     Gna=40
370     Gca=12
380     Gk=60
390     Gl=.1
400     Vna=55
410     Vca=65
420     Vk=-75
430     Vl=0
440     Deltid=-1               ! Id
450     Tauid=2000
460     Deltih=.5               ! Ih
470     Tauih=20000
480     K1=.0005                ! K accumulation kinetics
490     K2=.0005
500     RETURN
510     !
520 Init:  ! INITIALIZATION
530     V0=-40
540     V=V0
550     Id0=0
560     Ko=10
570     !
580     Il=Gl*(V-Vl)
590     GOSUB Pspike
600     GOSUB Pinf
610     GOSUB Ipoly
620     Ih0=-(Il+Ina+Ica+Ik)              ! Net current = 0 at start of simulation
630     T=T0
640     Spike_flag=0
650     RETURN
660     !
```

```
670 Iburs:  ! CURRENTS Id AND Ih
680   Id0=Id0*EXP(-Delt/Tauid)
690   Ih0=Ih0*EXP(-Delt/Tauih)
700   Id=Id0
710   Ih=Ih0
720   !
730   IF (Dv>0) AND (V>0) THEN Spike_flag=1          ! Detect a spike
740   IF (Dv>0) OR (Spike_flag=0) THEN RETURN
750   Id0=Id0+Deltid                                 ! Increment Id and Ih
760   Ih0=Ih0+Deltih
770   Spike_flag=0
780   RETURN
790   !
800 Ispike:  ! SPIKE CURRENTS
810   GOSUB Pspike
820   GOSUB Deltas
830   GOSUB Ipoly
840   RETURN
850   !
860 Pspike:  ! PARAMETER FUNCTIONS FOR SPIKE CURRENTS
870   !
880   X=EXP(.26*(V-2))
890   Y=X/(1+X)
900   Minf=Y^(1/3)
910   Taum=.5+4.5/(1+EXP((V+2)/5))
920   !
930   Hinf=1/(1+EXP((V+12)/3))
940   Tauh=10+10/(1+EXP(V/5))
950   !
960   X=EXP(.31*(V-10))
970   Y=X/(1+X)
980   Mminf=Y^(1/2)
990   !
1000  X=EXP(.31*(V-10))
1010  Y=X/(1+X)
1020  Taumm=1+2*Y
1030  !
1040  Hhinf=1/(1+EXP((V+20)/4))
1050  Tauhh=50+100/(1+EXP((V-25)/10))
1060  !
1070  X=EXP(.22*(V-7))
1080  Y=X/(1+X)
1090  Ninf=Y^(1/2)
1100  !
1110  X=EXP((V+35)/10)
1120  Y=X/(1+X)
1130  Taun=Y*(10+70/(1+EXP((V-10)/10)))
1140  !
1150  Jinf=.3+.7/(1+EXP(.26*(V+7)))
1160  Tauj=2000/(1+EXP((V-10)/5))
1170  !
1180  RETURN
1190  !
1200 Pinf: !
1210  M=Minf
1220  H=Hinf
1230  Mm=Mminf
1240  Hh=Hhinf
1250  N=Ninf
1260  J=Jinf
1270  RETURN
1280  !
1290 Deltas: ! ADJUST Delt TO KEEP PARAMETER CHANGES WITHIN LIMITS
1300  !
1310  Dm=(Minf-M)*(1-EXP(-Delt/Taum))
1320  Dh=(Hinf-H)*(1-EXP(-Delt/Tauh))
1330  Dmm=(Mminf-Mm)*(1-EXP(-Delt/Taumm))
1340  Dhh=(Hhinf-Hh)*(1-EXP(-Delt/Tauhh))
1350  Dn=(Ninf-N)*(1-EXP(-Delt/Taun))
1360  Dj=(Jinf-J)*(1-EXP(-Delt/Tauj))
1370  !
1380  IF (ABS(Dm)<.050) AND (ABS(Dh)<.050) AND (ABS(Dmm)<.050) AND (ABS(Dhh)<.05
0) AND (ABS(Dn)<.050) AND (ABS(Dj)<.050) THEN Delt2
1390  IF Delt<Delt_min*2 THEN Delt2
1400  Delt=Delt/2
1410  GOTO Deltas
1420  !
```

```
1430 Delt2: IF (ABS(Dm))>.01) OR (ABS(Dh))>.01) OR (ABS(Dmm))>.01) OR (ABS(Dhh))>.01
) OR (ABS(Dn))>.01) OR (ABS(Dj))>.01) THEN Delt3
1440   IF Delt>Delt_max/2 THEN Delt3
1450   Delt=Delt*2
1460   GOTO Deltas
1470   !
1480 Delt3: M=M+Dm
1490   H=H+Dh
1500   Mm=Mm+Dmm
1510   Hh=Hh+Dhh
1520   N=N+Dn
1530   J=J+Dj
1540   RETURN
1550   !
1560 Ipoly: ! CALCULATION OF CURRENTS
1570   !
1580   Ina=Gna*M^3*H*(V-Vna)
1590   Ica=Gca*Mm^2*Hh*(V-Vca)
1600   Ik=Gk*N^2*J*(V-Vk)
1610   Il=Gl*(V-Vl)
1620   RETURN
```

Kinetics of release as a tool to distinguish between models for neurotransmitter release

H. PARNAS AND I. PARNAS

The Otto Loewi Center for Cellular and Molecular Neurobiology and the Department of Neurobiology, The Hebrew University of Jerusalem, Israel

INTRODUCTION

The process of neurotransmitter release in synapses has been known to depend on extracellular Ca^{2+} ions for many years (del Castillo and Katz, 1954). More recently it became evident that the action of Ca^{2+} in promoting neurotransmitter release comes from inside the cell (Katz and Miledi, 1977; Llinás *et al.*, 1981). Based on these and similar experiments, the concept that Ca^{2+} is not only required, but is also sufficient to evoke release on its own, became widely accepted (Stockbridge and Moore, 1984; Simon and Llinás, 1985; Fogelson and Zucker, 1985). However, there is no conclusive evidence that Ca^{2+} is indeed the only limiting factor in the process of neurotransmitter release.

The assumption that Ca^{2+} is required and sufficient to start the release process is in fact the essence of the 'Ca^{2+} hypothesis' for release. An unavoidable conclusion from this hypothesis is that after an impulse, release should last as long as the Ca^{2+} concentration near crucial domains of release is above a certain level. Moreover, release after an impulse should last longer if more Ca^{2+} enters or if its removal is slowed (Parnas and Segel, 1984). In contrast, the time course of release after an impulse was found to be insensitive to variations in extracellular Ca^{2+} concentration (Datyner and Gage, 1980), to repetitive stimulation (Datyner and Gage, 1980; Barrett and Stevens, 1972; H. Parnas *et al.*, 1986a), or to other treatments that are known to modulate intracellular Ca^{2+} concentration (Matzner *et al.*, 1988) and its regulation. Therefore a conflict exists between predictions from the classical

Ca^{2+} hypothesis and experimental results pertaining to the time course of release.

Two main approaches were advanced for solving this conflict. The first remains in the boundaries of the classical Ca^{2+} hypothesis, but adds refinements associated with the spatio-temporal changes in intracellular Ca^{2+} concentration (Simon and Llinás, 1985; Fogelson and Zucker, 1985). In particular, these authors still assume that Ca^{2+} is the only limiting factor in the release process. Other factors, such as release sites or vesicles, are not only assumed to be ready and constant during release, but to be in excess. These authors developed detailed mathematical models to describe the entry of Ca^{2+} through patches of Ca^{2+} channels and away from the channels by diffusion.

Diffusion is the key process in keeping the time course of release short and insensitive to conditions that alter intracellular Ca^{2+} concentration.

A second approach questions the legitimacy of the foundation of the Ca^{2+} hypothesis, namely that Ca^{2+} is the only limiting factor in the release process. It suggests that while Ca^{2+} is certainly required for release, it is nevertheless unable to support evoked release by itself. This approach assumes that another factor, together with Ca^{2+}, accumulates during the natural stimulus, and that both this factor and Ca^{2+} are required to start the chain of events leading to release. In the case of the neuromuscular junction, the natural stimulus is membrane depolarization, and therefore this second factor, called S, must be produced by membrane depolarization and must rapidly disappear with membrane repolarization. Thus the classical Ca^{2+} hypothesis was extended, becoming the 'Ca^{2+}–voltage' hypothesis (H. Parnas *et al.*, 1986a; I. Parnas *et al.*, 1986). According to the Ca–voltage hypothesis, the amount of release or quantal content depends both on intracellular Ca^{2+} concentration and the amount of S, while the time course of release, after an impulse, depends mainly on the time course of disappearance of the membrane potential-dependent factor, S. Therefore, there is no need to correlate the time course of release to the time course of the increase and decrease in intracellular Ca^{2+} concentration after an impulse, and there is no more conflict between the hypothesis and the experimental results which show independence of the time course of release on conditions which manipulate intracellular Ca^{2+} concentration.

The most direct way to test the Ca^{2+} hypothesis is to monitor the changes in Ca^{2+} concentrations using Ca^{2+} indicators. However, even with the most sophisticated new developments in the field of Ca^{2+} indicators, the present spatio-temporal resolution is not refined enough to monitor changes in Ca^{2+} concentration near the Ca^{2+} channels and release sites. In the same way, there are no clues as to what that missing factor S could be. We therefore used a theoretical approach to distinguish between these two hypotheses.

In the present chapter, we show some of the main predictions from the two hypotheses. These predictions are compared with key experimental results, followed by a discussion as to the ability of these two models to cope with the experiments.

EXPERIMENTAL RESULTS

The essence of the Ca^{2+} hypothesis is that the time course of the release process should follow closely the temporal change in intracellular Ca^{2+} concentration. Therefore, we describe first the experimental procedure to measure the time course of release. Katz and Miledi (1965) showed that under conditions of low quantal content and low temperature, the normally short duration of release (1–2 ms at 20°C) is stretched to last about 8–10 ms. Moreover, most of the impulses do not elicit release, while others give rise to 1 quantum, and very rarely to 2 or 3 quanta. These single quanta vary in the delay that elapses between the stimulus and the appearance of the quanta. Katz and Miledi (1965) further showed that the origin of this delay is presynaptic. Therefore, a synaptic delay histogram reflects the probability of a single quantum being released at a given time after an impulse. Thus, it measures the kinetics of release and, as such, can reveal characteristic properties of the molecular events underlying the process of release.

Dependence of evoked release on membrane depolarization

Figure 1 (overleaf) shows that the time course of release is independent of the amplitude of a depolarizing pulse. In particular, release does not start earlier when following higher depolarization, even though at this range of depolarization, the influx of Ca^{2+} is larger and starts sooner.

Kinetics of release during facilitation

Synaptic delay histograms measured for two successive impulses (Fig. 1b and c) show complete overlap; thus we may conclude that the kinetics of release of the second impulse was not altered due to the residual Ca^{2+} from the first impulse. Similarly, release did not start earlier, even when the test pulse was administered after a train of pulses producing large facilitation (Parnas et al., 1989).

Dependence of the kinetics of release on intracellular Ca^{2+} concentration

It is expected that if the intracellular Ca^{2+} concentration is increased sufficiently, release should occur without a pulse, or at least start earlier after

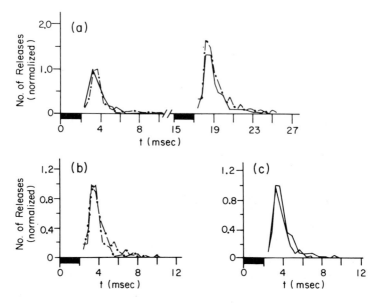

Fig. 1. Time course of release elicited by twin pulses at two levels of depolarization. (a) At left, delay histograms elicited by the first pulses, respectively, of −0.8 μA (---) and −1 μA (solid line). At right, the second pulses at the same depolarization. The procedure of normalization is as follows: Histograms elicited by the first pulses (left) are each normalized to its peak amplitude. Histograms elicited by the second pulses (right) are each normalized to the peak amplitude of the corresponding first pulse histogram. (b) Synaptic delay histograms elicited when both the first and the second pulses are at −0.8 μA. Histograms are each normalized to its peak amplitude. (c) Same as (b), except that pulse amplitude is −1 μA. (Redrawn from Parnas *et al.*, 1989.)

a pulse, as the critical level of Ca^{2+} will be reached sooner. When a Ca^{2+} ionophore was introduced (Fig. 2), release started with the same minimal delay, even though the quantal content increased five-fold after the ionophore treatment.

The conclusions formed from these types of experiments and others published in the literature (see earlier discussion) are that the time course of release is insensitive to changes in the entry of Ca^{2+} or to the level of intracellular Ca^{2+} concentration. Another important conclusion stems from the ionophore experiment. Even if Ca^{2+} is present presumably below the release sites, and in an appreciable concentration, synchronized release only starts after a depolarizing pulse. In summary, the predictions from the classical Ca^{2+} hypothesis do not match these experimental results.

Effect of temperature on the kinetics of release

The time course of release is very sensitive to temperature (Dudel, 1984).

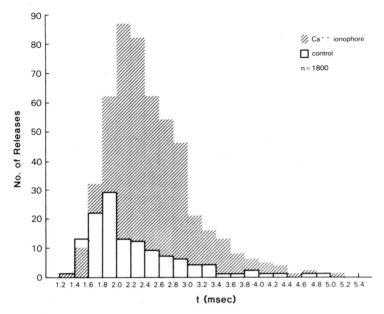

Fig. 2. Synaptic delay histograms before and after an application of a Ca^{2+} ionophore. *n* denotes the number of applied pulses. (J. M. Wojtowicz, H. L. Atwood, H. Parnas and I. Parnas, unpublished.)

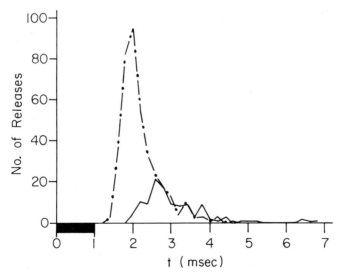

Fig. 3. Synaptic delay histograms at two temperatures (11°C, dashed line, 8°C, solid line). 1000 pulses of 1 ms duration and −1.2 µA amplitude were employed. The quantal contents are 0.38 for 11°C and 0.11 for 8°C. The bin size is 0.2 ms and the continuous lines were obtained by taking the middle time point of each bin. Normal Ringer+10^{-7} M TTX.

Release starts and reaches its peak sooner and terminates faster at higher temperatures (Fig. 3) and the Q10 can be as high as 4.

Ca²⁺ domain and diffusion theories and kinetics of release

We now turn to a group of theories which attempt to remain within the constraints of the Ca^{2+} hypothesis and yet account for the experimental results described above. First we present a detailed description of one such theory, the Fogelson–Zucker model (1985), followed by a general discussion of this approach.

According to the Fogelson–Zucker model (1985), Ca^{2+} enters through channels which are localized in patches in the membrane. As depolarization increases, more channels open, but the Ca^{2+} current through each channel, is smaller. Thus at a certain range of membrane depolarization, there may be an increase in the macrocurrent of Ca^{2+} while there is a decrease in the microcurrent through each channel. Once it has entered, the Ca^{2+} spatio-temporal distribution below each channel is mainly regulated by three-dimensional diffusion. The combination of patches of channels with three-dimensional diffusion results in the creation of restricted Ca^{2+} domains below and in close proximity to each channel. Thus an essential constraint is formed, namely the release sites at which Ca^{2+} acts to start the chain of events leading to release must be very close to the Ca^{2+} channel, within a distance at which the Ca^{2+} concentration is high. Aside from the kernel of the Ca^{2+} diffusion hypothesis described above, various representatives of the Ca^{2+} diffusion hypothesis differ in detail. However, the predictions outlined below result from this common kernel. The papers which originally presented the Ca^{2+} domain and diffusion theories mainly discussed predictions as to the amount of release under various conditions. We are presenting here predictions of these theories as to the time course of release.

Figure 4 depicts predicted synaptic delay histograms of twin pulses at two levels of depolarization. The most noticeable general result is that in contrast to experiments (Fig. 1), release occurs within the 1 ms pulse. Recall that in the experiments, even the minimal latency outlasted the 2 ms pulse duration. In Fig. 4 (in accord with Fogelson and Zucker), diffusion governs the kinetics of release, while the processes that follow and lead to exocytosis are fast. As a result, the synaptic delay histograms are insensitive to temperature, in contrast to the experimental results (Fig. 3).

Figure 5 shows simulation of the Fogelson–Zucker model in which the spatio-temporal distribution of intracellular Ca^{2+} is unaltered, but now a slow process links intracellular Ca^{2+} to exocytosis. Then, the synaptic delay histograms, as in the experiments, are sensitive to temperature. Also, the minimal latency is longer, and at least for the first pulse at the low depolarization, almost outlasts the length of the pulse (Fig. 5b). However,

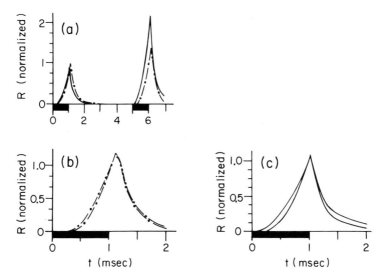

Fig. 4. Simulation of synaptic delay histograms (R) elicited by twin pulses at two levels of depolarization. The simulations are of the three-dimensional diffusion model of Fogelson and Zucker (1985) in which diffusion governs the kinetics of release. (a) Simulation of histograms (R) elicited by the first (left) and the second (right) pulses at low (-··-) and high (solid line) depolarization. Procedure of normalization as in (a) of Fig. 1. (b) R of the first and second pulse at the low depolarization. (c) R of the first and second pulse of the high depolarization. Note differences between the experimental (Fig. 1) and the simulated time course. (Redrawn from figure 9 of Parnas *et al.*, 1989.)

now the predicted delay histograms are even more sensitive than before (compare with Fig. 4) to the level of depolarization (Fig. 5a) and to the level of resting intracellular Ca^{2+} (Fig. 5b and c). A detailed description of the behaviour of the Fogelson–Zucker model under various conditions is given in Parnas *et al.* (1989).

In conclusion, the refined Ca theory, as does the classical one before (Parnas and Segel, 1984; Parnas *et al.*, 1986b), predicts in contrast to experiments that the time course of release is sensitive to variations in either influx of Ca^{2+} or its intracellular level prior to stimulation.

The Ca^{2+}–voltage hypothesis

It seems, therefore, that the Ca^{2+} hypothesis should be extended rather than refined. Recall from the experiments that even when the Ca^{2+} concentration below the release sites was quite high (Fig. 2), synchronized release started only when the membrane was depolarized. Such types of results provided the first clue as to which direction to extend the Ca^{2+} hypothesis, keeping in

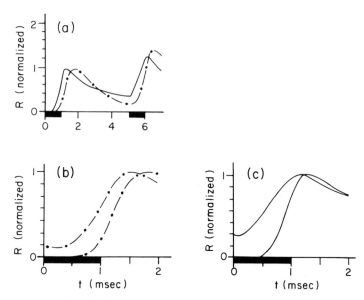

Fig. 5. Simulation of R according to the three-dimensional diffusion model when a slow process links Ca^{2+} to exocytosis. (a), (b) and (c), as in Fig. 4. (Redrawn from figure 13 in Parnas *et al.*, 1989.)

mind that Ca^{2+} is essential for release. Accordingly, we suggested (Dudel *et al.*, 1983; Parnas *et al.*, 1986a) that due to and during the depolarization produced by the action potential, two essential factors for release become available. The first is the increase in intracellular Ca^{2+} concentration due to the entry of Ca^{2+}, eventually at critical regions. The second is an additional factor, S, as yet unknown, which becomes available due to depolarization. Upon repolarization, S disappears with a certain time constant. While both Ca^{2+} and S determine the amount of transmitter released, it is the kinetics of S that governs the time course of release. We therefore extended the Ca^{2+} hypothesis to become the 'Ca^{2+}–voltage' hypothesis.

Formulation of the Ca–voltage hypothesis

A minimal model for the classical Ca^{2+} hypothesis and its refined version is:

$$n\,Ca + X \underset{k_{-2}}{\overset{k_2}{\rightleftharpoons}} (Ca^n\,X) \tag{1a}$$

$$(Ca^n\,X) + V \overset{k_3}{\to} L \tag{1b}$$

where X is some entity binding Ca^{2+} and starting the chain of events leading to release; k_2 and k_{-2} are rate constants; V stands for a vesicle and k_3 is the

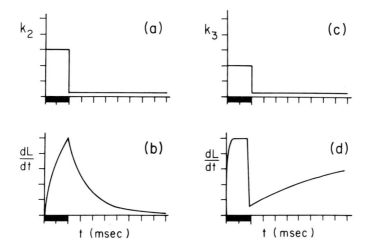

Fig. 6. Simulations of the expected synaptic delay histograms when depolarization increases either k_2 (b) or k_3 (d) in eqn (1a) and (1b), respectively. The time course of k_2 and k_3 are depicted in (a) and (b), respectively.

rate constant, lumping the many processes leading to exocytosis; n stands for the cooperative dependence of release L on Ca^{2+}.

In an attempt to account for an additional effect (aside from Ca^{2+} entry) of membrane depolarization in the release process, each one of the rate constants in the reactions that follow the entry of Ca^{2+} could be voltage-dependent. Figure 6 depicts the outcome of such a mode of extension. Indeed the time course of release depends, then, on membrane potential, but the peak of release is obtained prior to or at most at the end of the depolarizing pulse (Parnas et al., 1986a). This is in contrast to the experimental results where the peak of the synaptic delay histogram always follows the end of the impulse (Fig. 1).

An alternative way is that a second factor, S, accumulates during depolarization in parallel and independently to the entry and accumulation of Ca^{2+}. This factor, then, gradually inactivates upon repolarization. Thus,

$$T \underset{k_{-1}h}{\overset{k_1 d}{\rightleftharpoons}} S \tag{2a}$$

$$Ca^{2+} + S \underset{k_{-2}}{\overset{k_2}{\rightleftharpoons}} (CaS) \tag{2b}$$

In (2a), T stands for molecules in the membrane which are in an inactive form and cannot bind Ca^{2+} but can sense the electric field across the membrane. The rate of activation k_1^d increases with depolarization, while the rate k_{-1}^h of inactivation of S increases with hyperpolarization. Thus S differs from X in eqn (1) in the sense that its amount is not constant and at rest is rather low. Once S is produced, it binds Ca^{2+}, and starts, as in eqn (1), the chain of events leading to release. We show later that eqn (2) indeed enables the peak of the delay histogram to follow the end of the impulse. We must, however, attend another aspect of release before completing the model in eqn (2).

Cooperativity in release

The total amount of released transmitter (the quantal content) was shown to depend in a sigmoidal manner on extracellular Ca^{2+} concentration. This was interpreted to reflect a cooperative dependence of release on Ca^{2+} (Dodge and Rahamimoff, 1967). Later, other aspects of release, facilitation, were best accounted for quantitatively, if indeed a cooperative relation exists (Rahamimoff, 1968; Parnas and Segel, 1980; Parnas et al., 1982). The value of the cooperativity, n, obtained from the slope of a log release/log $[Ca^{2+}]_o$ plot, was suggested to be 4 in the frog (Dodge and Rahamimoff, 1967), but others suggested other values ranging from 1 to 5. Furthermore, the nature of cooperativity cannot be evaluated from measurements of total release. In the simplified scheme in (1), there are two alternatives for the cooperativity. In the first, n Ca^{2+} ions bind to one X (or S). Alternatively, one Ca^{2+} ion binds to X (or S), but n complexes (n(CaX) or n(CaS)) are required for release. From the experiment relating total release to extracellular Ca^{2+} concentration, it is impossible to distinguish between these two alternatives. Moreover, due to the masking effect of the resting intracellular Ca^{2+} concentration, the value of the measured slope of log release versus log $[Ca^{2+}]_o$ is an underestimate of the true cooperativity (Parnas and Segel, 1981; Parnas et al., 1982).

Parnas et al. (1986b) showed that the initial phase of the delay histogram reveals rather faithfully the value of n. This holds true both for (1), i.e. the Ca^{2+} hypothesis and for (2), the Ca^{2+}–voltage hypothesis. Thus, the slope of log dL/dt versus log t at the early phase of the histogram corresponds to n. The behaviour of this slope under various experimental conditions substantiated the choice of (2) over (1) (Parnas et al., 1986b). Finally, the experimental observation that the delay histogram rises in a sigmoidal fashion with time, and that the quantal content depends in a sigmoid manner on Ca^{2+} concentration, led to the conclusion that the cooperativity resides in the step where n complexes of CaS are required for release.

This last observation completes the list of details which are required to formulate a minimal model for the Ca–voltage hypothesis. Accordingly,

$$n(CaS) + V \overset{k_3}{\to} L \tag{2c}$$

Solution of (2) with the inclusion of the influx of Ca^{2+} through voltage-dependent Ca^{2+} channels, and removal of Ca^{2+} with a lumped group of processes which show saturative dependence on Ca^{2+}, is depicted in Fig. 7. It is clear that according to the Ca^{2+}–voltage hypothesis, similar to the experimental results, the kinetics of release is insensitive to the amplitude of the depolarizing pulse (Fig. 7a), and not to the level of intracellular Ca^{2+} concentration prior to stimulation (Fig. 7b and c). Moreover, dL/dt, as a function of time, also faithfully follows other experimental observations. It rises in a sigmoidal fashion in time, the minimal latency almost outlasts the duration of a 2 ms pulse and the histogram reaches its peak significantly after the end of the pulse. Moreover, according to (2), the delay histogram is sensitive to temperature.

As stated above, according to (2), dL/dt is insensitive to the amplitude of depolarization during the pulse. However, dL/dt is sensitive to the membrane potential prior to or immediately following the depolarizing pulse. Examination of (2a) shows that if k_{-1}^{h} is raised by post-pulse hyperpolarization, release will reach its peak and terminate sooner (I. Parnas *et al.*, 1986).

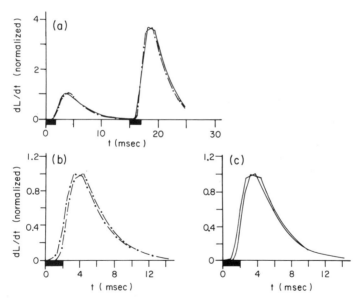

Fig. 7. Simulation of the time course of release of twin pulses at two levels of depolarization using the Ca–voltage hypothesis (eqn (2)). (a), (b) and (c) are as in Figs 4 and 5.

SUMMARY

In this chapter, we have presented combined theory and experiments, which together led to the suggestion and formulation of the Ca–voltage hypothesis for neurotransmitter release from neuromuscular junctions. The use of theory enabled us to choose between alternative possibilities and to select the route which best agrees with the experimental results. The outcome, the Ca^{2+}–voltage hypothesis, encompasses a wide scope of known experimental results in the field of the release of neurotransmitter from nerve endings. Moreover, the analysis presents constraints on what the as yet unknown S might be. The essence of the Ca–voltage hypothesis is that neurotransmitter release is not evoked by a change in a single parameter, Ca^{2+}; it requires the joined action of two factors that vary concomitantly as a result of the natural stimulus, which in the case of the neuromuscular junction is the depolarization produced by the action potential.

In other systems, where the natural stimulus triggering release is not an action potential, the cause for accumulation of a second limiting factor besides Ca^{2+} may obviously be one other than depolarization. Also, the nature of the second limiting factor may be different to S. In a system in which release is prolonged, the time course of release may reflect the time course of Ca^{2+} removal. But experimental results on such a system are lacking.

We believe that the rule that we have presented here in some detail, i.e. more than one limiting factor controls a fundamental biological process, is very advantageous. It diminishes the risk of 'mistakes' due to fluctuations in concentration of such an abundant element as Ca^{2+}. Furthermore, it is known that Ca^{2+} ions are involved in the control of many processes, even within a single cell. A mechanism by which two limiting factors are required, one common, the Ca^{2+}, but the second different for the various reactions, enables a more precise and differential control.

Acknowledgements: We thank Julie Anello for editing the manuscript, and for her skilled typing.

REFERENCES

Barrett, E. F. and Stevens, C. F. (1972) The kinetics of transmitter release at the frog neuromuscular junction. *J. Physiol., Lond.* **227**, 691–708.

Datyner, N. B. and Gage, P. W. (1980) Phasic secretion of acetylcholine at a mammalian neuromuscular junction. *J. Physiol., Lond.* **303**, 299–314.

del Castillo, J. and Katz, B. (1954) The effect of magnesium on the activity of motor nerve endings. *J. Physiol., Lond.* **124**, 553–9.

Dodge, F. A. Jr. and Rahamimoff, R. (1967) Cooperative action of calcium ions in transmitter release at the neuromuscular junction. *J. Physiol., Lond.* **193**, 419–32.

Dudel, J. (1984) Control of quantal transmitter release at frog's motor nerve terminals. I. Dependence on amplitude and duration of depolarization. *Pflugers Arch.* **402**, 225–34.

Dudel, J., Parnas, I. and Parnas, H. (1983) Neurotransmitter release and its facilitation in crayfish muscle. VI. Release is determined by both intracellular calcium concentration and depolarization of the nerve terminal. *Pflugers Arch.* **339**, 1–10.

Fogelson, A. L. and Zucker, R. S. (1985) Presynaptic calcium diffusion from various arrays of single channels. *Biophys. J.* **48**, 1003–17.

Katz, B. and Miledi, R. (1965) The measurements of synaptic delay and the time course of acetylcholine release at the neuromuscular junction. *Proc. R. Soc. Lond.* B **161**, 483–95.

Katz, B. and Miledi, R. (1977) Suppression of transmitter release at the neuromuscular junction. *Proc. R. Soc. Lond.* B **196**, 465–9.

Llinás, R., Steinberg, I. Z. and Walton, K. (1981) Relationship between presynaptic calcium current and post-synaptic potentiation in squid synapse. *Biophys. J.* **33**, 322–51.

Matzner, H., Parnas, H. and Parnas, I. (1988) Presynaptic effects of d-Tubocurarine on neurotransmitter release at the neuromuscular junction of the frog. *J. Physiol., Lond.* **398**, 109–21.

Parnas, H. and Segel, L. A. (1980) A theoretical explanation for some effects of calcium on the facilitation of neurotransmitter release. *J. theor. Biol.* **84**, 3–29.

Parnas, H. and Segel, L. A. (1981) A theoretical study of calcium entry in nerve terminals, with application to neurotransmitter release. *J. theor. Biol.* **91**, 125–69.

Parnas, H. and Segel, L. A. (1984) Exhaustion of calcium does not terminate evoked neurotransmitter release. *J. theor. Biol.* **107**, 345–65.

Parnas, H., Dudel, J. and Parnas, I. (1982) Neurotransmitter release and its facilitation in crayfish. I. Saturation kinetics of release and of entry and removal of calcium. *Pflugers Arch.* **393**, 1–14.

Parnas, H., Dudel, J. and Parnas, I. (1986a) Neurotransmitter release and its facilitation in crayfish. VII. Another voltage dependent process beside Ca entry controls the time course of phasic release. *Pflugers Arch.* **406**, 121–30.

Parnas, H., Parnas, I. and Segel, L. A. (1986b) A new method for determining cooperativity in neurotransmitter release. *J. theor. Biol.* **119**, 481–99.

Parnas, H., Hovav, G. and Parnas, I. (1989) The effect of Ca^{2+} diffusion on the time course of neurotransmitter release. *Biophys. J.* (in press).

Parnas, I., Parnas, H. and Dudel, J. (1986) Neurotransmitter release and its facilitation in crayfish. VIII. Modulation of release by hyperpolarizing pulse. *Pflugers Arch.* **406**, 131–7.

Rahamimoff, R. (1968) A dual effect of calcium ions on neuromuscular facilitation. *J. Physiol., Lond.* **195**, 471–80.

Simon, S. M. and Llinás, R. (1985) Compartmentalization of the submembrane calcium activity during calcium influx and its significance in transmitter release. *Biophys. J.* **48**, 485–98.

Stockbridge, N. and Moore, J. W. (1984) Dynamics of intracellular calcium and its possible relationship to phasic transmitter release and facilitation at the frog neuromuscular junction. *J. Neurosci.* **4**, 803–11.

Collective properties of insulin-secreting cells

ARTHUR SHERMAN AND JOHN RINZEL

National Institutes of Health, National Institute of Diabetes and Digestive and Kidney Diseases, Mathematical Research Branch, Bethesda, MD 20892, USA

INTRODUCTION

Pancreatic islets of Langerhans, composed primarily of insulin-secreting β-cells, are a rich system for the study of cellular signalling. In this chapter we will explore how electrical coupling by gap junctions enables the islet to function as a syncytium, and we will indicate both the unique features of the β-cell system and the similarities with other systems discussed in this conference.

At one time, the study of cells with excitable membranes coincided with neurobiology and muscle physiology. More recently the outlook has broadened to include endocrine and other cells, which have similar ionic channels and modulatory mechanisms. Indeed, the Chay–Keizer theory (Chay and Keizer, 1983) which first successfully accounted for many of the key features of β-cell electrical activity, was based on modifications of Hodgkin–Huxley (HH) kinetics (Hodgkin and Huxley, 1952). An important difference is that endocrine cells lack the complex geometry of axons and dendrites used by neurones for spatio-temporal signal processing at the individual cell level.

Like many neurones, such as the *Aplysia* neurosecretory R15 neurone and mammalian thalamic neurones, β-cells exhibit endogenous bursting oscillations (Fig. 1A and B). Like R15, the slow modulation of bursting appears to be controlled by the accumulation of intracellular calcium. We will make some comparisons of the mathematical models in the next section.

Calcium functions as part of a dual negative feedback system, involving both electrical activity and secretion. Calcium removal leads to depolarization and terminates the silent phase by deactivating calcium-activated

Cell to Cell Signalling: From
Experiments to Theoretical Models
ISBN 0–12–287960–0

61

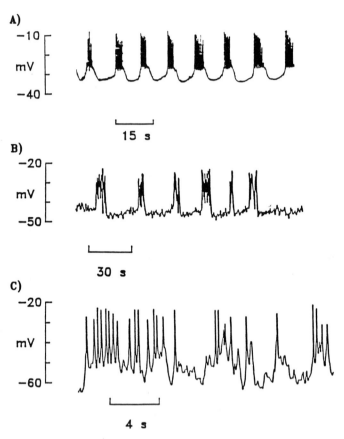

Fig. 1. Experimental recordings from individual β-cells in an intact islet of Langerhans (A), a reconstituted cluster of diameter 70 μm (B), and an isolated β-cell (C). (From Sherman *et al.*, 1988; originally, (A) is from Atwater and Rinzel, 1986, (B) and (C) from Rorsman and Trube, 1986.)

potassium (K–Ca) channels. This depolarization activates voltage-gated calcium and HH-like potassium channels and initiates spiking. The ensuing influx of calcium ultimately inhibits spiking as calcium accumulates. The connection with secretion is through glucose metabolism, which increases with glucose levels in the blood and provides the energy for calcium removal. As in many other secretory systems such as other endocrine cells, the cAMP relay signalling in *Dictyostelium discoideum*, and the neuromuscular synaptic junction, the entry of calcium is a trigger for secretion, in this case for insulin which reduces blood glucose levels over the course of many bursts. The Chay–Keizer class of models offers several possible explanations for how increase of glucose regulates electrical activity by increasing the relative duration of the active phase without affecting the spike amplitudes and frequencies. The key mathematical tools for obtaining this understanding are the separation into fast and slow processes and geometric phase plane analysis.

While some neurones, such as R15, can burst in isolation, isolated β-cells do not burst but show disorganized spiking (Fig. 1C). *In vivo* β-cells are found in islets of 10^3–10^4 cells coupled by gap junctions, and cultured cells spontaneously form coupled clusters. There are no reports of bursting in clusters smaller than 25–50 cells. Figure 1B shows bursting in a cluster of about 70 cells. Here we describe a first step toward developing a theory to understand the process by which this organized behaviour emerges. We propose that the irregular behaviour of a single cell is the result of stochastic openings and closings of the largest channels in the cell – in our treatment the K–Ca channels. In sufficiently large clusters of cells the single-channel conductance events are averaged by being shared through gap junctions over the combined capacitance (membrane area) of the entire cluster. This idea is based on a suggestion by Atwater *et al.* (1983) who found that the single-channel conductance of a K–Ca channel was comparable to the whole-cell K^+ conductance and concluded that the rest potential would be unstable unless channels could be shared.

At the end of this chapter we offer some speculation about why bursting may be beneficial to the main function of the β-cell, which is to regulate blood glucose levels by secreting insulin. Insulin secretion is pulsatile and under certain conditions has been shown to be closely correlated to the active phases of the bursts (Rosario *et al.*, 1986), although there are important dissociations as well (Atwater *et al.*, 1984). We note that many other secretory cells are pulsatile, such as the hourly oscillations of the GnRH system discussed in this volume. However, the bursting oscillations we discuss here (with a period of tens of seconds) appear to have the purpose of organizing the secretory process itself and not of optimizing the influence on target tissues, since the pancreas contains many islets which oscillate asynchronously *in vitro*. (It is not known whether islets synchronize *in vivo*.)

THE CHAY–KEIZER THEORY

The model

Chay and Keizer (1983) developed a successful dynamic model of β-cell electrical activity based on an earlier physiological model of Atwater *et al.* (1980). The model has evolved in detail over the years so we present the equations for a typical Chay–Keizer type model:

$$C_m \frac{dV}{dt} = -I_{Ca}(V) - I_K(V,n) - g_{K-Ca}(V - V_K) \tag{1}$$

$$\frac{dn}{dt} = \lambda \left[\frac{n_\infty(V) - n}{\tau_n(V)} \right] \tag{2}$$

$$\frac{dCa_i}{dt} = f(-\alpha I_{Ca}(V) - k_{Ca} Ca_i) \tag{3}$$

The independent variables are V, the transmembrane potential (inside minus outside); n, the fraction of HH-like K^+ channels open; and Ca_i, free intracellular calcium concentration.

In this section we consider the case of intracellular recording from a representative cell in an intact islet of Langerhans, and we take g_{K-Ca} to be an instantaneous function of Ca_i:

$$g_{K-Ca} = \bar{g}_{K-Ca} \frac{Ca_i}{K_d + Ca_i} \tag{4}$$

In the original Chay–Keizer model (Chay and Keizer, 1983, 1985) the kinetics for the inward calcium current, I_{Ca}, and the outward potassium current, I_K, were based on modifications of the HH model for squid sodium and potassium currents respectively. We have continued to follow the HH formulation in representing these currents as products of voltage-gated conductances with driving potentials that reverse at the Nernst potentials, but we have incorporated voltage-clamp data obtained by Rorsman and Trube (1986) for the β-cell. Details about the functional forms and parameter values may be found in Sherman et al. (1988).

Because the Ca^{2+} channel is an order of magnitude faster than the K^+ channel we let I_{Ca} be an instantaneous function of voltage. The 'time constant' for potassium channel activation, τ_n/λ, is in the range 15–20 ms and is comparable to the effective time constant for V. Calcium, however, is a slow variable because, f, the fraction of calcium which is free, is very small ($\sim 10^{-3}$). Slow oscillations of Ca_i modulate the bursts, while the fast V-n subsystem generates the spikes within a burst.

Figure 2A and B show the V, Ca_i and g_{K-Ca} time courses obtained by numerically integrating eqns (1)–(3), (4). In the low-voltage silent phase, I_{Ca} is small, the net flux of calcium is outward, g_{K-Ca} slowly decreases, and the cell slowly depolarizes. When the threshold for the voltage-gated Ca^{2+} channels is reached there is a rapid depolarization. This depolarization opens the voltage-gated K^+ channels, with a delay on the order of the time constant of eqn (2), and spiking oscillations of the active phase are established. These spikes bring Ca^{2+} into the cell which reactivates the K–Ca channels and eventually terminates the burst and repolarizes the cell. In this model the binding constant, K_d, of Ca^{2+} to the K–Ca channels is much larger than Ca_i, corresponding to the small fraction of open channels. Hence g_{K-Ca} is approximately proportional to Ca_i, as is revealed in Fig. 2B. Also observe that small changes in K^+ conductance can flip the cell between the active and

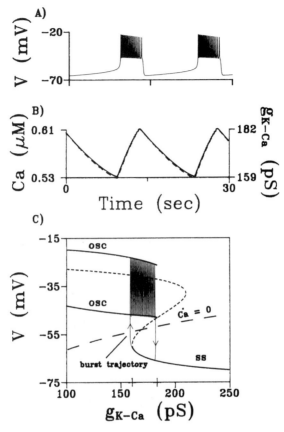

Fig. 2. Simulated bursting solution obtained by numerically integrating eqns (1)–(3), (4). (A) Voltage time course. (B) Time course of intracellular calcium (solid) and calcium-activated potassium conductance (dashed). (C) Projection of burst trajectory into V–$g_{K\text{-}Ca}$ phase plane. Direction of increasing time is indicated by arrows. Solution alternates between coexistent stable states of fast subsystem: low-voltage steady state (lower branch of Z-shaped 'ss' curve) and depolarized oscillation (amplitude indicated 'osc'). Unstable solutions of fast subsystem indicated by dashes. Intracellular calcium increases above and decreases below calcium nullcline thereby generating overall clockwise circulation.

silent phases. Adams and Benson (1985) point out that small currents modulate bursting in R15 neurones.

Other mechanisms have been proposed to account for bursting in β-cells. Some models replace $g_{K\text{-}Ca}$ by, or combine it with, an ATP-blockable potassium channel (K–ATP). Others use calcium inactivation of the calcium channel. See Keizer (1988) for a review, and also Chay and Cook (1988). All these models have in common the modulation of the bursts by slow feedback of Ca_i on some membrane channel. The analysis in the following subsection, and the general conclusions about coupling of β-cells presented in this paper hold for all models in this class.

Fast–slow analysis in the phase plane

Since g_{K-Ca} is a function of Ca_i it varies slowly relative to the time-scale of the spikes. We exploit this to gain additional insight into the mathematical mechanism for bursting by considering g_{K-Ca} momentarily as a parameter in the faster spike-generating subsystem, eqns (1)–(2). For some values of g_{K-Ca} this fast subsystem has multiple steady states; the Z-shaped curve (labelled 'ss') in Fig. 2C shows the dependence of the steady state on g_{K-Ca}. For the parameter values considered here, the hyperpolarized steady state (i.e. the lower branch of the Z-curve) is stable; the middle one is a saddle point and therefore unstable; and the upper steady state is an unstable spiral and, for a subset of g_{K-Ca} values, is surrounded by a stable limit cycle oscillation (maximum and minimum values are labelled as 'osc' in Fig. 2C). The existence of two stable states (here, one steady and the other oscillatory) for a given value of g_{K-Ca} is called 'bistability'. The saddle point plays the role of a threshold which separates the two stable states.

Bistability in the fast subsystem underlies the model's bursting behaviour. From the projection of the burst solution into this $V-g_{K-Ca}$ plane, we see that the lower steady state is slowly tracked during the silent phase, and the limit cycle oscillations around the depolarized level correspond to the spikes in the active phase. Moreover, this view shows how the active phase begins when the lower steady state coalesces with the saddle point (physiologically, when the silent phase potential reaches threshold). As the active phase proceeds the V-n subsystem oscillates rapidly and g_{K-Ca} increases; meanwhile, the threshold saddle point is slowly rising. When the minimum voltage of a spike meets the rising threshold, the active phase terminates. Note that the last few spikes of an active phase have longer interspike intervals because their V-n trajectories pass close to the saddle point and slow down there.

This phase plane treatment also shows clearly the narrow range of g_{K-Ca} values (tics on g_{K-Ca} axis) over which bursting occurs. Finally, the relative duration of the active phase depends on the position of the Ca_i-nullcline (labelled $\dot{C}a = 0$) which is obtained by setting the right hand side of eqn (3) to zero and expressing g_{K-Ca} in terms of Ca_i. The physiological significance of this will be taken up in the discussion.

Further details on this fast–slow mathematical dissection of the Chay–Keizer model can be found in Rinzel (1985). Similar arguments have been used by Hindmarsh and Rose (this volume) to formulate and understand a model for burst patterns in thalamic neurones, although their description of the fast subsystem does not include its oscillatory solutions. The approach has been generalized to classify and distinguish different types of bursting phenomena (Rinzel, 1987). For example, analysis of Plant's model (Plant, 1978) for the R15 bursting neurone reveals that the fast subsystem is not bistable (Rinzel and Lee, 1987). From this, one concludes that there must be two slow variables (e.g. a slowly activated Ca^{2+} current and Ca_i) to endogenously sweep the fast dynamics between oscillatory and non-oscillatory regions.

As in the β-cell, controversies have arisen about the membrane site for feedback of Ca_i; whether calcium activates I_{K-Ca} or inactivates I_{Ca} (Adams and Benson, 1985). Rinzel and Lee (1987) used the fast–slow treatment to construct modifications of Plant's model based on each of the two alternative calcium feedback mechanisms. The analysis further shows why interspike intervals can be low at the beginning as well as at the end of the active phase to explain the characteristic 'parabolic' pattern.

THE EMERGENCE OF BURSTING

Stochastic single-cell model

The theory presented above is appropriate for describing the behaviour of a cell in a perfectly synchronized, intact islet. We now modify the model to account for the disorganized spiking of a single isolated cell. Estimates for the unitary conductance of the K–Ca channel lie in the range 50–250 pS (Atwater et al., 1988), which is comparable to the whole-cell conductance. Therefore no more than a few of these channels are open on average, although there may be hundreds of K–Ca channels in the cell (Atwater et al., 1983).

The Hodgkin–Huxley formulation for ionic channels can be thought of as a continuum limit for a population of stochastically opening and closing channels so numerous that the macroscopic conductance is a smooth function of time. This approach is untenable for the β-cell, so we replace eqn (4) with the following:

$$g_{K-Ca} = \bar{g}_{K-Ca} \, p \tag{5}$$

where p is the fraction of channels open and is determined from a stochastic process such that

$$\langle p \rangle = \frac{Ca_i}{K_d + Ca_i} \tag{6}$$

This can be accomplished by describing the channel transitions as a two-state Markov process in which a closed channel opens with probablity per unit time $1/\tau_c$ and an open channel closes with probability per unit time $1/\tau_o$ where

$$\tau_o = \frac{Ca_i}{K_d}\tau_c$$

and τ_c is a fixed parameter. In our simulations we account for every channel event, but for purposes of computing p we do not need to know the state of every channel, only the total number of channels open and closed. We advance the solution in time using fixed steps of length Δt. Using the old value of p we update the values of V, n and Ca_i. The new value of Ca_i is used to compute τ_o. Since the channel events are independent, the probability that one of the open channels will close,

$$Pr\{N_c \to N_c + 1, N_o \to N_o - 1\}$$

is equal to $N_o \, \Delta t/\tau_o$. Similarly, the probability that one of the closed channels will open,

$$Pr\{N_o \to N_o + 1, N_c \to N_c - 1\}$$

is equal to $N_c \, \Delta t/\tau_c$. The new values of N_o and N_c give a new value of p, and the cycle is repeated. See Sherman et al. (1988) for implementation details.

In this new model calcium feedback operates only in the weakened sense that the mean of the distribution of open channels is a function of Ca_i

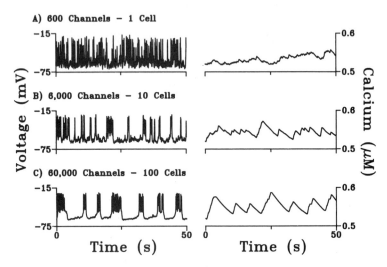

Fig. 3. Simulated V (left) and Ca_i (right) time courses obtained using the stochastic super-cell model. As the numbers of channels and cells increase the random spikes become organized into bursts. Calcium oscillations appear, allowing higher peak levels of Ca_i.

through the open time. If the number of channels is small then stochastic fluctuations can dominate, and thereby prevent regular bursting.

For our simulations we assumed that a cell has 600 K–Ca channels with a single-channel conductance of 50 pS for a maximal K–Ca conductance, $\bar{g}_{K-Ca}$, of 30,000 pS. Note that we do not choose p, but compute it as part of the solution. Its average value as determined by eqn (6) is about 0.005, which gives an average g_{K-Ca} of about 150 pS. (The variance, however, is enormous: g_{K-Ca} ranges from 0 to 500 pS. See Sherman et al., 1988, figure 7B.) We choose τ_c to be 1000 ms which means the average τ_o is about 5 ms. We give these values to convey some feeling for the quantities involved; the results do not depend qualitatively on the exact numbers used.

Figure 3A shows a sample solution of the stochastic differential eqns (1)–(3), (5). Compare with Fig. 1C. We see large-amplitude random spikes which may be viewed as bursts with only one spike and, occasionally, with a few spikes. Intracellular calcium (Fig. 3, right) does not oscillate as in the fully synchronized deterministic model, but rather drifts randomly because of channel fluctuations.

Coupled cells can burst

We now proceed to reconstruct the behaviour of clusters of β-cells using the above single-cell model as our atomic unit. In clusters or islets, β-cells are electrically coupled by gap junctions as shown by current injection (Eddlestone and Rojas, 1980), electron microscopy (Meda et al., 1984) and dye injection experiments in which several cells are labelled by injecting dye into one cell (Michaels and Sheridan, 1981; Meda et al., 1986). Suppose we have a cluster of N_{cell} cells with stochastic K–Ca channels, such that each cell is coupled symmetrically to an arbitrary set of other cells. The voltage in the jth cell satisfies

$$C_m \frac{dV_j}{dt} = -I_{ion}(V_j, n_j) - \hat{g} N_{o,j}(V_j - V_K) - \sum_{k \in \Omega_j} g_{jk}(V_j - V_k) \qquad (7)$$

where $\hat{g}$ is the conductance of a single K–Ca channel, $N_{o,j}$ is the number of open channels in the jth cell, $g_{jk} = g_{kj}$ is the coupling conductance between the jth and kth cells, Ω_j is the set of cells to which the jth cell is coupled, and $I_{ion}(V, n) = I_{Ca}(V) + I_K(V, n)$. Now let V be the average voltage of the cells:

$$V = \frac{1}{N_{cell}} \sum_{j=1}^{N_{cell}} V_j$$

Since the coupling is symmetric, V satisfies the equation

$$C_m \frac{dV}{dt} = -\frac{1}{N_{cell}} \sum_{j=1}^{N_{cell}} I_{ion}(V_j, n_j) - \hat{g} \frac{1}{N_{cell}} \sum_{j=1}^{N_{cell}} N_{o,j}(V_j - V_K) \tag{8}$$

Consider the extreme case in which the gap junction conductances become infinite ($g_{jk} \to \infty$); plausibly the entire cluster becomes perfectly synchronized ($V_j = V$, $j = 1, \ldots, N_{cell}$) since otherwise the coupling term in eqn (7) will become infinite. This is not realistic, but it is known that the bursts in islets are approximately synchronized, even in widely separated cells (Meda et al., 1984). Then let us *assume* synchrony and explore whether organized bursting emerges as the cluster size, N_{cell}, increases. Equation (8) becomes

$$C_m \frac{dV}{dt} = -I_{ion}(V, n) - \frac{\hat{g}}{N_{cell}} \sum_{j=1}^{N_{cell}} N_{o,j}(V - V_K) \tag{9}$$

An interpretation of this equation is that the effective single-channel conductance is $\hat{g}/N_{cell}$, i.e. the stochastic fluctuations decrease in magnitude as the cluster size increases. Indeed, when synchrony is assumed, a cluster is just a 'super-cell' with an enlarged surface membrane and pool of channels. A little algebra transforms eqn (9) into

$$C_m \frac{dV}{dt} = -I_{ion}(V, n) - \bar{g}_{K-Ca} p(V - V_K) \tag{10}$$

which is the same as eqns (1), (5) for the single-cell case, except that p is now the fraction of channels open in the entire cluster. Thus, a synchronous cluster can be simulated in exactly the same way as a single cell merely by multiplying the number of channels by N_{cell}. The results of such simulations with a sequence of increasing cluster sizes are shown in Fig. 3 and confirm the convergence to organized bursting. Inspection of the g_{K-Ca} time courses (Sherman et al., 1988, figures 7 and 9) reveals that the variance of g_{K-Ca} decreases and g_{K-Ca} stays closer to the mean determined by Ca_i. When the stochastic fluctuations are averaged over a large pool of channels, calcium oscillations emerge, and Ca_i is able to play a feedback role in modulating the bursts. It is perhaps not surprising that convergence occurs, given the Law of Large Numbers, but it was not guaranteed a priori that bursting would be obtained with modest numbers of cells when, it turns out, the variance is still quite large.

Another biological example in which the Law of Large Numbers provides a control mechanism is that of chick heart ventricular pacemaker cells. Clay and DeHaan (1979) found that the variance of the interbeat interval decreases as N_{cell}^{-1}.

DISCUSSION

Summary

Our long-term goals are to understand how bursting oscillations are generated in β-cells and what role bursting plays in insulin secretion. The deterministic Chay–Keizer (1983) model has provided a reasonable framework for explaining ionic mechanisms of bursting in a perfectly synchronized, intact islet of Langerhans. The model provided a link between metabolism and electrophysiology through an intracellular calcium-dependent K^+ (K–Ca) conductance to slowly modulate the bursts, as suggested by Atwater *et al.* (1980). Other versions of the model considered alternative ionic mechanisms, but all have essentially the same mathematical structure and are based on calcium feedback.

In order to unify the phenomena of disorganized spiking in individual cells and organized, synchronized bursting in coupled collections of cells we were led to consider a stochastic model. In our first approach to this problem we assumed synchrony and focused on the emergence of order. We call this the 'super-cell' model. Burst regulation by slow variation of calcium is disrupted in single cells and small cell clusters by sizeable stochastic fluctuations, but in sufficiently large clusters these fluctuations are averaged out and bursting can occur. Recently we have extended these results to show that in simulations based on eqn (7) synchrony is attained for sufficiently large, but finite, coupling conductance. The degree of organization in the bursts then depends on cluster size, as in the super-cell model.

It is widely believed that in β-cells metabolically sensitive K^+ channels determine the resting and silent phase potential. Patch clamp data support the existence of both the K–Ca and the K–ATP channels. On the other hand, there is scepticism over the role which either of these channels might play in bursting and setting the rest potential: the K–Ca channel is rarely open at physiological voltages and calcium levels, and the K–ATP channel is rarely open at the glucose levels where bursting occurs. Our stochastic models have shown that these rarely open channels can both play a role if they are shared among cells. Even if this theory is incorrect we believe that the only way to address the scepticism is to model the collective properties of β-cells.

Chay and Kang (1988) disregard both the K–Ca and K–ATP channels and propose a different stochastic model in which calcium inactivation of the calcium channels modulates the bursts. In this case the fluctuations of the next largest channel, the HH-like K^+ channel, dominate. The silent phase potential is determined by a large leak conductance with a reversal potential near that level. Their model, as ours, exhibits a transition to bursting with increasing cluster size, and thus demonstrates that it is not necessary for the stochastic element to be associated with a slow variable. They have not shown, however, that the occasional, but large, K–Ca channel fluctuations will not disrupt the bursts.

The stochastic models lend strong support to the hypothesis that bursting in β-cells is a collective phenomenon, achieved by channel sharing through electrical coupling. There remains a question as to whether coupling by gap junctions is sufficient to synchronize large clusters, since dye-coupled regions usually contain only three or four cells (Meda *et al.*, 1986). We do not know whether these connections percolate to form a network which can synchronize the entire islet or cluster. Possibly there is another class of gap junctions which does not pass dye. We hope that experimentalists will be motivated to test the theory quantitatively by recording the activity of clusters of different sizes. It also may be possible to examine clusters of fixed size but varying degrees of coupling – the extent of dye-coupled domains can be increased by prolactin (Michaels *et al.*, 1987). An alternative or additional coupling mechanism to gap junctions is diffusion of extracellular K^+ and Ca^{2+} among small iso-potential dye-coupled domains (Perez-Armendariz and Atwater, 1986).

The significance of bursting

We know that calcium is an essential trigger for secretion of insulin; we want to understand the role of bursting in organizing the electrophysiological response of the β-cell in order to deliver calcium to the secretory apparatus. Pancreatic β-cells respond to increased glucose with a graded increase in insulin secretion (Scott *et al.*, 1981). Parallel to this are the electrophysiological effects of glucose to modify the bursts by increasing the relative active phase duration (Beigelman *et al.*, 1977) and to increase average spike frequency (Scott *et al.*, 1981) without significantly affecting spike amplitudes (Beigelman *et al.*, 1977).

The separation of bursting into fast and slow processes has suggested that mechanisms for glucose sensing may be viewed as affecting the slow dynamics directly or indirectly (e.g. through the calcium-handling kinetics or calcium-sensitive conductances). To model glucose action Chay and Keizer (1983) assumed that the calcium removal rate, k_{Ca} in eqn (3), increases with glucose. Rinzel (1985) pointed out that when k_{Ca} is small the Ca_i nullcline (Fig. 2C) intersects the middle branch of the Z-curve near its left knee so that the burst trajectory spends much time in the silent phase. For large k_{Ca} an analogous argument shows why the active phase is long. Since the fast subsystem is unaffected by k_{Ca}, spike amplitude and instantaneous spike frequency are unaffected. Other glucose-sensing mechanisms have been suggested which affect the slow processes in different ways (Himmel and Chay, 1987). One alternative mechanism assumes that the Ca^{2+} binding constant, K_d, of the K–Ca channels decreases with glucose. This results in a similar electrical dose response to glucose as the k_{Ca} model but additionally predicts that average Ca_i increases with glucose.

We have also studied the glucose dose response of isolated cells and small clusters using the super-cell model. Although electrical activity and mean calcium influx increase in the same manner as in the deterministic models, randomly spiking cells do not realize the high calcium peaks found in simulations where the spikes are temporally organized into bursts (see Figs 2B and 3, right hand panels). This observation is suggestive that peak macroscopic calcium is significant for secretion and that bursting is necessary to achieve it. We caution, however, that the models have not yet addressed the question of how calcium influx interacts with other possible secretory co-factors in the cell.

There is considerable indirect evidence (Dawson *et al.*, 1984; Perez-Armendariz and Atwater, 1986) that intracellular calcium does oscillate in β-cells, but intracellular calcium has never been directly measured in bursting β-cells experimentally. The above discussion underlines the importance of making such measurements in order to distinguish between alternative models of glucose sensing and to understand the role of bursting in generating the glucose dose response of the β-cell.

REFERENCES

Adams, W. B. and Benson, J. A. (1985) The generation and modulation of endogenous rhythmicity in the *Aplysia* bursting pacemaker neurone R15. *Prog. Biophys. Molec. Biol.* **46**, 1–49.

Atwater, I., Gonçalves, A., Herchuelz, A., Lebrun, P., Malaisse, W. J., Rojas, E. and Scott, A. (1984) Cooling dissociates glucose-induced insulin release from electrical activity and cation fluxes in rodent pancreatic islets. *J. Physiol., Lond.* **314**, 615–27.

Atwater, I., Dawson, C. M., Scott, A., Eddlestone, G. and Rojas, E. (1980) The nature of the oscillatory behavior in electrical activity for pancreatic β-cell. *J. Horm. Metab. Res. Suppl.* **10**, 100–7.

Atwater, I., Li, M. X., Rojas, E. and Stutzin, A. (1988) Glucose reduces both ATP-blockable and Ca-activated K-channel activity in cell-attached patches from rat pancreatic B-cells in culture. *Biophys. J.* **53**, 145a (Abstr.).

Atwater, I. and Rinzel, J. (1986) The β-cell bursting pattern and intracellular calcium. In *Ionic Channels in Cells and Model Systems* (ed. R. Latorre), pp. 353–62. Plenum Publishing, New York and London.

Atwater, I., Rosario, L. and Rojas, E. (1983) Properties of calcium-activated potassium channels in the pancreatic β-cell. *Cell Calcium* **4**, 451–61.

Beigelman, P. M., Ribalet, B. and Atwater, I. (1977) Electrical activity of mouse pancreatic beta-cells II. Effects of glucose and arginine. *J. Physiol., Paris* **73**, 201–17.

Chay, T. R. and Cook, D. L. (1988) Endogenous bursting patterns in excitable cells. *Math. Biosci.* **90**, 139–53.

Chay, T. R. and Kang, H. S. (1988) Role of single channel stochastic noise on bursting clusters of pancreatic β-cells. *Biophys. J.* **54**, 427–35.

Chay, T. R. and Keizer, J. (1983) Minimal model for membrane oscillations in the pancreatic β-cell. *Biophys. J.* **42**, 181–90.

Chay, T. R. and Keizer, J. (1985) Theory of the effect of extracellular potassium on oscillations in the pancreatic β-cell. *Biophys. J.* **48**, 815–27.

Clay, J. R. and DeHaan, R. L. (1979) Fluctuations in interbeat interval in rhythmic heart-cell clusters. *Biophys. J.* **28**, 377–90.

Dawson, C. M., Atwater, I. and Rojas E. (1984) The response of pancreatic β-cell membrane potential to potassium-induced calcium influx in the presence of glucose. *Q. J. Exp. Physiol.* **69**, 819–30.

Eddlestone, G. T. and Rojas, E. (1980) Evidence of electrical coupling between mouse pancreatic β-cells. *J. Physiol.* **303**, 76–77P.

Himmel, D. and Chay, T. R. (1987). Theoretical studies on the electrical activity of pancreatic β-cells as a function of glucose. *Biophys. J.* **51**, 89–107.

Hodgkin, A. L. and Huxley, A. F. (1952) A quantitative description of membrane current and its application to conduction and excitation in nerve. *J. Physiol., Lond.* **117**, 205–49.

Keizer, J. E. (1988) Electrical activity and insulin release in pancreatic beta cells. *Math. Biosci.* **90**, 127–38.

Meda, P., Atwater, I., Gonçalves A., Bangham, A., Orci, L. and Rojas, E. (1984) The topography of electrical synchrony among β-cells in the mouse islet of Langerhans. *Q. J. Exp. Physiol.* **69**, 719–35.

Meda, P., Santos, R. M. and Atwater, I. (1986) Direct identification of electrophysiologically monitored cells within intact mouse islets of Langerhans. *Diabetes* **35**, 232–6.

Michaels, R. L. and Sheridan, J. D. (1981) Islets of Langerhans: dye coupling among immunocytochemically distinct cell types. *Science, N.Y.* **214**, 801–3.

Michaels, R. L., Sorenson, R. L., Parsons, J. A. and Sheridan, J. D. (1987) Prolactin enhances cell-to-cell communication among β-cells in pancreatic islets. *Diabetes* **36**, 1098–102.

Perez-Armendariz, E. and Atwater, I. (1986) Glucose-evoked changes in K^+ and Ca^{2+} in the intercellular spaces of the mouse islet of Langerhans. In *Biophysics of the Pancreatic β-Cell* (ed. I. Atwater, E. Rojas and B. Soria), pp. 31–51. Plenum Press, New York.

Plant, R. E. (1978) The effects of calcium + + on bursting neurons. A modeling study. *Biophys. J.* **21**, 217–37.

Rinzel, J. (1985). Bursting oscillations in an excitable membrane model. In *Ordinary and Partial Differential Equations* (ed. B. D. Sleeman and R. J. Jarvis), pp. 304–16. Springer-Verlag, New York.

Rinzel, J. (1987) A formal classification of bursting mechanisms in excitable systems. In *Mathematical Topics in Population Biology, Morphogenesis, and Neurosciences*, Lecture Notes in Biomathematics 71 (ed. E. Teramoto and M. Yamaguti), pp. 267–81. Springer-Verlag, New York.

Rinzel, J. and Lee, Y. S. (1987) Dissection of a model for neuronal parabolic bursting. *J. Math. Biol.* **25**, 653–75.

Rorsman, P. and Trube, G. (1986) Calcium and delayed potassium currents in mouse pancreatic β-cells under voltage clamp conditions. *J. Physiol.* **374**, 531–50.

Rosario, L., Atwater, I. and Scott, A. M. (1986) Pulsatile insulin release and electrical activity from single ob/ob mouse islets of Langerhans. In *Biophysics of the Pancreatic β-cell* (ed. I. Atwater, E. Rojas and B. Soria), pp. 413–25. Plenum Press, New York.

Scott, A., Atwater, I. and Rojas, E. (1981) A method for the simultaneous measurement of insulin release and B-cell membrane potential in single mouse islets of Langerhans. *Diabetologia* **21**, 470–5.

Sherman, A., Rinzel, J. and Keizer, J. (1988) Emergence of organized bursting in clusters of pancreatic β-cells by channel sharing. *Biophys. J.* **54**, 411–25.

The segmental burst-generating network used in lamprey locomotion: Experiments and simulations

S. GRILLNER, A. LANSNER,[a] P. WALLÉN, L. BRODIN, Ö. EKEBERG,[a] H. TRÅVEN[a] AND J. CHRISTENSON

The Nobel Institute for Neurophysiology, Karolinska Institute, Box 60400, S-104 01, Stockholm, Sweden
[a]NADA, The Royal Institute of Technology, Stockholm, Sweden

INTRODUCTION

The nervous system decides our ability to perceive, remember and act on the environment. The nerve cells of each species are connected by different types of synapses in neuronal networks forming rough templates, which can act as feature detectors to enable us to easily detect specific aspects of the environment by using the different senses. Experience can modify the strength of the synaptic connections and we can thereby learn to recognize and remember different stimuli. Correspondingly, preformed neural circuits are used to produce different motor acts ranging from withdrawal reflexes to voluntary movements.

In vertebrates and invertebrates fairly simple strategies in the neural organization have emerged. Networks of nerve cells within the nervous system serve as 'pattern generators'. Whenever such a network is activated, it will be responsible for activating the motoneurones, which control the different muscles used in the motor pattern and for their coordination. When each individual is born, a number of central pattern generators (CPG) are active or they can become activated by the appropriate stimuli, for example those generating breathing, sucking, swallowing, crying or simple withdrawal reflexes. As the nervous systems become more mature, other motor pattern generators may develop or become perfected as those used for walking or hand movements.

Cell to Cell Signalling: From
Experiments to Theoretical Models
ISBN 0–12–287960–0

The neural organization underlying locomotion has been particularly well studied and we will therefore consider this control system. The motor pattern during walking is generated by different pattern generators controlling each limb. The central network interacts with sensory signals activated by proximal joint movements (hip) or by the muscle force developed in extensor muscles (see Shik and Orlovsky, 1976; Stein, 1978; Grillner, 1981, 1985). The initiation and maintenance of activity in the locomotor CPG is controlled from locomotor areas in the midbrain and diencephalon. Much information about control principles and which parts of the nervous system are important have accumulated during the 1970s. However, the mammalian nervous system is so complex that it seemed impossible to reach an understanding about the detailed operation of, for instance, the CPGs underlying locomotion.

To be able to address this type of problem, we have turned to the simple nervous system of a lower vertebrate, the lamprey, which has comparatively few neurones and is amenable to experimental analysis (Rovainen, 1979; Grillner *et al.*, 1987). This is a reasonable approach since the general outline of the vertebrate nervous system is well preserved from lamprey to primates. This also applies specifically to the locomotor system. The lamprey CNS can be made to produce a rhythmic locomotor pattern even when its nervous system has been isolated (see Grillner *et al.*, 1987).

In this review we will briefly consider the control system for locomotion in lamprey and the neural network which has been established experimentally, and conjointly we will give some results from the computer simulations of this neural network.

INITIATION OF LOCOMOTION

The motor pattern of swimming (Wallén and Williams, 1984) can be studied in the isolated nervous system and the motor pattern can be elicited by:

(1) Specific brainstem areas normally activating the spinal locomotor circuitry via reticulospinal pathways (McClellan and Grillner, 1984; Kasicki and Grillner, 1986; Ohta and Grillner, 1989).

(2) By sensory input which will directly activate the spinal CPGs (McClellan and Grillner, 1983).

(3) By direct activation of the spinal pattern generator by adding excitatory amino acids activating *N*-methyl-D-aspartate (NMDA) (Grillner *et al.*, 1981) or kainate receptors (Brodin *et al.*, 1985).

The last method has been very useful experimentally, particularly for the cellular analysis of the neuronal interactions.

THE SEGMENTAL NETWORK PRODUCING LOCOMOTION

Experimental findings

The motor pattern in each segment constitutes a strictly alternating burst activity between the left and the right side. Each motoneurone is depolarized during ipsilateral burst activity and inhibited during contralateral activity (Fig. 1, Grillner *et al.*, 1987, 1988a,b). The inhibition is produced by contralateral interneurones (CCIN), which produce inhibitory postsynaptic potentials (IPSPs) utilizing glycine as neuronal transmitter (Buchanan, 1982). The excitation is produced by ipsilateral interneurones (EIN) which activate glutamate receptors (Buchanan and Grillner, 1987) and are active during the ipsilateral burst activity. Finally an ipsilateral inhibitory interneurone (LIN) is active in synchrony with EIN and it inhibits the ipsilateral inhibitory CCIN. Figure 1 shows the network interaction established experimentally and the discharge pattern of the different neurones.

Experimentally the network can be turned on by activation of the descending reticulospinal neurones from the posterior reticular nucleus of the brainstem (McClellan and Grillner, 1984), which are known to monosynaptically excite the spinal neurones of the network (Ohta and Grillner, 1989). The spinal network can also be turned on by two other procedures known to excite the different cellular elements: (1) sensory stimulation and (2) application of excitatory amino acids in the bath (Cohen and Wallén, 1980; Grillner *et al.*, 1981; McClellan and Grillner, 1983; Brodin *et al.*, 1985). It appears that a simple signal that excites the different neurones of the network can turn on rhythmic locomotor activity and the actual patterns will

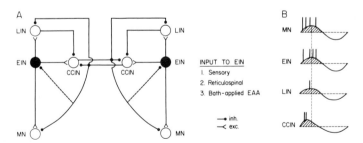

Fig. 1. Schematic diagram of the segmental burst-generating network in lamprey and the pattern of neuronal activity during locomotion. (A) Connections established by simultaneous recordings of identified pairs of neurones. EIN, excitatory interneurone; LIN, lateral interneurone; CCIN, inhibitory interneurone with crossed caudally projecting axon (inhibiting all types of neurones on the contralateral side); MN, motoneurone. Black spots indicate inhibitory synapses and open forks indicate excitatory synapses. The dashed line respresents a connection suggested by indirect evidence, but not yet verified with paired recordings. (B) Pattern of activity in the different types of neurones during fictive locomotion.

emerge by the inhibitory interaction between neurones. Essentially the different neurones of the network will tend to generate action potentials when not actively inhibited.

Let us first intuitively try to explain how a neural network with this pattern of interaction could function, with the premises given above. If one side initially starts to become active, the contralateral side will be inhibited through the CCIN and it could remain inhibited for ever if there were no mechanism to turn off the ipsilateral motor activity. The LIN, however, has this function. It has a high threshold for activation and it starts to spike only at the peak of the depolarization (Buchanan and Cohen, 1982). The LIN will inhibit the CCIN, which will then become silent and thereby all neurones on the contralateral side will become disinhibited. The contralateral neurones will then depolarize and a contralateral burst will start. The contralateral CCIN will then become active, which obviously will lead to an active inhibition of all neurones on the first side. In this way a simple alternating cycle of reciprocal burst activity will emerge. In summary, background excitation can activate neurones on one side leading to an active contralateral inhibition, the ipsilateral activity is switched off by LINs leading to a contralateral disinhibition and subsequent burst activation.

Simulation

With the findings and the interpretation discussed above it was natural to ask whether we could test the intuitive interpretation further by making a realistic simulation. We have used model neurones of an intermediate complexity (Lansner, 1986), in which each simulated neurone could be assigned an input resistance, and which had a soma and three different dendritic compartments. The inhibitory and excitatory synapses act by simulated conductance increases for ionic currents. If the membrane potential reaches a threshold value an 'action potential' is elicited which is not realistically simulated. After this event there is a long-lasting hyperpolarization which markedly reduces the excitability of the neurone and which serves to regulate the frequency of discharge during a constant excitatory synaptic drive.

The different neurones of the network of Fig. 1 have been assigned experimentally verified and/or realistic parameters (Grillner *et al.*, 1988a). The interneurones (EIN, CCIN, LIN) were simulated with three neurones of each kind on each side of the spinal cord, that is altogether $3 \times 3 \times 2$ neurones. The motoneurones which constitute the output, but which are not part of the pattern generator circuitry have not been included. They are assumed to be active in synchrony with EINs (Wallén and Lansner, 1984).

As the excitability of all the simulated neurones was increased to a level at which they tended to generate action potentials when not inhibited, the interneurones started to generate a patterned rhythmic activity (Fig. 2).

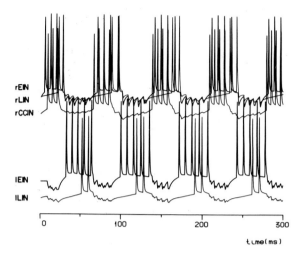

rEIN
rLIN
rCCIN

IEIN
ILIN

0   100   200   300

time(ms)

Fig. 2. Simulation of the neuronal network of Fig. 1A. The network consisted of three neurones of each type on both sides, i.e. a total of 18 neurones (MN not included). The patterns of activity of all neurone types are shown on the right (r) side, whereas on the left side only the EIN and LIN activities are shown. The after-hyperpolarization of all neurones had a time constant of 25 ms and the membrane time constant was 14 ms. The synaptic strengths between individual neurones were (in μmho): from CCIN to CCIN (contralateral) 0.03, to LIN 0.03, and to EIN 0.1; from EIN to CCIN (ipsilateral) 0.03, and to LIN 0.03. The network was turned on by a constant depolarizing current applied to all neurones.

Moreover, the pattern of activity of the different types of interneurones was similar to that of the real neurones. CCINs fire early in the excitatory half cycle, EINs throughout the half cycle and LINs start in the middle of the depolarizing plateau.

If the 'background excitation' is increased the burst rate will become enhanced. The simulated network covers a burst rate from 2–3 Hz up to 10 Hz, that is the normal upper part of the frequency range of swimming. Thus, it can be concluded that the network defined experimentally and discussed above could generate the burst pattern producing locomotion in the upper half of the normal frequency range (for the lower part see below).

In the preceding section we predicted that the LINs indirectly served to terminate the burst through their inhibitory effect on the CCINs. To test for that possibility (Fig. 3), we turned down the efficacy of the simulated synapse between LINs and CCINs to 0 and indeed the burst termination took a much longer time than with an effective LIN to CCIN synapse (Fig. 3A). This finding confirms the role of LINs for the burst termination.

However, of equal importance, alternating burst activity was continued but at a much reduced rate. There must thus be another factor, which turns out to be the decrease of spike frequency of the bursting neurones (EINs and CCINs). The after-hyperpolarization constitutes a very important factor in

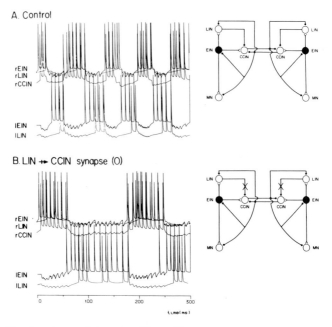

Fig. 3. Simulated test of the role of lateral interneurones in burst generation. (A) Control simulation with parameters as in Fig. 2. The connectivity is indicated in the right hand panel. (B) Simulation with the same parameters except for the lack of effective synapses between LINs and CCINs, as indicated with crosses in the right panel. Note the dramatic increase in burst duration.

this context. The after-hyperpolarization is one main factor in the frequency regulation of vertebrate neurones. If two action potentials follow in rapid succession as during a long depolarizing input, the after-hyperpolarization following the second action potential will become summated to the first after-hyperpolarization. The likelihood of a third action potential will then become lower. The summation of the hyperpolarizations is thus another putative burst terminating factor (see Grillner *et al.*, 1988b; Wallén *et al.*, 1989).

We would thus anticipate that if a neurone has a long-lasting after-hyperpolarization of large amplitude, there will be a large potential for summation and an early burst termination. Conversely, a small short-lasting hyperpolarization should lead to longer and more intense bursts. This supposition is tested in Fig. 4 in which the after-hyperpolarization of the simulated neurones were markedly reduced (Fig. 4B) and indeed the bursts became much longer and there is also a higher spike rate within each burst. We have thus identified two burst terminating factors: (1) the LINs (and similar interneurones) and (2) the summation of after-hyperpolarization.

A third related factor should also be mentioned. The frequency decrease of

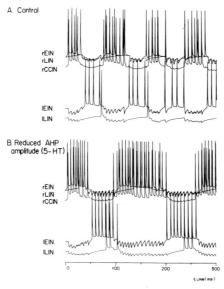

Fig. 4. Simulated test of the role of the postspike after-hyperpolarization (AHP) in burst generation. (A) Control simulation. (B) Simulation with identical parameters except for a reduced postspike AHP. Note the prolonged burst duration and the increase in firing frequency of all neurones. The phyiological effect of 5-HT is to reduce AHP and produce the motor pattern in B.

EINs due to the summation of the after-hyperpolarization will also lead to a reduction of the EIN-induced excitatory drive to CCINs, which will cause a further reduction in the discharge rate of the CCINs.

NEURAL MECHANISMS GENERATING SLOW SWIMMING

The network interaction discussed above can be produced by conventional synaptic transmission depending on kainate/quisqualate EPSPs and glycine inhibition. However, it appears that the normal cruising speed (0.3–3 Hz) is dependent on the additional activation of a different type of glutamate receptor, the NMDA receptor (Brodin *et al.*, 1985; Brodin and Grillner, 1985, 1986). If NMDA receptors are blocked pharmacologically, swimming at a slow steady rate is also blocked.

NMDA receptor ionophores have a different mode of operation than other excitatory ion channels, since they are not only transmitter-gated but also voltage-gated. An activation of NMDA receptors can induce pace-maker-like activity in isolated spinal neurones of the lamprey (Sigvardt *et al.*, 1985; Wallén and Grillner, 1987). Figure 5 shows experimentally recorded NMDA-induced oscillations. The interpretation of different experiments

NMDA induced TTX-resistant
oscillations

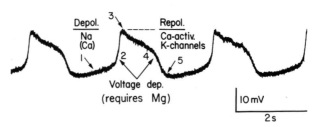

Fig. 5. Experimental recordings of NMDA-induced membrane potential oscillations in the presence of tetrodotoxin. The factors contributing to the intrinsic oscillatory activity were determined by ion substitution experiments in the *in vitro* preparation of the lamprey spinal cord. The depolarization is caused by a regenerative opening of the voltage-sensitive NMDA ionophore, resulting in a depolarizing current mainly carried by Na, and to some extent by Ca. The Ca entry will turn on Ca-activated K channels. The K current will cause a repolarization, which eventually will be aided by the voltage-sensitivity of the NMDA-ionophores. With successive buffering of intracellular Ca the hyperpolarizing current will decay and a new regenerative depolarization can occur.

have suggested that the NMDA receptor activation leads to an opening up of the NMDA ion channels at a threshold value; the more depolarized the neurones are, the more NMDA channels will become opened and a regenerative increase in open NMDA channels will take place. This will bring the membrane potential to a depolarized plateau. At this point there is an equilibrium between conventional voltage-dependent K channels, which will act in a hyperpolarizing direction, and depolarizing NMDA channels (in this experiment the Na^+ channels activated in the action potential are blocked with tetrodotoxin).

At the plateau the open NMDA channels will permit not only Na^+ but also Ca^{2+} ions to enter the cell. The Ca^{2+} entry will lead to a progressive increase of Ca^{2+} intracellularly, which will activate a Ca^{2+}-dependent K^+ channel, which will then cause a progressive hyperpolarization of the plateau level. At a certain point the membrane potential level will reach the level at which the NMDA channels will be closed again and this will cause a regenerative decrease of likelihood that NMDA channels should remain open. This event, together with a closing of voltage-dependent K^+ channels, will cause a rapid hyperpolarization to the most negative level in the pacemaker cycle. Since Ca^{2+} ions will not continue to increase the level (NMDA channels closed), the intracellular handling of Ca^{2+} ions will presumably lead to a progressive decrease of intracellular Ca^{2+} and a closing of the Ca^{2+}-dependent K^+ channels.

To test whether such an interpretation of this series of events is realistic, we have developed a new simulation model with more complex neurones in

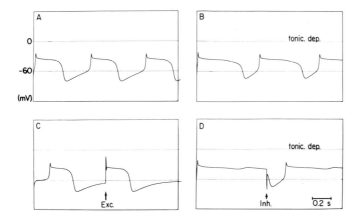

Fig. 6. Simulations of NMDA-induced, tetrodotoxin-resistant membrane potential oscillations. The model consists of a soma and three dendritic compartments. The following ion channels are included: NMDA ionophores with voltage-dependent opening properties; voltage-dependent K channels; voltage-activated Ca channels; Ca-activated K channels. No voltage-gated Na conductance was included, to simulate the effect of tetrodotoxin. The oscillations occur spontaneously and their frequency can be regulated by simulating current injection. A. Pacemaker-like oscillation in NHDA.   B. Effect of depolarization.   C. Effect of brief depolarizing pulse.   D. Effect of hyperpolarizing pulse. All effects resemble those obtained experimentally in Wallen and Grillner (1987).

which NMDA channels as well as Na^+, K^+ and Ca^{2+} channels are incorporated. Figure 6 shows a simulated cell with membrane potential oscillations similar to those observed experimentally. Manipulations of the simulated currents have in all cases mimicked the experimentally observed modifications. The simulation experiments with NMDA oscillations show that the interpretations of the cellular mechanisms responsible are indeed plausible.

These simulations are still of a preliminary nature. We are currently developing new and more realistic neuronal models. The next step will be to simulate the synaptic drive via NMDA receptors within the network. The subsequent steps will involve simulation of the intersegmental coordination, the steering signals from different brainstem neurones, and the interaction with sensory inflow.

REFERENCES

Brodin, L. and Grillner, S. (1985) The role of putative excitatory amino acid neurotransmitters in the initiation of locomotion in the lamprey spinal cord. I. The effects of excitatory amino acid antagonists. *Brain Res.* **360**, 139–48.

Brodin, L. and Grillner, S. (1986) Effects of magnesium on fictive locomotion induced by activation of *N*-methyl-D-aspartate (NMDA) receptors in the lamprey spinal cord *in vitro*. *Brain Res.* **380**, 244–52.

Brodin, L., Grillner, S. and Rovainen, C. M. (1985) NMDA, kainate and quisqualate receptors and the generation of fictive locomotion in the lamprey spinal cord. *Brain Res.* **325**, 302–6.

Buchanan, J. T. (1982) Identification of interneurons with contralateral, caudal axons in the lamprey spinal cord: synaptic interactions and morphology. *J. Neurophysiol.* **47**, 961–75.

Buchanan, J. T. and Cohen, A. (1982) Activities of identified interneurons, moto-neurons and muscle fibers during fictive swimming in the lamprey and the effects of reticulospinal and dorsal cell stimulation. *J. Neurophysiol.* **47**, 948–60.

Buchanan, J. T. and Grillner, S. (1987) Newly identified "glutamate interneurons" and their role in locomotion in the lamprey spinal cord. *Science, N.Y.* **236**, 312–14.

Cohen, A. and Wallén, P. (1980) The neuronal correlate of locomotion in fish. "Fictive swimming" induced in an in vitro preparation of the lamprey spinal cord. *Expl Brain Res.* **41**, 11–18.

Grillner, S. (1981) Control of locomotion in bipeds, tetrapods and fish. In *Handbook of Physiology, Motor Control* (ed. V. Brooks), pp. 1179–236. Waverly Press, Maryland.

Grillner, S. (1985) Neurobiological bases of rhythmic motor acts in vertebrates. *Science, N.Y.* **228**, 143–9.

Grillner, S., McClellan, A., Sigvardt, K., Wallén, P. and Wilén, M. (1981) Activation of NMDA receptors elicits "fictive locomotion" in lamprey spinal cord in vitro. *Acta Physiol. Scand.* **113**, 549–51.

Grillner, S., Wallén, P., Dale, N., Brodin, L., Buchanan, J. and Hill, R. (1987) Transmitters, membrane properties and network circuitry in the control of locomotion in lamprey. *Trends Neurosci.* **10**, 34–41.

Grillner, S., Buchanan, J. and Lansner, A. (1988a) Simulation of the segmental burst generating network for locomotion in lamprey. *Neurosci. Lett.* **89**, 31–5.

Grillner, S., Buchanan, J., Wallén, P. and Brodin, L. (1988b) Neural control of locomotion in lower vertebrates: from behavior to ionic mechanisms. In *Neural Control of Rhythmic Movements in Vertebrates* (eds A. H. Cohen, S. Rossignol and S. Grillner), pp. 1–40. John Wiley, New York.

Kasicki, S. and Grillner, S. (1986) Müller cells and other reticulospinal neurons are phasically active during fictive locomotion in the isolated nervous system of the lamprey. *Neurosci. Lett.* **69**, 239–43.

Lansner, A. (1986) Investigation into the pattern processing capabilities of associat-ing Nets. Techn. Report Royal Institute of Technology, Stockholm (TRITA-NA-8601; Doctoral thesis).

McClellan, A. and Grillner, S. (1983) Initiation and sensory gating of "fictive" swimming and withdrawal responses in an *in vitro* preparation of the lamprey spinal cord. *Brain Res.* **269**, 237–50.

McClellan, A. and Grillner, S. (1984) Activation of "fictive swimming" by electrical microstimulation of brainstem locomotor regions in an *in vitro* preparation of the lamprey central nervous system. *Brain Res.* **300**, 357–61.

Ohta, Y. and Grillner, S. (1989) Monosynaptic excitatory amino acid transmission from the posterior rhombencephalic reticular nucleus to spinal neurones involved in the control of locomotion in lamprey. *J. Neurophysiol.*, (in press).

Rovainen, C. M. (1979) Neurobiology of lampreys. *Physiol. Rev.* **59**, 1007–77.

Shik, M. L. and Orlovsky, G. N. (1976) Neurophysiology of locomotor automatism. *Physiol. Rev.* **56**, 465–501.

Sigvardt, K. A., Grillner, S., Wallén, P. and VanDongen, P. A. M. (1985) Activation of NMDA receptors elicits fictive locomotion and bistable membrane properties in the lamprey spinal cord. *Brain Res.* **336**, 390–5.

Stein, P. S. G. (1978) Motor systems, with special reference to the control of locomotion. *A. Rev. Neurosci.* **1**, 61–81.

Wallén, P. and Grillner, S. (1987) *N*-Methyl-D-aspartate receptor induced, inherent oscillatory activity in neurons active during fictive locomotion in the lamprey. *J. Neurosci.* **7**, 2745–55.

Wallén, P. and Lansner, A. (1984) Do the motoneurons constitute a part of the spinal network generating the swim rhythm in the lamprey? *J. exp. Biol.* **113**, 493–7.

Wallén, P. and Williams, T. L. (1984) Fictive locomotion in the lamprey spinal cord in vitro compared with swimming in the intact and spinal animal. *J. Physiol., Lond.* **347**, 225–39.

Wallén, P., Buchanan, J., Grillner, S. and Hökfelt, T. (1989) The effects of 5-hydroxytryptamine on the afterhyperpolarisation spike frequency regulation and oscillatory membrane properties in lamprey spinal cord neurons. *J. Neurophysiol.* **61**, 759–768.

Mathematical modelling of central pattern generators

G. B. ERMENTROUT[a] AND N. KOPELL[b]

[a]*Department of Mathematics, University of Pittsburgh, Pittsburgh, PA 15260, USA*
[b]*Department of Mathematics, Boston University, Boston, MA 02215, USA*

INTRODUCTION

Central pattern generators (CPGs) are neural networks that govern the stereotypic aspects of rhythmic motion, such as walking, swimming and running. Though they can be modulated by sensory feedback and higher brain structures, in many animals an isolated CPG is capable of autonomously generating spatio-temporal electrical signals similar to ones that guide the muscles in an intact animal. Thus, the question arises as to how the network self-organizes to produce the appropriate pattern of activity to the muscles. Many CPGs are thought to contain oscillating subnetworks or bursting neurones. In that case, the question becomes one of self-organization of relative phases of the oscillating subunits.

This paper reports on some of the mathematical problems that have arisen in an attempt to understand the structure and function of the swimming CPG of lamprey, an eel-like vertebrate. For more information on the biology of this preparation and the insights provided about the biological structure by the mathematics, see Kopell and Ermentrout (1988, 1989a) and the papers referenced therein. An elementary review article on modelling CPGs is given by Kopell (1988).

DESCRIPTION OF A CLASS OF MODELS

The swimming CPG of the lamprey is a neural network that lives in the spinal

Cell to Cell Signalling: From
Experiments to Theoretical Models
ISBN 0–12–287960–0

89

cord of the animal. A crude description of the organization of this network is that of a chain of oscillators, each of which is a local neural subnetwork (Grillner, 1975; Cohen *et al.*, 1982). The details of the segmental oscillators which comprise the CPG are largely unknown, and it is unlikely that they will ever be fully known. For this reason, we have chosen to approach the modelling from a very general point of view; we try to understand, within a very general class of models, the implications of the kinds of phenomenological data that can currently be obtained.

We assume there are segmental oscillators, each capable of autonomously and stably producing periodic output. More specifically, we assume that the kth oscillator can be modelled by a differential equation with an orbitally stable periodic solution:

$$\mathrm{d}X_k/\mathrm{d}t = F_k(X_k) \qquad (1)$$

where X_k lies in $\mathbf{R}^q$, $q \geqslant 2$. Equation (1) can be as complicated as is necessary and may model many neurones or a single neurone with many ionic channels. We will assume that oscillator k can communicate with its $2m$ nearest neighbours and that this coupling is additive. The coupled equations take the form:

$$\mathrm{d}X_k/\mathrm{d}t = F_k(X_k) + \sum_{j=1}^{m} [G_j^+(X_{k+j}, X_k) + G_j^-(X_{k-j}, X_k)] \qquad k = 1, \ldots, N \qquad (2)$$

Equation (2) is very general, and complete analysis appears to be impossible. Nevertheless, with a few very general hypotheses, it becomes possible to understand much about the qualitative behaviour, especially if the chain is long. The first hypothesis is that the coupling is not so strong as to cause the variables X_k to depart significantly from the periodic solution $Y_k(t)$. Thus, the state of each segmental oscillator can be described by a single variable, the phase $\theta_k(t)$, that denotes the position of the variables on the periodic solution. Such a hypothesis may be justified in at least two situations:

(i) Weak coupling; the interactions between the oscillators are small. The use of phase equations is then justified by the use of invariant manifold theory (Fenichel, 1971), which implies that the full set of equations has an invariant torus parameterized by the phases (Ermentrout and Kopell, 1984).

(ii) Strong attraction: the local oscillators, when perturbed, are rapidly pulled back to the periodic solution. Singular perturbation calculations are used to justify the use of phase models.

In addition, there is a third situation which, we conjecture, allows for the reduction to such phase models.

(iii) 'Distribution of effects': each local oscillator is composed of subunits which oscillate at phases distributed around the cycle, and which interact with analogous oscillators by means of pulses distributed around the cycle. This situation, which exists in some known CPGs (Friesen *et al.*, 1978), is far less understood; we have seen in numerical simulations that such distribution of effects can allow interactions strong enough so that, if they were concentrated in a small part of the cycle, they would disrupt the rhythmicity (Ermentrout and Kopell, 1989a).

Our second hypothesis is that the interaction terms depend only on the differences of the phases, i.e. the equations have the form

$$\mathrm{d}\theta_k/\mathrm{d}t = \omega_k + \sum_{j=1}^{m} [H_j^+(\theta_{k+j} - \theta_k) + H_j^-(\theta_{k-j} - \theta_k)] \quad k = 1, \dots, N \qquad (3)$$

The functions $H_j^\pm$ are 2π-periodic functions of their arguments. In case (i), it can be shown by standard averaging theory (Ermentrout and Kopell, 1989a) that new coordinates can be chosen on the invariant torus so that the coupling terms depend, to lowest order in the strength, only on the differences of the phases. The functions $H_j^\pm$ can be explicitly computed from the original eqn (2) by averaging (Ermentrout and Kopell, 1984, 1989a). Even with stronger coupling, equations of the form (3) can result, for example in phase models for which the interaction is dispersed around the cycle as in case (iii) (Ermentrout and Kopell, 1989b).

The final general hypothesis is motivated by an important property of the coupling of neural oscillators via chemical synapses. We shall say that a function G in eqn (2) is of diffusive type if, on the average over a cycle, $G(X,X)$ vanishes. Otherwise we shall say that it is synaptic (Kopell and Ermentrout, 1986; Kopell, 1988). Diffusive coupling arises from gap junctions and passive electrical coupling of tissue, as well as diffusion among chemical oscillators. Synaptic coupling may occur when there are chemical synapses or direct interactions between cells that do not depend on the difference between the two states. For diffusively coupled cells, it can be shown that the functions H in eqn (3) satisfy $H(0) = 0$; for synaptic coupling, $H(0) \neq 0$. Our third general hypothesis is that all of the functions H of eqn (3) satisfy $H(0) \neq 0$. We shall see that synaptic coupling forces the chain of oscillators to have non-zero phase-differences, which in turn allows the lamprey to swim in an undulating manner.

For most of the rest of the chapter, we will restrict our attention to the case of nearest-neighbour coupling, $m = 1$.

NEAREST-NEIGHBOUR COUPLING

We now describe three mathematical problems associated with eqn (3) with

$m = 1$. In each case, we give a brief description of the biological origin of the question.

Normal 'fictive swimming'

Lamprey and many other fish and fish-like animals swim by undulating through the water, with motion created by waves of muscular contraction that pass axially down the animal. These waves are governed by waves of electrical activity in the central pattern generator. The spinal cord of a lamprey, isolated from the rest of the animal, may be made to produce this electrical activity (Cohen and Wallén, 1980), and the behaviour of this activity in the isolated cord, called 'fictive swimming', is essentially the same as in an intact animal swimming in a stereotypic manner (Wallén and Williams, 1984). Recordings from the ventral roots, where processes from the motoneurones leave the spinal cord, show periodic bursts of activity. The periods of this activity are uniform throughout the cord, but phase-shifted. The intersegmental phase-lags are close to constant at about 1% per segment; in a lamprey, which has about 100 segments, this results in a travelling wave with a wave-number of about one body length. The adult can also be made to swim backwards; in that case, the phase-lags are similar but of opposite sign. Remarkably, the lags remain constant over a range of frequencies, which corresponds to a range of swimming speeds. Thus, these waves are not dispersive as are, for example, chemical waves, where the phase-lag is a function of the velocity of the wave.

The mathematical problem is to understand if the above behaviour can be modelled by the solutions of eqn (3), and under what hypotheses on the frequencies ω_k and the coupling functions.

Consider eqn (3) with nearest-neighbour coupling. We are interested in 'phase-locked' solutions, i.e. periodic solutions of eqn (3). It is easy to show that in this case, $d\theta_j/dt$ is a constant Ω, independent of j and t. Let $\varphi_j = \theta_{j+1} - \theta_j$. Suppose now that N is large. As $N \to \infty$, it can be shown that if $\omega_j \to \omega(x)$, where $x = j/N$, then eqn (3) tends to the following boundary value problem:

$$\Omega = \omega(x) + 2f(\varphi(x)) + \frac{1}{N} g(\varphi(x))_x \quad 0 < x < 1 \tag{4}$$

$$\Omega = \omega(0) + H^+(\varphi(0)) \tag{5}$$

$$\Omega = \omega(1) + H^-(\varphi(1)) \tag{6}$$

Here $2f(\varphi) = H^+(\varphi) + H^-(-\varphi)$ and $2g(\varphi) = H^+(\varphi) - H^-(-\varphi)$. This boundary value problem is complicated by the fact that the parameter Ω is unknown and determines the boundary conditions. Elsewhere (Kopell and Ermentrout, 1986, 1989b) we show that if $\omega(x)$ is not too far from constant, there is a unique formal solution to eqns (4)–(6) that selects the correct value of Ω, the phase-locked frequency. Furthermore, for each formal solution to eqns (4)–(6) there is an actual solution to eqn (3).

We now show the importance of chemical synaptic interactions for inducing fictive swimming in the lamprey. Suppose that there is no frequency gradient, i.e. $\omega(x) \sim$ constant. If the coupling is diffusive, then the unique solution to eqns (4)–(6) is $\varphi \equiv 0$. This does not correspond to swimming; the entire preparation simply oscillates synchronously. Small differences in frequency lead to phase-lags that lie around 0 and can take on either sign. On the other hand, if $H^{\pm}(0) \neq 0$, then $\varphi = 0$ does not solve eqns (4)–(6), since the boundary conditions will be violated. Instead, there is a solution $\varphi(x)$ that is constant except in a region near one of the ends of the medium. Depending on the sign of φ, this corresponds to forward or backward swimming. Significant variations from constant local frequency lead to non-constancy of the phase-lags as a function of k, at least for nearest-neighbour coupling.

An important observation is that the phase-lags are independent of the local frequencies of oscillation. More specifically, raising the local frequencies uniformly raises the emergent network frequency Ω by that amount without changing the phase relationships among the oscillators. Thus, one can get higher frequency (corresponding to faster swimming) while maintaining the wavelength.

In the special case that H^{+} is a multiple of H^{-}, the two boundary conditions are of opposite signs. Then the relative strengths of the forward and backward coupling determines if the constant part of the solution is positive or negative. Since the sign of the phase-lag specifies whether the wave goes forward or backward along the chain, such a case provides a simple mechanism for switching between forward and backward 'swimming': one need only tune the coupling strengths.

The reader may note that many of these results occur for coupling in only one direction (Stein, 1971). However, as we describe below, it is possible to entrain the lamprey spinal cord oscillators by periodically forcing the preparation at either end; this is impossible with unidirectional coupling. Thus, bidirectional coupling is a fundamental property of the lamprey CPG.

'Split baths' and large changes in local frequency

It is possible to induce different local frequencies in parts of the cord by exposing different parts of the cord to baths with different ionic concentrations. One mathematical question is to find out what the consequences of such manipulations would be in equations of the form (3), and how the consequences depend on the coupling functions. This was done elsewhere (Kopell and Ermentrout, 1989b), along with other generalizations of the work (Kopell and Ermentrout, 1986) described above.

We will consider a simple example. Suppose that $\omega(x) = \omega_0 + \alpha p(x)$, where $p(x) \geqslant 0$ is a bump centred at a point x_0 in the interior of the medium and having a maximum value of 1. If $\alpha < 0$, then the region in the neighbourhood of x_0 has a lower frequency than the surrounding cord, while if $\alpha > 0$, the same region has a higher frequency. For α of sufficient magnitude, the solution to eqns (4)–(6) undergoes a transition in behaviour from waves propagating in one direction to waves that propagate bidirectionally. It is shown by Kopell and Ermentrout (1989b) that if $f'' > 0$, then for α large enough and positive, the centre of the cord leads the rest of the cord in phase, and so waves propagate outward from the centre. If $f'' < 0$, then if $\alpha < 0$ is of sufficient magnitude, the centre of the cord acts as a wave sink; waves originate at the ends and collide in the centre. These two behaviours are mutually exclusive; for fixed f, this transition occurs for $\alpha > 0$ or $\alpha < 0$ but not both.

The few experiments with raised or lowered frequencies that have so far been done have uncovered an anomaly in our original guesses about the operation of the network. That is, the transition phenomenon described above appears to happen both for raised frequencies *and* for lowered ones. There are at least two sources of the possible discrepancy, both of which give rise to new and challenging mathematics problems. One is that the altered composition of the bath changes the coupling as well as the frequencies. Another is that long-distance fibres, which are known to exist and which have been left out of this class of equations, are capable of changing the output of that experiment.

Periodic forcing

There are stretch-sensitive neurones, thought to be the so-called 'edge cells' that are synaptically coupled to some of the neurones involved in the oscillations (Grillner *et al.*, 1984). Such mechanoreceptors provide the possibility of periodically forcing the chain, using a motor to move the last few segments of the cord, while the rest of the cord is pinned and so remains motionless (McClellan and Sigvardt, 1988; Sigvardt and Williams, in preparation). The effects of this forcing are believed to be local, i.e. only the oscillators closest to the forcing are directly driven (McClellan and Sigvardt, 1988; Sigvardt and Williams, in preparation). Thus, the resulting entrainment is due to the intersegmental coupling between the oscillators. The cord can be entrained to a range of frequencies by forcing at either end. This implies that there must be coupling between oscillators in both directions. (For unidirectional coupling, there is only one end at which forcing has any effect on the phase-lags throughout the chain.)

The mathematical question here is an inverse one: what information can be obtained, at least within the context of models of the form (3), from the phase-lags along the cord that can be measured during these entrainment experiments? To understand forcing of a single segment in a nearest-neighbour coupled chain, we return to (3) with $m=1$ and add a term for forcing (Kopell and Ermentrout, unpublished data). This term has the form $H(\omega_f t - \theta_m)$, where ω_f is the frequency of the forcing and $j=1$ or N. If the chain is phase-locked, then $d\theta_k/dt = \omega_f$ for all k and t. For simplicity of discussion and definiteness, we assume that the local frequencies are constant, say ω, and that $f'' > 0$. Suppose that we force the chain at $k=N$. Then eqn (3) gives rise to the following set of equations for the phase differences $\varphi_k = \theta_{k+1} - \theta_k$:

$$\omega_f = \omega + H^+(\varphi_1) \qquad (7)$$

$$\omega_f = \omega + H^+(\varphi_{k+1}) + H^-(-\varphi_k) \qquad k=1,\ldots,N-2 \qquad (8)$$

$$\omega_f = \omega + H^-(-\varphi_{N-1}) + H_f(\xi) \qquad (9)$$

where $\xi = \omega_f t - \theta_N$.

Equation (8) can be viewed as a discrete dynamical system: within the range in which H^+ is invertible, φ_{k+1} can be obtained from φ_k. This dynamical system can have fixed points, which must satisfy

$$\omega_f - \omega = H^+(\varphi) + H^-(-\varphi) \equiv 2f(\varphi) \qquad (10)$$

f is typically parabola-shaped in the region in which H^+ is monotone, and so outside some interval of values for ω_f, there are no critical points. If there are critical points, there are typically two, one stable and one unstable. Equation (7) may be interpreted as giving the initial condition for the dynamical system (8). With knowledge of the critical points, one can then predict the 'orbit' of the phase-lags along the chain (here k and not time is the variable). Locking occurs if eqn (9) can be satisfied. The larger the amplitude of H_f the larger the range of frequencies for which this is possible. For fixed H_f, for locking to occur, $|\omega_f - \omega - H^-(-\varphi_{N-1})| \equiv A(\varphi_{N-1})$ must be sufficiently small. If φ_1 is not in the basin of attraction of the stable fixed point, φ_{N-1} is generally very far from the value that minimizes $A(\varphi)$, making it very difficult to achieve locking unless the amplitude of H_f is very large. Furthermore, it is possible to compute the frequency values for which the initial φ is in the above basin of attraction, and it turns out that ω_f must satisfy $\omega_f > \Omega_L \equiv \omega + H^+(\varphi_L)$, where φ_L is the zero of $H^-(-\varphi)$ near $\varphi=0$. (Ω_L is one of the two possible frequencies taken by the unforced chain depending on the relative strengths of the coupling (Kopell and Ermentrout, 1986).) This provides an explicit bound on the range of possible forcing frequencies; a bound at the other end is given by the point at which eqn (9) cannot be satisfied. One also has from this a description of the trajectory for each possible ω_f. A similar analysis can be done with forcing at $k=1$, and in general this gives a different entrainment range and set of trajectories. The qualitative properties of the trajectories turn out to depend on the sign f'' and other properties of the coupling functions, as does the range of entrainment. Thus, data on the trajectories, which can be measured (at least crudely) in the lamprey spinal cord, constrain the possible ways in which the coupling can be done. Although all but some possibly anomalous data can be accounted for in this context, we are reserving any predictions until we understand better how the introduction of long-range fibres affects the conclusions of the analysis.

NON-NEAREST-NEIGHBOUR COUPLING

In the lamprey spinal cord, there is anatomical evidence that neurones in each segment have processes that extend locally for at least several segments. (This is in addition to the long-distance fibres.) If the local oscillators are roughly the size of a segment, this implies that the coupling of each oscillator is to several of its neighbours. A mathematical question is then: what difference, if any, does this make in the behaviour of the network?

The analysis of eqn (3) for $m > 1$ is much more complicated than for $m = 1$. So far, we have done work corresponding to the first question in the previous section, i.e. the ability to reproduce the stereotypic 'fictive swimming'; we have not studied the effects of large changes in frequency or of mechanical forcing. Even for the first question, the results are not as complete as for $m = 1$; in particular, we can no longer explicitly determine analytically the frequency Ω of the phase-locked ensemble. Also, the existence and uniqueness of a solution has not been fully established. The problem is not completely specified until one specifies how the edges are to be dealt with, e.g. by removing all terms referring to non-existent oscillators. It can be shown (Zhang, 1988; Kopell et al., 1989) that, for an open set of modifications of the end eqn (3), and away from the edges of the chain, there is a solution which approaches the solution to the algebraic equation

$$\Omega = \omega(x) + \sum_{j=1}^{m} (H_j^+(j\varphi) + H_j^-(-j\varphi)) \equiv \omega(x) + 2F(\varphi) \qquad (11)$$

for some value of Ω. Equation (11) is analogous to the 'outer' equation for (4), e.g. the equation obtained by letting $N \to \infty$. The methods used to show that (3) behaves like a solution to (11) involves exponential dichotomies for non-autonomous discrete systems (Palmer, 1988). In (4), the 'boundary' conditions allow us to determine the value of Ω, and matching between the inner and outer equation enables us to conclude this choice of Ω is unique. We have no explicit analogous conditions for multiple coupling. Recently, we have developed an iterative scheme that allows us to find Ω independently of the form of $\omega(x)$, if $\omega(x)$ is not too far from constant. Thus, we can predict the behaviour of the solution, at least away from the boundaries. This iterative scheme matches numerical calculations of the true solutions (and is much quicker), but it is unproven.

Using the outer eqn (11), we see that the multiple neighbour system behaves, at least away from the boundaries, like a modified version of a nearest-neighbour system, with the function $F(\varphi)$ replacing $f(\varphi)$ in the determination of most of the solution. One important effect is that local changes in frequency create smaller changes in phase-lags if there is multiple coupling, even if the coupling is normalized so that the total amount of the coupling is not increased; thus, the multiple coupling buffers against changes in local frequency and helps to insure that the phase-lags are constant. This can be understood by comparing the function F in eqn (11) with its analogue f in (3). For example, in the special case $f(\varphi) = A \cos \varphi$ for some A and $F(\varphi) = \Sigma_j A_j \cos(j\varphi)$, with A, A_j of the same sign, it can be seen from methods of trigonometric addition that $F(\varphi)$ is a much thinner function with a larger derivative in the relevant region. In addition, even in the absence of frequency differences, suitably normalized multiple coupling leads to smaller phase-lags than nearest-neighbour coupling, a phenomenon that is note-worthy, since the lags in the lamprey cord are only 1% per segment. This is more subtle to see and has not been rigorously proved, though it is clear from the numerics. Our intuitive explanation of it depends on the above heuristi-cally derived iterative scheme for the emergent frequency Ω of the locked chain, which in turn determines the lags away from the boundary according to (11) (see Zhang, 1988; Kopell et al., 1989).

REFERENCES

Cohen, A. H. and Wallén, P. (1980) The neuronal correlate to locomotion in fish: Fictive swimming induced in an in vitro preparation of the lamprey spinal cord. Expl Brain Res. **41**, 11–18.

Cohen, A. H., Holmes, P. J. and Rand, R. H. (1982) The nature of the coupling between segmental oscillators of the lamprey spinal generator for locomotion: a mathematical model. J. Math. Biol. **13**, 345–69.

Ermentrout, G. B. and Kopell, N. (1984) Frequency plateaus in a chain of weakly coupled oscillators I. SIAM J. Math. Anal. **15**, 215–37.

Ermentrout, G. B. and Kopell, N. (1989a) Oscillator death in systems of coupled neural oscillators. SIAM J. Appl. Math. (in press).

Ermentrout, G. B. and Kopell, N. (1989b) Multiple pulse interaction and averaging coupled neural oscillators. Preprint.

Fenichel, N. (1971) Persistence and smoothness of invariant manifolds for flows. Ind. Univ. Math. J. **21**, 193–226.

Friesen, W. O., Poon, M. and Stent, G. (1978) Neuronal control of swimming in the medicinal leech IV. Identification of a network of oscillatory interneurons. J. exp. Biol. **75**, 25–43. (There have been more cells discovered since 1978, but they have not been incorporated into a model network.)

Grillner, S. (1975) Locomotion in vertebrates: central mechanisms and reflex interactions. *Physiol. Rev.* **55**, 247–304.

Grillner, S., Williams, T. and Lagerback, P.-A. (1984) The edge cell, a possible intraspinal mechanoreceptor. *Science, N.Y.* **223**, 500–3.

Guckenheimer, J. and Holmes, P. J. (1983) *Nonlinear Oscillations, Dynamical Systems and Bifurcations of Vector Fields*. Springer, New York.

Kopell, N. (1988) Towards a theory of modelling central pattern generators. In *Neural Control of Rhythmic Movements in Vertebrates* (eds A. H. Cohen, S. Rossignol and S. Grillner). Wiley, New York.

Kopell, N. and Ermentrout, G. B. (1986) Symmetry and phaselocking in chains of weakly coupled oscillators. *Comm. Pure Appl. Math.* **39**, 623–60.

Kopell, N. and Ermentrout, G. B. (1988) Coupled oscillators and the design of central pattern generators. *Math. Biosci.* **89**, 14–23.

Kopell, N. and Ermentrout, G. B. (1989a) Structure and function in an oscillating neural network. In *Computational Neuroscience* (ed. E. Schwartz). MIT Press (in press).

Kopell, N. and Ermentrout, G. B. (1989b) Phase transitions and other phenomena in chains of coupled oscillators. *SIAM J. Appl. Math.* (in press).

Kopell, N., Zhang, W. and Ermentrout, G. B. (1989) Multiple coupling in chains of oscillators. Preprint.

McClellan, A. D. and Sigvardt, K. (1988) Features of entrainment of spinal pattern generators for locomotor activity in the lamprey spinal cord. *J. Neurosci.* **8**, 133–45.

Palmer, K. J. (1988) Exponential dichotomies, the shadowing lemma and transversal homoclinic points. *Dynamics Reported* **1**, 265–306.

Stein, P. S. G. (1971) Intersegmental coordination of swimmeret motorneuron activity in crayfish. *J. Neurophysiol.* **34**, 310–18.

Wallén, P. and Williams, T. (1984) Fictive locomotion in the lamprey spinal cord *in vitro* compared with swimming in the intact and spinal animal. *J. Physiol.* **347**, 225–39.

Zhang, W. (1988) Center manifolds for singularly perturbed discrete dynamical systems and an application to multiple coupling in chains of oscillators. PhD thesis, Northeastern University.

Universal learning mechanisms: From genes to molecular switches

YADIN DUDAI

Department of Neurobiology, Weizmann Institute of Science, Rehovot 76100, Israel

INTRODUCTION

Molecular and cellular analysis of learning is at the frontiers of modern neurosciences. It is practised on preparations ranging from simple invertebrates to mammalian brain (reviewed in Byrne, 1987; Dudai, 1989). As in other branches of molecular and cellular biology, this reductionist approach to learning is yielding an avalanche of publications containing intriguing results and interesting models. The findings even culminated in suggestions for a 'molecular grammar for learning' (Kandel, 1985). The purpose of this chapter is to review briefly some basic concepts of the molecular analysis of learning, and to illustrate with a set of experimental results the power as well as the limitations of the approach.

WHAT SHOULD WE EXPECT FROM MOLECULAR MODELS OF LEARNING?

Being so rewarding in terms of data, molecular analysis may lure its practitioners away from some of the cardinal problems of learning research. The question should, therefore, be posed: What do we expect to find by trying to reduce learning and memory to molecular processes? Any attempt to answer this question must start with a definition of the biological phenomenon which we try to explain.

Cell to Cell Signalling: From
Experiments to Theoretical Models
ISBN 0–12–287960–0

What is learning?

It is generally accepted that learning is an experience-dependent modification in behaviour, or in the potential to behave; the effect of purely developmental programmes, fatigue or pathology are excluded (Bower and Hilgard, 1981). This definition may satisfy psychologists or ethologists, but possibly not neurobiologists. A neurobiological definition should take into account the level of analysis with which neurobiology is concerned, namely, neurones, neuronal circuits and nervous systems.

Based on a detailed argumentation presented elsewhere (Dudai, 1989), I suggest that learning be defined in terms of a fundamental property of nervous systems, which is their ability to encode internal representations. Internal representations are structured bodies of the information possessed by the organism about the world and about the appropriate reactions to the world. Representations are, thus, neuronally encoded stylized versions of the world, capable of guiding behaviour. Internal representations are expected to vary tremendously in their complexity. Some are very simple. For example, a neuronal circuit subserving a reflexive limb withdrawal in response to pain, encodes a representation of 'no pain' or various intensities of pain, as well as the appropriate motor response. In the intact organism, this representation, although so simple, is still but one element in the complex structured body of information encoded by the nervous system *in toto*. Many internal representations are extremely complex, for example, language representations; we cannot yet even imagine how they are encoded in the neuronal hardware. It is important to note that the term 'representation' is used here in a very general sense, and is not restricted by the cognitive connotations usually associated with 'representations' in cognitive sciences and in mammalian neurophysiology. Learning is defined as an experience-dependent generation of enduring internal representations, and/or experience-dependent lasting modification in such representations. 'Enduring' and 'lasting' cover a physiological time window of seconds to a lifetime. Memory is the retention of these experience-dependent internal representations over time. The neurobiology of learning thus investigates experience-dependent modifications in neuronal substrates, which are relevant to the representational properties of these substrates.

Fundamental questions in learning research, and constraints on molecular and cellular models of learning

The aforementioned definition of learning directs our attention to a cardinal question, namely, how are representations encoded in the nervous system? This is a fundamental concern of the neurosciences in general, and hence is not unique to learning research. It is unlikely that a comprehensive account of learning would be provided without a detailed account of the representational codes involved. Alas, we know very little about representational codes. Considering the single-cell level, one can suggest two basic types of codes. One is a spike code in which the information is encoded in the timing and/or rate of spikes, or the rate of change in these parameters. The other is place code in which the information is encoded in the position of the neurone in the network. Place code is used, for example, by central sensory maps, be they projectional (topographical) or computational (centrally synthesized). Of course, the two codes are not mutually exclusive, and the same neurone may use both to encode different types of information (Konishi, in preparation).

In addition, it is highly likely that the nervous system encodes information not only in the activity of single units, but also in the coherent activity of populations of neurones. Such a code is an ensemble code in which the information is encoded in the collective activity of many distributed units, so that each unit plays only a minor role in the total output, and many different representations are superimposed in each unit. There is much theoretical support but only little experimental evidence for ensemble coding in the nervous system; the coding of saccadic eye movements by neurones in the superior colliculus is one recently described example (Lee et al., 1988).

Regardless of whether the system uses one or more of the above coding systems, one thing is clear: representations are encoded by the combination of the two fundamental properties of neuronal systems, namely, electrical activity and connectivity.

It follows that if we wish to understand the biological bases of learning, we should study enduring experience-dependent alterations in both neuronal activity and connectivity. However, to understand how the observed cellular changes ultimately result in a behavioural change, we must ultimately find out how the information is globally encoded in the activity and connectivity pattern of the concrete network(s) that subserve(s) the modifiable behaviour. By restricting our attention to only molecules and isolated cells, we may find mechanisms common to many types and instances of learning ('universal mechanisms'), but will still be unable to explain why the behaviour in question was altered in a specific way.

Figure 1 capitulates the above argument by depicting learning as a multilevel biological phenomenon. Specific representations are encoded by the activity and connectivity of circuits. The universal building blocks by which these representations are realized in the nervous system, are molecular and cellular mechanisms. Therefore, if we reduce learning to molecular and cellular terms, we address the universal mechanisms. At this level, the fundamental questions are: What types of mechanisms are used by the nervous system in order to achieve an enduring alteration in neuronal activity and connectivity? And could such mechanisms explain basic 'rules' of learning; for example, why does a certain sensory input result in learning

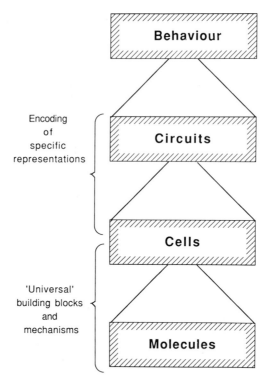

Fig. 1. Learning is a multilevel phenomenon, and its analysis can be performed at levels ranging from the molecular to the behavioural. The nervous system is expected to use universal molecular and cellular mechanisms to encode different representations in neuronal circuits. Molecular and cellular analysis can reveal such molecular and cellular universals, but ultimately, the representational code must be deciphered before the neuronal changes that lead to a modifiable behaviour are understood.

whereas another input does not, or why are some training procedures more effective than others?

AN EXAMPLE OF A MOLECULAR MODEL: HOW A SLUG AND A FLY IMPLICATE THE cAMP CASCADE IN ACQUISITION AND RETENTION

Keeping in mind the limitations of the molecular and cellular approach to learning, let us now illustrate its power. We will do so by addressing a set of studies which combine the neurogenetic approach in the fruit fly, *Drosophila*, with the cellular approach in the sea-hare, *Aplysia*. We will specifically address two questions: (a) Why is it that in classical (Pavlovian) conditioning, if the conditioned stimulus (CS) precedes the unconditioned stimulus (US) within an appropriate time window, conditioning is successful, whereas if the US precedes the CS, conditioning is unsuccessful? (A training schedule in which the CS precedes the US is termed 'forward conditioning', and a schedule in which the US precedes the CS is termed 'backward conditioning'. The constraints on the order and on the timing and stimuli are together termed the constraints of 'temporal specificity', and they make sense intuitively, because they are geared to detect causality.) (b) How is memory retained in the molecular systems that are modified during acquisition? We will address these questions briefly, one at a time.

The neurogenetic approach

Our experimental approach to the aforementioned questions was the following: first we wished to identify gene products that are crucial for learning more than for other physiological processes. This could be done by using genetics. In the neurogenetic dissection of learning, one induces mutations at random and searches for those mutations that disrupt learning and memory in a relatively specific manner. The appropriate mutants are then analysed and the molecular defect explored. The role of the defective gene product can then be investigated on the cellular level in cellular systems that subserve learning.

The organism best suited for neurogenetic dissection of learning is the fruit fly, *Drosophila melanogaster*. Its advantages for genetics and molecular biology need not be restated. Fruit flies, so it appears, are also quite intelligent, and can learn a variety of tasks that they deem important. For example, they learn to associate an odour with a punishment or a reward, or learn that a sexual partner is unresponsive and should be avoided. The neurogenetic dissection of learning in *Drosophila*, as well as the learning situations in which fruit flies excel, were recently reviewed in detail (Tully, 1987; Dudai, 1988). However, nobody is perfect, not even *Drosophila*: although excellent for genetics, fruit flies are not a convenient preparation for cellular analysis. Their central nervous system is dense and complex. Therefore, at present we must also use another preparation for the cellular analysis of the function of those gene products which are implicated in learning by the neurogenetic analysis. An excellent preparation for cellular analysis is *Aplysia* (Kandel and Schwartz, 1982).

Adenylate cyclase: a candidate convergence locus for the cellular codes of associative stimuli

Several single-gene mutations have been isolated in *Drosophila*, which affect learning and memory in a rather specific manner. One of these mutations is *rutabaga* (*rut*; Aceves-Pina *et al.*, 1983). *rut* is defective in acquisition and short-term memory in many different learning tests (Dudai *et al.*, 1984; Tully and Quinn, 1985). The sensory and motor behaviour of this mutant, excluding learning and memory, is astonishingly normal. In addition, *rut* is defective in adenylate cyclase activity (Dudai *et al.*, 1983; Livingstone *et al.*, 1984). Only a small fraction of the total adenylate cyclase activity is missing. This should probably be expected: it is difficult to conceive an apparently healthy mutant with gross defects in adenylate cyclase activity. The most interesting effect of the mutation is obliteration of the Ca^{2+}–calmodulin activation of adenylate cyclase (Dudai *et al.*, 1984; Livingstone *et al.*, 1984; Yovell *et al.*, unpublished data).

This lack of Ca^{2+}–calmodulin stimulation of the cyclase, when combined with results obtained independently by Kandel, Byrne and their colleagues on molecular mechanisms subserving learning in the gill and siphon or tail withdrawal reflexes in *Aplysia* (Kandel and Schwartz, 1982; Byrne, 1987), has led to the hypothesis that adenylate cyclase may function as a convergence locus for the molecular codes of the CS and US in classical conditioning. Adenylate cyclase plays a key role in mediating cellular information via the cAMP cascade; it does so by generating cAMP in response to an extracellular signal, i.e. neurotransmitter or hormone. In addition, the enzyme can be activated by Ca^{2+}, which functions here as an intracellular signal. Suppose the neurotransmitter and the Ca^{2+} are the cellular codes of the sensory stimuli used in classical conditioning; the lack of Ca^{2+} stimulation of the enzyme would not affect mediation of intercellular information by the cAMP cascade (hence the normal ongoing behaviour of the organism), but would damage the ability to associate stimuli. Furthermore, if the above hypothesis is correct, than the order of convergence of transmitter and Ca^{2+} on the cyclase might be expected to display temporal specificity, i.e. forward conditioning with activators that encode the CS and US, respectively, would be more efficient than backward conditioning.

To test this working hypothesis, one needs two experimental tools: (a) an experimental system to test the activation of adenylate cyclase by transient pulses of transmitter and Ca^{2+}, and by combinations of such pulses; (b) a system in which the molecular codes of the CS and US, respectively, are known.

A system for testing the activation of adenylate cyclase by transient pulses of activators, was developed by Yoram Yovell from our laboratory in collaboration with Eric Kandel and Tom Abrams (Yovell *et al.*, 1987). In brief, this is an electronically controlled apparatus in which a perfused membrane preparation housed in a microreaction chamber is flushed by pulses of the appropriate ligands. A system in which the cellular encoding of the CS and US was proposed, is the above-mentioned gill-and-siphon withdrawal reflex in *Aplysia*. The reflex can be classically conditioned by pairing a noxious stimulus to the head (US) with a mild tactile stimulus to the mantle skin (CS). Following training, the weak tactile stimulus to the skin comes to evoke an intense gill-and-siphon withdrawal response (CR) (Abrams, 1985). In this system, serotonin released by a facilitatory interneurone was proposed to be the cellular code of the US, and Ca^{2+} influx associated with an action potential in the sensory neurone was proposed to be the cellular code of the CS. Using the perfused membrane system, we were able to show that under certain *in vitro* conditions, if a pulse of Ca^{2+} precedes a pulse of serotonin, the amount of cAMP generated by adenylate cyclase from *Aplysia* is greater than if serotonin precedes the Ca^{2+} (Yovell and Abrams, 1988; Yovell *et al.*, unpublished data). This finding, though still preliminary, is in line with the hypothesis that Ca^{2+}-stimulated adenylate cyclase might serve as a convergence locus for the cellular representation of the CS and the US (Fig. 2). Moreover, it suggests that temporal specificity revealed in behavioural studies of classical conditioning, is a molecular property. Of course, network properties are not excluded.

Three caveats should be mentioned here: first, clearly the cAMP cascade is not the sole mechanism necessary for acquisition; *rut* can still learn up to 50–70% of normal in certain learning paradigms. Other molecular mechanisms must therefore function in parallel with the cAMP cascade during acquisition. Second, the *in vitro* results are encouraging, but they provide no

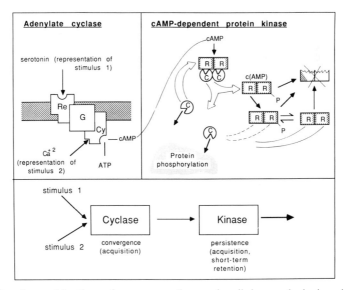

Fig. 2. A combination of neurogenetics and cellular analysis has led to a molecular model for the role of the cAMP cascade in acquisition and short-term memory in classical conditioning. According to this model, acquisition is based on coincidence detection and convergence of stimuli on the Ca^{2+}–calmodulin-stimulated adenylate cyclase. The resulting pulse of cAMP then persistently activates cAMP-dependent protein kinase. For further details, see text.

proof that adenylate cyclase indeed functions *in situ* as a convergence locus for a CS and US, neither in *Drosophila* nor in *Aplysia*. And third, *rut* is defective also in non-associative learning, and this could not be explained by the convergence hypotheses; other properties of the cAMP cascade might be critical for non-associative learning (Dudai, 1988).

cAMP-dependent protein kinase: a candidate retention locus for cellular memory

Let us now shift our attention to the second question posed above, namely, how is memory retained in molecular processes? Suppose adenylate cyclase indeed serves as a convergence locus in acquisition of classical conditioning; the question could then be asked, is the enzyme also a molecular locus for memory? In our experiments with the perfused-membrane system we could not find any evidence for persistence of activation of adenylate cyclase following a transient application of a transmitter or Ca^{2+}; the activity of the enzyme always returned to basal level immediately following the removal of the activating ligand.

The next station in the cAMP cascade, downstream from the cyclase, is cAMP-dependent protein kinase (cAMP-dPK). This enzyme has interesting properties that led to the suggestion, by several groups, that it might store molecular memory. The cAMP-dPK is composed of two types of subunits: a catalytic unit, C, and a regulatory unit, R. R inhibits C. Binding of cAMP to R causes dissociation of R from C and hence activation of the kinase (Fig. 2). R can be phosphorylated by C in an intramolecular process; this lowers the affinity of R for C. We have modelled the detailed kinetic properties of cAMP-dPK, performed extensive computer simulations with the model, and found that C can remain dissociated from R (and hence activated) for tens of minutes following a 1-s pulse of cAMP (Buxbaum and Dudai, 1989). In our model, the phosphorylation–dephosophorylation cycle of R plays a cardinal role in controlling the long-term activation of the enzyme.

Jimmy Schwartz and his colleagues found experimental evidence that the ratio of R to C decreases following learning in *Aplysia* (Greenberg *et al.*, 1987). They raised the possibility that R is degraded by a protease activated in learning. In both models, the kinase is thus suggested as a molecular retention device for short- and medium-term memory (Fig. 2). This device retains information before further cellular processes consolidate memory into a long-term process (Goelet *et al.*, 1986).

In conclusion, the following molecular model can be suggested for acquisition and short-term memory of classical conditioning (Fig. 2): Two stimuli (i.e. US and CS) converge on adenylate cyclase and activate it above a certain threshold. The resulting pulse of cAMP activates the cAMP-dPK for many minutes. Again, the experimental evidence from both *Drosophila* and

Aplysia show that other second messenger cascades function in parallel to the cAMP cascade in acquisition and short-term memory. Similar convergence and persistence loci should be expected, therefore, to be detected in other second messenger cascades as well (Dudai, 1989).

A combination of neurogenetics and cellular analysis has led us, thus, to a molecular model for some processes that take place on the molecular and cellular level during learning and at the early stages of memory. Again, we should not forget that the model can heuristically explain only pieces of the universal mechanisms employed by neuronal networks to modify their activity and connectivity. The concrete role of the latter in encoding the internal representation that underlies a given modifiable behaviour, must still be explored before the relevance of the molecular mechanisms to the behavioural change is understood.

REFERENCES

Abrams, T. W. (1985) Activity-dependent presynaptic facilitation: an associative mechanism in *Aplysia*. *Cell. molec. Neurobiol.* **5**, 123–45.

Aceves-Pina, E. O., Booker, R., Duerr, J. S., Livingstone, M. S., Quinn, W. G., Smith, R. F., Sziber, P. P., Tempel, B. L. and Tully, T. P. (1983) Learning and memory in *Drosophila*, studied with mutants. *CSH Symp. Quant. Biol.* **48**, 831–40.

Bower, G. H. and Hilgard, E. R. (1981) *Theories of Learning* (5th edn). Prentice-Hall, New Jersey.

Buxbaum, J. and Dudai, Y. (1989). A quantitative model for the kinetics of cAMP-dependent protein kinase (type II) activity: Long-term activation of the kinase and its possible relevance to learning and memory. *J. biol. Chem.* **264**, 9344–51.

Byrne, J. H. (1987) Cellular analysis of associative learning. *Physiol. Rev.* **67**, 329–439.

Dudai, Y. (1988) Neurogenetic dissection of learning and short-term memory in *Drosophila*. *Ann. Rev. Neurosci.* **11**, 537–63.

Dudai, Y. (1989) *The Neurobiology of Learning: Concepts, Findings, Trends*. Oxford University Press, Oxford (in press).

Dudai, Y., Zvi, S. and Segel, S. (1984) A defective conditioned behavior and a defective adenylate cyclase in the *Drosophila* mutant *rutabaga*. *J. Comp. Physiol.* **155**, 569–76.

Dudai, Y., Uzzan, A. and Zvi, S. (1983). Abnormal activity of adenylate cyclase in the *Drosophila* memory mutant *rutabaga*. *Neurosci. Lett.* **42**, 207–12.

Goelet, P., Castellucci, V. F., Schacher, S. and Kandel, E. R. (1986) The long and the short of long-term memory – a molecular framework. *Nature* **322**, 419–22.

Greenberg, S. M., Castellucci, V. F., Bayley, H. and Schwartz, J. H. (1987). A molecular mechanism for long-term sensitization in *Aplysia*. *Nature* **329**, 62–5.

Kandel, E. R. (1985) Steps towards a molecular grammar for learning: explorations into the nature of memory. In *Medicine, Science and Society* (ed. K. J. Isselbacher), pp. 555–604. Wiley, New York.

Kandel, E. R. and Schwartz, J. H. (1982) Molecular biology of learning: modulation of transmitter release. *Science, N.Y.* **218**, 433–43.

Lee, C., Rohrer, W. H. and Sparks, D. L. (1988) Population coding of saccadic eye movements by neurons in the superior colliculus. *Nature* **332**, 357–60.

Livingstone, M. S., Sziber, P. P. and Quinn, W. G. (1984) Loss of calcium/calmodulin responsiveness in adenylate cyclase of *rutabaga*, a *Drosophila* learning mutant. *Cell* **37**, 205–15.

Tully, T. (1987) *Drosophila* learning and memory revisited. *Trends Neurosci.* **10**, 330–5.

Tully, T. and Quinn, W. G. (1985) Classical conditioning and retention in normal and mutant *Drosophila melanogaster*. *J. comp. Physiol.* **157**, 253–77.

Yovell, Y., Kandel, E. R., Dudai, Y. and Abrams, T. W. (1987) Biochemical correlates of short-term sensitization in *Aplysia*: temporal analysis of adenylate cyclase stimulation in a perfused-membrane preparation. *Proc. natn. Acad. Sci. U.S.A.* **84**, 9285–9.

Yovell, Y. and Abrams, T. W. (1988) Order dependence in the activation of adenylate cyclase by serotonin and Ca^{2+}/calmodulin: a possible molecular mechanism for associative learning in *Aplysia*. *Soc. Neurosci. Abst.* **14**, 910.

Neural networks: From neurocomputing to neuromodelling

J. DEMONGEOT, T. HERVÉ, F. BERTHOMMIER AND
O. FRANÇOIS

*TIM3-IMAG & ICP-INPG, Université J. Fourier de Grenoble,
Faculté de Médecine, 38 700 La Tronche, France*

INTRODUCTION

Our purpose in this chapter is to give an introduction to neurocomputing and
neuromodelling based on some recent papers published at the frontier
between these domains. The appearance of synthetic books (MacGregor,
1987; Anderson and Rosenfeld, 1988) bringing together basic papers or first
references in these fields proves that there is now an interesting convergence
between techniques coming from cellular automata theory (Demongeot *et
al.*, 1985; Arbib, 1987) or from solid-state physics (Bienenstock *et al.*, 1986),
and modelling problems coming from neurophysiology and experimental
psychology, essentially from the fields of perception and recognition (Buser
and Imbert, 1987; Heierli *et al.*, 1987; Patuzzi and Robertson, 1988;
Piccolino, 1988; Silvera, 1988). We will present successively some recent
theoretical results about the convergence of neuroalgorithms and about
memory of neural networks. Then we will show how numerical experiments
of neurocomputing can be characterized by a small number of parameters.
Finally, we will apply these results and techniques to neuromodelling, by
addressing problems of neural signal coding and contrast exhaustion (of a
sound in a noise), based on lateral inhibition in a network. The example given
concerns the auditory system, but the reinforcement of contrasts appears also
in vision and in taste.

Cell to Cell Signalling: From
Experiments to Theoretical Models
ISBN 0–12–287960–0

NEUROCOMPUTING: THEORETICAL BASIS

Following the first paper published about Boolean neurocomputing (McCulloch and Pitts, 1934), we deal here with a large number of elementary subunits (called 'neurones' in the following) having only two possible states, 0 and 1; these neurones are, in general, located at the vertices of a network, which can be regular – a square B in the integer lattice $\mathbf{N}^d$ – or not – any finite graph G in the continuous space $\mathbf{R}^d$, where the dimension d in general is equal to 2. Then we define the neighbourhood $V_{i,j}$ of each neurone (i,j) (where i and j denote the coordinates of the corresponding site in B) by the layer of the nearest neighbours, plus in general the layer of the nearest neighbours of these ones. Moreover, each neurone (i,j) can receive an external input $i_{i,j}(k)$ at time k; we determine the new output state $o_{i,j}(k)$ of the neurone (i,j), by calculating a potential function U for the configuration $C(k)$ having state 1 in the neurone (i,j), states $o_{m,n}(k-1)$ in each neurone (m,n) of the neighbourhood $V_{i,j}$, and state 0 elsewhere in the rest of B:

$$U(C(k))= \sum_{(m,n)\in V_{ij}} w_{m,n,i,j}(k-1)\, o_{m,n}(k-1)+i_{i,j}(k-1)-S_{i,j}(k-1)$$

where $w_{m,n,i,j}(k-1)$ denotes the synaptic weight between neurones (m,n) and (i,j) (depending on the concentration of neurotransmitter present in the synapse at time $(k-1)$; to simplify, we will denote this weight by $w_{m,n}(k-1)$ when there is no ambiguity); $S_{i,j}(k-1)$ denotes a threshold the neurone (i,j) has to exceed in order to fire. This threshold depends in general on the past history of the neurone and its received stimulations (see later for an example).

The transition rule of change of state is then very simple:

(a) in a deterministic network, we decide that $o_{i,j}(k)$ is equal to 1, if
$U(C(k))/(1+U(C(k)))>1/2$;

(b) in a stochastic network, we decide that $o_{i,j}(k)=1$ with the probability
$U(C(k))/(1+U(C(k)))$.

The main problem is how the network converges, from a certain initial configuration $C(0)$ and after a certain transient, during which the configuration $C(k)$ (represented by all states $o_{i,j}(k)$) fluctuates and reaches a steady state. Many papers have described convergence in the deterministic network (see the study of the attractors of the dynamics for example in Kerszberg and Mukamel (1988)). We shall only give here recent results concerning the stochastic case, where we observe asymptotically a unique invariant measure (ergodic behaviour). The convergence has been studied when the randomness only concerns:

(a) the order in which the transition rule is successively applied to the

neurones; if they are chosen randomly, with a deterministic transition rule, the convergence has been proved in Fogelman *et al.* (1987) when the evolution of the *w*s allows the building of a Lyapunov function;
(b) the inputs $i_{ij}(k)$ (Cottrell, 1988), also by using a Lyapunov function;
(c) the transition rule (Demongeot and Fricot, 1986; Demongeot, 1987; Demongeot and Tchuente, 1987);
(d) both the order of the sequence of neurones and the transition rule (Geman and Geman, 1984).

If there is no convergence nor ergodicity, we can meet phase transitions as in solid-state physics (Kaneko and Akutsu, 1986).

Another important theoretical problem concerns the memory capacity of a neural network (Morris and Wong, 1988). For example, Seifert (1988) proves that the recall memory capacity of the network is considerably increased by the presence of a diagonal term $w_{ij}(k-1)o_{ij}(k-1)$ in the potential U; an asymptotic (when the size B tends to infinity) formula for this capacity shows that, if there is no diagonal term in U, the capacity exponentially decreases to 0, when the length of the sequence of configurations to recall is increasing to infinity.

NEUROCOMPUTING: SIMULATIONS

Many experiments have been processed on classical computers or on devoted parallel machines (Soucek and Soucek, 1988), by using only one or a succession of neural network layers (Fogelman *et al.*, 1987; Gorman and Sejnowski, 1988; Kawahara and Irino, 1988; Irie and Miyake, 1988; Funahashi, 1989). The interest in replacing soft sequential simulations by hard parallel processing has been emphasized in many surveys (Lippmann, 1987; Cowan and Sharp, 1988a: Decamp and Amy, 1988; Grossberg, 1988; Kohonen, 1988; Sejnowski *et al.*, 1988). Here we want only to dwell on the possibility of characterizing the stationary state reached by a neural network and of replacing the sequence of a large number of iterations by only a small number of parameters (Gerstein *et al.*, 1985; Hervé *et al.*, 1989). For that purpose, we have to introduce the notion of random field: let us consider spike trains $M(i,j)$, representing the sequence of inputs received by the neurone (i,j); we can build a succession of binary matrices M_k such that each term $M_k(i,j)$ is 1 if there is a spike in the interval $[k\Delta t,(k+1)\Delta t]$ of the train $M(i,j)$, else $M_k(i,j)$ is 0. Figure 1 shows this transformation for four spike trains. The originality of this representation lies in the idea of considering the succession of the matrices M_k like the successive realizations of a random field, defined on the finite set $B=[1, \ldots,n] \times [1, \ldots,n]$.

A random field is entirely defined by the interaction potential of its configurations, which defines the parameters of its associated Gibbs proba-

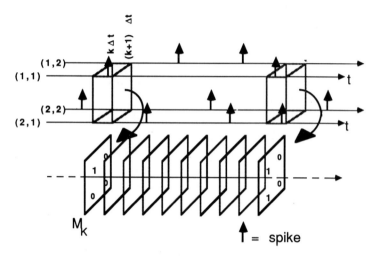

Fig. 1. The sequence of the (2×2) Boolean matrices represents the input activity of four fibres, considered to be like the successive realizations of a random field defined on four sites.

bility measure P; this measure is defined very simply, on each configuration C of 0s and 1s in the neurones of the network, by the formula:

$$P(\{C\}) = \exp(U(C))/Z$$

where Z is a normalization constant and U is defined by

$$U(C) = \sum_{D < C} J_U(D)$$

where J_U is called the interaction potential and '$D < C$' signifies that the 1s of the configuration D are included in the 1s of C.

The estimation of the interaction potential is made by the Gibbs sampling technique (Geman and Geman, 1984; Demongeot *et al.*, 1985). It suffices to simulate the network for a sufficiently long time and to calculate the empirical frequency $\underline{P}(C)$ of each configuration C; then we have, for the estimates $\underline{U}$ and $\underline{J}_U$: $\underline{U}(C) = \mathrm{Log}\,\underline{P}(C) - \mathrm{Log}\,\underline{P}(\varphi)$ and by applying the classical Moebius formula:

$$\underline{J}_U(C) = \sum_{D < C} (-1)^{|C \backslash D|} \underline{U}(D)$$

where $|C \backslash D|$ denotes the number of 1s being in the configuration C and not in D. Among all the possible configurations, let us call: φ the configuration whose sites are all in the state 0, n-uplet any configuration all sites, but n,

Fig. 2. Spatial distribution of the singleton and pair potentials of an output random field induced by a homogeneous Gaussian input random field (i.e. corresponding to the same gaussian law for the intervals between successive 1s of the M_ks). The brilliancy of a red site is proportional to its singleton potential (simulations have been made in parallel on a TNode 40 FPS). The green bars have a surface proportional to the value of the pair potential corresponding to its extremities (a threshold has been used here in order to eliminate all sufficiently large positive pair potentials and keep only 'cold' interactions).

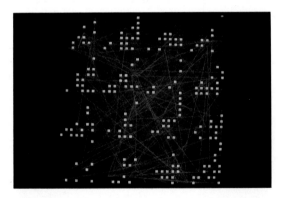

Fig. 3. Spatial distribution of the singleton and pair potentials of an output random field induced by an inhomogeneous input random field.

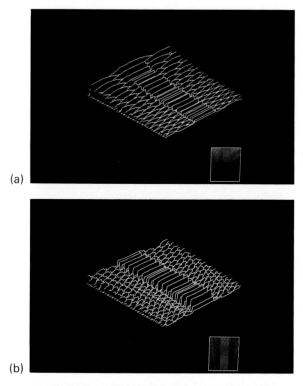

Fig. 5.(a) Transient activity of a 1-dimensional network (the time axis is orthogonal (coming toward the observer) to the 1-dimensional network and the activity on the vertical axis corresponds to the normalized mean number of spikes per unit of time). (b) Stationary activity of the same network; the exhaustion of the central activity can be seen, corresponding to a noised low frequency sound (670 Hz).

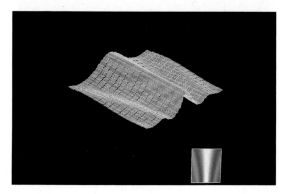

Fig. 6. Correlation function between the activity of the central neurone and that of neurones in its neighbourhood. The observed behaviour has been also described in other sensory systems such as vision, olfaction and taste.

which are in the state 0. With N sites, there will be N possible 1-uplets (called singletons), $\binom{N}{2}$ 2-uplets (called pairs) ,.., $\binom{N}{n}$ n-uplets.

The interaction potential J_U verifies: $J_U(\varphi) = 0$ and the $J_U(D)$s are called, respectively, singleton (resp. pair) potentials, if Ds are singletons (resp. pairs). The Gibbs measure is then parametrized by the $2^{|B|}$ values taken by J_U. Very often, it is sufficient to know the $J_U(C)$s on configurations C having their 1s only in the neighbourhood $V_{i,j}$ of a neurone (i,j): that reduces considerably the number of parameters necessary to parametrize the asymptotic Gibbs measure representing the stationary behaviour of the network. In Fig. 2, we give a coloured representation of singleton and pair potentials: it summarizes 16 000 parallel iterations after the transient phase and allows qualitative visualization of the response of the network.

The ws used for the simulations above followed the rule of lateral inhibition: they evolved with the Hebbian rule described in the next part of the chapter by remaining activatory at short distance and inhibitory after. In general the ws converge, as in Figs 2 and 3. This convergence appears simultaneously with that of the Gibbs probability measure representing the frequency of occurrence for the configurations. If there is no input, we can prove that, after convergence of the ws, the observed asymptotic random field has the same potential U as the potential of the transition rule defined earlier in this chapter (Demongeot $et\ al.$, 1985), if the simulation is sequential. If it is synchronous and the ws symmetric, S and i being equal to 0, we have:

$$U(C) = \log \left(\sum_{D < B} \exp \left(\sum_{(m,n) \in D;(i,j) \in C} w_{m,n,i,j} O_{m,n} O_{i,j} \right) \right)$$

APPLICATIONS TO NEUROMODELLING

The application of neurocomputing to neuromodelling implies the existence of a realistic model for coding the spike trains on the fibres delivering inputs to an intermediary neural network located, for example, in a sensory system. This problem has been studied in the past by many people (Johnson, 1974; Holden, 1976; Sampath and Srinivasan, 1977; Hoppensteadt, 1986). In the case of the auditory nerve activity, the mean firing rate of discharge of a fibre can be easily related to the tonotopic coding of the information (Hervé and Demongeot, 1988), but this coding is severely limited because of the weak dynamic range due to the refractory period of the discharges; as the interspike interval is greater than 1 ms, the mean firing rate is slower than 1 kHz and there is some evidence that this concept is not completely efficient in representing the speech signal (10 kHz bandwidth). The second coding (known as the temporal coding), lies in the density function of the interspike intervals, the spikes being supposed to occur following a renewal point process. This idea has been used, for example, in Bishop *et al.* (1964), Segundo *et al.* (1966), Schroeder and Hall (1974) and Donhouède (1984), contrary to other approaches assuming, for example, a partially deterministic coding (Lestienne and Strehler, 1987; Colbert and Levy, 1988). In order to take into account both kinds of coding, we propose to represent the activity of N neurones, i.e. N spike trains, by a binary random field defined on N sites, in the following two steps:

(1) each spike train is first sampled at the frequency $1/\Delta t$, so that each spike (represented by a Dirac pulse) falls in an interval $[k\Delta t,(k+1)\Delta t]$; Δt is shorter than the refractory period and the Boolean variable $M_k(i,j)$ is equal to 1 if a spike is emitted during the interval $[k\Delta t,(k+1)\Delta t]$, otherwise it is equal to 0;
(2) the probability measure of the interspike intervals is distributed on $[T_R, \infty]$, where T_R denotes the value of the refractory period, by translating the geometric law of parameter p.

Figure 4 shows the histogram of the position of spikes emitted in response to stimulation after a refractory period. The different types are obtained by changing p and T_R and by randomizing T_R in the case presented at the top right of Fig. 4. The smaller histograms correspond to experiments and the larger ones to simulations of the model, showing a good qualitative fit in each of the four cases.

After modelling the inputs, we have used the model of neural network presented above. This model can be described more specifically as follows:

(a) time sampling period: Δt; $t = k\Delta t$.
(b) neurone (i,j) output: $o_{i,j}(k) = 1 \langle = \rangle$ occurrence of a spike at time $k\Delta t$, according to eqn (1):

$$o_{i,j}(k) = \mathbf{H}\left[\sum_{(m,n)\in V_{i,j}} W_{m,n}(k-1).o_{m,n}(k-1) + i_{i,j}(k-1) - S_{i,j}(k-1) \right]; \tag{1}$$

$$\mathbf{H}(t) = 1 \text{ if } t > 0, \text{ else } \mathbf{H}(t) = 0$$

where the set of neurones (m,n), denoted $V_{i,j}$, is the excitatory/inhibitory neighbourhood of the neurone (i,j) and $w_{m,n}$ denotes the weight of the synapse between the neurone (m,n) and the neurone (i,j).

(c) $S_{i,j}(k)$ is the threshold of the neurone i,j, according to eqn (2):

$$S_{i,j}(k+1) = \min(S_{max}, \max(S_{min}, S_{i,j}(k) - \sum_{(m,n)\in V_{i,j}} W_{m,n}(k)o_{m,n}(k) - i_{i,j}(k))), \tag{2}$$

$$\text{if } o_{i,j}(k) = 0 = S_o, \text{ if } o_{i,j}(k) = 1$$

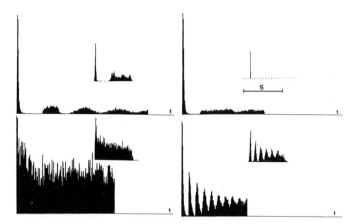

Fig. 4.   Post-stimulation histograms (large: simulations, small: experiments).

(d) short-term plasticity rule: $w_{m,n}(k) = w_{m,n}(k_0)$ as long as $k_0.\Delta t \leqslant k.\Delta \leqslant k_0.\Delta t + 3$ ms, so that the weights change every 3 ms, $w_{m,n}(k)$ keeping the same sign as $w_{m,n}(0)$ chosen following experimental data (Heierli *et al.*, 1987), according to eqns (3)–(5).

$$\left| w_{m,n}(k) \right| = \min(0, \left| w_{m,n}(k-1) \right| + \Omega \left[i_{m,n}(k-1), \quad o_{i,j}(k) \right] \quad w_{m,n}(k-1)) \tag{3}$$

$\Omega(0,0) = \Omega(0,1) = 0$, $\Omega(1,0) = -\beta$, $\Omega(1,1) = N_f$, for excitatory synapses $(w_{m,n}(0) > 0)$

$$\tag{4}$$

$\Omega(0,0) = \Omega(0,1) = 0$, $\Omega(1,0) = N_f$, $\Omega(1,1) = -\beta$, for inhibitory synapses $(w_{m,n}(0) < 0)$

$$\tag{5}$$

The above plasticity rules lead to a learning, which can be represented as in Fig. 5, where, after a transient phase, we see the occurrence of an exhaustion of the normalized activity on the central part of the network, allowing the vowel-feature extraction (Irino and Kawahara, 1988) (the sites are represented in dimension 1 orthogonally to the time axis coming toward the observer). Figure 6 shows the classical 'dog' profile of the correlation function between activities of the central neurone and of the neurones of its neighbourhood (supposed here to be equal to the whole set of neurones *B*) (Berthommier *et al.*, 1989).

CONCLUSION

Neuromodelling is still very influenced by Boolean neurocomputing. In order to be more realistic, the models should be improved by:

(1) taking into account a plasticity based on the kinetics of neutrotransmitters (cf. the classical papers by Hubel *et al.* (1977) and by Wernig *et al.* (1980), and more recently by Aoki and Siekevitz (1988), Gottlieb (1988), Nakai *et al.* (1988) and Nicoll (1988));

(2) studying more complex, integrated functions (Axelrad *et al.*, 1987; Pratap *et al.*, 1988; Trelease, 1988);

(3) studying the attractors of the dynamics even in case of non-stationarity (Babloyantz and Destexhe, 1986; Degn *et al.*, 1987; Labos, 1987; Lagerlund and Sharbrough, 1988);

(4) comparing qualitatively the behaviour of a simulated network to that of a network studied with voltage-sensitive dyes (Blasdel and Salama, 1986);

(5) generating spike trains with continuous differential models having bursting properties (Martiel and Goldbeter, 1987; Chay and Cook, 1988; Goldbeter and Moran, 1988; Llinás, 1988; see also the paper by Hindmarsh and Rose in this volume);

(6) including ideas coming from artificial intelligence in order to be able to model cognitive phenomena (McLelland and Rumelhart, 1986; Schwab and Nusbaum, 1986; Cowan and Sharp, 1988; Demongeot *et al.*, 1988; Feldman, 1988; Schwartz, 1988);

(7) adapting the timing in neural networks, for example to the desensitization kinetics of the neuroreceptors (Rose and Hindmarsh, 1985; Desmond and Moore, 1988; Martiel, 1988);

(8) taking into account experimental magnetic recording (Papanicolaou *et al.*, 1988).

These improvements in neuromodelling will lay the foundations of a field that will take advantage of the ever-increasing simulation abilities of neurocomputing.

REFERENCES

Anderson, J. A. and Rosenfeld, E. (1988) *Neurocomputing. Foundations of research.* MIT Press, Cambridge, MA.

Aoki, C. and Siekevitz, P. (1988) Plasticity in brain development. *Sci. Am.* **259**, 34–43.

Arbib, M. A. (1987) *Brains, machines, and mathematics*, 2nd edn. Springer Verlag, New York.

Axelrad, H., Bernard, C., Cottrell, M. and Giraud, B. (1987) The use of an artificial neural network to analyse the informational transfer properties of a simplified model of the cerebellar cortex. *IEEE 1st Ann. Conf. Neural Networks*, IV59–IV65.

Babloyantz, A. and Destexhe, A. (1986) Low-dimensional chaos in an instance of epilepsy. *Proc. Natl Acad. Sci. U.S.A.* **83**, 3513–17.

Berthommier, F., Demongeot, J., Schwartz, J. L., Escudier, P. and Hervé, T. (1989) Auditory processing in a post-cochlear stochastic neural network: reinforcement of spectral contrasts based on spike synchrony with application to vowel spectrum processing. *Proc. Eurospeech 89* (in press).

Bienenstock, E., Fogelman Soulie, F. and Weisbuch, G. (1986) *Disordered Systems and Biological Organization*, NATO ASI Series, Vol. 20. Springer Verlag, Berlin.

Bishop, P. O., Levick, W. R. and Williams, W. O. (1964) Statistical analysis of the dark discharge of lateral geniculate neurons. *J. Physiol.* **170**, 598–612.

Blasdel, G. G. and Salama, G. (1986) Voltage-sensitive dyes reveal a modular organization in monkey striate cortex. *Nature* **321**, 579–85.

Buser, P. and Imbert, M. (1987) *Neurophysiologie Fonctionnelle*. Hermann, Paris.

Chay, T. R. and Cook, D. L. (1988) Endogenous bursting patterns in excitable cells. *Math. Biosci.* **90**, 139–53.

Colbert, C. M. and Levy, W. B. (1988) What is the code? *Neural Networks* **1**, S1, 246.

Cottrell, M. (1988) Stability and attractivity in associative memory networks. *Biol. Cybernetics* **58**, 129–39.

Cowan, J. D. and Sharp, D. H. (1988a) Neural nets and artificial intelligence. *Daedalus* 85–122.

Cowan, J. D. and Sharp, D. H. (1988b) Neural nets. *Q. Rev. Biophys.* **21**, 365–427.

Decamp, E. and Amy, B. (1988) *Neurocalcul et réseaux d'automates*. Editions EC2, Paris.

Degn, H., Holden, A. V. and Olsen, L. F. (eds) (1987) *Chaos in Biological Systems*, Life Sciences Series, Vol. 138. Plenum Press, New York.

Demongeot, J. (1987) Stochastic automata. In *Dynamical Behaviour of Cellular Automata* (eds F. Fogelman *et al.*), pp. 83–95. Manchester University Press, Manchester.

Demongeot, J. and Fricot, J. (1986) *Random Fields and Renewal Potentials*, NATO AS1 Series F, Vol. 20, pp. 71–84. Springer Verlag, Berlin.

Demongeot, J. and Tchuente, M. (1987) Cellular Automata Theory. *Encyclopedia of Physical Science & Technology*, Vol. 4, pp. 464–72. Academic Press, New York.

Demongeot, J., Goles, E. and Tchuente, M. (eds) (1985) *Dynamical Systems and Cellular Automata*. Academic Press, London.

Demongeot, J., Hervé, T., Rialle, V. and Roche, C. (1988) *A.I. and Cognitive Sciences*. Manchester University Press, Manchester.

Desmond, J. E. and Moore, J. W. (1988) Adaptive timing in neural networks: the conditioned response. *Biol. Cybernetics* **58**, 405–15.

Donhouède, B. (1984) Le codage nerveux du message acoustique. Thèse Sciences, Grenoble.

Feldman, J. A. (1988) Neural representation of conceptual knowledge. Preprint, University of Rochester.

Fogelman Soulie, F., Robert, Y. and Tchuente, M. (1987) *Automata Networks in Computer Science*. Manchester University Press, Manchester.

Funahashi, K. (1989) On the approximate realization of continuous mappings by neural networks. *Neural Networks* (in press).

Geman, S. and Geman, D. (1984) Stochastic relaxation, Gibbs distributions, and the Bayesian restoration of images. *IEEE Trans. Pattern Anal. & Machine Intell.* **6**, 721–41.

Gerstein, G. L., Perkel, D. H. and Dayhoff, J. E. (1985) Cooperative firing activity in simultaneously recorded populations of neurons: detection and measurement. *J. Neurosci.* **5**, 881–9.

Goldbeter, A. and Moran, F. (1988) Dynamics of a biochemical system with multiple oscillatory domains as a clue for multiple modes of neuronal oscillations. *Eur. Biophys. J.* **15**, 277–87.

Gorman, R. P. and Sejnowski, T. J. (1988) Analysis of hidden units in a layered network trained to classify sonar targets. *Neural Networks* **1**, 75–90.

Gottlieb, D. I. (1988) Les neurones GABAergiques. *Pour la Science* **126**, 44–51.

Grossberg, S. (1988) Nonlinear neural networks: principles, mechanisms, and architectures. *Neural Networks* **1**, 17–62.

Heierli, P., de Ribaupierre, F. and de Ribaupierre, Y. (1987) Functional properties and interactions of neuron pairs simultaneously recorded in the medial geniculate body of the cat. *Hearing Res.* **25**, 209–25.

Hervé, T. and Demongeot, J. (1988) Random field and tonotopy: simulation of an auditory neural network. *Neural Networks* **1**, S1, 297.

Hervé, T., Dolmazon, J. M. and Demongeot, J. (1987) Neural network in the auditory system: influence of the temporal context on the response represented by a random field. *IEEE-ICASSP Proceedings*, 161–4.

Hervé, T., Dolmazon, J. M. and Demongeot, J. (1989) Random field and neural information: a new representation for multi-neuronal activity. *Proc. Natl. Acad. Sci. U.S.A.* (in press).

Holden, A. V. (1976) Models of the stochastic activity of neurones. *Lecture Notes in Biomaths*, Vol. 12. Springer Verlag, Berlin.

Hoppensteadt, F. C. (1986) *An Introduction to the Mathematics of Neurons.* Cambridge University Press, Cambridge.

Hubel, D. H., Wiesel, T. N. and Le Vay, S. (1977) Plasticity of ocular dominance columns in monkey striate cortex. *Phil. Trans. R. Soc. Lond.* B **278**, 377–409.

Irie, B. and Miyake, S. (1988) Capabilities of three-layered perceptrons. Preprint, ATR Auditory & Visual Perception Res. Lab.

Irino, T. and Kawahara, H. (1988) Vowel-feature extraction from cochlear vibration using neural networks. Preprint, NTT BRL.

Johnson, D. H. (1974) The response of single auditory-nerve fibers in the cat to single tones. PhD thesis, MIT.

Kaneko, K. and Akutsu, Y. (1986) Phase transitions in two-dimensional stochastic cellular automata. *J. Phys. A* **19**, L69–L75.

Kawahara, H. and Irino, T. (1988) Introduction to saturated projection algorithm for artificial neural network design. Preprint, NTT Basic Res. Lab.

Kerszberg, M. and Mukamel, D. (1988) Dynamics of simple computer networks. *J. Stat. Phys.* **51**, 777–95.

Kohonen, T. (1988) An introduction to neural computing. *Neural Networks* **1**, 3–16.

Labos, E. (1987) Chaos and neural networks. *Life Sci. Series* **138**, 195–206.

Lagerlund, T. D. and Sharbrough, F. W. (1988) Computer simulation of neuronal circuit models of rhythmic behavior in the electroencephalogram. *Comput. Biol. Med.* **18**, 267–304.

Lestienne, R. and Strehler, B. L. (1987) Time structure and stimulus dependence of precisely replicating patterns present in monkey cortical neuronal spike trains. *Brain Res.* **437**, 214–38.

Lippman, R. P. (1987) An introduction to computing with neural nets. *IEEE ASSP Magazine*, 4–22.

Llinás, R. (1988) The intrinsic electrophysiological properties of mammalian neurons. *Science, N.Y.* **242**, 1354–63.

MacGregor, R. J. (1987) *Neural and Brain Modeling.* Academic Press, San Diego.

Martiel, J. L. and Goldbeter, A. (1987) Origin of bursting and birhythmicity in a model for cAMP oscillations in *Dictyostelium* cells. *Lecture Notes in Biomaths*, **71**, 244–55.

McCulloch, W. S. and Pitts, W. (1943) A logical calculus of the ideas immanent in nervous activity. *Bull. Math. Biophys.* **5**, 115–33.

McLelland, J. L. and Rumelhart, D. E. (1986) *Parallel Distributed Processing.* MIT Press, Cambridge, Mass.

Morris, R. J. T. and Wong, W. S. (1988) A short term neural network memory. *SIAM J. Comp.* **17**, 1103–18.

Nakai, K., Sasaki, K., Matsumoto, M. and Takashima, K. (1988) Effects of furosemide on the resting membrane potentials and the transmitter-induced responses of the *Aplysia* ganglion cells. *Tohoku J. Exp. Med.* **156**, 79–90.

Nicoll, R. A. (1988) The coupling of neurotransmitter receptors to ion channels in the brain. *Science, N.Y.,* **241**, 545–51.

Papanicolaou, A. C., Wilson, G. F., Busch, C., De Rego, P., Orr, C., Davis, I. and Eisenberg, H. M. (1988) Hemispheric asymmetries in phonological processing assessed with probe evoked magnetic fields. *Int. J. Neurosci.* **39**, 275–82.

Patuzzi, R. and Robertson, D. (1988) Tuning in the mammalian cochlea. *Physiol. Rev.* **68**, 1009–82.

Pratap, R., Nampoori, V. P. N. and Varghese, L. (1988) Invariant characterization of neural systems. *Int. J. Neurosci.* **39**, 245–51.

Rose, R. M. and Hindmarsh, J. L. (1985) A model for a thalamic neuron. *Proc. R. Soc. London B* **225**, 161–93.

Sampath, G. and Srinivasan, S. K. (1977) Stochastic models for spike trains of single neurons. *Lecture Notes in Biomaths,* Vol. 16. Springer-Verlag, Berlin.

Schwab, E. C. and Nusbaum, H. C. (1986) *Pattern Recognition by Humans and Machines.* Academic Press, Orlando.

Schwartz, J. T. (1988) The new connectionism: developing relationships between neuroscience and A.I. *Daedalus* 123–42.

Schroeder, M. R. and Hall, J. L. (1974) Model for mechanical to neural transduction in the auditor receptor. *J. Acoust. Soc. Am.* **55**, 1055–60.

Segundo, J. P., Perkel, D. H. and Moore, G. P. (1966) Spike probability in neurones: influence of temporal structure in the train of synaptic events. *Kybernetik* **3**, 67–82.

Segundo, J. P., Tolkunov, B. F. and Wolfe, G. E. (1976) Relation between trains of action potentials across an inhibitory synapse. Influence of presynaptic irregularity. *Biol. Cybernetics* **24**, 169–79.

Seifert, B. G. (1988) A Hopfield type network with prodigious memory. Preprint, Oxford.

Sejnowski, T. J., Koch, C. and Churchland, P. S. (1988) Computational neuroscience. *Science, N.Y.* **241**, 1299–306.

Soucek, B. and Soucek, M. (1988) *Neural and Massively Parallel Computers.* J. Wiley, New York.

Trelease, R. B. (1988) Connectionism, cybernetics, and the cerebellum. *AI Expert,* 30–6.

Wernig, A., Pecot-Dechavassine, M. and Stöver, H. (1980) Sprouting and regression of the nerve at the frog neuromuscular junction in normal conditions and after prolonged paralysis with curare. *J. Neurocytol.* **9**, 277–303.

Part 2
Morphogenesis and development

The evolution of morphogens

JOHN TYLER BONNER

Department of Biology, Princeton University, Princeton, NJ 08544, USA

Everyone accepts the idea that development in general and pattern formation in particular are mediated by key chemical signals in the embryo which control the events. These 'morphogens' have been the mainstay for modellers of development, and the search for the identification and biochemical reactions related to morphogen production and activity has been a central pursuit of experimental developmental biologists since the early work of Hans Spemann. The interchange between the experimentalists and the theoreticians has been vigorous and fruitful, yet we all have a feeling that the ultimate goal of understanding the details of the mechanics of development is still far away. To some extent the difficulty is due to being unable to decide what actual chemicals in the organism are the morphogens of the equations, but more important is the uneasy feeling that there are so many morphogens and development is so complicated that we may never see it in any kind of idealized clarity. We all secretly pine for a simple picture, and envy those, like Charles Manning Child, who saw everything in terms of metabolic gradients, yet at the same time we automatically reject any such easy, unifying ideas because they seem to conflict with our daily experience in the laboratory, where we are constantly faced with the complexity of even our simplest developmental systems.

One way in which we might decrease the difficulty of the problem is to examine it from a different and unconventional point of view. I will do this here by looking at development in terms of the evolution of morphogens. By this I mean we will make a comparative survey of the simplest and smallest morphogens to the most complicated involving large molecules. It cannot be assumed with any confidence that this is a genuine evolutionary series, but it is certainly true that in primitive organisms the low molecular weight morphogens play a significant role and in more advanced ones we more often

Cell to Cell Signalling: From
Experiments to Theoretical Models
ISBN 0–12–287960–0

find the larger morphogens. Perhaps it would be more accurate to describe the discussions to come as a survey of morphogens from small and simple to large and complex. But since there is a very rough correlation between this trend and the evolution of the developing organisms, it may not be completely unreasonable to call it the evolution of morphogens. Also, as we shall see, the increase in molecular weight in the morphogen is only a small part of the story, because besides this there are the questions of how the morphogen is transmitted from one part of an embryo to another and of how the chemical signal is received so that it can produce a response. And finally, there is the question of whether the response is one that produces a genuine pattern.

First I shall consider small, volatile morphogens. Sometimes these are general in their effects and perhaps would more appropriately be called signal molecules rather than strict morphogens, but the line between the two is thin. Next I shall discuss non-volatile molecules in the molecular weight range of hundreds to many thousands, which include proteins. In each of my examples I will examine the method of transport of the molecule, the kind of receptors that are associated with it, and whether or not it produces a pattern. The organisms that we will consider will range from cellular slime moulds to social insects, although there will be a strong bias towards the former.

VOLATILE MORPHOGENS

In the last few years volatile morphogens have been of special interest to me. It is reasonable to make the assumption that substances which are normally produced by the metabolism of a developing organism might during the course of evolution be put to some use in directing or influencing development. It is simply a matter of taking advantage of the resources that are already present. During normal metabolism there are a number of volatile substances given off in small quantities, but CO_2 and NH_3 are characteristically abundant. CO_2 is, of course, the direct product of aerobic respiration and NH_3 comes from the degradation or deamination of proteins and other nitrogen-containing molecules in the cells.

I would like to show how the cellular slime moulds have made use of NH_3 to guide their developing stages to the ideal location for the formation of the mature fruiting body. This role of NH_3 could hardly be considered one of pattern formation, but it is involved in directed cell movements which are certainly an integral part of development. But first let me very briefly describe the development of one species of slime mould, *Dictyostelium discoideum*.

In asexual development a single spore of *D. discoideum* produces an amoeba that can, upon feeding and frequent cell division, produce a multicellular organism. After the food supply of bacteria has been exhausted a starvation reaction takes over in which many amoebae aggregate into central collection points to form a multicellular cell mass. This migrating slug, which moves at roughly 1 mm/h is positively phototactic and also very sensitive to temperature gradients. Furthermore, if two slugs happen to be close to one another, they will be mutually repelled by negative chemotaxis. It is known that the aggregation to the multicellular stage is caused by positive chemotaxis of the individual amoebae to a pulse gradient of cyclic-AMP, and that the same system seems to be operating within the slug so that the internal amoebae are moving in the same direction, towards the tip of the slug which results in the overall migration of the slug. It is presumed that these chemo-, photo-, and thermotaxes of the entire slug all see to it that spores are formed in an ideal location for dispersal which is different from the ideal feeding area. These are soil organisms and the bacteria in the soil are consumed in moist regions where the individual, delicate amoebae can feed, but spores are best dispersed near the surface of the soil where they are supported on a delicate stalk of dead cells and can be disseminated by touching passing insects and other invertebrates. In this primitive organism, development is under the direct influence of the environment so that through taxes different stages of development will be located in different places in the soil.

We have found that NH_3 increases the rate of movement of amoebae and by applying more NH_3 to one side of a cell mass than the other, we can cause it to turn away from the high concentration of NH_3. We also have excellent evidence that NH_3 production is stimulated by light, and since the migrating slug concentrates directional light on the distal side of itself by means of the 'lens effect', that side moves more rapidly and therefore the slug migrates towards the light. Finally, we have shown that NH_3 production increases with increased temperature and from this we postulate that for negative thermotaxis (which occurs in temperature gradients below the growth temperature of the amoebae, as shown by Whitaker and Poff, 1980) the cells on the warmer side of the slug move faster and therefore the slug moves away from heat. If the temperature gradient is above the growth temperature, then we speculate that the NH_3 becomes so high on the warmer side of the slug that the cells become inhibited in their role of movement, and therefore the cooler side moves faster, producing a positive thermotaxis.

Let us now consider these observations (and speculations) in terms of the properties of NH_3. First it should be noted that there can be small regional differences in NH_3 production; NH_3 liberation must be extremely sensitive to minute changes in light and temperature. Secondly, there is the question of how NH_3 moves, and clearly diffusion must be solely involved; very rapid in the air and more slowly within the cell masses. Thirdly, there is the question of how it is received in the cells and how it has its effect on rate of movement. NH_3 has the property of being both lipid and water soluble, so it can pass through cell membranes, but as soon as it enters the cytoplasm it will disassociate into NH_4^+ and cause the internal pH of the cell to rise. It is known from the work of others on leucocytes that they respond more rapidly to chemical gradients and that high pH increases the activity of contractile proteins (for reviews see Simchowitz and Cragoe, 1986).

To summarize, NH_3 is the simplest of molecules; a product of normal cell chemistry which diffuses with great speed easily penetrating cells and has the effect of making the cell cytoplasm more alkaline. This, in turn, affects the rate of movement of cells which results in oriented movement. There is no localization of specific receptors; all cells can receive the NH_3 signal.

(Let me add a note here. Animal embryos have mass movements at early stages, in particular gastrulation, and neurulation in vertebrates, and it is well known that early embryos excrete nitrogen as NH_3, while later in development it is packaged in the form of urea. It would be particularly interesting to know if the localization of NH_3 production correlates with the regions of extensive cell movement.)

It is also thought that NH_3 may play a role in pattern formation in slime moulds, for NH_3 will increase the percentage of spores, and again it is assumed (but the evidence is controversial) that this effect is due to a regional difference of internal pH (for a review see Schaap, 1986). I will return to factors which affect stalk–spore ratios presently.

There are other volatile morphogens and of these perhaps ethylene is the best known. But like many of the growth hormones in higher plants, ethylene seems to have multiple functions. It is produced by the plant as a natural metabolite and encourages fruit ripening and leaf abscission by stimulating various enzymes, and it has an interaction with auxin, giberellin and kinetin in stimulating cell growth. Perhaps the important point to stress here is that it is thought that although ethylene is a small, volatile substance, there is a specific receptor to which it binds (reviewed in Roberts and Tucker, 1985). If this is indeed the case, then ethylene has advanced in the degree of the specificity of its action. It is, however, a difficult binding to study, and we know nothing of the distribution of the receptors within the plant.

LARGER MORPHOGENS (IN ORGANISMS LACKING GAP JUNCTIONS)

We have now advanced to the consideration of larger molecules that are incapable of significant diffusion in the gas phase and first we will examine those morphogens which are hydrophilic, that is they cannot penetrate the cell membrane, but attach to specific protein receptors on the cell surface. Such morphogens can vary widely in their size, from small molecules with molecular weights in the low hundreds to large proteins with molecular weights in the high thousands. Since we are considering organisms without gap junctions, let us examine briefly cyclic-AMP as a morphogen in the cellular slime mould, *Dictyostelium discoideum*.

To begin with, cyclic-AMP serves two main functions as a morphogen: it is responsible for the chemotactic orientation of the amoebae and it is involved in the pattern of differentiation of stalk cells and spores. In its former role, as we saw earlier, it complements NH_3 which acts as a chemokinetic agent. There are specific receptors in the cell surface which appear in abundance during the aggregation and later stages and that, through a series of steps, mobilize actin in the leading edge of each amoeba.

One of the most interesting aspects of slime mould chemotaxis during aggregation and later stages is the system of transport. It is clear that a simple diffusion gradient can orient the amoebae so they will move up the gradient, but in normal development there is a much more sophisticated system. This is the pulse-relay, where, if a cell is hit with a puff of cyclic-AMP, one of the results is the stimulation of an adenyl cyclase within the cell and the synthesis and release of additional cyclic-AMP into the surroundings. The cyclic-AMP therefore acts by diffusion in the external aqueous environment either as one large gradient or more effectively by a series of small gradients generated in succession through a line of cells by this pulse relay. The latter allows orientation to occur over great distances before large centres are formed, and probably makes it possible for orientation to occur within a migrating cell mass or slug.

Cyclic-AMP also plays a significant role in pattern formation in slime moulds, that is, the regulation of the ratio of stalk cells to spores. There are a number of other known morphogens besides cyclic-AMP and NH_3 which seem to interact in interesting ways to produce patterns. But before examining this, let me briefly discuss other more general aspects of morphogens.

There are many animal hormones, such as the mammalian steroid hormones, that are lipid soluble, but relatively few similar examples of developmental morphogens. This is a pity because so much is known about how steroid hormones turn genes on and off. In the water mould *Achyla* the sexual organs are produced by hormones that pass from plants of one sex to the other, and hormone A of the female, which causes the differentiation of the male hyphae and orients the hyphae towards the female, is a steroid. Also some of the insect developmental hormones are lipid soluble, such as the juvenile and the moulting hormone (ecdysone). In these cases the morphogen hormones penetrate readily through the cell membranes and no doubt combine with an internal receptor which then passes into the nucleus and combines to specific sites in the DNA of the chromosomes to turn on specific genes. It is interesting to speculate why some chemical signals combine with receptors on the surface, while others combine with internal receptors. Is this a fortuitous consequence of the solubility properties of the signal substance or is it a difference that has important functional implications where lipid-soluble signal molecules can do things that the insoluble ones cannot and vice versa?

GAP JUNCTIONS AND SIGNAL TRANSMISSION

In all the cases given so far we have assumed that there is free diffusion of the morphogens. (This is true of the insect developmental hormones too, which diffuse into the haemolymph.) Recent work has shown that some morphogens can only spread through the gap junctions between cells, which are found in higher animals but not in slime moulds. They only allow molecules of molecular weights slightly over a thousand to pass through, and there have been no estimates to what degree this affects the rates of diffusion from cell to cell. The reasons for thinking gap junctions are important comes from the work of N. B. Gilula and his associates (for a brief review see Gilula *et al.*, 1988) who have blocked them with antibodies to specific gap-junction proteins. By blocking the junctions in this way different sorts of developmental anomalies can be produced and these adverse effects are presumably due to preventing key morphogens from passing from cell to cell. By analysing such abnormalities it might be possible to get some indirect view of the role that gap-junction morphogens play in normal development. However, those studies must be combined with the fact that the gap-junction coupling between cells does not remain constant throughout development, but becomes progressively restricted as development proceeds. This leads to the further question of how the topography of these restrictions is controlled; what is the mechanism of the determination of their pattern? Perhaps it is specified in some way by the morphogen itself, or some other morphogen.

A number of morphogens are polypeptides or proteins. These will be too large to penetrate gap junctions and therefore they must diffuse in the extracellular matrix and bind surface receptors on the target cells. There will be a severe limit on their rate of transport simply because macromolecules diffuse so slowly. (It would take about an hour for a molecule of a molecular weight of 300 to diffuse a millimetre, while a protein would take one or more days depending on its size.) The best known examples of such morphogens are the growth factors in vertebrates, for example the epidermal growth factor and the nerve growth factor which stimulate growth of specific cell types.

PATTERN

Thus far we have looked into the mechanical properties of morphogens, that is, their size, their transport, and their reception. Now we come to the central issue of pattern formation. (One important facet of pattern is oriented cell movement.)

Pattern without form

In the simplest kind of pattern that can be achieved the ratios of units dividing the labour are kept constant, but the units do not come together into a cohesive shape. One of the best examples of this is found in social insects when the ratio between the number of individuals in different castes will remain constant despite the size of the colony. Furthermore, it can recover its proper porportions if one of the castes is decimated. For instance, among the termites there are some species found where the ratio of soldiers to workers remains about 4% as the colony increases in size over a period of 15 years (Wilson, 1971). If the soldiers are artificially removed, Light (1942–43) showed that in the subsequent moults of the nymphs, new soldiers will appear and soon regain the fixed ratio. Furthermore, Light was able to prevent this restoration of the ratio by feeding the nymphs a paste made from mashed soldiers, and he concluded that the ratio was maintained by the soldiers producing a pheromone which inhibited the formation of new soldiers at moults, and that this inhibition was responsible for maintaining the fixed ratio.

Here we have a case of constant proportions but clearly the mass of termites does not produce any fixed form or shape. Moreover, it is interesting that the mechanism is by the chemical inhibition of one caste by another. There is a splendid parallel in the cellular slime moulds. A number of workers beginning with Oyama *et al.* (1983) have shown that if the cells of a developing cell mass are separated and placed in a shaking or rotating suspension, it prevents them from forming slugs, but the cells, under the right conditions, will form disorganized clusters for which the ratio of prestalk to prespore cells remains normal and constant, even though all morphogenesis has been inhibited. It is possible to artificially change the ratio by beginning the experiment with largely prestalk cells or largely prespore cells and the proper ratio of about 80% prespore cells takes over after some hours.

Recently Akiyama and Inouye (1987) have repeated this kind of experiment and rigorously examined an old observation of Raper (1940) of whether prestalk cells could be converted to prespore cells when some of the prespore cells have been removed. The answer is clearly 'yes' and they suggest that the simplest hypothesis to explain the results is that the prespore cells somehow prevent prestalk cells from becoming prespores. In other words, the prespore cells produce an inhibitor – much like the termite soldiers – that prevents prestalk cells from changing to prespore. (There is a sociobiological point of interest here. The stalk cells that support the spore mass are dead, and therefore they are altruists which help in the better dispersal of the spores. They are, however, being forced into this non-reproductive mode, according to Akiyama and Inouye's hypothesis, by the selfish spores. In this case the parallel in social insects is between reproductives and the neuter workers, that in many cases are prevented from becoming fertile by inhibitor pheromones given off by the reproductives.)

The only problem with this simple model is that there are at least four candidates for proportion-producing morphogens (for reviews see Gerisch, 1987; Schaap, 1986). Weijer and Durston (1985) have some evidence that the control may be under the influence of cyclic-AMP and its hydrolysis products (such as adenosine). It is also known from many studies that DIF, a small molecule discovered by Kay and his colleagues, not only stimulates the formation of stalk cells, but turns on stalk-specific genes. It has also been shown that NH_3 stimulates the formation of spore cells. Finally cyclic-AMP is involved in the formation of both cell types, and there is evidence that NH_3 turns on specific spore genes. All these substances not only act directly but there is the further complication that they may interact by inhibiting the action of one or more of the other substances.

If we return to normal development, the spore–stalk ratio is fixed and the 20% prestalk cells reside in the tip, while the 80% prespore cells make up the bulk of the posterior position of the migrating slug. There have been many studies on the causes of this distribution and the general view is that two factors are involved; one is the chemotactic gradient of cyclic-AMP inside the slug which pulls the more responsive prestalk cells to the anterior end, and the other is that if a cell happens to be in a particular region, it will be influenced by that environment and become a prestalk cell if it is anterior, and a prespore cell if it is posterior. As the migration stage enters into the final fruiting or culmination stage, all the prespore cells turn into spores, and the prestalk cells secrete the central cylinder of cellulose, and then enter into the top of the cylinder they have created, become large and vacuolate with cellulose walls, ultimately dying. And this is a simple organism!

REGIONALIZATION OF RECEPTORS

To return to a more general problem, and one of fundamental importance, we must ask how it is that often the pattern is produced by the receptors to a specific morphogen existing in a specific region in the embryo. In this way, different regions will differentiate into different structures. The question is, how are these large protein receptors placed in particular parts of the embryo, and not in others?

If we look at all multicellular organisms, there seems to be two principle ways this localization is achieved. One is by cell lineage. In the classic case of mosaic eggs, following fertilization there is a redistribution of macromolecules within the one-celled zygote which is then divided up into compartments by successive cell cleavages. One must ask how the determinants were distributed in the egg, and one of the most interesting possibilities comes from the work of Jaffe (1966) who has shown that the polarization of eggs will result in a flow of calcium ions that in turn generates electrical currents, and these currents are of sufficient magnitude to produce differential electrophoresis of charged macromolecules within the cell.

The other way in which receptors are localized is no doubt the more prevalent one among multicellular organisms. One of the most important functions of morphogens is to serve as signal substances which turn on genes, and those genes can be the progenitors of the receptor proteins. But, in order to avoid having all the genes turned on everywhere, the distribution of the stimulating morphogen must have a pattern. Here we have a plethora of ideas of how these could be produced in a non-uniform fashion, starting with C. M. Child's metabolic gradients, and more recently including all the models related or parallel to those involving reaction-diffusion dynamics.

There is an interesting implication to what I have just said. It is that in many organisms there are two rounds of morphogen activity: the first which produces the regionalization of receptors and the second in which the receptors are in place and ready to respond to morphogens to produce some sort of mature, differentiated pattern. In the first round the responsibility for the pattern lies with the non-uniform distribution of a morphogen, in the second the morphogen can be (but need not be) uniformly distributed, for the pattern is already there by virtue of the localization of the receptor molecules. This means that in early development the diffusing morphogen plays a key role in the pattern formation, but later that role is taken over by receptors. The timing of these two processes could certainly overlap, and even be different for different parts of a complex embryo at any one time.

CONCLUSION AND SUMMARY

If one looks at the activities of morphogens from an evolutionary point of view, it seems reasonable to assume that the most primitive condition is where the morphogens produce ratios without patterns, as we saw in termites and slime moulds. The next step might have been to produce form along with ratios, and this also is governed by morphogens which affect cell orientation (and cell adhesion as well, a matter I have not touched on in this short essay). In this sculptured cell mass the next evolutionary step would be for morphogens to set up gradients and other patterns which directly turn on genes, and in the most advanced state of these gene products there could be protein receptors for specific morphogens.

Another evolutionary sequence can be found in the size and complexity of the morphogen, where in the simplest case the molecules are volatile and have no receptors, while the larger and more advanced ones do.

Also there is an evolutionary trend in the means of transport of the morphogens. In the most primitive case it is by diffusion, by far the most prevalent means of transport in all developing organisms, but gap junctions are used in some advanced and refined instances of morphogen transport.

One final evolutionary trend might be towards an increase in the number of morphogens of the various kinds so that any one developing organism might have a veritable symphony of chemical signals all playing together in exquisite harmony.

I hardly know what suggestions to make to the mathematical modeller from all this, but I desperately feel that his powers are needed. It is never helpful to stress how complicated things are. What we need are some crystalline insights to clear the fog I have created.

REFERENCES

Akiyama, Y. and Inouye, K. (1987) Cell-type conversion in normally proportioned and prestalk-enriched populations of slug cells in *Dictyostelium discoideum*. *Differentiation* **35**, 83–7.

Gerisch, G. (1987) Cyclic AMP and other signals controlling cell development and differentiation in *Dictyostelium*. *A. Rev. Biochem.* **56**, 853–79.

Gilula, N. B. *et al.* (1988) *Science* **230**, 1439.

Jaffe, L. (1969) On the centripetal course of development, the *Fucus* egg, and self-electrophoresis. *Dev. Biol. Suppl.* **3**, 83–111.

Light, S. F. (1942, 1943) The determination of the castes of social insects. *Quart. Rev. Biol.* **17**, 312–26; **18**, 46–63.

Oyama, M., Okamoto, K. and Takeuchi, I. (1983) Proportion regulation without pattern formation in *Dictyostelium discoideum*. *J. Embryol. exp. Morph.* **75**, 293–301.

Raper, K. B. (1940) Pseudoplasmodium formation and organization in *Dictyostelium discoideum*. *J. Elisha Mitchell Sci. Soc.* **56**, 241–82.

Roberts, J. A. and Tucker, G. A. (1985) *Ethylene and Plant Development*. Butterworths, Boston.

Schaap, P. (1986) Regulation of size and pattern in the cellular slime molds. *Differentiation* **33**, 1–16.

Simchowitz, L. and Cragoe, E. J. Jr (1986) Regulation of human neutrophil chemotaxis by intracellular pH. *J. biol. Chem.* **261**, 6492–6500.

Weijer, C. J. and Durston, A. J. (1985) Influence of cyclic AMP and hydrolysis products on cell type regulation in *Dictyostelium discoideum*. *J. Embryol exp. Morph.* **86**, 19–37.

Whitaker, B. D. and K. L. Poff (1980) Thermal adaptation of the thermosensing and negative thermotaxis in *Dictyostelium*. *Expl Cell Res.* **128**, 89–93.

Wilson, E. O. (1971) *The Insect Societies*, p. 114. Harvard University Press, Boston.

Positional information and prepattern in the development of pattern

LEWIS WOLPERT

Department of Anatomy and Developmental Biology, University College and Middlesex School of Medicine, London W1P 6DB, UK

During development the differentiation of cells results in the well-defined patterns that constitute organs such as limbs or eyes. The process whereby this patterning occurs is related to, but different from, the process whereby cells become different – cell differentiation. The same cell types – muscle, cartilage, and so on – are present in the forelimb and hindlimb of vertebrates yet the pattern and form of these structures can be very different. In evolution of the vertebrates it is essentially the patterning process which makes giraffes, fish and man take on different shapes. There may well be differences between the cells that make up the skeletomuscular system of these different animals but these differences do not account for the different patterns. Again, there is no cell type that man has which is lacking in the chimpanzee; the difference lies in the spatial organization of the cells.

Change in form, morphogenesis, is a process that again is related to pattern formation but different from it. Change in form during development results from the forces generated by the cells within the embryo. In many cases these are contractile forces generated by the cytoskeletal components. The patterning process specifies which cells will generate the forces. Thus patterning must precede all morphogenetic movements. This view stands in contrast to proposals that the mechanics itself generates patterning (e.g. Murray and Maini, in this volume).

The basic idea of positional information is that there is a cell parameter which is related to the cells' position in the developing system (Wolpert, 1969, 1981b). It is as if there is a coordinate system. This parameter, positional value, is then interpreted by the cells in terms of cell differentiation. There is thus an uncoupling between patterning and cell differentiation. There is no

Cell to Cell Signalling: From
Experiments to Theoretical Models
ISBN 0–12–287960–0

133

necessary intrinsic link between a particular set of positional values and a particular cellular differentiation (Wolpert and Stein, 1984). In principle, any cell differentiation could result from interpretation. It also means that the interactions between the cells are involved in the specification of cell states and not any particular cellular differentiation. Some support for such a view comes from studies on muscle differentiation. A single gene can be used to transfect cells and so cause them to differentiate as muscle (Davis *et al.*, 1987). Different positional values could activate this gene.

Mechanisms based on positional information should be contrasted with those based on induction (Wolpert, 1989). Here the interactions are directly linked to cellular differentiation. One group of cells induces a particular pathway of differentiation in an adjacent group, as for example, in the induction by the endoderm of the early amphibian embryo, of mesoderm in the adjacent tissue (Gurdon, 1987).

Some of these considerations also apply to mechanisms based on pre-pattern. Here, too, there is a specification of cell state that precedes cell differentiation but in this case, instead of a monotonic relation between position and cell parameter, there may be a wave-like distribution of values so that the same values are repeated (Nagorcka, 1989) (Fig. 1). It essentially provides a mechanism for generating repeated patterns, and if differentiation only occurs above a threshold then a repeated pattern will develop. Because of the relationship between the distribution of the cell states or morphogen and the observed pattern, such a prepattern should be called an isomorphic prepattern to emphasize the relationship. This distinguishes it from a positional information model where there is no obvious relation between the cell states or morphogen. The pattern arises entirely from the interpretation process.

An important feature of pattern formation that is related to positional

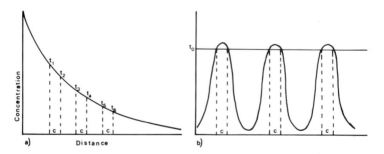

Fig. 1. Diagrams to illustrate two different mechanisms to specify three regions of cell differentiation into, say, cartilage. In (a) the mechanism is based upon positional information provided by a concentration gradient. The cartilage regions are specified by the six thresholds t_1-t_6. In (b) the cartilage is specified by a prepattern mechanism with a single threshold t_0. In this case all the cartilage regions are equivalent.

information is the concept of non-equivalence: cells of the same differentiation class may differ in some intrinsic property such as positional value (Lewis and Wolpert, 1976). Thus the early digits of the chick embryo wing may be made up of cartilage, but the cartilage in the different digits differs with respect to some other property. This non-equivalence permits the cartilage cells in the different digits to follow different, developmental pathways, such as, for example, different growth programmes. There is good evidence that cartilage is, in fact, non-equivalent. That is also probably true for connective tissue in general. However, there is good evidence that embryonic muscle cells and endothelial cells are equivalent. For example, embryonic muscle cells from the cervical region of the chick embryo can provide a normal musculature for the wing, but the character of the vertebrae of the cervical region is determined very early on and is expressed even when grafted to a quite different location (Wolpert, 1988).

Persuasive evidence for positional information, specifically the idea that cells have a parameter positional value, comes from grafting experiments in insects (French, 1985). In general terms when epidermal regions are grafted to different sites, there is intercalary growth as if to replace the missing positional values. This technique thus provides a biological assay of positional value. Many of the results obtained by this procedure can be accounted for by the polar coordinate model (Bryant et al., 1981). It is important to recognize that the intercalation is related only to the cell's position and not the nature of the cell's differentiation.

POSITIONAL INFORMATION AND INTERPRETATION

A mechanism based on positional information could, in principle, generate an infinite variety of patterns. If each cell has a positional value corresponding to its position, as in, say a sheet of cells, then if the cells contain an appropriate mechanism for interpretation it is possible to generate any imagined pattern. The pattern arises from the interpretation mechanism. The question then arises as to whether such an interpretation mechanism is, in fact, plausible and what evidence there is for it.

In principle the interpretative mechanism requires that there be a complete specification of the behaviour of every cell in every position. If cells are to differentiate at different positions into cell type A, then all those positions must be 'listed' as activating cell differentiation type A. This could be achieved by sites adjacent to the gene which lead to its activation or suppression. This list could be very long if the number of cells is large and seems a cumbersome and inelegant mechanism. But what appears inelegant to biologists may be beautiful to the cell. It is also worth remembering that all positional fields are quite small, the maximum linear dimensions rarely being greater than 50 cells (Wolpert, 1981b). Further, it may be possible to economically specify the behaviour of contiguous cells that undergo similar differentiation (Wolpert, 1985). Cells can be specified in blocks. There is even some evidence in one case at least that a listing of cell types occurs. This may be the situation in the control of the pattern of sensory bristles on the thorax of flies (Ghysen and Dambly-Chaudiere, 1988). There are 11 such bristles on the thorax and these can be removed, often in pairs by a series of mutants on the *achaete–scute* complex. Molecular analysis has shown that the phenotype of most scute alleles can be correlated with their location on the chromosome. These different sites on the chromosome may be activated at appropriate positional values and thus may correspond to the 'list' of positional values necessary for interpretation. It is argued that the regional control of expression need not be very precise if lateral inhibition is involved; one bristle inhibits the development of another in its immediate neighbourhood.

Another problem is how the positional value is read and recorded by the cell prior to its interpretation. If the initial positional specification is due to variation in concentration of a morphogen then some kind of threshold mechanism must be invoked. This could be based, for example, on a positive feedback loop (Lewis *et al.*, 1977) or on covalent phosphorylation of a protein, or on some other cooperative event. In general terms the study of thresholds has received little attention (but see Meinhardt, 1982).

The most impressive evidence that a gradient in a morphogen does indeed specify pattern comes from the work of Nüsslein-Volhard on early insect development. She has demonstrated that the protein coded for by the *bicoid* gene is present in the egg in a concentration gradient and altering this gradient alters the boundary between the head and thorax (Driever and Nüsslein-Volhard, 1988).

LIMB DEVELOPMENT

We have proposed a model for pattern formation in chick limb development based on the interpretation of positional information (Wolpert, 1981a). Positional values are thought to be specified in a region at the tip of the limb known as the progress zone which is itself specified by the overlying apical ectodermal ridge. Position along the antero-posterior axis would be specified by a signal from the polarizing region at the posterior margin of the progress zone. The signal may be the concentration of a diffusible morphogen, possibly retinoic acid (Tickle *et al.*, 1985; Thaller and Eichele, 1987). The position of the cartilaginous elements would then be determined by thresholds with respect to the concentration of the morphogen. Thus, when an additional polarizing region is grafted to the anterior margin of the limb bud the pattern of digits is *4 3 2 2 3 4* which is to be compared with the normal pattern *2 3 4*. Grafting the polarizing region at a variety of different positions along the antero-posterior axis, in general, gives results that can be accounted for by the morphogen concentration model. A graft, for example, to the middle of the limb typically gives *2 3 4 4 3 3 4*. In spite of the success of the model there are good reasons to question its validity. The most telling experiments involve disaggregation of the mesodermal cell, followed by reaggregation and placing the aggregate in an ectodermal jacket. Without a discrete polarizing region moderately good digits form (Patou, 1973) and this suggests that another mechanism, such as one generating an isomorphic prepattern, is involved. Such mechanisms have, in fact, been put forward by Wilby and Ede (1975) and Newman and Frisch (1979).

The two types of mechanism make rather specific predictions with respect to the relationship between the number of elements along the antero-posterior axis and the width of the limb bud. We have tested these predictions with respect to the development of the humerus. For prepattern mechanisms the development of an additional element requires fitting in an additional wavelength by widening the bud. For the digits, where the wavelength would be relatively short, only a modest widening would be required and Tickle and Stein (personal communication) have found that about 20% is sufficient for an extra digit. However, for the humerus, which would correspond to a single wave, limb width would probably have to increase substantially, possibly doubling, in order to accommodate another peak and so duplicate the humerus.

Polarizing region grafts which generate additional elements are always associated with limb bud widening (Smith and Wolpert, 1981). However, the rate of widening is such that widening would not occur early enough for duplication of the humerus to occur if it were necessary to substantially widen the limb at the time when the humerus is being laid down. Thus the prepattern model predicts that it will be very difficult to obtain duplication of the humerus following a graft of the polarizing region to the anterior margin.

By contrast a mechanism for specifying the humerus by positional information makes rather different predictions. While it is well recognized that the concentration of the morphogen following a graft of the polarizing region would be very dependent on the width of the bud (Smith and Wolpert, 1981; Wolpert and Hornbruch, 1981), the model requires that when a polarizing region is placed at different positions along the antero-posterior axis, elements must be duplicated or eliminated in the region between the grafted host polarizing regions (Fig. 2). For the digits, this is, in general, true (Wolpert and Hornbruch, 1981). The results from grafting an additional polarizing region to the anterior margin before the humerus is laid down do not conform to this prediction (Wolpert and Hornbruch, 1987). None of the grafts eliminated the humerus and only 10% of the grafts gave a duplicated humerus even when the graft was well to the anterior of the bud. In this position the limb bud only widened by about 15%.

Our interpretation of the results is that the formation of the cartilaginous elements involves both an isomorphic prepattern together with a positional

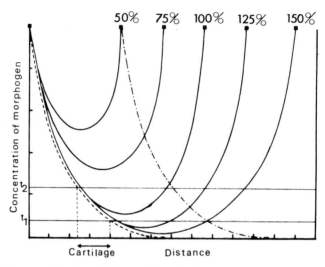

Fig. 2. The specification of the cartilage in the humerus by a diffusible signal from the polarizing region is shown by the dashed line and thresholds t_1 and t_2. The solid lines show the predicted concentration profiles when an additional polarizing region is placed at different positions along the antero-posterior axis.

signal (Wolpert and Stein, 1984). The prepattern would essentially specify the elements and the positional signal would provide positional information thus making the elements non-equivalent and lead to the detailed differences in form of the elements. An example of such a highly localized change in form of the elements, is the protrusion at the distal end of the tibia leading to its capture of the distal epiphysis of the fibula (Archer *et al.*, 1983).

If the prepattern were based on a reaction diffusion mechanism then the waveform has to change from a single peak in the most proximal region to two peaks in the middle arm to give a radius and ulna, and then to three peaks to give the digits. These changes are not due to changes in the geometry of the limb bud, the overall shape of the cross-section at the time when the humerus is laid down is similar to that when the radius and ulna are being specified. Moreover, quite significant changes in geometry of the cross-section have no effect on the pattern (Wolpert and Hornbruch, in preparation). The change in the waveform could result from the changing proximo-distal positional value altering the parameters of the reaction diffusion system (Wolpert and Stein, 1984).

The analysis above has been in terms of specifying cartilage in the appropriate regions. It may be more appropriate to think in terms of specifying where cartilage should not form, that is, inhibiting cartilage differentiation. This is based on the observation that the progress zone in dense culture forms a sheet of cartilage (Cottrill *et al.*, 1987). It is as if the cells in the progress zone will form cartilage unless they are prevented from doing so. One factor is an inhibitory signal from the ectoderm (Solursh, 1984), though removal of the dorsal ectoderm still permits normal limb development (Martin and Lewis, 1986).

We are currently exploring reaction diffusion models to generate the isomorphic prepattern. A key feature in this model is an inhibitory signal from the ectoderm.

MECHANICS AND PATTERN IN THE LIMB BUD

As pointed out in the introduction, changes in form are concerned with the forces generating the change in shape. It was assumed that whenever a change in form occurred it was preceded by a patterning process which specified which cells would generate the forces (cf. Gustafson and Wolpert, 1967). One of the features of cartilage development in the limb bud is that the cells of the future cartilaginous elements condense – that is they come closer together. Oster *et al.* (1983) have proposed that this condensation and the pattern of the cartilaginous elements could result from cell tractions or changes in the extracellular matrix (Oster *et al.*, 1985). The essence of such models is that in principle they require no patterning process involving either a prepattern or positional information, the pattern being generated purely by mechanical interactions between the cells and the matrix. This approach, while interesting, is somewhat odd. For the very cells that condense give rise to cartilage whereas the other cells in the bud do not (Rooney *et al.*, 1984). Yet the model largely ignores the factors determining cartilage differentiation and concentrates only on the form of the condensations. There are, then, only three possible explanations for the congruence of cartilage differentiation with the mechanical condensations: (i) they are completely independent and miraculously correspond; (ii) the condensations cause cartilage differentiation; (iii) the cells that are specified as cartilage by a patterning mechanism undergo condensation. There is some evidence relating cell shape to cartilage differentiation; in general flattened cells do not differentiate into cartilage (Zanetti and Solursh, 1986). Thus, if the condensation resulted in the rounding up of the cells, mechanics and differentiation would be linked. Even so such a mechanical model is not capable, on its own, of specifying the more detailed shapes of the cartilaginous elements. There are also studies *in vitro* suggesting that a signal from the ectoderm plays a major role in inhibiting cartilage formation in the periphery of the limb and that this does not act via cell shape (Gregg *et al.*, 1989).

A more critical test of the model is based on the temporal differences between (ii) and (iii). A patterning model suggests specification before condensation, whereas with the mechanical model specification occurs only at the time of the mechanical interactions. So, if it can be shown that specification occurs prior to condensation the mechanical model must be rejected. To do just this we have constructed double anterior limbs at stages well before condensation of the humerus occurs (Fig. 3) and found that in a significant number of cases two elements develop proximally (Wolpert and Hornbruch, in preparation). The overall form of the limb is unaltered and the mechanical model thus cannot account for this result. The result is best interpreted in terms of early specification of the humerus long before condensations develop, and that condensation is an early feature of cartilage differentiation. This interpretation is also consistent with our results obtained by constructing limb buds, at stages well before condensation, which are doubled along the dorso-ventral axis. The cartilaginous pattern develops autonomously and is, in general, unaffected by the geometry.

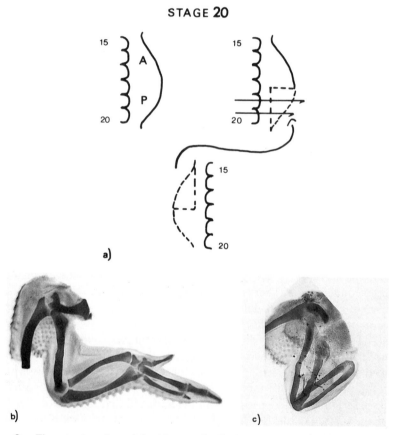

STAGE 20

Fig. 3. The construction of double anterior limbs by the operation in (a) leads to two humeri developing proximally followed by two radii (c). A normal wing is shown in (b).

CONCLUSIONS

Prepatterns which generate repeated equivalent structures together with positional information which makes the structures non-equivalent and gives them a positional value could provide a potent system for generating patterns. It requires little effort to see how the segmental body structures of vertebrates or feather patterns could be generated in this way. From an evolutionary point of view it makes a great deal of sense. The prepattern mechanism generating a series of similar structures is 'easier' to evolve and requires few thresholds. By contrast, positional information requires many thresholds and while the positional values may be easy to generate, their interpretation is, from the point of view of evolution, probably more complex.

Both prepattern and positional information require cell to cell interactions. There is little evidence in vertebrates and insects, at least, that lineage mechanisms are of any importance (Wolpert, 1989).

It is curious that the distance over which cell to cell interactions are thought to occur seems with time, to become less and less. There are very few well-documented cases in which signalling occurs over more than 10–20 cell diameters. Local cell to cell interactions have always been central to intercalation in the polar coordinate model (Bryant *et al.*, 1981). Recent studies on insect development, too, have emphasized local interactions (Ingham, 1988; Tomlinson, 1988), though evidence for the role of graded morphogens is also well established (Driver and Nusslein-Volhard, 1988).

The studies on early insect development (Ingham, 1988) have shown us how combining genetics, development and molecular biology has revealed a much more complex set of interactions between both cells and genes than was previously thought. The relevance of this complexity to vertebrate systems such as the limb, and to positional value in particular, remains to be determined.

REFERENCES

Archer, C. W., Hornbruch, A. and Wolpert, L. (1983) Growth and morphogenesis of the fibula of the chick embryo. *J. Embryol. exp. Morph.* **75**, 101–16.

Bryant, S. V., French, V. and Bryant, P. J. (1981) Distal regeneration and symmetry. *Science, N.Y.* **212**, 993–7.

Cottrill, C. P., Archer, C. W., Hornbruch, A. and Wolpert, L. (1987) The differentiation of normal and muscle-free distal limb bud mesenchyme in micromass culture. *Devl Biol.* **119**, 143–51.

Davis, R. L., Weintraub, H. and Lassar, A. B. (1987) Expression of a single transfected cDNA converting fibroblasts to myoblasts. *Cell* **51**, 987–1000.

Driever, W. and Nüsslein-Volhard, C. (1988) The *bicoid* protein determines position in the drosophila embryo in a concentration-dependent manner. *Cell* **54**, 95–104.

French, V. (1985) Positional maps and cellular interactions in insect development. In *Molecular Determinants of Animal Form* (ed. G. M. Edelman), pp. 455–75. Alan Liss, New York.

Ghysen, A. and Dambly-Chaudiere, C. (1988) From DNA to form: the achaete–scute complex. *Genes & Development* **2**, 495–501.

Gregg, B. C., Rowe, A., Brickell, P., Devlin, C. J. and Wolpert, L. (1989) Ectodermal inhibition of cartilage differentiation in micromass culture of chick limb bud mesenchyme in relation to gene expression and cell shape. *Development* (in press).

Gurdon, J. B. (1987) Embryonic induction-molecular prospects. *Development* **99**, 285–306.

Gustafson, T. and Wolpert, L. (1967) Cellular movement and contact in sea urchin morphogenesis. *Biol. Rev.* **42**, 442–98.

Ingham, P. W. (1988) The molecular genetics of embryonic pattern formation in *Drosophila*. *Nature* **335**, 25–34.

Lewis, J. H. and Wolpert, L. (1976) The principle of non-equivalence in development. *J. theor. Biol.* **62**, 479–90.

Lewis, J. H., Slack, J. M. W. and Wolpert, L. (1977) Thresholds in development. *J. theor. Biol.* **65**, 579–90.

Martin, P. and Lewis, J. (1986) Normal development of the skeleton of chick limb buds devoid of dorsal ectoderm. *Devl Biol.* **118**, 233–46.

Meinhardt, H. (1982) *Models of Biological Pattern Formation*. Academic Press, London.

Nagorcka, B. (1989) Wavelike isomorphic patterns in development. *J. theor. Biol* **137**, 127–62.

Newman, S. A. and Frisch, H. L. (1979) Dynamics of skeletal pattern formation in developing chick limb. *Science, N.Y.* **205**, 662–8.

Oster, G. F., Murray, J. D. and Harris, R. K. (1983) Mechanical aspects of mesenchymal morphogenesis. *J. Embryol. exp. Morph.* **78**, 83–125.

Oster, G. F., Murray, J. D. and Maini, P. (1985) A model for chondrogenic condensations in the developing limb: the role of extracellular matrix and cell tractions. *J. Embryol. exp. Morph.* **89**, 92–112.

Patou, M. P. (1973) Analyse de la morphogénèse du pied des Oiseaux à l'aide de mélanges cellulaires interspécifiques. I. Etude morphologique. *J. Embryol. exp. Morph.* **29**, 175–96.

Rooney, P., Archer, C. and Wolpert, L. (1984) Morphogenesis of cartilaginous long bone rudiments. In *The Role of the Extracellular Matrix in Development* (ed. R. Trelstad), *Symp. Soc. Devl Biol.*, Vol. 42, pp. 305–22. Alan Liss, New York.

Smith, J. C. and Wolpert, L. (1981) Pattern formation along the anteroposterior axis of the chick wing: the increase in width following a polarizing region graft and the effect of X-irradiation. *J. Embryol. exp. Morph.* **63**, 127–44.

Solursh, M. (1984) Ectoderm as a determinant of early tissue pattern in the limb bud. *Cell Differentiation* **15**, 17–24.

Tickle, C., Lee, J. and Eichele, G. (1985) A quantitative analysis of the effect of all-*trans*-retinoic acid on the pattern of limb development. *Devl Biol.* **109**, 82–95.

Thaller, C. and Eichele, G. (1987) Identification and spatial distribution of retinoids in the developing chick limb bud. *Nature* **327**, 625–8.

Tomlinson, A. (1988) Cellular interactions in the developing *Drosophila* eye. *Development* **104**, 183–94.

Wilby, O. K. and Ede, D. A. (1975) A model generating the pattern of cartilage skeletal elements in the embryonic chick limb. *J. theor. Biol.* **52**, 47–59.

Wolpert, L. (1969) Positional information and the spatial pattern of cellular differentiation. *J. theor. Biol.* **25**, 1–47.

Wolpert, L. (1981a) Pattern formation in limb morphogenesis. *Fortschritte der Zoologie* **26**, 141–52.

Wolpert, L. (1981b) Positional information and pattern formation. *Phil. Trans. R. Soc. B* **295**, 441–50.

Wolpert, L. (1985) Positional information and pattern formation. In *Molecular Determinants of Animal Form* (ed. G. M. Edelman), pp. 423–33. Alan Liss, New York.

Wolpert, L. (1988a) Craniofacial development: summing up. *Development* **103** (Suppl.), 249–89.

Wolpert, L. (1988b) Stem cells: a problem in asymmetry. *J. Cell Sci. Suppl.* **10**, 1–9.

Wolpert, L. (1989) Positional information revisited. *Development Suppl.* (in press).

Wolpert, L and Hornbruch, A. (1981) Positional signalling along the anteroposterior axis of the chick wing. The effect of multiple polarizing region grafts. *J. Embryol. exp. Morph.* **63**, 145–59.

Wolpert, L. and Hornbruch, A. (1987) Positional signalling and the development of the humerus in the chick limb bud. *Development* **100**, 333–8.

Wolpert, L. and Stein, W. D. (1984) Positional information and pattern formation. In *Pattern Formation* (ed. G. M. Malacinski and S. V. Bryant), pp. 2–21. Macmillan, New York.

Zanetti, N. C. and Solursh, M. (1986) Epithelial effects on limb chondrogenesis involve extracellular matrix and cell shape. *Devl Biol.* **113**, 110–8.

The early amphibian embryo: Evidence for activating and for modulating or self-limiting components in a signalling system that underlies pattern formation

JONATHAN COOKE

Laboratory of Embryogenesis, National Institute for Medical Research, Mill Hill, London NW7 1AA, UK

INTRODUCTION

In the development of most animal types, the egg comes equipped with very little of the large-scale spatial information that will form a necessary part of the description of the embryo or larva a few days later. Indeed in mammalian development, that which leads to the most highly spatially and functionally organized body of all, self-polarization and then self-organization accomplish the whole pattern formation, starting from the slightest of spatial gradations that depend upon circumstances literally surrounding the embryo rather than being intrinsic to its structure or biochemistry (Smith, 1985). Self-amplifying, self-stabilizing physiological processes of an unknown nature then lead to establishment of a field or map of developmental tendencies for the regions of the body, within a sheet of tissue that is some hundreds or thousands of cells in extent and growing rapidly meanwhile (Eyal-Giladi, 1985). Such totally 'epigenetic' development takes longer than the average time, for embryos generally, to achieve an advanced organization, and is the minority condition. Child (1941), however, was a proponent of the idea that it represented the primitive nature of multicellular development, which is retained as a pronounced ability to regenerate typical body form by isolated tissue or asexual, propagative 'buds' from many marine organisms. Typically the module of sexual reproduction, the zygote, has achieved pre-specialization for extra rapid and reliable development, being equipped with at least a minimal pre-structure in the form of a gradient or an eccentric localization of some sort.

Cell to Cell Signalling: From
Experiments to Theoretical Models
ISBN 0–12–287960–0

These ideas would fit with the expected properties of plausible model mechanisms for development of regulated 'positional information' (Wolpert, 1971), in the form of graded concentration landscapes of critical metabolites or intercellular signal molecules across the tissue that is to give rise to embryo or regenerate (Gierer and Meinhardt, 1972; Meinhardt, 1982 for review). Many of these models can be seen, at least mathematically, as special cases of a generic class pioneered by Turing (1952), though they best account for development of 'whole body pattern' when engineered to produce primarily single gradient systems (i.e. a 'chemical wavelength' close to the dimensions of the tissue that is to be organized into a series of unique domains), whereas Turing was primarily modelling re-iterating wave-like patterns of a morphogen concentration in space (i.e. 'prepatterns' for periodically repeated structure such as occurs in the bodies of radially symmetrical forms or in the late-developing surface details of many animals). These formalisms, represented simply in Fig. 1, can be spoken of as 'reaction–diffusion' models for the generation of biological pattern, relying on short-range activation with autocatalysis followed by longer-range self-inhibitory or modulatory signalling. The logic of simple cases of the single gradient type is evident from the diagram and legend.

That most developments from eggs utilize a structurally pre-formed polarity or reference locality is not surprising. Simulation has revealed that although these reaction systems can be so adjusted as to break symmetry reliably and progress spontaneously towards pattern, the stability and reliability of the morphogen landscape achieved are improved, and the time taken to regulate to completion diminished if the requirement to break symmetry from spatial homogeneity and localize the area for production of the activating morphogen is circumvented by such minimal pre-structuring. Single gradient systems best model the early phase of embryogenesis for two reasons. The problem at this stage is one of partitioning the tissue spatially into rather a few (though more than two) zones, of particular relative extents, in each of which a particular developmental sequence then uniquely occurs, and which together make up 'the body'. Although growing greatly during later differentiation, this body plan is first established while the whole tissue is of a particular ($<$ 1 mm) size-range. Secondly, in at least one dimension of spatial organization, the body plans of all embryo types that form by intercellular signalling show positive regulatory capacity, even if only for a restricted time in development. By this is meant that the morphogen system can adjust the outcome of pattern formation so as to found territories of the right *relative* extents (i.e. to allocate tissue in the proper proportions for form) across a range of sizes for individual eggs or embryos of the species. Multi-peak (i.e. original Turing-type) prepattern systems can only plausibly maintain absolute control over peak number for small (3–6) numbers of peaks in constant-sized tissue fields. They cannot adjust the territory size required by each peak in relation to any significant variation in the overall

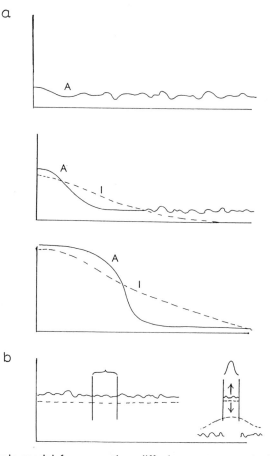

Fig. 1. A simple model for a reaction–diffusion system producing regulated morphogen concentration gradients. (a) The system in normal operation. A cell state leading to production of morphogen A (activator), which is autocatalytic and slowly spreading via intercellular signalling, ultimately leads to production of a second morphogen I (inhibitor or modulator). I is more rapidly spreading, probably by a more passive, diffusion-like process. Cells receiving I while themselves only at early stages of the A-producing pathway of development, fail to become activated. There may be a continuous, damped negative feedback relationship between I concentration and level of A production. Morphogen gradients as shown can be stable, and regulated against tissue-size variation to some degree. Coherent polarity of the system would be assured, in different organisms, by external physiological gradation or by the inherent structural localization of a spontaneously activating region (see first text section). (b) Possible response to the experimental situation of simultaneous, homogeneous exposure to the signal that transmits activation (see last text section). Due to lack of a tissue 'sink' for I in this condition, a typical gradient profile and in particular the normal 'upper boundary' levels of position value (see a) may not be able to develop unless a piece of the tissue is subsequently isolated into surroundings lacking I (right).

extents of developing individuals, unless by an accessory mechanism, lacking biological plausibility because it does not grow out of the primitive components of the system, that feeds back information about overall tissue size onto the local variables that describe effective diffusion rates or relative production rates of the interacting morphogens. Single gradient systems *can* perform the elements of such scale regulation to preserve the completeness of pattern more naturally, though the limitations with which they do this has led to the proposal that a more multi-stage but still integrated mechanism, of the same general form, underlies primary pattern formation in higher embryos in biological reality (e.g. Cooke, 1983; Pate, 1984).

In what follows, I outline the geometrical conditions attending specification of the body plan in amphibian embryos, and then describe certain experimental phenomena regarding whole embryos that are to be accounted for. I finally survey some experimental evidence that both activating and inhibitory or modulatory signal species are involved in the real mechanism of spatial diversification. This evidence comes from studies on the cellular responses, in tissue explants and in whole embryos, to a signalling molecule called a 'mesoderm-inducing factor' (MIF), whose molecular relationships and properties make it a strong candidate to be a primary activating component in the natural mechanism.

THE FROG EMBRYO: SPATIO-TEMPORAL PATTERN IN MESODERM AND IN GASTRULATION

In the embryo of *Xenopus*, the clawed frog, development is fairly rapid and the polarity or orientation for pattern is established at the outset as an asymmetry of unknown nature within the yolky, downward hanging part of the giant cell that is the egg (Gerhart *et al.*, 1981; Vincent *et al.*, 1986). This egg region will itself contribute directly the endoderm, or gut-lining structure, to the body. The next stage of pattern elaboration occupies the 100–10,000 cell stage (blastula), during very rapid cleavages of the egg, but without true growth and essentially without intervention of the embryo's own genomic activity, and lasts a few hours. It is only plausible to suppose that the position-related diversity brought about by the events with which we are concerned is beginning to be made permanent, as differential settings for gene activity, towards the very end of this period as gastrulation is about to commence, or even later. The essential interaction is the induction, by a system of intercellular signalling originating from the endoderm, of a territory specified to be mesoderm, lying as a broad belt of tissue round the equator of the embryo which has meanwhile become a hollow spherical wall of cells. The mesoderm will undergo a programme of differentiations as the middle layer of the final three-layered body – including axial skeleton and musculature, kidney, mesenchymes, heart – but before this it engages in a programme of mechanical activities that create the forces driving it into its definitive position between the other two layers. This is the process of gastrulation. By induction is meant the diversion of a zone of tissue, adjacent to the endodermal source of initiating signals, from the 'default programme' of differentiation that it had as ectoderm (epidermis etc.). Tissue lying beyond the zone of effective induction, i.e. nearer the uppermost pole of the original egg and opposite the yolky mass, remains specified as ectoderm at gastrulation.

A variety of experiments reveal that the mesoderm territory has already a considerably detailed and autonomously maintained spatial pattern of characters at the very onset of gastrulation, that is, after a few hours of inductive signalling. Gradations of physiological state (and possibly in gene transcriptional activity) are manifest as remarkable adherence to a temporal schedule, whereby mesoderm destined for particular body positions in undisturbed development, and coming from particular positions within the pre-gastrulation mesoderm field, begins new behaviours at particular time-points in relation to the original start of development. The completeness and normal proportions of the body plan of differentiated structures that will arise in the mesoderm are thus prefigured, many hours beforehand, by a particular spatio-temporal schedule of mechanical activities within that tissue. All the evidence is that this spatio-temporal schedule with which new locomotory-adhesive behaviours are begun is a direct expression of a pattern of 'pre-specification' for the regionality of the later body, already painted in broad brushstrokes in the mesodermal field after its first induction. The geometry of this process is depicted in Fig. 2. What opportunities are there for self-regulating systems of intercellular signalling to occur in relation to the endodermal sources of initiating signals at this early stage, that would make possible the drawing of a fate map for the body that is already reflected in differences of property in the mesoderm?

It is not possible here to discuss the geometrical complexities of deriving two dimensions of final anatomical organization, the medial-to-lateral or dorsal-ventral and the head-to-tail, from the belt-shaped zone of tissue that is first recruited into mesodermal specification in the blastula. We can only say that there are certain types of early disturbance which reveal an *inability* of the pattern-forming system to adjust regulatively, because the initiating preconditions are set up in a way that loses this property after the first hour or so of development (see Scharf and Gerhart, 1980; Cooke, 1985, 1987), and other experimental disturbances which are followed by evidence for mutual, long-range adjustment interactions such as would be predicted by theories of the kind mentioned in the last section (see Cooke, 1981, 1983). It is in the rapid events, at least partly mechanical in nature, that lead development past the point where the overall boundary conditions for pattern formation can be further adjusted, that *Xenopus* development differs from that of mammals and probably birds (Eyal-Giladi, 1985). Pattern formation in the induced

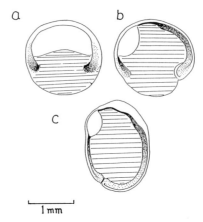

1 mm

Fig. 2. Sectional views, in the plane of bilateral symmetry, of the events of mesoderm induction and of gastrulation. (a), (b) and (c) show longitudinal sectional views of initial, mid-gastrular and closing stages, respectively, in the process whereby mesoderm that was specified before gastrulation changes cell behaviour and actively involutes to form a new, middle cell layer. Graded density of shading for the mesoderm in each diagram represents the pre-organization, in relation to distance from initial sources of induction in the vegetal region, whereby mesoderm cells from successively higher in the marginal zone commence the change in their behaviour that causes involution at successively later times. Positions of cells within the pre-involution mesoderm are thereby significantly related to their expected final position in the body plan. In *Xenopus* where involution spans 3 or 4 h only, pre-organization of these properties has largely occurred before its onset. In slower-developing forms the progressive organization of pre-involuted material may continue after gastrulation is underway.

mesoderm starts from a given asymmetry, whereby a sector of that part of the egg from which signals are originating – and a sector occupying a particular relative extent in the system – is set to produce an initiating condition (probably a particular signal species, see later) that gives rise specifically to mesoderm of head and dorsal axial positional character. If any disturbance during the first hour after fertilization, or surgical intervention thereafter, causes an abnormal balance of spatial extents as between this sector and the remainder of the egg where a different type of mesoderm is specified, then the body pattern will be qualitatively incomplete or quantitatively imbalanced in terms of tissue types and structure.

There is much evidence suggesting that subsequent pattern formation within the mesoderm territory is largely based on a dynamic mechanism of interaction between the sectors of mesoderm of these two types. They are not themselves directly specified to be the precursors of particular mesodermal structures, but could instead interact as source and sink zones that set up a gradient with respect to a further morphogen signal, that polarizes mesodermal development (Dale and Slack, 1987) and sets the local 'rate' with which progression to determination is occurring (Cooke, 1983, 1985). It is on the profile of *this* gradation that the spatially ordered and proportioned allocation to types of differentiation partially depends, but because the relative extent of the zone of 'source'-type tissue is not positively regulated, the gradient itself, and hence the pattern of development controlled by it, can be incomplete or imbalanced in profile.

All around the marginal zone, but especially in the sector that produces the head and dorsal axial midline, the head-to-tail dimension of body organization is related to what in geographical terms we would call 'latitude' in the blastula; that is, to distance in cellular terms from the origin of inducing signals at the endoderm–mesoderm junction. Lineage labelling and grafting procedures reveal that the induction process spreads through an appreciable number of cells, progressively across the blastula period, recruiting them into mesodermal specificity. Thus, for this dimension of the body plan, we must postulate that cells experiencing conditions far from the site of the first transmission of inductive stimuli, which are necessarily also the conditions obtaining after the whole set of signalling interactions has been in progress for a longer time, achieve specification as more posterior members of the body plan. This is first revealed as the absolute time in development when the change in cell social and mechanical properties that leads to participation in gastrulation is actively commenced, this being later for more posterior-specified cells. The dorsal quadrant of the marginal zone of a beginning gastrula reveals itself, when cultured in isolation, to be already a highly spatio-temporally diversified 'morphogenetic organ' (see Keller and Danilchik, 1988), and can give rise to an essentially complete axial pattern sequence (Cooke, 1989).

As mentioned earlier, there are aspects of the patterned induction system that *do* show a capacity for pronounced regulatory interactions, that tend to normalize the quantitative allocations of tissue to structures and are active across distance within the multicellular (10^3–10^4 cells) embryo. The most striking concerns the relative extent of the whole mesoderm itself, considered as an allocation within a two-part pattern when the marginal zone is marked off as mesoderm from the remaining, still ectodermally pre-specified tissue. But there is also adjustment of territory sizes within mesoderm. These phenomena will be described in turn.

Normally some 40% of the descendants of the total original non-endodermal tissue have become specified as mesoderm by the neurula stage, when essentially all the primary mesoderm reveals its identity by having occupied a middle cell layer. Allowing for reduced division rate in mesoderm cells after they gastrulate, it appears that about half of the territory that would have been competent to be induced as mesoderm has in fact been so induced. This is largely by position-dependent pattern formation rather than by random local assignment to mesoderm followed by co-aggregation of induced cells to the marginal zone. When blastulae in early stages of mesoderm induction are equipped microsurgically with either abnormally extensive or abnormally restricted terrains of competent ectodermally pre-specified tissue, in relation to their normally structured endodermal source of inducers, it is found that the absolute position of the boundary of induction is adjusted so as to preserve a normal *relative* allocation into mesoderm and ectoderm in the later embryo (Cooke, 1989). The system can, so to speak, solve the pattern problem with respect to one frontier (cf. the slime mould slug, this volume). Furthermore, the body plans produced within such abnormally large or small mesoderms are not imbalanced with respect to dorsal anterior versus other structure, as are the earlier-mentioned ones resulting from imbalances in the endodermal initiating pattern of inducers. Further experimental observations lead to the view that the whole non-endodermal territory of the blastula is, in fact, invaded by inductive stimuli. The meso-ectodermal boundary does not mark the upper limit to their spread through its wall, but is to be regarded instead as the position of some response threshold within a unified pattern. The differentiation of fully normal epidermis, as well as the positions of some special ectodermal structures, almost certainly requires signalling at pre-gastrula stages just as does the specification and patterning of mesoderm.

There are strong indications that the extent of a territory within the body pattern can be adjusted downward if an abnormal supernumerary edition of that territory is developing elsewhere in the embryo, in continuity with it through the same tissue layer. This has been described for somite muscle and for kidney development in the amphibian mesoderm (Cooke, 1983 – though there are some problems with secure interpretation of those observations), but is most evident for head and dorsal midline structures. If two sets of the latter structures are arranged to be developing in one embryo, i.e. the whole body is bipolar for pattern, then pronounced interactions occur as follows (Cooke, 1981, 1987). If the duplicate axes are developing near-parallel to one another within the wall of the embryo, so that the tissue distance interconnecting them on at least one side is small, they each act to exaggerate the other's scale of construction, leading to a more than two-fold over-allocation to dorsal and anterior structure in the embryo's mesoderm as a whole. If, on the other hand, they are developing at near-opposite meridians of the embryo, so that the tissue distance between midlines (thus origins for the specifying process) is maximal, they act mutually to scale down the sizes of cellular allocations and thus minimize the over-allocation in total. These interactions occur without influence upon the cell cycle, hence upon the total tissue space or cell number available. One is reminded of the situation modelled in certain of the more complex reaction–diffusion schemata (Meinhardt, 1982, 1986) where emerging territories in a pattern, or cell types, 'help' one another's progress via local mechanisms while at the same time competing or inhibiting examples of themselves that are more distant. The behaviours that have been surveyed above almost formally require that self-limiting, or negative feedback signal components with considerable mobility in the tissue are at play.

RESPONSES TO A MESODERM-INDUCING PROTEIN: EVIDENCE FOR MODULATING OR INHIBITORY SIGNALS THAT FOLLOW ACTIVATION IN INDUCTION

There are now at hand two purified proteins that offer good evidence of playing roles in the natural induction of the mesoderm territory during *Xenopus* development (Kimelman and Kirschner, 1987; Smith *et al.*, 1988). Both are at least related to proteins, of different families, first defined as 'growth factors' having effects upon the cell cycle or upon matrix secretion in cultured cells from differentiated stages of the life cycle. They are intercellular signals, released from the cells of origin and acting at the surfaces of cells competent to respond by binding to receptors. The pathways of onward transduction (second messenger systems) and modes of action within the target cell are not understood yet even for their above functions, and are quite obscure here where they effect fundamental changes in cell specification, from ectodermal to other fates. The most novel feature of this response to them is diagnostic of an early developmental phenomenon, but unique to my knowledge in the history of defined cellular responses to defined signalling factors in culture. The earliest, sudden and readily visible response, a cell behavioural one as in natural gastrulation, occurs at a precise cellular 'age' counting from the start of development. This is particular to the type of mesoderm that has been specified, regardless of the much more variable earlier timepoint, and length of time, for experimental exposure to the factors (Cooke and Smith, 1989). This directly verifies the conclusion, from observation of normal and experimentally disturbed whole embryos, that early embryonic cells are temporally organized in a way that can time responses, as well as time-gate the period of competence to respond, to stimuli.

Work with one of these factors, *Xenopus* XTC cell-derived mesoderm-inducing factor (XTC-MIF), which appears to activate production of the 'dorsal anterior' member of the original pair of mesodermal states (see previous section), offers a variety of as yet indirect evidence that the set of downstream intercellular signals that it evokes includes those that modulate, limit or downregulate the progress of the initial response, in a way that mediates pattern formation. Large collections of cells, exposed to various concentrations of the soluble factor as pieces of the normal tissue explanted from the competent ectoderm of embryos, develop a range of the tissue types and structures of normal mesoderm. Theoretical difficulties exist with the idea that cells could be organized to make several different, yet direct

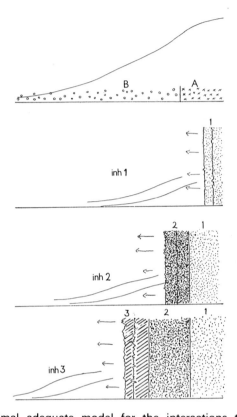

Fig. 3. A minimal adequate model for the interactions that polarize and proportion the body plan in amphibian mesoderm. Events in the first cell cycle establish the relative extents of two zones around the meridians of the egg, in which occur the specification of two initial mesoderm states, A and B. This is represented linearly, in side view, to show events that set up one side of a bilaterally symmetrical system. A is proposed to act as source, and B as sink, for a further morphogen that will then produce a gradient profile as shown. This gradient is not directly read as positional information in the sense of Wolpert (1971), but instead acts to rank the intrinsic rates of development towards the determined state in cells, from fast (at right) to slow. Cells meanwhile have access to an intrinsic 'clock' or time structure of a more absolute sort; that is, they 'age' in a consistent and structured way during the hours of early development. The rank ordering in space, and proportioning, of the zones for mesodermal states that together constitute the body (e.g. 1–4), is then brought about by a cascade of inhibitory interactions as follows. State 1 develops as the default option in any cells that are developing fast enough, i.e. that arrive at the point of determination while 'young' enough. Cells far advanced towards state 1 produce an inhibitor of entry to that state that spreads rapidly (see Fig. 1). If cells that would otherwise be developing towards state 1 experience enough of this inhibitor beforehand, they are diverted and progress instead towards state 2. Progression towards or into state 2 involves production of a new specific, laterally diffusing self-inhibitor, which after a while causes diversion of the character of cell determination to the next option, state 3, as commitment spreads across the system. This system is in

responses to different concentrations of one signal, and also with the idea that stable morphogen gradients could be of a shape that offers information for the setting of several response thresholds across a space in one operation (see Pate, 1984 and Lacalli and Harrison, 1978 for relevant theoretical treatments). We expect therefore that the apparent diversity of response to MIF represents the different outcomes of a more complex set of interactions triggered by the variable initiating conditions (e.g. morphogen concentration, length of exposure) that we can impose. The following evidence is relevant to the idea of second signals.

Injected into the central cavity of the embryo, the factor causes there the abnormal specification of a sheet of mesoderm throughout the blastula wall. But this mesoderm never attains the fully dorsal and anterior character that can be obtained in cultured explants of isolated but naive competent tissue. It is as if the embryo's own system of mesoderm induction meanwhile primes the whole tissue with the second inhibitory signal, instrumental in normal pattern control, to an extent that is avoidable when an explant is isolated. The MIF concentration ranges necessary to specify even the cell types that *are* seen in this ectopic mesoderm in the whole embryo are 10–50-fold elevated in relation to those that are effective in *in vitro* isolates.

Mesoderm specified by microinjection into the blastocoel as above, that is within the first hour of its response and has not yet made the first mesoderm-specific cell behaviours, *can* upgrade its position value to the most 'dorsal anterior' in the normal body if removed as a small (200 cell) graft and implanted into a region of a host embryo that normally forms very different (ventral and posterior) mesoderm. When such grafts are made, whole supernumerary axial body plans are induced to form, meaning that such implanted cells have attained the 'organizer' status that only one small geographical region of the normal embryo's mesodermal territory has (Spemann and Mangold, 1924; Cooke et al., 1987). It can be seen from Fig. 1 that a likely property of the reaction–diffusion kinetics of self-stabilizing positional morphogen gradients, is that the biologically abnormal homogeneous and synchronous activation of a sheet of tissue may lead to a build-up of the consequent inhibitory signal to an extent that prevents, everywhere,

---

effect a linked series of reaction–diffusion subsystems, each operating to regulate the position of one frontier, as in Fig. 1. Size-regulatory features exist because the rate of concentration build-up of each inhibitor acts as a size sensor of the system. The primary gradient does not itself need a well-regulated profile as it is not directly positional information (for further discussion see Cooke, 1983). Limitations on regulation of pattern exist, however, because of the gating role of developmental rate in cells, and the non-regulating nature of the early system that has set the extent in the tissue of the 'source' state, A. If there are too few A cells, and the gradient therefore has an abnormally low upper boundary concentration at equilibrium, no cells may develop fast enough to enter state 1 as the default option. State 2 then becomes the new upper boundary state and a dorso-anteriorly truncated, incomplete pattern results.

the emergence of the upper levels of activation that the normal system leads to. A piece of such kinetically 'trapped' tissue, however, removed to surroundings that allow it to be relieved of inhibitor, may be able to resume the development of an activation peak with consequent control of a morphogen landscape leading to typical pattern.

A particular low concentration of the factor *in vitro* is able to cause all the cells of a dissociated suspension, from the ectodermal region of an early embryo, to make certain responses that we define as an early step in mesoderm specification. Under conditions of prolonged cell dissociation, these changes are stable once induced, but are not followed by any further steps resulting in currently definable mesodermal cell differentiations. When explants of the same cells but with their normal tissue organization are exposed *in vitro* even to much higher concentrations of the factor, an episode of pattern formation in effect results, whereby much mesoderm (including defined cell types) differentiates, but much tissue is left in – or perhaps has reverted to – a state of epidermal specification. It is as if a second, modulatory signal that normally acts as the inhibitor substance in a reaction–diffusion formalism, requires cell contact or tissue arrangement for its production or else its effective receipt by cells.

The precise time within normal gastrulation that ectopically induced mesoderm makes its 'gastrulation' behaviours is a measure of its regional character within the head-to-tail sequence of a normal body (see earlier section). Using this measure, we have been able to establish that neither the concentration of the injected (MIF) signal nor the particular 'age' at which the blastula tissue has first experienced it constitutes the crucial spatial variable that sets position value for mesoderm normally. Spatial gradations with respect to these parameters may exist in the marginal zone in normal induction, but do not in themselves constitute the positional information. Regional character of artificially induced mesoderm cannot be varied detectably by varying these two parameters over wide ranges (200-fold concentration, 3.5 h in 'age' of first exposure in the blastula period), provided that an extensive sheet of competent tissue has been synchronously exposed to a homogeneous concentration of the factor as is necessarily the case after these injections. We are reminded again of the predicted inability of artificially, homogeneously activated versions of certain modelled reaction–diffusion morphogen systems to break symmetry and achieve spatial diversification.

There are advantages to proposing a multi-stage reaction–diffusion process, setting up a sequence of frontiers between zones that will be occupied by the different elements of the pattern of differentiation. These advantages can be seen in a way which is independent of the formal logic, i.e. whether serial activation or induction of successive cell states is proposed, or only a relationship whereby each state inhibits itself to give way to its successor. Either may finally be found to operate when the system is understood at the molecular level, though the principle of self-limitation for each state in relation to the extent of tissue would seem necessary to account for size regulation. Figure 3 displays features of a model for spatial diversification in the amphibian mesodermal territory along these lines, that accounts for most of the phenomena mentioned here, including the limitations on regulative ability as well as the documented capacity for interactions after surgical disturbance. For further details, the experimental papers by Cooke, especially Cooke (1983), and by Dale and Slack (1987) as well as a theoretical treatment by Pate (1983) should be consulted.

REFERENCES

Child, C. M. (1941) *Patterns and Problems of Development*. University of Chicago Press, Chicago.

Cooke, J. (1981) Scale of body pattern adjusts to available cell number in amphibian embryos. *Nature* **290**, 775–8.

Cooke, J. (1983) Evidence for specific feedback signals underlying pattern control during vertebrate embryogenesis. *J. Embryol. exp. Morph.* **76**, 95–114.

Cooke, J. (1985) Early specification for body position in mes-endodermal regions of an amphibian embryo. *Cell Diff.* **17**, 1–12.

Cooke, J. (1987) Dynamics of the control of body pattern in the development of *Xenopus laevis*. IV. Timing and pattern in the development of twinned bodies after re-orientation of eggs in gravity. *Development* **99**, 417–27.

Cooke , J. (1989) *Xenopus* mesoderm induction: evidence for early size control, and partial autonomy for pattern development by onset of gastrulation. *Development* **106**.3, 519–29.

Cooke, J. and Smith, J. C. (1989) Gastrulation and larval pattern in *Xenopus* after blastocoelic injection of a *Xenopus*-derived inducing factor; experiments testing models for the normal organisation of mesoderm. *Devl Biol.* **131**, 383–400.

Cooke, J., Smith, J. C., Smith, Emma, J. and Yaqoob, M. (1987) The organisation of mesodermal pattern in *Xenopus laevis*: experiments using a *Xenopus* mesoderm-inducing factor. *Development* **101**, 893–908.

Dale, L. and Slack, J. M. W. (1987) Regional specification within the mesoderm of early embryos of *Xenopus laevis*. *Development* **100**, 279–95.

Eyal-Giladi, H. (1985) The gradual establishment of cell commitments during the early stages of chick development. *Cell Diff.* **14**, 245–56.

Gerhart, J., Ubbels, G., Black, S., Hara, K. and Kirschner, M. (1981) A re-investigation of the role of the grey crescent in axis formation in *Xenopus laevis*. *Nature* **292**, 511–16.

Gierer, A. and Meinhardt, H. (1972) A theory of biological pattern formation. *Kybernetik* **12**, 30–9.

Keller, R. and Danilchik, M. (1988) Regional expression, pattern and timing of convergence and extension during gastrulation of *Xenopus laevis*. *Development* **103**, 193–210.

Kimelman, D. and Kirschner, M. (1987) Synergistic induction of mesoderm by FGF and TGFβ and the identification of an mRNA coding for FGF in the early *Xenopus* embryo. *Cell* **51**, 869–77.

Lacalli, T. C. and Harrison, L. G. (1978) The regulatory capacity of Turings model for morphogenesis, with application to slime moulds. *J. theor. Biol.* **70**, 273–95.

Meinhardt, H. (1982) *Models of Biological Pattern Formation*. Academic Press, London.

Meinhardt, H. (1986) Hierarchical inductions of cell states: A model for segmentation in *Drosophilia*. *J. Cell Sci.* (Suppl.) **4**, 357–81.

Pate, E. (1984) The organiser region and pattern regulation in amphibian embryos. *J. theor. Biol.* **11**, 387–96.

Scharf, S. R. and Gerhart, J. C. (1980) Determination of the dorsal-ventral axis in the egg of *Xenopus laevis*: complete rescue of UV-impaired eggs by oblique orientation before first cleavage. *Devl Biol.* **79**, 181–98.

Smith, J. C., Yaqoob, M. and Symes, K. (1988) Purification, partial characterisation and biological effects of the XTC mesoderm-inducing factor. *Development* **103**, 591–600.

Smith, L. J. (1985) Embryonic axis orientation in the mouse and its correlation with blastocyst relationships to the uterus. *J. Embryol. exp. Morph.* **89**, 15–35.

Spemann, H. and Mangold, H. (1924) Uber induktion von embryonenanlagen durch implantation artfremder organisatoren. *Roux arch.* **100**, 599–638.

Turing, A. M. (1952). The chemical theory of morphogenesis. *Phil. Trans. R. Soc. B* **237**, 37–72.

Vincent, J.-P., Oster, G. F. and Gerhart, J. C. (1986) Kinematics of grey crescent formation in *Xenopus* eggs: the displacement of subcortical cytoplasm relative to the egg surface. *Devl Biol.* **113**, 484–500.

Wolpert, L. (1971) Positional information and pattern formation. *Curr. Topics Devel.* **6**, 183–223.

Pattern formation mechanisms – a comparison of reaction–diffusion and mechanochemical models

J. D. MURRAY[a] AND P. K. MAINI[b]

Centre for Mathematical Biology, Mathematical Institute, 24–29
St Giles', Oxford OX1 3LB, UK
[a]Current address: Applied Mathematics FS-20, University of
Washington, Seattle, WA 98195, USA.
[b]Now at: Department of Mathematics, University of Utah, Salt
Lake City, UT 84112, USA.

INTRODUCTION

The elucidation of the underlying mechanisms that occur during morphogenesis is of fundamental importance in developmental biology. From the
homogeneous mass of dividing cells of the early blastula, emerge the rich and
varied range of pattern and structure observed in animals. Little is known
about the mechanisms involved in such development. Although genes must
play a key role, genetics says little about the actual mechanisms which
produce structure as an organism develops from the egg to its adult form.

Morphogenesis consists of a complex series of chemical, mechanical and
electrical interactions and an identification and understanding of these has
been the focus of much experimental and theoretical research. Here we
briefly compare the two main mathematical models for morphogenetic
pattern formation, namely reaction–diffusion systems and mechanochemical
models, both of which produce patterns which closely resemble those
observed during several developmental processes.

First we outline the theory behind each mechanism and then we focus on
the application of these models to only two problems in development,
namely, the patterning in the limb that presages cartilage formation and the
patterning of feathers and scales. These models have been applied to a

remarkable array of problems and the reader is referred to the papers listed for other examples. These applications rely on the models' ability to generate spatially heterogeneous solutions. We conclude with a brief discussion of some of the successes, failures and drawbacks of the models.

THE MODEL MECHANISMS

In his seminal paper on the chemical basis of morphogenesis, Turing (1952) showed that a system of reacting and diffusing chemicals (morphogens) could evolve from an initial homogeneous steady state to a spatially heterogeneous steady state of morphogen concentration. This behaviour is called diffusion-driven instability. If we assume cell differentiation occurs above a certain threshold in chemical concentration, then the landscape of chemical pattern will be reflected in a spatial pattern of cell differentiation.

Numerous reaction–diffusion schemes have since been proposed (for example, Gierer and Meinhardt, 1972; Schnakenberg, 1979) with different reaction kinetics. The general form of the equations for a two-species reaction diffusion system is

$$\partial u / \partial t = D_1 \nabla^2 u + f(u,v) \tag{1}$$

$$\partial v / \partial t = D_2 \nabla^2 v + g(u,v) \tag{2}$$

where $u(\mathbf{x},t)$ and $v(\mathbf{x},t)$ denote morphogen concentrations at time t and spatial coordinate $\mathbf{x}$, D_1 and D_2 are positive diffusion coefficients, ∇^2 is the Laplacian operator and $f(u,v)$ and $g(u,v)$ describe the non-linear reaction kinetics.

The fundamental idea underlying the formation of such chemical prepatterns can be couched in terms of short-range activation and long-range inhibition and is most easily seen in the Gierer–Meinhardt (1972) model kinetics. One of the morphogens, say u, may be thought of as an activator. It is produced locally, for example by autocatalysis, and diffuses slower than the other chemical, v, which plays the role of inhibitor. If the parameters of the system lie in the appropriate domain – the Turing domain (see Murray, 1982) – this interaction can lead to spatially patterned peaks of activator chemical concentration. This chemical prepattern provides the bauplan for cell differentiation and is the basis for Wolpert's positional information theory (see, for example, Wolpert, 1981) in which cells interpret the chemical prepattern and differentiate accordingly.

Reaction–diffusion mechanisms have been widely studied both mathematically and experimentally. The mathematical analysis has concentrated on the pattern-producing abilities of such systems (see, for example, the books by Britton, 1986; Murray, 1989, the former of which is more mathematical with the latter blatantly applications oriented). Equation systems like (1) and

(2) can produce a large range of spatially heterogeneous patterns, spiral patterns and travelling wave solutions. Much experimental work has been involved in trying to identify the morphogens so as to provide some experimental justification for a reaction–diffusion basis of morphogenesis. So far this has met with limited success.

In the 1980s a fundamentally different approach to cell patterning was proposed by Oster, Murray and their colleagues (for example, Oster *et al.*, 1983; Murray and Oster, 1984a,b; Oster *et al.*, 1985). They modelled the mechanochemical interaction of a major class of cells, known as fibroblasts, with the substratum, the extracellular matrix (ECM), on which they move. The Oster–Murray approach consists of three equations which describe, respectively, cell and matrix conservation and the mechanical equilibrium between cell and matrix. Each equation describes the physico-chemical processes occurring in the cell–matrix milieu. An important difference between the mechanochemical (MC) and reaction–diffusion (RD) models is that the constituent variables and properties involved in the MC mechanisms are all well known and the parameters in principle measurable. The assumptions behind each equation are the following.

Cell equation

The general form of the cell equation is

$$\partial n / \partial t = - \nabla \, \boldsymbol{J} + F(n) \tag{3}$$

where $n(\boldsymbol{x},t)$ is the cell density at time t and position $\boldsymbol{x}$ and $\boldsymbol{J}$ is the cell flux. The second term on the right hand side models cell mitosis, a reasonable form of which is logistic growth

$$F(n) = rn(N - n) \tag{4}$$

where N is a typical homogeneous cell density and rN the linear mitotic rate of cells.

Several factors contribute to the cell flux term, for example,

(a) Convection. Cells may ride passively on the ECM, pulled by the forces exerted by their neighbours. This contributes a term of the form

$$\boldsymbol{J}_c = n \partial \boldsymbol{u} / \partial t \tag{5}$$

where $\boldsymbol{u}(\boldsymbol{x},t)$ is the displacement of a material point initially at $\boldsymbol{x}$.

(b) Haptotaxis. Cells move by attaching their filopodia to adhesive sites on the external substratum. It is well known from *in vitro* studies that cells tend to move up a gradient in adhesive site density (see references in

Oster *et al.*, 1983). This is known as haptotaxis and may be modelled, in its simplest form, by

$$J_{\text{h}} = \alpha n \nabla \rho \tag{6}$$

where $\rho(x,t)$ is the ECM density and α is the (positive) constant coefficient of haptotaxis.

(c) Chemotaxis. Movement of cells up a gradient in some chemoattractant. This type of behaviour is common in many types of cell (for example, in the slime mould *Dictyostelium discoideum*) and may be modelled, in its simplest form, by

$$J_{\text{ch}} = \alpha' n \nabla \chi \tag{7}$$

where $\chi\ (x,t)$ is the concentration of chemoattractant and α' is the positive constant coefficient of chemotaxis. Note that inclusion of chemotaxis would necessitate having an equation to describe the behaviour of the chemoattractant $\chi(x,t)$.

(d) Diffusion. Cells may move randomly. This contributes a term

$$J_{\text{D}} = - D \nabla n \tag{8}$$

where D is the positive diffusion coefficient.

Various other physical processes which might be involved in development can also be incorporated: a full pedagogical discussion is given in the book by Murray (1989).

Matrix (ECM) equation

The matrix moves by convection caused by forces exerted by cells and its density is described by the equation

$$\partial\rho/\partial t = - \ \nabla.(\rho\partial u/\partial t) \tag{9}$$

In situations where the cells are secreting matrix, this equation has to be augmented by a matrix source term on the right hand side.

Mechanical force balance equation

The cell–matrix milieu is modelled as a linear, viscoelastic material with low Reynolds number which implies inertial terms are negligible in comparison with viscous terms and the material is in mechanical equilibrium. The equation of motion is thus of the form

$$\nabla.\sigma + \rho F = 0 \tag{10}$$

where σ is a stress tensor and ρF is an external body force which comes from the surrounding tissue. As well as the viscoelastic contribution from the matrix to σ, there are several other factors to be taken into account, for example:

(a) Cell traction. *In vitro*, it is known that mesenchymal (fibroblast) cells exert large traction forces which deform the matrix (Harris *et al.*, 1980). At high cell densities the traction force per cell decreases due to contact inhibition.

(b) Osmotic pressure. In certain cases, for example, in the developing limb bud, the ECM is composed of materials, such as glycosaminoglycans which can exert high osmotic pressure.

In some cases the cell–matrix composite is tethered by elastic-type fibres to some external layer and this must be accounted for in the body force ρF.

Investigations of the pattern-forming capabilities of such MC models (see, for example, Murray and Oster, 1984a,b; Perelson *et al.*, 1986; Maini and Murray, 1988) show that they can produce stable spatially heterogeneous patterns and exhibit travelling wave solutions similar to those obtained from the RD approach. The study of MC models are only at a preliminary stage and, in view of their tensor structure, it is almost certain that they can produce richer pattern structures and a wider range of phenomena than RD systems. From MC theory, the patterning process can also be couched in terms of local short-range activation, initiated by the cell traction, and long-range inhibition but where these are now in terms of forces. A simple detailed intuitive and mathematical description of lateral inhibition models in development is given in Murray (1989).

From a mathematical point of view the onset of a spatially heterogeneous pattern is broadly similar in both models. As a certain parameter increases beyond a critical value, the uniform steady state loses stability and evolves to a spatially heterogeneous pattern. Several parameters can play the role of this bifurcating parameter, for example the relative rates of various chemical reactions or the ratio of diffusivities in RD systems and the cell traction parameter in the MC models. Scale and geometry of the domain also play an important role. The larger the scale, the richer are the possible patterns.

Let us now consider two specific practical problems each mechanism has addressed.

CHONDROGENESIS

The vertebrate limb has played a major role in embryological and evolutionary studies not only because of its morphological diversification but also because it is one of the easier systems to study experimentally. The mechanism that initiates condensations is still unknown. Figure 1 illustrates the process of development of the early limb bud. Cells are produced in the progress zone, the area at the tip of the limb bud, and as they move out of this area they condense and these cell aggregations eventually form cartilage. As development proceeds the skeletal pattern becomes more complex as the limb bud grows, widens and flattens.

The RD approach to explain this spatial patterning phenomenon is essentially the following. As cells leave the progress zone they secrete morphogens which initially form a uniform distribution. As one of the model parameters passes through a bifurcation value the uniform state becomes unstable and evolves to a morphogen concentration pattern with a single central peak. Cells in the centre of the limb bud interpret this peak as the cue to become pre-cartilage cells. As the scale and geometry of the limb bud change, the RD mechanism can produce more complex patterns which can account for the later condensation patterns.

There are several MC scenarios. That in Oster *et al.* (1985) has the cells move out of the progress zone, secrete hyaluronic acid (HA), a highly osmotic glycosaminoglycan, which swells the limb bud and initially keeps the cells apart. Prior to condensation, it is known that cells secrete hyaluronidase, the enzyme which degrades HA. The model suggests that this could

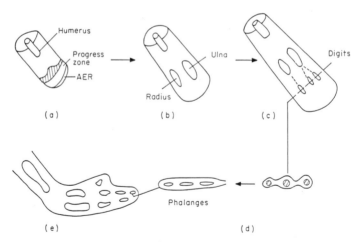

Fig. 1. A scenario showing a possible pattern of bifurcations according to the MC model leading to skeletal pattern formation in the chick limb. (a) Cells move through the progress zone, which is adjacent to the apical ectodermal ridge (AER), aggregate and condense to form the humerus. (b) When a critical cross-sectional ellipticity is reached, a bifurcation occurs leading to the radius and ulna. (c) The asymmetry induced in the limb geometry by cell tractions leads to a bifurcation on the ulna and a focal condensation on the radius. This leads to the digits. (d) The limb bud is now flattened and the digits bifurcate longitudinally to give the phalanges. (e) Sketch of the full chick limb skeletal pattern.

lead to partial osmotic collapse of the internal limb bud which brings cells in closer proximity to their neighbours. The short-range traction forces then take over and can induce a bifurcation wherein the uniform distribution of cells evolve to one where cells aggregate in the centre. These cells then condense to form cartilage. As with the RD system, scale and geometry effects can account for subsequent pattern.

Although the mathematical theory behind pattern formation is similar for both mechanisms, the biological scenarios are quite different. RD relies on a number of discrete steps which effectively form an 'open loop' system (which is notoriously unstable):

(i) The formation of a prepattern of morphogen concentration.
(ii) The interpretation of this morphogen concentration by cells.
(iii) A change in limb scale and geometry leading to the more complex structures.

The MC approach relies on fewer steps. The cellular pattern is formed directly by the process of aggregation – not a prepattern. This cell aggregate has the ability to deform the geometry of the limb bud thus inducing subsequent bifurcations (Oster *et al.*, 1983). Since mechanical deformation can alter the transcription of cells the MC model is not an open-loop system; it has the ability for self-correction. The RD approach does not address the question of the change in geometry and scale of the developing limb bud.

Several experiments have been performed to try and test the above mechanisms. From the RD viewpoint the first essential step is to identify the morphogens. To date this has met with only limited success. Calcium and retinoic acid (RA) have been proposed as two possible candidates but the evidence is controversial. Indeed, in the case of RA, experiments by Kochar *et al.* (1984) suggest that its effects may be due to the observation that RA greatly changes the ability of cells to synthesize HA. This tends to substantiate an MC approach.

The main body of experimental work centres around grafting experiments. For example, if cells from the region known as the zone of polarizing activity (ZPA), which is located in the posterior part of the limb bud, are grafted onto the anterior of the limb, the skeletal elements are duplicated. It is known that such grafting experiments result in abnormal widening of the limb bud (Smith and Wolpert, 1981). Both the above mechanisms would thus predict an increase in richness of pattern due to the increase in scale.

Alberch and Gale (1983) treated developing frog limb buds with colchicine, a mitotic inhibitor. They found that the resulting limb usually had one or more digits missing and was smaller in size than a normal limb. This, again is consistent with both the above mechanisms – a decrease in size resulting in a decrease in pattern. Further results and a comprehensive theory of limb development has been given by Oster *et al.* (1988).

A major criticism of RD theory is that it is very sensitive to initial and boundary conditions and to parameter values. This is obviously a major drawback to any mechanism proposed for a robust developmental process such as chondrogenesis. Arcuri and Murray (1986) partially answered this criticism when they showed that with fixed concentrations on the boundary, RD systems became less sensitive. In the limb bud, the sleeve of HA already provides such a fixed boundary condition in the MC models.

Recent graft experiments by Wolpert (personal communication) seem to indicate that there is indeed a prepattern formed before any variation in cell densities are evident. This would certainly be a major point in favour of a RD theory.

SKIN ORGAN FORMATION

Vertebrate skin forms many specialized structures, such as hair, scales, feathers and glands. A widely studied problem of considerable interest is the generation of feather germs in chicks (see, for example, Davidson, 1983). These structures are distributed across the surface of the animal in a characteristic and regular fashion. As with other organs the developmental processes involved are not understood. This is currently a problem of considerable interest since its solution would be a milestone in understanding embryogenesis.

The chick skin consists of an epidermal layer of columnar, non-motile cells overlying a dermal layer composed of motile, mesenchymal (fibroblast) cells in a collagenous extracellular matrix. Each feather germ, or primordium, consists of a thickening of the epidermis, called a placode, and a condensation of dermal cells, called a papilla. Davidson (1983) found that in the chick, initially a primordium became visible in the centre of the dorsal midline. A row of primordia then appeared along the dorsal midline. Subsequent rows of primordia then formed on either side of this initial row at intermediate points: the pattern spread laterally forming a hexagonal array.

Perelson et al. (1986) solved a MC model in one dimension and showed that, as cell traction increased, an initially uniform distribution of cells could evolve to a pattern of aggregations. This pattern of cells produces a strain field which causes cells in neighbouring rows to aggregate at interdigitating points, thus leading to hexagonal structure as shown in Fig. 2.

It is not easy to reconcile RD theory with such sequential pattern formation. The mechanism would have to rely on forming a prepattern, to which cells become competent to respond in a sequential manner. The process by which pattern forms on the dorsal midline suggests some prepattern mechanism which determines the initial central condensation. However, Davidson (1983) took a section of chick pteryla on which pattern was forming and cut ahead of the 'wave' of laterally expanding primordia. He found that primordia formed in the skin in front of the cut but their

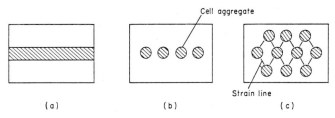

Fig. 2. Scenario for hexagonal pattern formation in the chick back according to MC theory. (a) Section of the chick back illustrating uniform cell density along the dorsal midline. As these cells mature, their traction increases and the uniform steady state loses stability and evolves to a heterogeneous spatial pattern of cell aggregations (b). This line of aggregates imposes a strain field causing aggregation along neighbouring rows at intermediate points leading to hexagonal patterning (c).

position was random compared to the previous row behind the cut. This argues against a prepattern theory.

Both RD and MC theories make predictions on the spacing between primordia. The RD prediction is in terms of diffusion coefficients and reaction rates for chemicals which have yet to be identified and therefore such predictions cannot be tested. The MC approach predicts that decreasing the total cell number will increase the spacing (Oster *et al.*, 1983). Davidson (personal communication) irradiated a section of dorsal pteryla, decreasing the total cell number, and found that the spacing did indeed increase.

Both the above mechanisms simply concentrate on the dermal pattern. Dermal–epidermal recombination experiments strongly suggest that there is interaction between both layers. Nagorcka *et al.* (1987) proposed a hybrid model as a mechanism for scale pattern formation in reptiles. Their model consisted of a two-component RD system of epidermal origin coupled to a cell traction (MC) model of dermal origin. They showed that such a mechanism could produce a pattern consisting of two different sized scales, with correspondingly different interscalar spacings, superimposed on each other. Their patterns closely resemble those observed in certain reptiles such as the armadillo. Coupled MC systems could also produce such patterns.

DISCUSSION

In the last two sections we focused on the ability of RD and MC systems to generate spatially heterogeneous solutions and showed how they could be considered as possible mechanisms for certain developmental processes. They have also been proposed as possible mechanisms for several other developmental phenomena, for example, butterfly and moth wing patterns (Murray, 1981) in the case of RD theory and, in the case of MC theory, movements of epidermal cell sheets during the process of gastrulation (Odell *et al.*, 1981), contraction waves on the surface of vertebrate eggs just after fertilization (Cheer *et al.*, 1987; Lane *et al.*, 1987) and others (see, for example, Murray, 1989). In a broader sense, RD theory has been proposed as the mechanism which accounts for certain regeneration processes, for example in the alga *Acetabularia* (Goodwin *et al.*, 1985) where calcium is proposed as the morphogen: quantitative experimental evidence is presented to support this hypothesis. RD has definitively been shown to be the mechanism in the case of spiral wave phenomena in slime mould *Dictyostelium* (see, for example, the brief review by Tyson and Murray, 1989). MC theory has been applied to the process of wound healing (Murray *et al.*, 1988): qualitative comparison of the theory with experiment is encouraging. The list of proposed applications is now very large.

An important feature of all of these models is the effect of geometry and scale. Oster *et al.* (1988) used such ideas to help make more precise the notion of developmental constraints in the vertebrate limb. They showed that, with only a limited number of allowable condensation patterns, one could use the effects of scale and geometry to 'construct' the entire limb. Their results have important consequences for evolution, which suggest that mutation can only give rise to a limited variety of different limb skeletal structure from a normal limb.

Although both theories have very rich pattern formation structure, clearly both suffer from major drawbacks. The existence of morphogens is, in most cases, still a controversial issue. The problem of how cells interpret a prepattern cannot begin to be addressed until they are identified. RD mechanisms are of 'open-loop' type, that is, once the prepattern is laid down, the eventual morphogenetic structure is fixed – there can be no feedback from the environment to modify the pattern. Such open-loop systems are unstable and seem an unreliable mechanism for many developmental processes which are, in the main, robust and self-correcting.

MC theory is based, to a large extent, on known cellular properties from *in vitro* studies. It is a controversial issue as to whether certain of these properties are exhibited by cells *in vivo*. In some cases of skin organ formation it is unclear if cells do actually form aggregations.

The ability of a model to mimic patterns observed in a developmental process is only the first, but *necessary* step to its candidacy as a possible mechanism. A model must be able to make testable predictions. From this point of view many predictions of RD theory are hard to verify because they are in terms of properties of morphogens yet to be identified, whereas those of MC theory are based on parameters which are measurable.

The processes that occur in development are clearly highly complex. Mathematical models are, necessarily, based on gross simplification. However, they do tend to focus on particular elements of the process and help determine the role played by such elements in the overall process. The role of theory is to provide a conceptual framework for the experimenter and to make predictions which encourage, or provoke, experimentation. From this point of view, mathematical models for development are fulfilling their role and will certainly play a significant part in the understanding of one of the major problems in science.

Acknowledgements: P.K.M. would like to thank the Science and Engineering Research Council of Great Britain for its financial support through a grant to the Centre for Mathematical Biology, University of Oxford.

REFERENCES

Alberch, P. and Gale, E. (1983) Size dependency during the development of the amphibian foot. Colchine induced digital loss and reduction. *J. Embryol. exp. Morph.* **76**, 177–97.
Arcuri, P. and Murray, J. D. (1986) Pattern sensitivity to boundary conditions in reaction–diffusion models. *J. Math. Biol.* **24**, 141–65.
Britton, N. F. (1986) *Reaction–Diffusion Equations and their Applications to Biology.* Academic Press, New York.
Cheer, A., Nuccitelli, R., Oster, G. F. and Vincent, J.-P. (1987) Cortical activity in vertebrate eggs I: The activation waves. *J. theor. Biol.* **124**, 377–404.
Davidson, D. (1983) The mechanism of feather pattern development in the chick. I. The time determination of feather position. II. Control of the sequence of pattern formation. *J. Embryol. exp. Morph.* **74**, 245–73.
Gierer, A. and Meinhardt, H. (1972) A theory of biological pattern formation. *Kybernetik* **12**, 30–9.
Goodwin, B. C., Murray, J. D. and Baldwin, D. (1985) Calcium: the elusive morphogen. In *Acetabularia. Proc. 6th Intern. Symp. on Acetabularia* (eds S.
Harris, A. K., Ward, P. and Stopak D. (1980) Silicone rubber substrata: A new wrinkle in the study of cell locomotion. *Science, N.Y.* **208**, 177–9.
Kochar, D. M., Penner, J. D. and Hickey, T. (1984) Retinoic acid enhances the displacement of newly synthesized hyaluronate from cell layer to culture medium during   early phases of chondrogenesis. *Cell Diff.* **14**, 213–21.
Lane, D. C., Murray, J. D. and Manoranjan, V. S. (1987) Analysis of wave phenomena in a morphogenetic mechanochemical model and an application to post-fertilisation waves on eggs. *IMA J. Math. appl. Med. Biol.* **4**, 309–31.
Maini, P. K. and Murray, J. D. (1988) A nonlinear analysis of a mechanical model for biological pattern formation. *SIAM J. appl. Math.* **48**, 1064–72.
Meinhardt, H. (1982) *Models of Biological Pattern Formation.* Academic Press, London.
Murray, J. D. (1981) On pattern formation mechanisms for Lepidopteran wing patterns and mammalian coat markings. *Phil. Trans. R. Soc. Lond. B* **295**, 473–96.
Murray, J. D. (1982) Parameter space for Turing instability in reaction diffusion mechanisms: a comparison of models. *J. theor. Biol.* **98**, 143–63.
Murray, J. D. (1989) *Mathematical Biology.* Springer-Verlag, Berlin.
Murray, J. D. and Oster, G. F. (1984a) Generation of biological pattern and form. *IMA J. Math. appl. Med. Biol.* **1**, 51–75.

Murray, J. D. and Oster, G. F. (1984b) Cell traction models for generating pattern and form in morphogenesis. *J. Math. Biol.* **19**, 265–79.

Murray, J. D., Maini, P. K. and Tranquillo, R. T. (1988) Mechanochemical models for generating biological pattern and form. *Phys. Rep.* **171**, 59–84.

Nagorcka, B. N., Manoranjan, V. S. and Murray, J. D. (1987) Complex spatial patterns from tissue interactions – an illustrative model. *J. theor. Biol.* **128**, 359–74.

Odell, G. M., Oster, G. F., Burnside, B. and Alberch, P. (1981) The mechanical basis of morphogenesis I: Epithelial folding and invagination. *Devl Biol.* **85**, 446–62.

Oster, G. F., Murray, J. D. and Harris, A. K. (1983) Mechanical aspects of mesenchymal morphogenesis. *J. Embryol. exp. Morph.* **78**, 83–125.

Oster, G. F., Murray, J. D. and Maini, P. K. (1985) A model for chondrogenic condensations in the developing limb: the role of extracellular matrix and cell tractions. *J. Embryol. exp. Morph.* **89**, 93–112.

Oster, G. F., Shubin, N., Murray, J. D. and Alberch, P. (1988) Evolution and morphogenetic rules. The shape of the vertebrate limb in ontogeny and phylogeny. *Evolution* **42**, 862–84.

Perelson, A. S., Maini, P. K., Murray, J. D., Hyman, J. M. and Oster, G. F. (1986) Nonlinear pattern selection in a mechanical model for morphogenesis. *J. Math. Biol.* **24**, 525–41.

Schnackenberg, J. (1979) Simple chemical reaction systems with limit cycle behaviour. *J. theor. Biol.* **81**, 389–400.

Smith, W. R. and Wolpert, L. (1981) Pattern formation along the anteroposterior axis of the chick wing: the increase in width following a polarizing graft and the effect of X-irradiation. *J. Embryol. exp. Morph.* **63**, 127–44.

Turing, A. M. (1952) The chemical basis of morphogenesis. *Phil. Trans. R. Soc. Lond.* *B* **237**, 37–72.

Tyson, J. J. and Murray, J. D. (1989) Cyclic-AMP waves during aggregation of *Dictyostelium* amoebae. *Development* (in press).

Wolpert, L. (1981). Positional information and pattern formation. *Phil. Trans. R. Soc. Lond. B* **295**, 441–50.

The budding of membranes

GEORGE F. OSTER[a] AND HSIAO-PING H. MOORE[b]

[a]Departments of Biophysics, Entomology and Zoology,
University of California, Berkeley, CA 94720, USA
[b]Department of Physiology and Anatomy, University of
California, Berkeley, CA 94720, USA

INTRODUCTION

There are two general mechanisms cells use to signal one another. Small molecules and ions can travel between neighbouring cells via gap junctions. This route limits the diversity of messages for the gap junction cannot pass molecules larger than about 1 kDa. Thus more elaborate signalling must proceed by the process of exocytosis and endocytosis, which is not limited to small molecules, nor to diffusion driven transport. During secretory communication, the signalling cell releases the messenger molecule by packaging it in a vesicle and then fusing the vesicle with its plasma membrane. The receiving cell internalizes the signal molecule by binding it to a surface receptor and packaging it into an endocytic vesicle. A variant on this is so-called fluid phase pinocytosis, wherein external fluid is constitutively internalized in vesicles whose contents reflect the concentrations in the extracellular fluid. Thus the packaging of molecules into vesicles is a universal feature of intra- and inter-cellular communication. Nevertheless, despite its ubiquity, little is known about the mechanisms of vesicle formation.

We present here a proposal for how lipid bilayers bud off vesicles. In particular, we are interested in how vesicles bud from the trans-most cisternae of the Golgi apparatus; however, the same physical mechanism may operate in other circumstances involving membrane deformation. For example, viruses have evolved a way to subvert the process of exocytosis and endocytosis to aid their own reproduction, and we shall propose a mechanism for how this occurs. Our focus is on the *physics* of vesicle budding; that is, we address the question: What are the forces that deform a lipid bilayer into a

Cell to Cell Signalling: From
Experiments to Theoretical Models
ISBN 0–12–287960–0

171

vesicle? Our thesis is that *a bilayer can be bent by releasing its own stored elastic energy.* We do not presume to identify a unique chemical mechanism for, as we shall see, several chemical means can lead to the same bending forces.

The chapter is organized as follows: first we briefly summarize the various forces that determine membrane conformation. Then we present a molecular model for vesicle formation and apply it to several phenomena that involve membrane budding. Finally we discuss some more general aspects of membrane conformation and organelle morphogenesis. We have used side-bar 'boxes' to segregate out certain mathematical aspects of the discussion so as not to discourage less mathematically inclined readers. A more complete discussion of the ideas contained herein can be found in Oster *et al.* (1989).

MEMBRANE FORCES

Before discussing organelle shape and presenting our model for membrane bending, we briefly summarize some of the physics underlying membrane conformation.

Membranes are held together by the opposition of tensile and compressive forces

A lipid bilayer is held together largely by the hydrophobic forces at the interface between the aqueous solvent and the lipid leaflet. This 'force' is not an attraction between the lipid molecules, but is generated by the attractive hydrogen bonding forces between the polar water molecules: the water 'squeezes' the lipids out of solution. That is, the water molecules create a tangential stress at the lipid–water interface – the interfacial tension, τ (dyne/cm = erg/cm^2) (Hill, 1960; Berry, 1971; Tanford, 1980; Miller and Neogi, 1985).

Opposing the compressive solvent force are the repulsive forces between the lipid molecules, which constitute an interfacial, or surface, pressure, π (dyne/cm). These include the electrostatic repulsion between similarly charged lipid head groups as well as steric repulsions between lipid head and tail groups (Cevc and Marsh, 1987). In addition, if the membrane is charged, or if it sustains a transmembrane potential, counterions will accumulate near the interfacial surface. These counterions repel each other and create an 'ion pressure', π_I, which opposes the interfacial tension. Thus the balance of tangential forces at the interface between the solvent and each bilayer leaflet can be described by a net interfacial stress, σ, defined as the sum of the interfacial pressure and tension forces (Fig. 1):

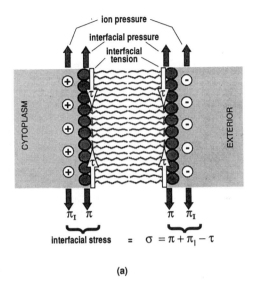

(a)

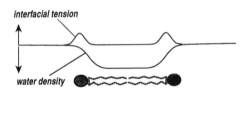

(b)

Fig. 1. (a) The stress, σ, at the interface between the lipid head groups and the aqueous solvent is determined by the difference between the interfacial tension, τ, generated by the solvent and the interfacial pressure generated by the lipid head groups, π. If some lipids are charged, or if the membrane sustains a transmembrane potential, then counterions will be drawn into the neighbourhood of the interface. These counterions will develop a pressure, π_I, which also opposes the interfacial tension. Thus the net stress on each leaflet is $\sigma_{C,L} = \pi + \pi_I - \tau$, where the subscripts C and L refer to the cytoplasmic and lumenal (or exoplasmic) face, respectively. (b) A schematic of the water density profile, ρ, across the membrane. The hydrophobic lipid tails exclude water while the hydrophilic head groups associate with water molecules. The interfacial tension, τ, is proportional to the square of the water gradient (Box 1). Therefore, any surface perturbation that distorts the water gradient will strongly influence the interfacial tension.

$$\sigma = \pi + \pi_I - \tau \tag{1}$$

where σ is the interfacial stress, π is the interfacial pressure, π_I is the ion pressure and τ is the interfacial tension.

Box 1. The molecular basis of surface tension

The interfacial tension is defined as the excess of the tangential component of the stress p_T, over the bulk hydrostatic pressure, p:

$$\tau \equiv \int_{-\infty}^{\infty} [p - p_T(z)]dz \qquad (B1.1)$$

where the integration is carried out over a line perpendicular to the interface. p_T can be obtained from the intermolar potential, $u(r)$, and the pairwise radial distribution function, $g(r,\rho)$, where ρ is the fluid density (Hill, 1960). Indeed, it can be shown that the interfacial tension is proportional to the square of the gradient of the water density near the interface (Rowlinson and Widom, 1982; Davis and Scriven, 1982):

$$\tau = c(\nabla \rho)^2 \qquad (B1.2)$$

where c can be computed from $u(r)$ and $g(r,\rho)$. Thus anything that affects the local *gradient* of water molecules will heavily influence the interfacial tension. For example, the presence of hydrated glycoproteins, or of charged groups which attract counterions, will alter ρ as well as $g(r,\rho)$, and thus strongly affect the local interfacial tension (Willows and Hatschek, 1923).

Upsetting the balance of forces bends the membrane

Equation (1) applies to both leaflets of the bilayer. If the interfacial stress in one leaflet changes while that in the other leaflet remains the same, a bending moment (M) will be generated across the membrane:

$$M = \delta \, \Delta\sigma = \delta \, \{ [\pi_E - \pi_C] - [\tau_E - \tau_C] \} \qquad (2)$$

where δ is the membrane thickness, $\Delta\sigma$ is the difference in surface stress, $[\pi_E - \pi_C]$ is the difference in interfacial pressure and $[\tau_E - \tau_C]$ is the difference in interfacial tension. Thus any effect that alters the interfacial pressures or tension differentially in one leaflet will bend the membrane (Fig. 2). There are many physical effects that can bring about changes in the interfacial forces. For example, the composition and shape of lipids and membrane proteins, including charges on head groups, alter the interfacial pressure. Osmotic hydration of glycolipids and proteins near the leaflet interface can disrupt the local water structure and thus decrease the interfacial tension (a review of many such interfacial effects is given in Oster et al., 1989). Below we shall propose a particular mechanism for membrane deformation based on the notion that a molecular surfactant intercalates into one leaflet of the bilayer and alters the interfacial forces.

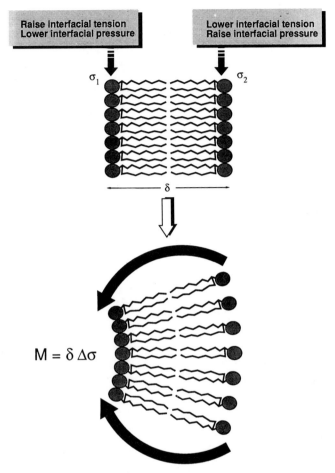

Fig. 2. The membrane will buckle inwards (towards the cytoplasm) if (1) the interfacial pressure (including the ion pressure) increases, and/or the interfacial tension decreases on the cytoplasmic face; (2) the interfacial tension increases and/or the interfacial pressure decreases on the lumenal (external) leaflet.

A MODEL FOR VESICLE BUDDING

Vesicles budding from bilayers commonly have a molecular 'coat'. For example, coated pits budding from the plasma membrane, and secretory granules budding from the trans-Golgi, are coated with a complex of clathrin and several clathrin-associated proteins. We propose that molecular coatings are generally the proximal agent for vesicle formation. That is, we hypothesize that (Fig. 3):

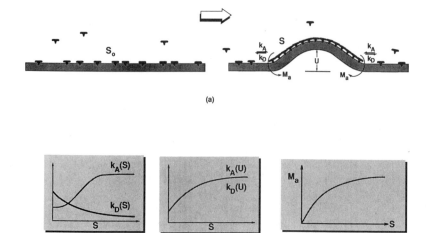

Fig. 3. The surfactant model for the onset of vesicle formation. Surfactant at concentration S_o diffuses on the cytoplasmic face of the membrane. Vesiculation initiates when S commences to crosslink into a coat. This generates an active bending moment, M_a, by altering the balance of interfacial tension and pressure in the cytoplasmic leaflet according to eqn (3). This bending moment initiates a deformation, U, according to eqn (B2.1). The crosslinking of S to form a cytoplasmic coat proceeds according to eqn (B2.2). The shaded insert panels show the qualitative shape of the constitutive relations governing the bending moment, $M_a(S)$, and the association and dissociation rates, $k_A(S,U)$ and $k_D(S,U)$, respectively.

(1) A molecular coating, **S**, condenses onto the cytoplasmic leaflet of the bilayer and cross-links into a coherent layer.

(2) **S** acts as a 'surfactant', changing the interfacial stress on the cytoplasmic leaflet. That is, **S** generates an active bending moment, M_a:

$$M_a(S) = \delta \, \Delta\sigma = \delta \, \{[\pi_E - \pi_C(S)] - [\tau_E - \tau_C(S)]\} \qquad (3)$$

where the subscripts E and C refer to the external (or lumenal, in the case of internal organelles) and cytoplasmic faces, respectively.

Numerical solutions to the model equations (see Box 2, eqns (B2.1) and (B2.2)) are presented in Oster *et al.* (1989). The essential points are:

(1) The surfactant model predicts that vesicle formation is an autocatalytic process: once initiated, it accelerates; this is because bending the membrane lowers the free energy for further intercalation of surfactant and so produces even greater bending. Moreover, vesicle initiation is favoured in regions of a bilayer that are highly strained where the splay

---

Box 2. The onset of budding

We can use the model to study the onset of vesicle formation as follows (Oster *et al.*, 1989). We introduce the active bending moment into the equation for the buckling of an elastic plate (Landau and Lifshitz, 1959);

$$\mu\frac{\partial U}{\partial t} = B\left(\frac{d^2U}{dr^2} + \frac{v}{r}\frac{dU}{dr}\right) + M_a(S) - k_m U \tag{B2.1}$$

| Displacement rate | Elastic bending resistance | Active bending moment | Lumenal elasticity |

where B is the bending modulus ($\approx 10^{-12}$ dyne cm), m is the surface viscosity of the bilayer ($\approx 10^{-6}$ dyne s/cm), and n is Poisson's ratio (≈ 0.5). The last term represents the bending resistance of the vesicle contents (proteins and/or lumenal matrix).

The aggregation of the surfactant can be modelled by the diffusion and cooperative binding of S on the membrane:

$$\frac{\partial S}{\partial t} = D(S)\left(\frac{d^2S}{dr^2} + \frac{1}{r}\frac{dS}{dr}\right) + k_A(U, S)S_0 - k_D(U, S)S \tag{B2.2}$$

| Aggregation rate of surfactant | Diffusion of surfactant on the membrane | Crosslinking of surfactant | Dissociation of surfactant |

where S_o is the concentration of surfactant in the cytoplasm, and the constitutive relations $D(S)$, $k_A(U,S)$ and $k_D(U,S)$ are shown in Fig. 3.

---

of the lipid heads reduces the free energy barrier to intercalation of S. Thus, for example, the edges of the Golgi cisternae should be more susceptible to vesiculation than planar regions.

(2) Crosslinking of S is essential to localize a bending moment sufficiently large to commence vesiculation.

(3) Why doesn't S crosslink into one large patch? It turns out that the model equations predict that vesicles space themselves apart by a characteristic distance determined by the diffusion limited binding of surfactant to the growing patches of S.

The model equations in Box 2 describe only the instability that initiates vesicle formation. To follow the formation of a vesicle further we must resort to a different formulation, capable of following large deformations. We have constructed a finite element shell model to describe the progression of vesicle budding up to the point where the neck of the vesicle fuses to separate the vesicle from the parent membrane (the actual fusion event is a separate physical question that cannot be dealt with by finite element methods). These simulations show that (Oster *et al.*, 1989):

(1) Because the lipid bilayer is almost incompressible, *it is not possible to bend the membrane from a planar bilayer into a spherical cap*. That is, simulations of a budding vesicle with an incompressible bilayer do not admit solutions with spherical geometry. The elastic bending modulus for the bilayer can be expressed as a function of the surface compressibilities (Evans, 1974):

$$B = \delta^2 \left(\frac{K_C K_L}{K_C + K_L} \right) \tag{4}$$

where K_C and K_L are the area moduli of the cytoplasmic and lumenal leaflets, respectively. Since a surfactant that reduces K_C allows the cytoplasmic leaflet to expand, this will reduce the hoop stresses that enforce the mushroom geometry. This supports the surfactant hypothesis since relaxing K_c permits a spherical vesicle to evolve.

(2) The bending moment generated by the surfactant produces the 'necking' of the vesicle that brings the cytoplasmic leaflets into contact.

(3) The contents of a vesicle contribute a normal force which aids in vesicle budding. Without an internal pressure, a spherical vesicle is difficult to achieve.

Figure 4 shows a typical simulation of the vesicle bud under the bending moment of a surfactant cap. Note that while the process is autocatalytic in its beginning stages, it is ultimately self-limiting since, as the membrane curves into a spherical vesicle, the free perimeter first increases, then decreases as the vesicle necks down.

Budding of vesicles from the Golgi

We have applied the model to study several phenomena involving vesicle formation. For example, consider the budding of vesicles from the trans-Golgi cisternae. These vesicles carry secretory proteins to the cell surface, and fall into at least two classes: constitutive and regulated vesicles (Moore *et al.*, 1988). At the moment of budding, both types of vesicles are coated on the cytoplasmic leaflet, the regulated vesicles by clathrin and its associated proteins, the constitutive vesicles by a less characterized molecular coat (Orci *et al.*, 1986). While the functional role of these coats has not been determined, we propose the following hypothesis.

According to the postulates of our model, these molecular coats are surfactants which bend the membrane by altering the balance of interfacial stresses. That is, the coating on constitutive vesicles is the agent for vesicle budding, and it acts by lowering the interfacial tension on the cytoplasmic leaflet. The resulting bending moment bends the membrane until the bilayer is brought into apposition at the vesicle neck. At this point, membrane fusion

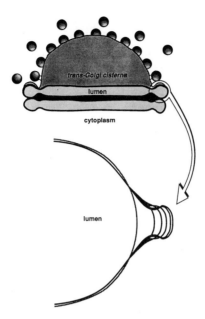

Fig. 4. A large deformation finite element model was used to model the budding of a vesicle from the edge of the trans-Golgi cisterna. The sequence of shapes shown arise from imposing a growing cap of bending moments. Because of the hoop stresses developed as the membrane bends the budding vesicle cannot achieve a spherical shape unless the interfacial tension in the outer leaflet is relaxed and/or a large internal normal force is supplied by the vesicle contents. (From Oster *et al.*, 1989.)

must come into play, a process which is probably mediated by a protein fusagen, and which we do not address in the model.

In the case of regulated vesicles the surfactant is probably not clathrin itself, but one of the associated proteins that link clathrin to the cytoplasmic leaflet (Fig. 5a). Clathrin is necessary to crosslink the surfactant and focus the bending moment. The clathrin-associated proteins insert into the cytoplasmic leaflet only, and their surfactant effect arises either from the counterions they attract, or from the disruption of the water gradient at the interface (the interfacial tension is proportional to the square of the solvent density; c.f. Box 1). It is worth mentioning that the only ATPase activity associated with coated vesicle formation appears to be the enzymatic step that removes the coat once the vesicle has budded from the parent membrane (Rothman and Schmid, 1986).

Proteins affect the surface pressure of phospholipid monolayers (Wiedmer *et al.*, 1978; Llerenas, 1984), and so a mechanically equivalent possibility for budding of regulated granules is that, instead of weakening the interfacial tension on the cytoplasmic face, a vesicle could be formed by weakening the

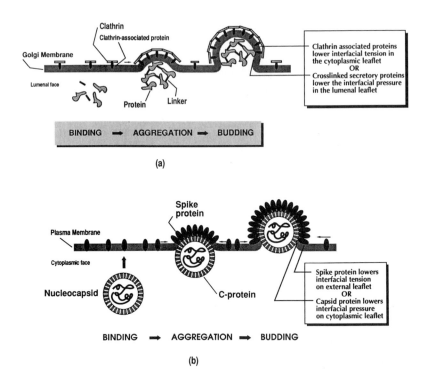

Fig. 5. (a) Model for the formation of a secretory granule. Clathrin crosslinks surfactant proteins which relaxes the interfacial tension on the cytoplasmic leaflet. This causes the membrane to buckle and commence wrapping around the proteins of the incipient vesicle. Alternatively, the proteins which crosslink to form the vesicle core (Chung *et al.*, 1988) probably associate with a membrane receptor. This association can lower the surface pressure on the lumenal face, leading to a bending moment equivalent to that of reducing the surface tension on the cytoplasmic face. The receptor could also be the trigger for aggregation of the cytoplasmic surfactant. (b) Budding of an enveloped virus from the plasma membrane. The spike proteins diffuse in the plane of the membrane until they encounter a C-protein on a capsid. This nucleates the subsequent binding of spikes to C-protein sites. The growing patch of spikes relaxes the interfacial tension in the outer leaflet which generates a membrane bending moment according to eqn (4), and forces the membrane to wrap around the capsid. Alternatively, the association of the cytoplasmic tail of the spike protein with the capsid C-protein can lower the surface pressure on the cytoplasmic face, leading to an equivalent bending moment.

surface pressure on the lumenal face. Since it is necesssary to crosslink the surfactant so as to localize the bending moment, vesicle budding could as well be generated by proteins crosslinking and associating with the lumenal leaflet lipids. That is, the surface pressure culd be reduced by the association of secretory proteins with secondary 'linker' proteins. Indeed, Chung *et al.* (1988) have isolated a family of multivalent binding proteins that appear to crosslink secretory proteins in the incipient secretory granules of the trans-Golgi. Moreover, in the absence of protein synthesis cells continue to secrete via the constitutive pathway, but regulated secretion ceases (Brion and Moore, 1989). It is an attractive notion that the same process that aggregates secretory proteins in the trans-Golgi also drives the budding process that packages them for export.

Budding of viruses from the plasma membrane

Enveloped viruses escape from their host cell's membrane by a budding process that resembles vesicle formation in reverse (Fig. 5b). Simons and his co-workers propose that, for togaviruses such as Semliki Forest virus, the association of the viral spike protein, which resides mostly on the external leaflet of the plasma membrane, with sites on the virus' capsid surface, allows the capsid to wrap the membrane around itself (Simons et al., 1982; Metsikko and Simons, 1986; Simons and Fuller, 1987). However, they do not address the problem of how this bending force is generated. We propose that the mechanisms of vesicle formation and virus budding have a similar physical basis. There are two possibilities.

(1) The spike protein is a surfactant: its close association with the external leaflet of the plasma membrane reduces its interfacial tension. When the cytoplasmic tail of a spike protein binds to the capsid, aggregation of spikes can commence: when free spikes diffuse into the vicinity of the bound spike, they bind to the capsid and/or their glycoprotein head associates with the bound spike. The growing aggregation localizes the bending moment each spike generates so that the membrane is forced to wrap around the capsid. The surface activity of the spike protein can arise from several of its structural properties. Recall that the magnitude of the interfacial tension varies as the square of the water density gradient (cf. Box 1). Thus the hydration of the sugars that becloud the spike's external protein structure can disrupt the solvent density profile near the interface and reduce the interfacial tension. Alternatively, the counterions attracted by the spike charges could increase the ion pressure near the interface (eqn (2)).

(2) The cytoplasmic tail of the spike protein binds to the C-protein on the capsid. The surface pressure of the associated proteins can be lower than that of the two separately. Therefore, binding of the capsid to the spike can reduce the surface pressure on the cytoplasmic leaflet. This has the same mechanical effect as reducing the interfacial tension on the external leaflet – both produce a bending moment that will wrap the capsid in the membrane.

There are several observations that support our model for virus budding:

(a) Viruses seem to bud from a 'hill' on the plasma membrane, suggesting a weakening of the tension in the external leaflet.

(b) When entering the cell, viruses frequently bind to microvilli. Subseqently, they move to the base of the microvilli where a coated pit forms to ingest them. These raise two questions: (i) How does the bound virus move to the microvillus base? (ii) Why does a coated pit form preferentially at the base of the microvilli?

Our model suggests the answers. (i) Binding of the virus spike to the cell surface locally reduces the interfacial tension in the neighbourhood of the binding site. Thus a tension gradient is set up between the base and the shaft of the microvillus; this will induce a convective flow of lipid toward the base (i.e. the 'Marangoni effect') which carries the bound virus down the shaft (Girault and Schiffrin, 1986). (ii) Because of the high curvature of the membrane at the base of the microvilli the lipids are splayed. This lowers the free energy for intercalation of clathrin-associated protein, one of the cell's surfactants, that initiates coated pit formation.

The thermal ratchet model

The model we propose depends on specific molecular interactions to alter the balance of forces in the membrane. There is another model that deserves to be considered for it is implied in many discussions of vesicle and viral budding. Consider the following situation. The viral capsid first binds to a single spike protein. The membrane is constantly undulating under the influence of thermal fluctuations, while other spike proteins are diffusing into the region surrounding the bound spike. Eventually, the following coincidence can occur: a freely diffusing spike protein finds itself very near the bound spike at the moment that a membrane fluctuation of sufficient amplitude occurs such that its cytoplasmic binding site is carried close enough to the capsid to bind. Thus the membrane is held in its deformed configuration, and the process can repeat itself. In this fashion the membrane can seemingly wrap itself around the capsid by 'capturing' the thermal fluctuations of the membrane, the force for bending the membrane coming from thermal energy. We call this model the 'thermal ratchet' mechanism. The thermal fluctuations of the membrane are opposed by the membrane's bending modulus; however, as the membrane bends, the modulus actually decreases (Leibler et al., 1987). This model has a certain appeal, for it requires only the binding specificity between spike and capsid. Clearly, the same idea can be applied to the formation of coated vesicles.

The question is: are thermal fluctuations large enough, and do they occur quickly enough, and in concert with the random arrival of a spike protein to wrap the capsid? This is a difficult problem, which we will address in a subsequent study. Preliminary calculations suggest that thermal fluctuations will not do the job. For example, the frequency spectrum of the membrane 'flicker phenomenon' in erythrocytes follows a (frequency)$^{-4/3}$ law, with a mean square amplitude of about 8×10^{-2} μm – about 10 times the membrane thickness. This is a generous estimate, for fluctuations of this magnitude are characteristic of the broad, flat expanse of erythrocyte membrane, rather than the much smaller domain involved in virus budding. At the scale of Golgi budding, this seems even more unlikely. Moreover, using the bending modulus of a red blood cell membrane, one can calculate that the ratio of the energy of bending to thermal energy is at least several thousand. Therefore, the cell must provide a positive force to bend the membrane. Since there is no evidence for ATPase activity associated with vesicle budding (although an ATP-dependent enzyme is necessary to strip the clathrin coat from coated vesicles), the elastic energy stored in the membrane itself provides a readily available source of bending force.

The instability trap model

There is a variant of the thermal ratchet model that is more likely, and which is based on the amplification of instabilities in the bilayer. A lipid bilayer can be modelled as a viscoelastic film sandwiched between two aqueous reservoirs (Dimitrov and Jain, 1984; Gallez and Coakley, 1986). A fundamental property of such films is that they have become unstable under a variety of perturbations and asymmetries, such as transmembrane potentials, surface charges or lipid composition. When such an instability is excited, the bilayer commences to bend away from its initial planar geometry, and wave-like disturbances grow and propagate along the surface. These instabilities are observed in a number of experimental situations. For example, when erythrocytes are heated above the denaturation temperature of the spectrin lattice that reinforces the plasma membrane, surface waves grow along the cell rim which can reach amplitudes sufficient to fragment the cell (Gallez and Coakley, 1986). These observations suggest that interfacial instabilities can give rise to major geometric deformations of a bilayer. While these deformations are excited by random thermal perturbations, they are driven by interfacial tension; i.e. they use the stored elastic energy in the membrane itself. If the viral spike protein, acting as a local surfactant, weakens the interfacial tension in one leaflet (probably the outer) of the plasma membrane, this could excite a bending instability whose amplitude is easily large enough to bring a second spike protein into cytoplasmic contact with the capsid. This resembles the thermal ratchet model, but unstable bending modes generate the membrane displacements rather than thermal fluctuations. We might call this variant the 'instability trap'. Note that this mechanism is not contradictory to our surfactant hypothesis: the surfactant may be necessary to excite the bending instability and to relax the outer leaflet so that the planar bilayer can wrap around the spherical capsid. The difference between the ratchet mechanisms and the surfactant model is that the former do not require the crosslinking of the surfactant coat to generate a large bending moment – this comes from the unstable bending mode. Rather the crosslinking serves to trap the membrane deformations; thus it is still necessary to capture the bilayer excursion, by binding a spike protein to the virus capsid, or in the case of a coated pit, by crosslinking clathrin. While possibly applicable to the virus budding situation, this model looks problematical for the budding of constitutive vesicles from the Golgi because of the size and curvature constraints. Both the instability trap model and the thermal ratchet model require more detailed calculations to determine their domain of applicability; we shall report on them in a subsequent publication.

Finally, we note that there is no evidence for ATPase activity associated with vesicle budding, although an ATP-dependent enzyme is necessary to strip the clathrin coat from coated vesicles. This lends further support to the hypothesis that the elastic energy stored in the membrane itself provides the bending force.

DISCUSSION

Budding of vesicles from membranes is the key mechanical step in the intracellular trafficking of proteins between organelles, in the import of extracellular substances into the cell, and in the escape of viruses from the cell interior. In this paper we have discussed the budding of vesicles from membranes. Our thesis is that the energy to bend the membrane is stored in the membane itself. This stored elastic energy is released by the action of protein surfactants that change the balance of forces – the interfacial tensions and interfacial pressures – between the leaflets of the bilayer. In order to form a vesicle the surfactants must crosslink to localize and focus the bending moments. Thus vesicle formation should always be associated with a crosslinked coating on one leaflet of the bilayer. Two examples we have discussed are the budding of secretory granules and constitutive vesicles from the trans-Golgi network, and the budding of enveloped viruses from the plasma membrane. Of course, we do not mean to imply that ATP hydrolysis is unnecessary for vesicle formation – any of the coat assembly steps might involve such activity – however, the bending step itself does not require ATP hydrolysis.

This strategy of storing elastic energy for later release is a trick the cell employs in other situations. For example, certain secretory proteins (e.g. mucin, chromogranin, hyaluronic acid) are stored in secretory vesicles whose ionic conditions maintains them at their isoelectric point. When the vesicle fuses with the plasma membrane, the ionic environment suddenly changes, and the proteins swell osmotically and explosively, and expand into the extracellular space (Verdugo *et al.*, 1987). Another example is the storage of elastic energy in cytoskeletal fibres. The curlicue shape of depolymerizing microtubules suggests that microtubules can exist in a number of elastic strain states. This could come about as follows. Tubulin monomers are not in a conformation that permits polymerization; however, when a monomer binds GTP it undergoes a conformational change that removes the steric hindrance and permits it to polymerize. Subsequently, when the bound GTP is hydrolysed to GDP, the monomer attempts to recoil to its free conformation, but cannot because of the constraints of its bonds to neighbouring subunits in the bundle. Thus the intact tubule contains stored elastic energy. This elastic energy can be released to perform useful work, such as moving chromosomes during mitosis – a process that does not seem to require further GTP hydrolysis (Koshland *et al.*, 1988). Alternatively, since the internal bonds of an intact tubule are subject to an elastic strain, triggering the release of this strain energy could accelerate depolymerization – a process that proceeds much faster than polymerization. A similar scenario may well apply to actin polymerization. Recent studies indicate that, like tubulin, actin polymers can exist in a number of helical states; therefore, the stored elastic energy in an actin fibre could be used, for example, by actin-severing proteins, such as gelsolin, which are able to sever fibres without the aid of ATPase activity. A similar process might accompany assembly of clathrin networks: the clathrin cage could store elastic energy which, when released, would create a bending moment for vesiculation; however, no ATPase activity or conformational change has been associated with clathrin assembly. Thus the cell need not 'pay as you go' with regard to mechanical energy utilization, any more than it must with regard to chemical energy; it can live on 'credit' by releasing previously stored energy, just as it does with metabolically stored chemical energy.

Finally, we point out that our discussion of vesicle formation touches on a larger puzzle concerning cell structure. The internal membranous organelles of cells fall into three categories: the extended bilayer of the plasma membrane, the folded 'pita bread' cisternae of the rough endoplasmic reticulum, lysosomes, and the tubular networks formed by the endoplasmic reticulum and the trans-Golgi network (Griffiths and Simons, 1986; Terasaki *et al.*, 1986; Dabora and Sheetz, 1988; Lee and Chen, 1988). What are the forces that shape these organelles, and how is their morphogenesis related to their biochemical functioning? We shall address these questions in a subsequent investigation.

Acknowledgements: This research was supported by NSF Grant No. MCS-8110557 to G.F.O., and NIH Grant No. GM35239, NSF Grant No. DCB8451636 and Juvenile Diabetes Foundation Grant No. 187381 to H.-P. M. The authors would like to thank Paul Janmey for a critical reading of the manuscript.

REFERENCES

Berry, M. V. (1971) The molecular mechanism of surface tension. *Phys. Educ.* **7**, 79–84.

Brion, C., Moore, H.-P. H. (1989) The regulated, but not the constitutive, secretory pathway requires new protein synthesis. *Nature* (submitted).

Cevc, G. and Marsh, D. (1987) *Phospholipid Bilayers*. Wiley, New York.

Chung, K.-N., Walter, P., Aponte, G. and Moore, H.-P. H. (1988) Molecular sorting in the secretory pathway. *Science, N.Y.* **243**, 192–7.

Dabora, S. and Sheetz, M. (1988) The microtubule-dependent formation of a tubulvesicular network with characteristics of the ER from cultured cell extracts. *Cell* **54**, 27–35.

Davis, H. and Scriven, L. E. (1982) Stress and structure in fluid interfaces. *Adv. Chem. Phys.* **49**, 357.

Dimitrov, D. and Jain, R. (1984) Membrane stability. *Biochim. biophys. Acta* **779**, 437–68.

Evans, E. (1974) Bending resistance and chemically induced moments in membrane bilayers. *Biophys. J.* **14**, 923–31.

Gallez, D. and Coakley, W. (1986) Interfacial instability at cell membranes. *Prog. Biophys. molec. Biol.* **48**, 155–99.

Girault, H. and Schiffrin, D. (1986) Charge effects on phospholipid monolayers in relation to cell motility. *Biochim. biophys. Acta* **857**, 251–8.

Griffiths, G. and Simons, K. (1986) The trans Golgi network: sorting at the exit site of the golgi complex. *Science, N.Y.* **234**, 438–43.

Hill, T. (1960) *Introduction to Statistical Thermodynamics*. Addison Wesley, Reading, MA.

Hotani, H. (1984) Transformation pathways of liposomes. *J. molec. Biol.* **178**, 113–20.

Koshland, D., Mitcheson, T. and Kirschner, M. (1988) Polewards chromosome movement driven by microtubule depolymerization *in vitro*. *Nature* **331**, 499–504.

Landau, L. and Lifshitz, E. (1959) *The Theory of Elasticity*. Pergamon, London.

Lee, C. and Chen, L. B. (1988) Dynamic behavior of endoplasmic reticulum in living cells. *Cell* **54**, 37–46.

Leibler, S., Lipowsky, R. and Peliti, L. (1987) Curvature and fluctuations of amphiphilic membranes. In *Physics of Amphiphilic Layers*. Springer-Verlag, New York.

Llerenas, E. (1984) The molecular interaction between F-actin and lecithin in a phospholipid monolayer system. *Bol. Estud. Med. Biol., Mex.* 3333–9.

Metsikko, K. and Simons, K. (1986) The budding mechanism of spikeless vesicular stomatitis virus particles. *EMBO J.* **5**(8), 1913–20.

Miller, C. and Neogi, P. (1985) *Interfacial Phenomena*. M. Dekker, New York.

Moore, H.-P. H., Orci, L. and Oster, G. (1988) Biogenesis of secretory granules. *Protein Transfer and Organelle Biogenesis* (eds R. Das and P. Robbins). Academic Press, New York.

Orci, L., Glick, B. and Rothman, J. (1986) A new type of coated vesicular carrier that appears not to contain clathrin: its possible role in protein transport within the Golgi stack. *Cell* **46**, 171–84.

Oster, G., Cheng, L., Moore, H.-P. H. and Perelson, A. (1989) Budding of vesicles from the Golgi. *J. theor. Biol.* (in press).

Rothman, J. and Schmid, S. (1986) Enzymatic recycling of clathrin from coated vesicles. *Cell* **46**, 5–9.

Rowlinson, J. and Widom, B. (1982) *Molecular Theory of Capillarity*. Clarendon Press, Oxford.

Simons, K. and Fuller, S. (1987) The budding of enveloped viruses: A paradigm for membrane sorting? In *Biological Organization: Macromolecular Interactions at High Resolution* (eds R. Burnett and H. Vogel). Academic Press, New York.

Simons, K., Garoff, H. and Helenius, A. (1982) How an animal virus gets into and out of its host cell. *Scient. Am.* **246**(2), 58–66.

Tanford, C. (1980) *The Hydrophobic Effect: Formation of Micelles and Biological Membranes*. Wiley, New York.

Terasaki, M., Chen, L. and Fujiwara, K. (1986) Microtubules and the endoplasmic reticulum are highly interdependent structures. *J. Cell Biol.* **103**, 1557–68.

Verdugo, P., Aitken, M., Langley, L. and Viillalon, M. (1987) Molecular mechanism of product storage and release in mucin secretion. II. The role of extracellular calcium. *Biorheology* **24**, 625–33.

Wiedmer, T., Brodbeck, U., Zahler, P. and Fulpius, B. (1978) Interactions of acetylcholine receptor and acetylcholinesterase with lipid monolayers. *Biochim. biophys. Acta* **506**, 161–72.

Willows, R. S. and Hatschek, E. (1923) *Surface Tension and Surface Energy and their Influence on Chemical Phenomena* (3rd edn). P. Blakiston's Son & Co., Philadelphia.

Pattern formation and the activation of particular genes

HANS MEINHARDT

Max-Planck-Institut für Entwicklungsbiologie, 7400 Tübingen, FRG

INTRODUCTION

The development of an adult organism with all its differentiated cells in complex but precise spatial arrangement from a single cell, the fertilized egg, is one of the most spectacular events in living systems and a central issue of contemporary investigations. Most information about how pattern formation is achieved has been derived from perturbations of normal development, for instance, by removal of some parts, by their transplantation into an unusual environment or by mutagenesis of genes involved in this process. The subsequent changes in pattern formation provide an inroad into the regulatory properties of the underlying pattern-forming system. Important concepts have been developed in this way, for instance that of the morphogenetic gradient (Morgan, 1904), the embryonic organizer (Spemann and Mangold, 1924) or positional information (Wolpert, 1969).

By considering hypothetical mechanisms (models) possible molecular interactions can be found which are compatible with the observations. If the models are formulated in a mathematically precise way, the danger that the postulated and the actual properties of a model disagree is much reduced. The mathematical formulation requires equations for the production rates of the molecules involved as functions of the other molecules present, for their movement (e.g. diffusion) and removal. The total time course can be calculated by a numerical integration of these coupled partial differential equations. In this way the regulatory properties of the models can be compared with the experimental observations.

We have proposed that the following mechanisms are essential steps in development:

Cell to Cell Signalling: From
Experiments to Theoretical Models
ISBN 0–12–287960–0

(i) Patterned distributions of substances are generated by the interaction of at least two substances. A short-ranging substance enhances its own production and a long-ranging substance acts antagonistically to this autocatalysis (Gierer and Meinhardt, 1972).

(ii) Cells obtain stable differentiated states by gene activation involving self-enhancement and competition between alternative genes. This leads to a situation in which a cell has to make a choice between a few alternative cell states. This choice can be under control of a pattern generated by mechanism (i) (Meinhardt, 1978).

(iii) Spatially ordered sequences of cell states are formed if such locally exclusive cell states activate each other on long range. This leads to a controlled neighbourhood of cell states. Discontinuities can be repaired by intercalation (Meinhardt and Gierer, 1980).

(iv) After separation into distinct domains, the borders between two domains (or the intersections between two such borders) are the organizing regions for the formation of substructures, for instance of legs and wings (Meinhardt, 1983a,b).

These mechanisms will be discussed in more detail and a comparison with insect and especially *Drosophila* development should show what the models can explain and how the elements can be linked together such that the formation of highly complex patterns becomes possible in a reproducible way.

GENERATION OF POSITIONAL INFORMATION BY AUTOCATALYSIS AND LATERAL INHIBITION

Pattern formation from almost homogeneous initial conditions is by no means unique to living systems. The formation of high sand dunes (despite the permanent redistribution of the sand by the blowing wind) or of sharply contoured rivers (despite the fact that the rain is homogeneously distributed) are good examples. Common in these inorganic pattern formations are that small deviations from a homogeneous distribution have a strong positive feedback such that the deviation grows further. We have proposed that primary embryonic pattern formation is accomplished similarly by the coupling of a short-range self-enhancing (autocatalytic) process with a long-range effect which acts antagonistically to the self-enhancement (Gierer and Meinhardt, 1972; Gierer, 1981; Meinhardt, 1982, see also Segel and Jackson, 1972). A simple molecular realization would consist of an 'activator' whose autocatalysis is antagonized by a long-ranging 'inhibitor'. The production rate of the activator (a) and inhibitor (h) can be formulated by the following set of couplet differential equations:

$$\frac{\partial a}{\partial t} = \frac{\rho a^2}{h(1 + ka^2)} - \mu a + \frac{\partial^2 a}{\partial x^2} + \rho_0 \tag{1a}$$

$$\frac{\partial h}{\partial t} = \rho a^2 - vh + \frac{\partial^2 h}{\partial x^2} \qquad (1b)$$

Figure 1 shows numerical solutions of these equations. The simulations demonstrate that the interaction eqn (1) has properties which are basic for the explanation of biological pattern formation: a pattern emerges whenever the extension of the field exceeds a critical size. At the critical size, determined by the range of the activator, a high activator concentration is formed at one end of the field. In this way, monotonic activator and inhibitor distributions result. Therefore, such a system can provide positional information (Wolpert, 1969): the activator and/or inhibitor with their graded distributions can act as morphogen and activate different genes in different parts of the developing organism, as will be discussed below in more detail. If the ranges of the activator and inhibitor are much smaller than the total field, periodic structures will emerge.

An important feature of many developing systems is their capacity for

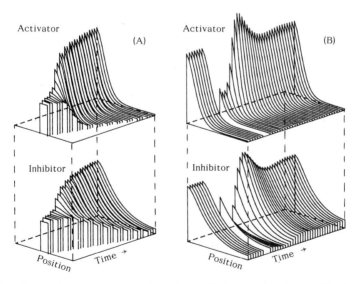

Fig. 1. Generation and regeneration of a graded concentration profile. Assumed is an autocatalytic substance, the activator, and its highly diffusible antagonist, the inhibitor (eqn (1a,b)). (A) Assumed is a linear array of cells (to enable a space–time plot) which grows at both margins. Random fluctuations are sufficient to initiate pattern formation if the extension of the field has surpassed a critical size. At this critical size, the high activator concentration appears at a marginal position since a central maximum would require space for two slopes which is not available at the critical size. (B) Pattern regulation. With removal of the activated region, the inhibitor-producing region is removed too. The remnant inhibitor decays and a new maximum regenerates via autocatalysis and the gradients become restored.

regeneration: the removal of some parts of an embryo are compensated by pattern regulation. This is a property of the activator–inhibitor system since, for instance, after removal of the activated region, the remnant inhibitor decays and a new maximum is formed via autocatalysis (Fig. 1).

The model also accounts for a very puzzling observation, the induction of new structures at ectopic positions by unspecified manipulations, for instance the induction of a second embryo after implantation of denatured tissue in a developing frog egg (Waddington *et al.*, 1936). In the model, the inhibitor concentration decreases with increasing distance from the activator maximum. A second autocatalytic centre can be triggered if the inhibitor concentration is further reduced by an unspecific manipulation, for instance by poisoning, leakage through a wound or destruction by irradiation. The resulting maximum is self-regulating and thus largely independent of the mode of induction and indistinguishable from an activator maximum induced by the application of the real activator. Whether the second maximum leads to an independent embryo or to a fused symmetric malformation depends on the total available space.

The saturation term $1/(k + a^2)$ in eqn (1a) has influences on the emerging pattern which are important for the modelling of many pattern-forming systems. Due to this saturation, the local activator increases and thus the competition among neighbouring cells becomes limited. Therefore, more cells remain active, although at a lower level. This allows, for instance, that the activated region can have a certain proportion in relation to the total size or that it can obtain a stripe-like extension. Stripes are very frequent in developing systems. For instance, the domains of activity of the *engrailed* or the *fushi-tarazu* genes in *Drosophila* (see below) are stripes. They have an extension of about one and four cells respectively in the antero-posterior but about 100 cells in the dorso-ventral direction. Or, to give another example, the organization of a higher organism requires at least two global pattern-forming processes, one determining the antero-posterior (AP) and another one determining the dorso-ventral (DV) axis. In insects, as a rule, the latter is much shorter than the former. Thus, the organization of the DV axis requires not a point-like source but a stripe-like source to generate the gradient, extending along the whole egg from the anterior to the posterior pole. This stripe of high concentration will determine the ventral side of the embryo. Systems with a saturation at low activator concentrations can generate such patterns without the problem that the activated stripes decay into separate patches.

Two different organizing regions at opposite sides of a field or two complementary stripe systems can be formed by two activator–inhibitor systems if they are coupled by a crossreaction between the two inhibitors. A high concentration of one system will appear at the region of the least inhibitor concentration, i.e. at the minimum of the other system. Figure 2 shows a simulation of the formation of two interdigitating periodic patterns.

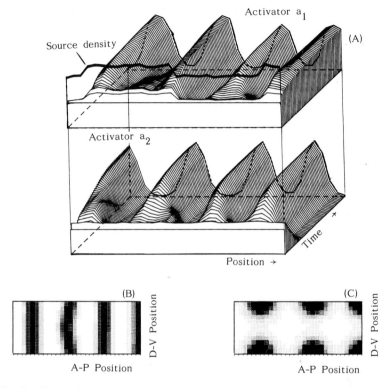

Fig. 2. Formation of complementary periodic pattern. Assumed are two activator–inhibitor systems (only the activator distributions are shown) which are
coupled by a crossreaction of the two inhibitors. To simulate the influence of the
gap genes on the pair-rule genes (see Figs 6 and 8), the pattern of the a_1-activator
is initiated by a modulation of the source density ρ (eqn (1a,b)). The elevated
region of ρ is too large for a single maximum, thus the region of high ρ forms two
peaks. An additional maximum is formed centrally in the region of low ρ values.
The resulting periodic pattern has a much higher spatial frequency than the ρ
pattern but the number of peaks is controlled by the number of elevations in the ρ
pattern as long as the regions of high and low ρ values are not too large or not too
small. Otherwise more than two a_1-peaks per ρ peak or only a single peak would
be formed. (B) Stripe formation in a two-dimensional array of cells. If autocatalysis saturates at low activator concentrations ($k>0$ in eqn (1a)), more cells
become activated, although at a lower level. A stripe-like pattern is especially
stable since, in this case, non-activated cells are nearby into which the inhibitor
can diffuse. This is essential for the simulation of the stripe-like activation pattern
of *Drosophila* segmentation genes (see Fig. 8). (C) In contrast, if the extension of
the maxima is controlled by diffusion, patches instead of stripes are formed.

A special feature of such a system is that a pattern can also be formed if one of the systems is absent due to a mutation. This is in contrast to the behaviour of a system based on the lateral activation mechanism. As discussed below, both types of regulation seem to occur.

GENE ACTIVATION UNDER CONTROL OF POSITIONAL INFORMATION

In higher organisms the patterns generated by reaction–diffusion mechanisms are necessarily transient. Since the time required for communication among the cells of a field increases quadratically with the mean dimension of the field, pattern formation must take place in a small assembly of cells (or, as is the case in early insect development, within a single cell). This requires that at an appropriate stage the cells make use of the positional information so that they become determined for a particular pathway. Afterwards the cells maintain this determination whether or not the evoking signal is still present. We have proposed that the formation of two or more alternative stable states is based on genes that feed back directly or indirectly on their own activation (Meinhardt, 1978). Once turned on by an external signal, such a gene will remain active due to the positive feedback. In addition, a competition must exist between the alternative genes (feedback loops) such that only one of the alternative loops can win this competition. Thus, pattern formation and gene activation have much in common: both processes may depend on autocatalysis. While in spatial pattern formation, the antagonist must have a long range, i.e. act at alternative positions, in gene activation the antagonist may be less specific in that it acts on all alternative genes.

By analogy to the activator–inhibitor system for spatial pattern formation (eqn (1)) the following equations describe a possible mechanism for activation of particular genes (Meinhardt, 1978, 1982):

$$\frac{dg_i}{dt} = \frac{c_i g_i^2}{r} - \mu g_i + m_i \quad (i = 1,2,...) \tag{2a}$$

$$\frac{dr}{dt} = \sum c_i g_i^2 - vr \tag{2b}$$

In this view, each gene product (g_i) has an autocatalytic feedback on its own production but produces and reacts upon the repressor r (a direct repression of the alternative gene products would be possible as well). The last term in eqn (2a), m_i, describes the influence of the positional information. Several types of couplings are conceivable. An example will be discussed below. Due to the competition, within one cell only one of the genes can be active (Fig. 3A). Thus, even if the evoking signal(s) are smoothly graded, the resulting

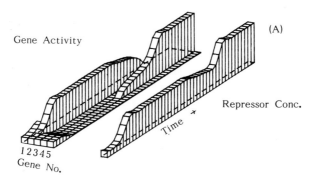

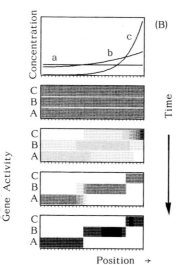

Fig. 3. Formation of cell states and space-dependent gene activation. (A) A set of genes (1,2...5) is assumed whose products feed back on the activation of the corresponding genes. In addition, all genes compete with each other by repressor (eqn (2a,b)). This has the consequence that only one of the genes can be active within one cell. In this simulation, random fluctuations have been decisive in that gene no. 2 initially wins this competition. By an external signal, gene no. 4 becomes activated. Full activation results from autoregulation and is accompanied by the repression of the previously active gene. (B) Spatially restricted cell activation under the influence of a multi-gradient system. Assumed are the genes *A*, *B* and *C* regulated as described above. Gene *A* is activated in a position-independent manner (line *a*) while *B* and *C* are activated by the gradients *b* and *c*, respectively (m_i in eqn (2a)). Shown is the initial, two intermediate and the final pattern of gene activities. At a particular position the gene with the dominating external signal wins the competition. Sharply confined regions of gene activities emerge despite the fact that the initiating signals are graded (compare also with Fig. 7).

pattern will produce sharply bordered regions. In each of them only one of the alternative genes is active.

The postulated autocatalysis can be indirect. For instance, if two genes, a and b, exist which inhibit each other, together they form an autocatalytic system since, for instance, an increase of a leads to a decrease of b which, in turn, leads to a further increase of a. Thus, if only a choice between two alternative states is required, both self-enhancement and competition, can result from two genes which inhibit each other (eqn (3), Meinhardt, 1982).

$$\frac{da}{dt} = \frac{c_a}{k+b^2} - \mu a \tag{3a}$$

$$\frac{db}{dt} = \frac{c_b}{k+a^2} - \nu b \tag{3b}$$

A fully characterized system of this type is the Lambda-phage with its bistable switching between the lytic and the lysogenic phases (Ptashne et al., 1980).

GENE ACTIVATION IN MULTI-GRADIENT SYSTEMS

The patterned activation of genes becomes especially simple if several gradients with different slopes are present which spread out from the same source region. If, for instance, each of these substances activate a particular gene and if the absolute concentrations are appropriate, in regions close to the source the steepest gradient will be dominating and the corresponding gene will become activated. Further away, a shallower gradient will be dominating. Even a position-independent activation can be used. It will be the dominating signal at very large distances. Figure 3B shows a simulation of three genes under the influence of the two gradients.

According to these mechanisms of gene activation, the sharpness of the regions of gene activation result from the competition among the alternative genes but not from sharp thresholds. This has led to the prediction that if one of these genes is missing due to a mutation, the domains of activity of neighbouring genes will extend into the corresponding region (Meinhardt, 1985). Meanwhile, this prediction has been proven to be correct for the Krüppel gene of Drosophila (Jäckle et al., 1986; Gaul and Jäckle, 1987, see Fig. 7).

THE FORMATION OF SELF-REGULATING SEQUENCES BY MUTUAL ACTIVATION OF CELL STATES

In the mechanism described above, the cells do not communicate directly with each other to obtain the correct state of determination. They measure (or compare) local concentrations. This leads to an activation of genes which depends only on these local signals, not on the gene activation in neighbouring cells. Therefore, a mismatch in the neighbourhood of cell states cannot be repaired when the evoking signals are gone. An example for a system which is unable to perform such regulation is the sequence of segments of an insect embryo. For instance, by some manipulations, a second abdomen is formed at the anterior pole (see below). Frequently a pronounced discontinuity remains between the normal structures and the additional abdomen. No tendency exists to intercalate the intervening structures. In other developmental systems, for instance in the pattern formation within insect segments, it appears that not a global signal but a direct communication between adjacent cells is involved to obtain a particular determination. We have proposed that in such systems a sequence of cell states is generated by a mutual long-range activation of cell states (Meinhardt and Gierer, 1980). Neighbouring cell states need each other for mutual stabilization while local exclusiveness assures that the two states do not merge.

Such a system also has the capability to generate stripes. Stripes have necessarily a very long border between cells of different cell states. A particular cell state is close to cells of the other type. This facilitates the mutual support of the two cell states by diffusable substances. A stripe-like arrangement is therefore especially stable.

The mutual activation scheme can comprise more than two states. For instance, state 1 can activate state 2, state 2 can activate state 3 and so on until state n. The system can be cyclic in that state n activates state 1. Equation (4) describes a mutual activation scheme (Meinhardt and Gierer, 1980).

$$\frac{\partial g_i}{\partial t} = \frac{c_i\, g_i'^2}{r} - \alpha g_i + D_{gi}\, \frac{\partial^2 g_i}{\partial x^2} \tag{4a}$$

with

$$g_i' = g_i + \delta^- s_{i-1} + \delta^+ s_{i+1}$$

$$\frac{\partial s_i}{\partial t} = \gamma(g_i - s_i) + D_s \frac{\partial^2 s_i}{\partial x^2} \tag{4b}$$

$$\frac{\partial r}{\partial t} = \sum c_i\, g_i'^2 - \beta r \tag{4c}$$

This system is able to generate self-regulating sequences. If cell states are juxtaposed which are usually not neighbours, the missing cell states become intercalated, restoring in this way a normal neighbourhood of cell states (Fig. 4). This need not necessarily be the natural sequence of structures. A

(A) The reaction scheme

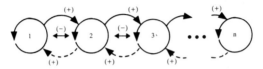

(B) Formation of a sequence

Position

(C) Intercalary regeneration

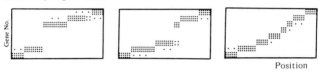

Position

(D) Intercalary regeneration with polarity reversal

Position

(E) Experimental observation

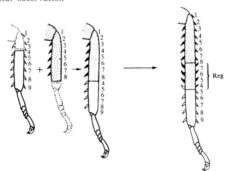

Fig. 4. Generation of sequences of structures and their repair. (A) Reaction scheme: by autocatalysis and mutual short-range repression, cell states are formed (see Fig. 3). These cells activate each other on long range. (B) Computer simulation with eqn (4) in a growing array of cells. Subsequent structures are added whenever sufficient space is available. (C) If, by a manipulation, groups of cells become juxtaposed which are usually not neighbours, the pattern discontinuity is smoothed out. (D) This intercalation can lead to a polarity reversal (after Meinhardt and Gierer, 1980). (E) An experimentally induced pattern discontinuity and its repair: graft experiment with parts of the femur segment of cockroach legs (Bohn, 1970). A large distal fragment, denoted arbitrarily 4–9, is

biological system in which, after manipulation, a correct neighbourhood but a non-natural pattern can be formed, is the intrasegmental pattern of insect legs (Fig. 4E). Since missing structures can be added later on, this mechanism allows the formation of a fine-grained pattern out of a coarse pattern. However, to form a polar and periodic pattern such as segments, at least three elements per segment must be laid down initially in order to allow intercalation in a defined polarity (see below).

BORDERS BETWEEN DOMAINS AS ORGANIZING REGIONS

After a separation of a field into distinct domains, the borders between these domains can be used to organize the domains themselves. Imagine an A and a P domain with a common border. If only in the A domain a certain diffusible substance, the morphogen m, is produced, m will obtain a graded distribution in the P domain (Fig. 5A). Many other schemes are conceivable. If A- and P-specific pathways are simultaneously required for m production, such production will take place only in a zone of $A–P$ overlap. Such a mechanism requires that the border is not completely sharp. Or, in one domain, an m precursor may be produced and in the other the final product. Thus, production can take place only in the boundary region because further away, either no precursor molecules are present or no capability exists to produce the morphogen (Fig. 5B). If domains are generated by the mutual activation mechanism, the diffusible substance spreading out, e.g. from the A domain in order to stabilize the P domain has in the P domain a graded distribution and could supply positional information as well.

To determine a particular location of a structure such as a leg, a wing or an eye, a single border is insufficient since a border has a long extension. In contrast, intersections between two borders, one resulting from antero-posterior (AP) and the other from dorso-ventral (DV) patterning, determine unique points. Moreover, for any reasonable AP and DV pattern, these intersections will occur in pairs (Fig. 5C). Both of these intersections have opposite handedness, they are mirror-images. Thus, the model accounts for the pairwise initiation of these structures with a defined handedness and orientation in respect to the main body axes. Many experiments with limb formation in vertebrates and insects can be accounted for under this assumption (Meinhardt, 1982, 1983a,b).

This model has the regulatory features demanded by the polar coordinate

---

grafted on a large proximal stump (1–8). After one or two moults, the pattern discontinuity is repaired by intercalary regeneration (7–5 Reg.). The spines of the intercalate have a reversed orientation, which indicates a polarity reversal. The regulatory system is able to restore a normal neighbourhood of structures. However, in this situation the system is unable to restore the natural pattern. A simulation of this process is shown in D.

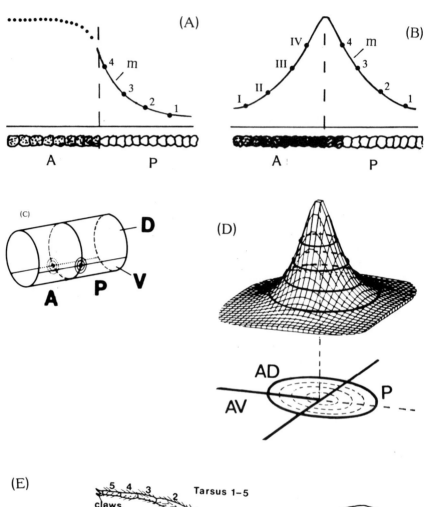

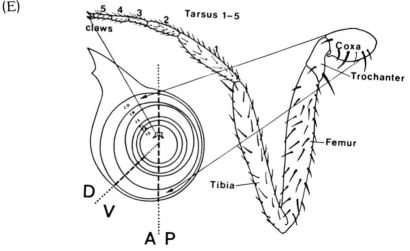

model (French *et al.*, 1976) since the requirement of two intersecting borders is equivalent to the requirement of a complete circle of four quadrants or three sectors. The boundary model is molecularly feasible and accounts for primary initiation of appendages as well as for regeneration. In addition, it is based on borders which have been determined by independent experiments.

EARLY INSECT DEVELOPMENT AS AN EXAMPLE

For several reasons the insect embryo is a convenient system for the analysis of pattern formation. The sequence of segments may be regarded, in a first approximation, as a linear pattern. Many manipulations are known which alter the sequence of segments in a reproducible way, for instance, centrifugations, ligation of the egg and UV-irradiation (see Sander, 1976 for review). Different species develop in different ways, but it is likely that the underlying mechanisms are similar. Especially for *Drosophila* many mutations are known which alter segmentation in a characteristic manner (Fig. 6; Nüsslein-Volhard, 1977; Nüsslein-Volhard and Wieschaus, 1980; Schüpbach and Wieschaus, 1986). The corresponding genes have been subdivided into four classes:

(i) Coordinate genes. A mutation leads to large-scale changes in the pattern. Decisive is the genotype of the mother, not that of the embryo, indicating that the primary positional information is set up by the mother during development of the oocyte. An example is the

---

Fig. 5. Boundaries and intersection of boundaries as organizing regions for legs and wings. (A) If a morphogen *m* is produced in one domain and diffuses into a neighbouring domain it forms an exponential gradient and is thus appropriate to generate positional information. (B) If the production of a morphogen *m* requires the collaboration between two different cell types situated in two neighbouring regions (*A* and *P*) the morphogen production is restricted to the boundary region. If *m* is diffusable it obtains a symmetrical distribution. The local concentration is a measure for the distance from the border and thus appropriate as positional information. The symmetrical pattern can determine nevertheless very asymmetric structures if both cell types react in a different manner (1–4 versus I–IV). In vertebrate limb formation, only one cell type seems to respond (Meinhardt, 1983a). The resulting pattern, the sequence of digits, is very asymmetric. (C) Borders generated by an antero-posterior (A/P) and a dorso-ventral pattern (D/V) form pairwise intersections (concentric circles). If a cooperation of the two pairs of these compartments is required, morphogen production becomes restricted to these two points. (D) A point-like source leads to a cone-shaped morphogen concentration, appropriate to initiate, for instance, the primordium of an insect leg, the imaginal disk. (E) The cone-shaped morphogen distribution is appropriate to determine the proximo-distal sequence of leg structures in concentric rings within the disk (Schubiger, 1968).

mutation *bicaudal* which causes a symmetrical malformation with a second abdomen instead of head and thorax.

(ii) Gap genes. If mutated, the sequence of segments is interrupted by gaps about seven segments wide. The gaps are overlapping.

(iii) Pair rule genes. Mutations cause the deletion of either the even- or the odd-numbered segment borders. The remaining pattern maintains polarity.

(iv) Segment polarity genes. Mutations lead to deletions in every segment, the remaining structures are symmetrical.

The early development of insects is unique in that first a rapid division of nuclei takes place. Later, these nuclei migrate to the egg periphery and it is only then that the nuclei become separated by cell walls. At this stage, a cell is not only determined as to which segment it belongs but also whether its progeny will populate the anterior or posterior compartment (Garcia-Bellido

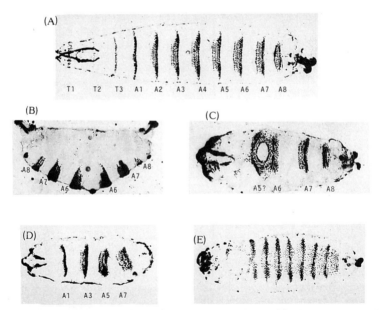

Fig. 6. Examples of the four classes of mutations affecting segmentation in *Drosophila* (Nüsslein-Volhard, 1977; Nüsslein-Volhard and Wieschaus, 1980). (A) Wild type embryo. (B) The coordinate mutant *bicaudal*: a mirror-symmetrical embryo with two abdomens. (C) *Krüppel*, a gap mutant: a single region of about seven segments is missing (*T1–A4*). (D) *Hairy*, a pair-rule mutant: homologous pattern elements around every second segment border are deleted; the remaining denticle belts indicate the maintenance of the polarity. (E) *Gooseberry*, a segment polarity mutant. Certain pattern elements are deleted in every segment, the remaining pattern of each segment is duplicated. The polarity of the segments is lost. (Negatives of dark-field photographs, kindly provided by Christiane Nüsslein-Volhard, after Meinhardt, 1986.)

et al., 1973; Steiner, 1976). Thus, a very fine-grained pattern formation must have taken place in this short time interval of 3–4 h after egg deposition.

MATERNALLY SUPPLIED POSITIONAL INFORMATION

Many classical experiments with early insect embryos (see Sander, 1976) are explicable under the assumption that a gradient is produced at the posterior egg pole by autocatalysis and long-range inhibition (Meinhardt, 1977). In this model, a high morphogen concentration would lead to abdominal structures, low concentrations to head structures. This model accounts, for instance, for the general instability which exists in many insect species at the anterior pole such that after some unspecific treatments posterior structures often develop. According to the model, at the anterior pole the inhibitor concentration is low. Any further (unspecific) reduction can lead to the onset of autocatalysis and thus to the formation of a second activator peak there. This, in turn, would lead to a high morphogen production and to the formation of a second abdomen instead of a head and a thorax. Figure 6B shows such a malformation based on a genetic defect.

AT LEAST THREE *AP* POSITIONAL INFORMATION SYSTEMS EXIST IN *DROSOPHILA*

Recent genetic investigations in *Drosophila* have shown that in addition to this positional information system which has its high point at the posterior pole, the *oskar-nanos* system (Lehmann and Nüsslein-Volhard, 1986) at least two other systems exist, the *bicoid* (Driever and Nüsslein-Volhard, 1988a,b) and the *torso* system (Schüpbach and Wieschaus, 1986). The *bicoid* system has its high point at the anterior pole and is responsible for anterior structures. It is not generated in a dynamic way as outlined above but by the formation of a local source. The oocyte is fed by nurse cells. The latter produce the *bicoid* RNA (Berleth *et al.*, 1988). This RNA becomes transported to the anterior pole of the oocyte and immobilized there. Local production of the *bicoid* protein leads to the set-up of the *bicoid* gradient. Thus, the dynamically regulated pattern-forming event (including autocatalysis) does not consist of the generation of a gradient but in the separation into two different cell types, the nurse cell and the oocyte. The formation of the *bicoid*-gradient is more of the border-type mentioned above (for a more detailed modelling of this system, see Meinhardt, 1988).

The decisive morphogenetic molecule of the *torso* system presumably has a symmetrical distribution with a narrow peak at the anterior as well as at the posterior pole and controls the formation of the terminal structures of the embryo. Dominant mutations are known (Klinger *et al.*, 1988) which behave as if the morphogen has a high concentration everywhere in the field. A possible explanation in terms of the model would be that an inhibitory action is abolished in these mutants such that activation can occur everywhere.

CONTROL OF SEGMENTATION

The existence of the three classes of zygotic genes which affect the segmentation either by forming large gaps, by deletion of structures in a two segment interval or by deletions within every segment indicates that segmentation is achieved in a stepwise hierarchical manner. The model I have proposed (Meinhardt, 1985, 1986) can be summarized as follows: under the influence of the maternally supplied positional information a space-dependent activation of the gap genes takes place which leads to a separation into three to four cardinal regions plus anterior and posterior marginal regions (Fig. 7). The borders between these cardinal regions provide the scaffolding for the first truly periodic pattern, the pattern of pair-rule gene activities. The latter can be described as the repetition of (at least) four cell states – to be called 1, 2, 3 and 4. Each 1 2 3 4 cycle forms two segments in the adult organism. An individual segment consists of the reiteration of three or four cell states, to be called S, A and P (or S, A_1, A_2 and P). A segment border is formed whenever P and S cells are juxtaposed. In the following I would like to discuss what these models can explain, where changes have been necessary to cope with more recent observations and where the models are still insufficient.

At the beginning it should be discussed of what a segment, the basic unit, consists. Based on this, an attempt should be made to reconstruct the events which lead to this pattern.

THE THREE-FOLD SUBDIVISION OF SEGMENTS

Segments clearly resemble a periodic structure. The segments proper alternate with segment borders. The simplest periodic structure would consist of the alternation of two specifications, for instance A and P, such as suggested by the early subdivision of the thoracic segments into anterior and posterior compartments. However, in a sequence ...$APAPAP$... neither the polarity of the segments nor position of the segment border is determined ($AP/AP/AP/$ or $A/PA/PA/P$). For that reason I have proposed (Meinhardt, 1982, 1984) that segments consist of the alternation of at least three states ...$P/SAP/$ SAP/S.... A segment border is formed whenever S and P cells are contiguous. A sequence of three states has a defined polarity. This model predicts that, if one of the three states is absent due to a mutation, the segment polarity will be lost and a symmetrical pattern will emerge. For instance, a loss of A would lead to a pattern ...$S/P/S/P/S/$... i.e. to a symmetrical

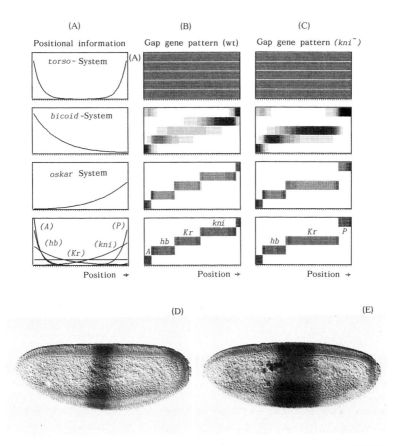

Fig. 7. Positional information and gap-gene activation in *Drosophila*. (A) Three *A/P positional* information systems exist (Nüsslein-Volhard *et al.*, 1987), the *torso-*, the *bicoid-* and the *oskar*-system. *torso* on its own is assumed to activate genes responsible for the most posterior structures (*P*), in combination with *bicoid* the most anterior structure (*A*); *oskar* activates *knirps* (*kni*), *bicoid* *hunchback* (*hb*) (Tauz, 1988) while *Krüppel* (*Kr*) has a position-independent activation (Gaul and Jäckle, 1987). (B) The gap genes are assumed to compete with each other. Thus, the locally dominating activation is decisive as to which gene wins the competition at a particular position (see Fig. 3). The initial, two intermediate and the final pattern of gap-gene activities are shown. (C) If, for instance, the *knirps* gene is lost by a mutation, the neighbouring domains of gap-gene activity extend into the region in which normally *knirps* would be active. (D) For comparison, distribution of *Kr*-protein in the wild type (dark region) and (E) in the *kni*-mutation. The more posterior extension is clearly visible (photographs kindly supplied by Dr Ulrike Gaul, see Gaul and Jäckle, 1987). A simulation of mutations of the positional information systems can be found elsewhere (Meinhardt, 1988).

pattern with the same number of denticle bands (*S*) but twice the normal number of segment borders. This corresponds to the mutation *patch* (Nüsslien-Volhard and Wieschaus, 1980).

More recent experiments indicate a four-fold subdivision (Martinez-Arias *et al.*, 1988). How can a loss of one of four cell states lead to a symmetrical pattern? For simplicity, let us call the four cell states ...*abcdabcd*.... After a mutation of a gene responsible for instance, for the cell state *c* a sequence ...*ab/dab/d*... remains (the slash indicates the mismatch in the sequence of cell states). Pattern regulation which leads to a situation that every cell is juxtaposed to a normal neighbour can proceed either by intercalation of *a*-cells resulting in a sequence ...*aba**d**abab*... (the intercalated cell state is written in bold face), or by degeneration of *d*-cells which leads to the sequence ...*abab*.... In both cases, the polarity of the pattern is lost. Both modes seems to occur. According to Martinez-Arias *et al.* (1988) the genes *engrailed, naked, patch* and *wingless* (...*ENPWENPE*...) are involved in a four-fold subdivision. If *naked* is missing, an extra band of *wingless* is intercalated (...*EWPWEWPW*...). In contrast, if *wingless* is missing, after a period of initial expression, the *engrailed* gene is turned off and a symmetrical structure (...*NPNPNP*...?) remains.

THE FOUR-FOLD SUBDIVISION OF DOUBLE SEGMENTS

The pair-rule mutations indicate a temporary formation of a periodic structure with a repeat length of two segments. Although eight pair-rule genes are known, I have proposed (Meinhardt, 1986) that this pattern can be regarded essentially as two binary patterns of cell states, to be called ...*13131*... and ...*24242*.... Both sequences are superimposed on each other with a phase shift. The phase shift determines the polarity of the pattern. The sequences $1_23_41_23_4$ and $_21_43_21_43$ have, of course, opposite polarities. The genes *hairy* and *runt* are involved in the generation of the first, *even-skipped* and *ftz* in the generation of the second pattern (Frasch and Levine, 1987; Ingham and Gergen, 1988).

The pattern of pair-rule gene activities has different regulatory features in comparison with the pattern of segment polarity genes. A loss of one element does not lead to an intercalation as described above. From the point of view of the models it came as a surprise that genes which have a complementary pattern of expression, for instance *fushi-tarazu* and *even-skipped*, do not depend on each other as expected from a lateral activation model and one gene can show a stripe-like expression even if the gene responsible for the complementary stripes is lost. Therefore, this pattern formation can be better described under the assumption of activator–inhibitor systems coupled by cross-interactions of the inhibitors and in which the autocatalysis saturates at low activator concentrations. As mentioned, this can lead to complementary stripe formation (see Fig. 2).

The double segment pattern, i.e. the repetitive *1234*-sequence, must give rise to the pattern corresponding two single segments. This requires that the same segment polarity gene becomes activated at two different positions of the pair-rule pattern. For instance, cell state 1 induces cell state A, 2 induces S, and 1 and 2 together induce P; *3* and *4* induce the second set of APS states in an analogous way. This model accounts for essential features of the pair-rule mutants. If one of these four pair-rule controlled cell states is lost, the remaining three would still determine a polar pattern. If, for instance, cell state *1* is lost, I expect that the remaining pattern $...4_{1}234_{1}234...$ induces a pattern $...S_{AP}SAP/S_{AP}SAP... = ...SSAP/SSAP...$ (the deleted pattern is written as subscript) i.e. every second segment (SAP) remains normal. The same segment border would be deleted if the cell state 2 from the pair-rule pattern was lost since the remaining $41_{2}341_{2}34...$ pattern would lead to $...SA_{P/S}AP/SA_{P/S}AP... = SAAP/SAAP....$ A mutation of cell states *3* and *4* would lead to similar patterns, only that instead of the odd- the even-numbered segment borders would be affected. Thus, for each segment two frames of deletions exist, in basic agreement with the observation (Nüsslein-Volhard and Wieschaus, 1980). In the case of $/SSAP/SSAP/$, a broader denticle band (SS) is expected, such as is observed in *hairy*. The other type expected, $...SAAP/SAAP...$, should carry a small denticle band (S) and a larger naked region (AAP). Such a pattern is observed in weak alleles of *even-skipped*.

CONTROL OF THE PAIR-RULE GENE PATTERN BY THE GAP GENES

The omission of segments in the gap mutants indicates that the gap-gene pattern controls the formation of pair-rule gene activities. The gaps caused by mutated gap genes are overlapping. I have proposed that the overlap does not result from overlapping domains of the gap genes but from the organizing properties of the borders of the gap domains. For simplicity let us denote the domains in which the gap genes *hunchback*, *Krüppel* and *knirps* are active with A, B, and C. If no B region would be formed due to a mutation, neither an A/B nor a B/C border would be formed. Thus, if the B domain is lost, half of the A region and half of the C region would be lost in addition to the B region. This has led to the prediction (Meinhardt, 1985) that the gap domains are half as large as the gaps formed if a corresponding gene is lost. Meanwhile, this prediction has turned out to be basically correct (Knipple *et al.*, 1985; Gaul and Jäckle, 1987). The model also predicts correctly that the *hunchback* and the *knirps* gap are complementary to each other since in the first, the A/B border, in the latter, the B/C (and possibly a C/D) border is lost. Also in agreement with the model is the observation that in a *hunchback/knirps* double mutant, almost all periodicity of the pair-rule genes is lost (Frasch and Levine, 1987) since in this case all borders are lost including those with the marginal regions. Unexpected was the finding that normal expression of a gap gene is located in the anterior part of that region which is lost if the corresponding gene is mutated. For instance, the *Krüppel* (*Kr*) region extends from *T1* to *A1* while the *Kr*-gap extends from *T1* to *A5*. Thus, the *Kr* (*B*) region is presumably required to organize the whole *knirps* (*C*) region but is not required to organize the *hunchback* (*A*) region.

In order to activate the periodic pair-rule pattern by the essentially non-periodic gap-gene pattern, a particular pair-rule gene must become activated at different positions in the gap-gene pattern. Since, for instance, the *Krüppel* domain has an extension of at least three segments (Knipple *et al.*, 1985), most pair-rule genes must become activated at two different positions within a single cardinal region. The gap-gene pattern must not only trigger the seven stripes of the pair-rule genes but also control the phase shift between the *hairy/runt* pattern on the one hand and the *even-skipped/fushi-tarazu* pattern on the other (in the terminology used above, the ...*1313*... and the ...*2424*... pattern) in order to transmit the polarity.

The following model is able to generate a precise number of stripes and the two periodic patterns with the correct phase shift: the cardinal regions *A*, *B*, *C* activate the pair-rule genes 2, 4 and 2 respectively. Each of these broad stripes resolve into two smaller stripes (see Fig. 2). In contrast, the *A/B* border and the *B/C* border activate genes 3 and 1, respectively. Thus, for instance, around the *A/B* border, a pattern ...*234*... emerges. A complete simulation is given in Fig. 8. Although basic features are correctly reproduced, this model has several problems which are not yet solved. The borders on both sides of a cardinal region play the same role. The unilateral long-ranging influence of a particular cardinal region on the next posterior cardinal region is not yet incorporated. Further, in some gap mutants, very broad stripes are formed which do not resolve into individual stripes, indicating that the pair-rule pattern needs a strong trigger to resolve but larger coherent patches are possible as well (for a model which accounts for this feature but which fails to describe the phase shift see Edgar *et al.*, 1989).

FORMATION OF LEGS AND WINGS AT THE INTERSECTION OF COMPARTMENT BORDERS

Many observations with leg and wing formation in insects can be accounted for by the boundary model outlined above (Meinhardt, 1983b). In holometabolous insects – such as *Drosophila* – the appendages are formed during pupation from nests of cells, the so-called imaginal disks. The borders

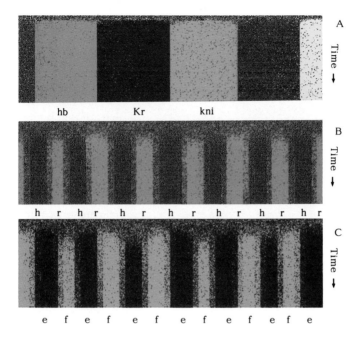

Fig. 8. Simulation of the activation of the gap and pair rule genes as a function of time. The density of dots with a particular colour is a measure for the activity of a particular gene. Initially, the genes are active in a position-independent manner (random dot pattern at the top of each panel). Shown are the activation of (A) the gap genes *huchback (hb)*, *Krüppel (Kr)*, *knirps (kni)*, (B) the *hairy/runt* pattern and (C) the *even-skipped/ftz* pattern. A, B and C are separate plots of a single simulation. After spatial restriction of gap gene expression, i.e. after formation of the cardinal regions (see Fig. 7), the pattern of pair rule gene activity emerges. In this simulation, it has been assumed that *Kr* (dark blue in A) enhances the *ftz* activation (light blue in C) while *hb* and *kni* enhance that of *even-skipped* (dark blue in C). The resulting broad bands resolve into two stripes (see Fig. 2). Conversely the *hb/Kr* border leads to the activation of *runt* (green) while the *Kr/kni* border activates *hairy* (red). With this initiation, a second stripe system is formed which has a defined phase shift in respect to the other. This phase shift contains the polarity information. For comparison, distribution of *Krüppel* (D) and of *fushi-tarazu* (E) protein in early embryos are shown overleaf.

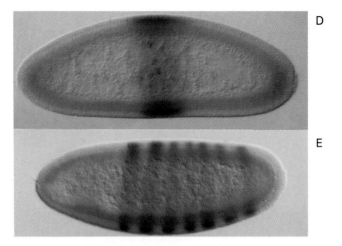

D

E

Fig. 8 (*cont'd*) Distribution of *Krüppel* (D) and of *fushi-tarazu* (E) protein in early embryos (photographs kindly provided by U. Gaul and H. Jäckle, see Gaul and Jäckle, 1987; Carroll and Scott, 1985).

postulated and their intersections are well known. An early event in the pattern formation of appendages is the clonal separation into an anterior (A) and a posterior (P) compartment (Garcia-Bellido et al., 1973; Steiner, 1976). With the completion of the blastoderm, a cell is programmed that its progeny form either anterior or posterior parts of a leg or a wing but never contribute to both parts. Somewhat later, a similar clonal separation takes place in the dorso-ventral dimension. A particular cell contributes only to a leg or to a wing. A further step is the formation of dorsal and ventral compartments of legs and wings respectively. It should be mentioned that the borders required by the models must not necessarily be borders of clonal restriction.

Involved in this clonal separation is the gene *engrailed* (Lawrence and Morata, 1976). A stripe-like pattern of expression becomes detectable soon after cellularization (Kornberg et al., 1985; Fjose et al., 1985), i.e. much earlier than any disk structure could be present. The width of the stripe is one cell only. No disk can be present at this early stage and this observation is incompatible with the assumption that existing disks become subdivided into an anterior and a posterior compartment. It supports, however, the assumption that the imaginal disk is formed around previously formed borders. According to the model of segmentation discussed above, segmentation results from the repetition of at least three cell states, S, A and P. Thus, the formation of the $A–P$ border is part of the antero-posterior pattern formation and not a special pattern-forming event in disks or for the formation of disks.

The primordia of leg segments are concentrically arranged in the disks: the distal structures (tarsi and claws) in the centre, the proximal structures (coxa and thorax) more at the periphery. The appropriate positional information to generate such concentric arrangement of structures would be a cone-shaped morphogen distribution since the contour lines of a cone-shaped distribution are concentric circles (see Fig. 5C,D). The most distal structures – the claws of the legs or the tip of the wings – are formed at the intersection of the compartment borders, in obvious agreement with the model proposed.

A strong support for the hypothesis that, on the one hand, the juxtaposition of differently determined tissues (A and P) is required for leg induction and, on the other, that these two tissues are separated by a third tissue (S) is provided by an experiment with cockroaches. Bohn (1974) has removed small stripes of ectodermal tissue from the anterior region of a thoracic segment. After one or two moults he observed the formation of an additional leg. This is not some sort of regeneration since, under normal circumstances, a leg would never have been formed at this position. In addition, this supernumerary leg has the reversed handedness (for instance, a left limb would be formed on the right side) and the opposite orientation in respect to the body axis (posterior leg side points anteriorly). According to the model, by such an operation a portion of the S region is removed. After wound healing, the P region of one segment becomes juxtaposed to the A region of the following segment. The result is an additional *P–A* confrontation with a reversed antero-posterior polarity (*P* becomes anterior). Since the dorso-ventral axis remains normal, this polarity change is accompanied by a change in the handedness (Meinhardt, 1984).

CONCLUSIONS

Pattern formation in higher organisms can be regarded as a hierarchical process in which the following processes play a key role: (i) primary pattern formation by autocatalysis and long-ranging inhibition; (ii) formation of stable cell states by direct or indirect autoregulation of genes accompanied by competition of alternative genes; (iii) mutual long-range stabilization of cell states to achieve a controlled neighbourhood; and (iv) boundaries between regions generated by these mechanisms obtain organizing properties for the formation of substructures, such as legs and wings. Many of the very detailed observations available for the early insect and especially of *Drosophila* development are explicable in a straightforward manner under these models. Discrepancies between models and observations now become most interesting since they uncover elements not yet fully understood. So far, these discrepancies result from questions about how these basic mechanisms have to be linked and not because the basic mechanisms proposed are inadequate. I hope that the interactions between models and experiments provide an inroad for a better understanding of development.

REFERENCES

Berleth, T., Burri, M., Thoma, G., Bopp, D., Richstein, S., Frigerio, G., Noll, M. and Nüsslein-Volhard, C. (1988) The role of localization of *bicoid* RNA in organizing the anterior pattern of the *Drosophila* embryo. *EMBO J.* **7**, 1749–56.
Bohn, H. (1970) Interkalare Regeneration und segmentale Gradienten bei den Extremitäten von *Leucophaea*-Larven (Blattari). I. Femur und Tibia. *Wilhelm Roux's Arch.* **165**, 303–41.
Bohn, H. (1974) Extent and properties of the regeneration field in the larval legs of cockroaches (*Leucophaea maderae*). I. Extirpation experiments. *J. Embryol. exp. Morph.* **31**, 557–72.
Driever, W. and Nüsslein-Volhard, C. (1988a) A gradient of *bicoid* protein in *Drosophila* embryos. *Cell* **54**, 83–93.
Driever, W. and Nüsslein-Volhard, C. (1988b) The *bicoid* protein determines position in the *Drosophila* embryo in a concentration dependent manner. *Cell* **54**, 95–104.
Edgar, B. A., Odell, G. M. and Schubiger, G. (1989) A genetic switch, based on negative regulation, sharpens stripes in *Drosophila* embryos. *Devl Genetics* (in press).
Fjose, A., McGinnis, W. J. and Gehring, W. J. (1985) Isolation of a homoeo box-containing gene from the engrailed region of *Drosophila* and the spatial distribution of its transcripts. *Nature* **313**, 284–9.

Frasch, M. and Levine, M. (1987) Complementary patterns of *even-skipped* and *fushi tarazu* expression involve their differential regulation by a common set of segmentation genes in *Drosophila*. *Genes Develop.* **1**, 981–95.

French, V., Bryant, P. J. and Bryant, S. V. (1976) Pattern regulation in epimorphic fields. *Science, N.Y.* **193**, 969–81.

Garcia-Bellido, A., Ripoll, P. and Morata, G. (1973) Developmental compartmentalization of the wing disk of *Drosophila*. *Nature, New Biol.* **245**, 251–3.

Gaul, U. and Jäckle, H. (1987) Pole region-dependent repression of the *Drosophila* gap gene *Krüppel* by maternal gene products. *Cell* **51**, 549–55.

Gierer, A. (1981) Generation of biological patterns and form: Some physical mathematical, and logical aspects. *Prog. Biophys. molec. Biol.* **37**, 1–47.

Gierer, A. and Meinhardt, H. (1972) A theory of biological pattern formation. *Kybernetik* **12**, 30–9.

Ingham, P. and Gergen, P. (1988) Interactions between the pair-rule genes *runt, hairy, even-skipped* and *fushi tarazu* and the establishment of the periodic pattern in the *Drosophila* embryo. *Development* (Suppl.), **104**, 51–60.

Jäckle, H., Tautz, D., Schuh, R., Seifert, E. and Lehmann, R. (1986) Cross-regulatory interactions among the gap genes of *Drosophila*. *Nature* **324**, 668–70.

Klinger, M., Erdelyi, M., Szabad, J. and Nüsslein-Volhard, C. (1988) Function of *torso* in determining the terminal anlagen of the *Drosophila* embryo. *Nature* **335**, 275–7.

Knipple, D. C., Seifert, E., Rosenberg, U. B., Preiss, A. and Jäckle, H. (1985) Spatial and temporal pattern of *Krüppel* gene expression in early *Drosophila* development. *Nature* **317**, 40–4.

Kornberg, T. I., Siden, I., O'Farrell, P. and Simon, M. (1985) The *engrailed* locus of *Drosophila*: *In-situ* hybridisation of transcripts reveals compartment-specific expression. *Cell* **40**, 45–53.

Lawrence, P. A. and Morata, G. (1976) Compartments in the wing of *Drosophila*: A study of the *engrailed* gene. *Devl Biol.* **50**, 321–37.

Lehmann, R. and Nüsslein-Volhard, C. (1986) Abdominal segmentation, pole cell formation, and embryonic polarity require the localized activity of *oskar*, a maternal gene in *Drosophila*. *Cell* **47**, 141–52.

Martinez-Arias, A., Baker, N. E. and Ingham, P. W. (1988) Role of segment polarity genes in the definition and maintenance of cell states in the *Drosophila* embryo. *Development* **103**, 157–70.

Meinhardt, H. (1977) A model of pattern formation in insect embryogenesis. *J. Cell Sci.* **23**, 117–39.

Meinhardt, H. (1978) Space-dependent cell determination under the control of a morphogen gradient. *J. theor. Biol.* **74**, 307–21.

Meinhardt, H. (1982) *Models of Biological Pattern Formation*. Academic Press, London.

Meinhardt, H. (1983a) A boundary model for pattern formation in vertebrate limbs. *J. Embryol exp. Morph.* **76**, 115–37.

Meinhardt, H. (1983b) Cell determination boundaries as organizing regions for secondary embryonic fields. *Devl Biol.* **96**, 375–85.

Meinhardt, H. (1984) Models for positional signalling, the threefold subdivision of segments and the pigmentation pattern of molluscs. *J. Embryol. exp. Morph.* **83** (Suppl.) 289–311.

Meinhardt, H. (1985) Mechanisms of pattern formation during development of higher organisms: A hierarchical solution of a complex problem. *Ber. Bunsenges. Phys. Chem.* **89**, 691–9.

Meinhardt, H. (1986) Hierarchical inductions of cell states: a model for segmentation in *Drosophila*. *J. Cell Sci.* (Suppl.) **4**, 357–81.

Meinhardt, H. (1988) Models for maternally supplied positional information and the activation of segmentation genes in *Drosophila* embryogenesis. *Development* **104** (Suppl.), 95–110.

Meinhardt, H. and Gierer, A. (1980) Generation and regeneration of sequences of structures during morphogenesis. *J. theor. Biol.* **85**, 429–50.

Morgan, T. H. (1904) An attempt to analyse the phenomena of polarity in tubularia. *J. exp. Zool.* **1**, 587–91.

Nüsslein-Volhard, C. (1977) Genetic analysis of pattern formation in the embryo of *Drosophila melanogaster*. *Wilhelm Roux's Arch.* **183**, 249–68.

Nüsslein-Volhard, C., Frohnhöfer, H. G. and Lehmann, R. (1987) Determination of anteroposterior polarity in *Drosophila*. *Science, N.Y.* **238**, 1675–81.

Nüsslein-Volhard, C. and Wieschaus, E. (1980) Mutations affecting segment number and polarity in *Drosophila*. *Nature* **287**, 795–801.

Ptashne, M., Jeffrey, A., Johnson, A. D., Maurer, R., Meyer, B. J., Pabo, C. O., Roberts, T. M. and Sauer, R. T. (1980) How the lambda repressor and Cro work. *Cell* **19**, 1–11.

Sander, K. (1976) Specification of the basic body pattern in insect embryogenesis. *Adv. Ins. Physiol.* **12**, 125–238.

Schubiger, G. (1968) Anlageplan, Determinationszustand und Transdeterminationsleistungen der männlichen Vorderbeinscheibe von *Drosophila melanogaster*. *Wilhelm Roux's Arch.* **160**, 9–40.

Schüpbach, T. and Wieschaus, E. (1986) Maternal-effect mutations altering the anterior–posterior pattern of the *Drosophila* embryo. *Roux's Arch. Dev. Biol.* **195**, 302–17.

Segel, L. A. and Jackson, J. L. (1972) Dissipative structure: an explanation and an ecological example. *J. theor. Biol.* **37**, 545–54.

Spemann, H. and Mangold, H. (1924) Über Induktion von Embryonalanlagen durch Implantation artfremder Organisatoren. *Wilhelm Roux's Arch. Entw. mech. Org.* **100**, 599–638.

Steiner, E. (1976) Establishments of compartments in the developing leg imaginal discs of *Drosophila melanogaster*. *Wilhelm Roux's Arch.* **180**, 9–30.

Tautz, D. (1988) Regulation of the *Drosophila* segmentation gene *hunchback* by two maternal morphogenetic centres. *Nature* **332**, 281–4.

Waddington, C. H., Needham, J. and Brachet, J. (1936) Studies on the nature of the amphibian organizing centre. III. The activation of the evocator, *Proc. R. Soc. Lond. B* **120**, 173–90.

Wolpert, L. (1969) Positional information and the spatial pattern of cellular differentiation. *J. theor. Biol.* **25**, 1–47.

Bifurcations, harmonics and the four colour wheel model of *Drosophila* development

BRIAN C. GOODWIN[a] AND STUART A. KAUFFMAN[b]

[a]*Developmental Dynamics Research Group, Department of Biology, The Open University, Milton Keynes, MK7 6AA, UK*
[a]*Department of Biochemistry and Biophysics, University of Pennsylvania, PA 19104–6059, USA*

INTRODUCTION

Our aim in this chapter is to discuss the relationship between the dynamically changing spatial patterns of gene activity and the emergence of phenotypic patterns, both normal and mutant, in early *Drosophila* development. Evidence accumulated over the past decade has identified at least five major sets of genes which regulate two fundamental processes: organization of the embryo into periodic segments, and specification of segment identity. The five major sets of genes are the maternal effect and zygotic genes which play a role in establishing the antero-posterior and dorso-ventral axes; the gap genes which are each necessary for normal development of a contiguous set of segments in the embryo; the pair-rule genes which are required in overlapping sets in double segment-spaced zones; the segment-specific genes which appear to establish segment boundaries and proper polarity within segments; and the homeotic genes which cause transformations of one segment or its adult derivatives to those of another segment (Nüsslein-Volhard and Wieschaus, 1980; Jäckle *et al.*, 1986; Nüsslein-Volhard *et al.*, 1987; Akam, 1987).

We wish to explore two basic questions concerning this system. First, what type of mechanism is involved in generating the observed spatial and temporal pattern of expression of these genes? A detailed analysis of the currently available data leads to a particular class of model that is based upon a bifurcation sequence of periodic functions that define the spatial order of gene expression. Second, granted that some mechanism achieves the

Cell to Cell Signalling: From
Experiments to Theoretical Models
ISBN 0–12–287960–0

213

observed spatial and temporal pattern of expression of the genes in the early embryo, how does the embryo *make use* of those patterns to control periodic segmentation and segment identity? It is proposed that the relevant variables are ratios of phase-shifted gene products over different periodic domains, resulting in what we call the four colour wheel model. We first examine the evidence.

GENE EXPRESSION PATTERNS ARE SPATIAL HARMONICS

In recent years, among the experimental observations that drew attention most forcefully to the possibility of a harmonic sequence of spatial modes in segmentation were those reported by Nüsslein-Volhard and Wieschaus (1980) in their now classic study of the *Drosophila* segmentation genes. Three major consequences emerged from this elegant piece of research.

(1) Segmentation is a result of a hierarchical progression of spatial subdivisions of the embryo, proceeding from global to local order in accordance with the conclusions of classical embryology.

(2) The totally unexpected pair-rule mutants, which revealed a double-segment periodicity prior to the establishment of the final segmented pattern, pointed to spatial frequency doubling as an important aspect of the periodicity-generating process underlying segmentation.

(3) The emergence of mirror-symmetric patterns in mutations affecting all four levels of spatial patterning (maternal, gap, pair-rule, and segment-polarity mutations) suggested that such duplications arise naturally (generically) from perturbations to the system causing morphogenesis.

In this part of the chapter we consider the implications of the first two consequences, amplified by more recent studies on the molecular biology of segmentation.

The use of molecular probes to observe spatial patterns of segmentation and other gene transcripts has now uncovered an unprecedented wealth of detail about the dynamics of the pattern-generating process in *Drosophila*. It is impossible not to be struck by certain characteristics of these patterns. For example, whole-mount studies of embryos at the late syncytial blastoderm stage using fluorescein-labelled antibodies to *fushi-tarazu*, (*ftz*) protein (Carroll and Scott, 1986) revealed the now familiar seven-striped zebra banded pattern. The regularity of the waveform and the curvature of the bands in relation to that of the embryo are precisely the characteristics expected of a standing-wave solution or *eigenfunction* of a Turing (1952) reaction–diffusion or other field equation defined on this geometry.

The only difference between the expected eigenfunction patterns on a full egg shape and the actual pattern of *ftz* is that bands should occur throughout the domain, whereas the embryonic pattern has a large empty domain anteriorly and a smaller one posteriorly. However, Edgar *et al.* (1986) showed that two extra bands of gene transcripts are observed in this anterior domain if protein synthesis is inhibited by cycloheximide 1 h prior to the normal period of *ftz* banding (cycle 14). This suggests that there is a global periodic pattern in the normal embryo, but *ftz* transcription is repressed in the anterior (and presumably the posterior) domain by a local signal. The embryonic pattern of *ftz* thus appears to be the combined result of a global periodicity together with another spatial pattern with an anterior and posterior localization; i.e. the actual pattern looks like a superposition of spatial modes of a morphogenetic field, as suggested by Edgar *et al.* (1986), following the earlier standing-wave harmonic model of Kauffman *et al.* (1978).

Molecular probes for the segmentation gene transcripts have in recent years uncovered the dynamics of the process leading to relatively stable patterns such as the seven pair rule stripes. A beautiful example is the study of the pair-rule genes *even-skipped* (*eve*) and *ftz* by Macdonald *et al.* (1986). Transcripts of each gene first show up in single broad bands during stage 13, *eve* showing a more pronounced antero-posterior gradient than *ftz*. Both show also a dorso-ventral gradient, transcript intensity being initially higher ventrally (Fig. 1).

Towards the end of the 13th cleavage cycle, these broad single bands begin to split in two, then into four during interphase of the 14th cycle, and by the middle of stage 14A both show the distinct seven-stripe zebra pattern of the pair-rule genes, phase-shifted with respect to one another. The posterior band is broad, failing to split. *Eve* later undergoes a further progression to 14 stripes when lines of cells posterior to the initial stripes are activated to produce transcripts. Thus, both *eve* and *ftz* pass through an approximate spatial frequency doubling bifurcation sequence with the exception that four bands progress to seven rather than eight bands, thence to 14 in *eve*. Other genes, such as *paired* (Kilchherr *et al.*, 1986), also develop 14 bands at cellular blastoderm by band-splitting. This process is initiated ventrally.

All of these observations on the spatial and temporal patterns of pair-rule gene expression are extremely suggestive of a harmonic sequence of field solutions defined on a particular geometry, the distorted ellipsoid of the *Drosophila* embryo. What is observed is a sequence in which successive solutions show spatial frequency doubling, the wavelength decreasing by one half at each step.

This sequence of spatial patterning is qualitatively similar to that proposed by Kauffman *et al.* (1978) for the subdivision of the *Drosophila* egg into progressively finer spatial domains. This model made use of harmonic solutions of a morphogenetic field equation (Turing's reaction–diffusion model), doubling the spatial periodicity at each step and superimposing the standing-wave harmonics to get sequential spatial subdivisions into alternatively committed domains.

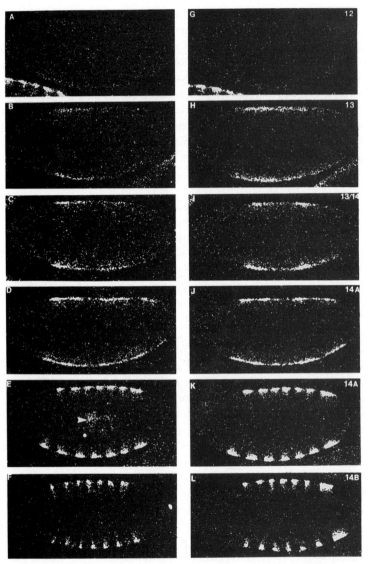

Fig. 1. Harmonic sequences of transcript patterns of *even-skipped* (left) and *fushi-tarazu* (right) in the development of the pair-rule *striping* zebra stripe pattern. (After Macdonald *et al.*, 1986, with permission.)

A large body of detailed mathematical work on the general class of Turing and other field models has accumulated. In particular, recent work has analysed both the expectations of linearized models, and the important consequences of several fully non-linear models. The general implications of these non-linear models is that the particular bifurcation sequences expected on linear theory can be modified, and that the final spatial patterns which arise can be relatively insensitive to perturbations of initial conditions. Thus the class of models is more powerful and robust than initially thought.

Arcuri and Murray (1986) have carried out particularly careful studies of the linear model and non-linear versions. The general feature of any Turing model is that a sequence of standing-wave patterns arise in which more and more peaks and troughs 'fit' into a domain as particular parameters are changed. Two major parameters enter the theory. The first corresponds to the ratio of the diffusion constants of the molecular variables in the system. The second can be taken to be any of the three meanings: strength of reaction terms, absolute size of diffusivities, or linear size of the domain.

The authors showed that for large ratios of the diffusion constants, a sequence of solutions undergoing spatial frequency doubling at each step of the process is indeed possible in solutions of Turing's equations in response to a continuous change in a bifurcation parameter. Stages in such a process are shown in Fig. 2. Thus a typical behaviour of a reaction–diffusion model, for proper parameter values, is to display a period-doubling bifurcation sequence as a parameter is tuned.

Evidence from transcript and genetic data that the gap and segment polarity genes, as well as the zygotic genes of the antero-posterior group such as *caudal*, all progress through similar frequency-doubling bifurcation sequences but at characteristic times will be presented in a more extended analysis by Goodwin and Kauffman (1989). The result is that at every moment of development there is a hierarchy of wave patterns that the embryo can use to specify pattern on progressively smaller spatial frames. Although they pass through similar harmonic sequences, the different genes are functional over restricted sets of spatial domains, many of their transient

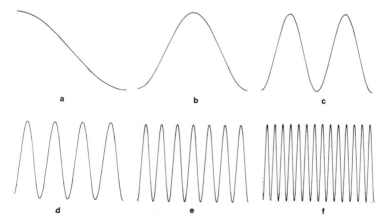

Fig. 2. The sequence of harmonics of the type shown by many segmentation gene transcripts, from a half-wave to fourteen cycles.

harmonics being genetically invisible. The picture that emerges from this analysis is that development can be analysed into a spatially and temporally nested set of field patterns generated by gene products and other variables. These are all coupled together to give a complex, robust system that passes through a set of ordered transients, generating spatial structure on progressively smaller wavelengths as higher harmonics arise. The exact nature of the couplings and the field variables other than gene products are the focus of active research and will be discussed at greater length in the papers of Goodwin and Kauffman (1989) and Kauffman and Goodwin (1989).

Harrison *et al.* (1988) have demonstrated the applicability of a two-stage hierarchical Turing process to pattern formation in *Acetabularia*, while Lacalli *et al.* (1988) have simulated *Drosophila*-type bands using a one-dimensional Turing model. A full three-dimensional simulation of robust pair-rule type banding has been achieved by Hunding *et al.* (1989) using a hierarchial model of dynamically nested Turing systems.

MUTATIONS CREATING LONGITUDINAL DELETIONS AND MIRROR-SYMMETRIC DUPLICATION

The next question we address concerns the way the embryo makes use of these dynamically changing spatial harmonics to generate detailed spatial structure. An important clue to this is provided by the deletions and mirror-symmetric duplications that are characteristic of mutants on different wavelengths.

Consider first the maternal effect gene *bicaudal*. The spectrum of phenotypes seen includes fully symmetric double abdomen embryos, with the axis of mirror symmetry as far anterior as the second abdominal segment. But it is critical to stress that the exact position of mirror symmetry varies from fly to fly, and may be in the third to seventh abdominal segment (Nüsslein-Volhard, 1977; Mohler and Wieschaus, 1986). In addition, the symmetry position may occur at any point within an abdominal segment. Thus, no crisp position is repeatably the locus about which symmetry occurs. Further, the mirror-symmetry axis is typically oblique in the dorsal ventral axis, with more ventral segments than dorsal segments (Gergen *et al.*, 1986). In addition to the symmetric double abdomen phenotype, asymmetric embryos, with fewer anterior mirror duplicate segments than posterior segments are common. A yet weaker phenotype is the headless embryo in which the head and progressively larger portions of the thorax are missing, but with no apparent onset of mirror symmetry. Thus, arranged as a series, progressively more of the head and thorax are missing, and beyond some critical extent of segment *deletion*, mirror-symmetric segment *duplication* arises.

Bicaudal embryos also exhibit an additional striking phenomenon. It is not uncommon to find embryos in which the *dorsal* aspect is a symmetrical bicaudal on the left and right halves of the embryo, with posterior spiracles on left and right halves at the anterior pole. But ventrally, one side is a mirror-symmetric double abdomen phenotype, while the other, perhaps left side, is a *headless* phenotype. This astonishing form means that if one considers a transect from dorsal to ventral along the left side near the anterior pole of such an embryo the segment phenotype will jump discontinuously from posterior abdomen dorsally, to thorax ventrally. This phenotype is found in other insects. For example, van der Meer (1984) induced bicaudal phenotypes in *Callosobruchas* and recovered embryos with a *longitudinal stripe* of bicaudal segmental pattern elements embedded within a normal left or right half embryo. Obviously we must ask what kind of system can readily account for such a strikingly discontinuous pattern in the embryo. Clearly, it is not easy to do so on the basis of a stable monotonic morphogen gradient high at one end of the embryo and low at the other (Meinhardt, 1986). Very steep gradients would have to be postulated across the observed discontinuity. We shall shortly propose that anterior–posterior position is measured by assessing one or more phase angles. As we shall see, this hypothesis naturally leads to the existence of a *phase singularity* (Winfree, 1984). Crossing such a singularity leads to sudden jumps in measured phase even though only small alterations in the underlying 'morphogen' concentrations occur. Thus, our model will easily accommodate such discontinuities.

The same type of phenomenon as that described for *bicaudal* is observed for mutations in the gap genes *Krüppel* and *hunchback*, the pair-rule genes *runt* and *even-skipped*, and the segment polarity genes, but on the different wavelengths that characterize their phenotypic effects. This pattern of deletions and mirror-symmetric duplications thus emerges as a generic property of the process whereby spatial patterns of gene products give rise to the segmental phenotype in *Drosophila*. To see how this arises, consider first the simplified case of pattern specification by the products of two genes within one of the segmentation categories, for example *hairy* and *eve*, belonging to the set of pair-rule genes. At Stage 14 these both have the spatial pattern shown schematically in Fig. 2(e), but they are phase-shifted relative to each other by about 90°. Since each cycle is a repeat of the next, it is sufficient to consider the pattern over one double-segment domain.

Plotting one variable (gene product) against the other over a cycle results in a closed curve, shown as a circle in Fig. 3(a). This is just a polar coordinate description of spatial state, familiar from the model of French *et al.* (1976). It is reasonable to assume that differences of spatial state specified by gene products can be resolved only within finite concentration ranges, so that states are quantized. These are labelled R, O, Y, G, B, V, the primary colours of the rainbow obtained by quantizing the spectrum. This colour chart or wheel specifies the discrete states which are assumed by tissue in different regions of the double-segment domain defined by the pair-rule genes: it defines a map of tissue specificity space (TSS), as described by Winfree (1984). At the centre of this wheel is a point where all the colours meet, giving white or no colour and corresponding to a state which cannot be interpreted. This defines a singularity which, because of the quantization assumption, has a finite extent (the hatched region). Mutations alter the shape of the curve that normally surrounds this singularity. Fig. 3(b) presents an example in which a mutation results in a distorted curve of gene products which passes through the singularity, with the result that two pattern elements (G and B) are deleted, the sequence being V-R-O-Y instead of V-R-O-Y-G-B (assuming V as the start). A stronger mutation can result in a further distortion of the curve so that it lies entirely to one side of the singularity (Fig. 3(c)). In this case there is a mirror-image duplication of the pattern, the sequence being R-O-Y-O-R.

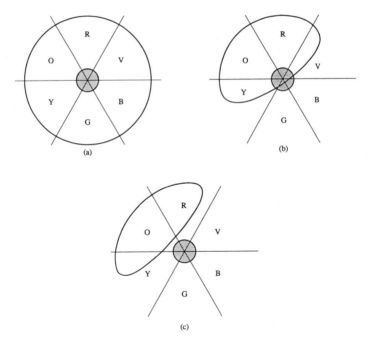

Fig. 3.   Distortions of the colour wheel in 2-space showing the normal pattern (a), deletions when the wheel transects the singularity domain a dead-zone (b), and mirror-symmetric duplication when the distortion carries the wheel across the singularity (c).

If this description is extended to the case of three gene products within a category, phase-shifted relative to one another and defining spatial state, then the analysis must be conducted in three dimensions. For any genotype there will be a closed curve in 3-space which defines state in TSS over a spatial domain corresponding to one cycle of gene products. Quantization of this TSS results in a finite set of discriminable states (shown again as 6 colours) and there is again a singularity where all the colours come together to produce an undefined state. However, in this case the singularity is a line rather than a point (Fig. 4(a)) and the quantized states are like the sections of

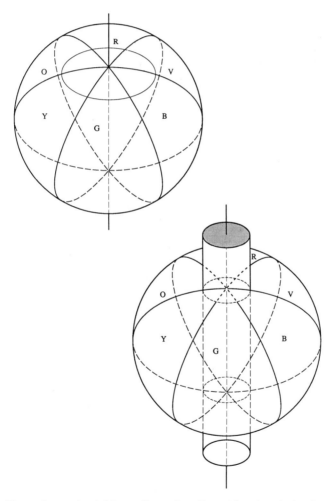

Fig. 4. The colour wheel (three-dimensional) and its singularity (one-dimensional) in a quantized tissue specificity space with six states (R=red, O=orange, Y=yellow, G=green, B=blue, V=violet) for both on line and a domain singularity.

an orange. Extending the singularity to a finite domain results in a cylinder of undefined states (Fig. 4(b)). For N genes in a particular category, the description requires N dimensions and the dimensionality of the singularity is N-2 (Winfree, 1984). Each of the four major categories of segmentation genes requires such a description, resulting in four colour wheels each of which defines spatial state over domains that extend from the whole A–P axis to the single segment.

PATTERN ELEMENT DELETION AND MIRROR DUPLICATION ARE GENERIC

The analysis given in Fig. 3 showing how deletions and mirror duplications come about as a result of distortions of the colour wheels by mutations can now be generalized. Over each of the pattern domains of the segmentation genes, the model yields four critical results.

1. Pattern element deletions arise when the spatial wave forms are modifed such that part of the cycle passes through the dead zone.
2. Mirror symmetric duplications occur when the wave pattern of the gene products are modified such that the entire cycle falls on one side of the dead zone. This does not require cell death and intercalary regeneration.
3. Continuous deformation of the normal cycle in TSS from that which surrounds the dead zone to that which does not surround it leads *first* to pattern element deletion and *next* to mirror symmetric pattern duplication.
4. There will be variability of the mirror symmetry because there is no constraint on the exact manner in which the deformed cycle crosses the singularity.

These are all topological consequences of the basic assumptions about the way in which spatial state is specified by periodic wave-forms, as noted by Kauffman (1984) and Russell (1985) in two-dimensional analyses. They are precisely the properties observed in pattern element deletions and duplications on all length scales.

Since the basic phenomena of deletion of pattern elements and duplication of pattern elements occurs at all length scales along the egg, and exhibits fundamentally similar features, we propose that position along the entire anterior posterior axis of the embryo is specified *simultaneously* by measuring phase angle, or 'colour' on at least four separate colour wheels. The first colour wheel, representing the longest wavelengths, gives crude overall information about position along the entire AP axis. The crudeness is a necessary consequence of supposing that cells can only measure differences in concentration beyond some minimal range, hence that any colour wheel must be quantized into minimum discriminable sectors. Maternal genes such as *bicaudal* (Nüsslein-Volhard, 1977) and zygotic plus maternal genes such as *caudal* (Macdonald and Struhl, 1986) presumably are the constituents of the first colour wheel. The second colour wheel represents information from the gap genes such as *Krüppel, hunchback, knirps, giant*, and perhaps *tailless*. This second colour wheel passes through about two cycles along the full egg length, hence refines the positional precision yielded by the first colour wheel. The third colour wheel is based on the phase offset patterns of the pair-rule genes, and the fourth colour wheel is derived from the segment polarity genes. Jointly, the information tells any cell in the AP axis what pattern element to make in the larval cuticle. We note that this formal model is, in effect, a more precise statement of Gergen *et al*'s (1986) concept that in each cell position is specified by measuring ratios of segment polarity, pair-rule and perhaps gap-gene products.

This four colour wheel model, therefore, specifies the positional identity of each cell *combinatorially*, not merely in the sense that each phase angle on each colour wheel is itself a ratio combination of many genes, but in the further sense that position is specified simultaneously on at least four colour wheels. A further issue is critical. Each wheel must have a phase singularity threading through it, hence a dead-zone. Thus, it is natural to postulate that if the morphogen cycle on *any* of the four colour wheels lies in the dead-zone, that cell will make no pattern element. Therefore, the four colour wheel model asserts that to make a pattern element, a cell must have a colour from each of the four wheels. Naturally, the model predicts that a deformation of the cycle of morphogen values on one colour wheel such that a cell has the 'wrong' colour from that wheel but correct colours from the remaining wheels will lead to a homeotic-like transformation which alters the pattern made to that appropriate for a different longitudinal region of the embryo. Such transformations are precisely what are observed, although not usually thought of as homeotic, in the duplication phenotypes at all length scales which we have described.

Gergen *et al.* (1986) have noted that these features appear difficult to account for on the present common hypothesis that developmental commitments in *Drosophila* are made on a compartmental basis. The problem is that a compartmental polyclone (Garcia-Bellido, 1975; Lawrence and Morata, 1976, 1977) is thought to be an *equivalence* class of cells with an identical developmental commitment. Yet mirror symmetry boundaries do not seem to respect compartmental boundaries. On the other hand, the data demonstrating compartmental phenomena are sound and the existence of some form of heritable determined state in imaginal tissues is absolutely clear (Gehring, 1966). The tension between these two sets of phenomena must be resolved, and we return to it below.

A 'WAVE PARTICLE DUALITY' ARISING FROM JOINT PARTIAL AND ORDINARY DIFFERENTIAL EQUATIONS GOVERNING PATTERN SPECIFICATION

Control of pattern specification and developmental commitments must eventually be viewed in terms of dynamical systems. The complexity of transcribed RNA in *Drosophila* indicates that on the order of 15,000 distinct sequences are generated. These transcripts, their translation products and further metabolic products, together with cis acting regulatory loci at the DNA and perhaps other levels, comprise the genetic regulatory system within each cell. Many of the molecular variables in the regulatory system are confined within the cell that creates them. A subset of macromolecular and small metabolic variables pass between cells. In addition, physical properties such as strain fields couple neighbouring cells.

The natural description of processes which involve transport of molecules or forces between cells is in terms of partial differential equations. Those processes which are confined to a network of interactions within one cell may often be approximated as ordinary differential equations. It may clarify language to think of the variables which pass between cells as the 'positional variables' in the entire coupled system, while those which are confined to act within one cell as the 'interpretive' mechanism. In a loose and inaccurate analogy to the wave–particle duality, the positional variables are the 'waves', while the attractors of the ordinary differential equation system in each cell are the 'particles'.

Within this general framework, what is a map? And what is interpretation? Imagine a tissue which, like imaginal disks, grows from a small initial anlagen. If the positional variables reliably assume some spatially heterogeneous but stable distribution, we might call that the reliable establishment of positional information in the tissue. Imagine, in such a context, that each cell utilizes the values of the positional variables within that cell coupled to the remaining components of the network, and flows from an initial state to a single stable attractor. Then each cell has 'interpreted' the positional information in a unique way and an ordered tissue arises with each cell in a specific state according to the 'map'.

Then is the tension noted by Gergen *et al.* (1986) real or apparent? Apparent, we think. In the picture of coupled partial and ordinary differential equations, the four colour wheel model corresponds to the waves arising from the partial differential equations, while the assessment of colour wheel phase to assume developmental commitments corresponds to the interpretation step arising from the ordinary differential equation system in each cell. In this description, cell state is specified by a combination of values that come from different colour wheels so that individual cells share developmental commitments with particular groups of cells and differ from others. We have noted that any colour wheel can be distorted by mutation so that the cycle in tissue specificity space lies entirely to one side of the singularity region, and that this distortion can occur in any position so that any plane of mirror symmetry is possible. This altered pattern together with the undisturbed colour wheels specifies new boundaries of cell specificities and heritable cell states. Hence there is no incompatibility in having a plane of mirror symmetry that falls anywhere within a normal compartment.

The themes of linked partial and ordinary differential equations in a system with graded positional signals and alternative heritable cellular commitments in the whole embryo seems likely to carry over to later development. For example, the Turing model of Kauffman *et al.* (1978) hypothesized increasingly complex morphogen waveforms on the whole embryo and later on the imaginal disks. We now know that such complex waveforms are real and might control both colour wheels and binary commitments. There is no reason not to suppose that in the later development of the imaginal disks similar colour wheels arise, determine the locations of compartmental boundaries according to the natural wave patterns of the eigenfunctions of the Laplacian (Kauffman *et al.*, 1978), and persist in each disk to constitute positional information within the disk. For example, mirror symmetry induced by *engrailed* mutants in the wing (Lawrence and Morata, 1976) are remarkably similar in this picture to those induced by the segment polarity mutants in general in the larval hypoderm.

We note, finally, that the supposition that similar genetic systems to those which underlie the four colour wheels in the early embryo may arise and generate persistent standing gradient patterns in the imaginal disks during larval life accounts for a long-standing failure of familiar models of epimorphic pattern regulations (e.g. French *et al.*, 1976). All these familiar models explain duplication and regeneration by cell loss and juxtaposition of normally non-adjacent 'positional values', followed by intercalary filling in of missing positional values and hence intercalary regeneration of missing positional elements. A necessary implication of this view is that complementary fragments of a field must regenerate the same structure. If one fragment duplicates, the other must regenerate. No model of positional 'smoothing' by intercalation will allow both fragments of one disk, or one field, to *regenerate* (Goodwin, 1976; Kauffman, 1984; Winfree, 1984). Yet just such observations have been reported, notably by Karpen and Schubiger (1981), as well as Kauffman and Ling (1980). Such regeneration by both fragments of a field is most naturally accounted for by presuming that the field variables which specify position actively recreate missing maxima and minima, not only by diffuse smoothing of discontinuities, but by Turing or other now familiar mechanisms of symmetry breaking and pattern formation. The supposition that active pattern-forming mechanisms set up different waveforms in the early embryo, and similar systems act in the later disks, provides a uniform picture of *Drosophila* pattern formation.

REFERENCES

Arcuri, P. and Murray, J. D. (1986) Pattern sensitivity to boundary and initial conditions in reaction–diffusion models. *J. Math. Biol.* **24**, 141–65.

Akam, M. (1987) The molecular basis for metameric pattern in the *Drosophila* embryo. *Development* **101**, 1–22.

Carroll, S. B. and Scott, M. P. (1986) Zygotically active genes that affect the spatial expression of the *fushi tarazu* segmentation gene during early *Drosophila* embryogenesis. *Cell* **45**, 113–26.

Edgar, B. A., Weir, M. P., Schubiger, G. and Kornberg, T. (1986) Repression and turnover pattern of *fushi tarazu* RNA in the early *Drosophila* embryo. *Cell* **47**, 747–54.

French, V., Bryant, P. J. and Bryant, S. V. (1976) Pattern regulation in epimorphic fields. *Science, N.Y.* **193**, 969–81.

Garcia-Bellido, A. (1975) Genetic control of wing disc development in *Drosophila*. In *Cell Patterning*, Ciba Foundation Symposium No. 29, pp. 161–182. Amsterdam: Elsevier.

Gehring, W. (1966). Übertraging und Änderung der Determinationsequalitäten in Antennenscheiben-Kulturen von *Drosophila melanogaster*. *J. Embryol. Exp. Morphol.* **15**, 77–111.

Gergen, J. P., Coulter, D. and Wieschaus, E. (1986) *Segmental Pattern and Blastoderm Cell Identities. Gametogenesis and the Early Embryo*, pp. 195–220. Alan R. Liss, New York.

Goodwin, B. C. (1976) *Analytical Physiology of Cells and Developing Organisms*. Academic Press, London.

Goodwin, B. C., and Kauffman, S. A. (1989) Spatial harmonics and pattern specification in early *Drosophila* development. Part 1. Bifurcation sequences and gene expression. *J. theoret. Biol.* (submitted).

Harrison, L. G., Graham, K. T. and Lakowski, B. C. (1988) Calcium localization during *Acetabularia* whorl formation: evidence supporting a two-stage hierarchical mechanism. *Development* **104**, 255–62.

Jäckle, H., Tautz, D., Schuh, R., Seifert, E. and Lehmann, R. (1986) Cross-regulatory interactions among the gap genes of *Drosophila*. *Nature* **324**, 668–70.

Karpen, G. H. and Schubiger, G. (1981) Extensive regulatory capabilities of a *Drosophila* imaginal disk blastema. *Nature* **294**, 744–7.

Kauffman, S. A. (1984) Pattern generation and regeneration. In *Pattern Formation* (eds G. M. Malacinki and S. Bryant), pp. 73–101. Macmillan, New York.

Kauffman, S. A., and Goodwin, B. C. (1989) Spatial harmonics and pattern specification in early *Drosophila* development. Part II. The four color wheel model. *J. theoret. Biol.* (submitted).

Kauffman, S. A. and Ling, E. (1980) Regeneration in complementary wing disc fragments of *Drosophila melanogaster*. *Devl Biol.* **82**, 238–57.

Kauffman, S. A., Shymko, R. M. and Trabert, K. (1978) Control of sequential compartment formation in *Drosophila*. *Science, N.Y.* **199**, 259–70.

Kilchherr, F., Baumgartner, S., Bopp, D., Frei, E. and Noll, M. (1986) Isolation of the *paired* gene of *Drosophila* and its spatial expression during early embryogenesis. *Nature* **321**, 493–9.

Lacalli, T. C., Wilkinson, D. A., and Harrison, L. G. (1988). Theoretical aspects of stripe formation in relation to *Drosophila* segmentation. *Development* **104**, 105–114.

Lawrence, P. A. and Morata, B. (1977) The early development of mesothoracic compartments in *Drosophila*: An analysis of cell lineage and fate mapping and an assessment of methods. *Devl Biol.* **56**, 40–51.

Macdonald, P. M. and Struhl, G. (1986) A molecular gradient in early *Drosophila* embryos and its role in specifying the body pattern. *Nature* **324**, 537–44.

Macdonald, P. M., Ingham, P. and Struhl, G. (1986) Isolation, structure, and expression of *even-skipped*: a second pair-rule gene of *Drosophila* containing a hameo box. *Cell* **47**, 721–34.

Meinhardt, H. (1986) Hierarchial inductions of cell states: a model for segmentation in *Drosophila*. *J. Cell Sci.* (Suppl.) **4**, 357–81.

Mohler, J. and Wieschaus, E. F. (1986) Dominant maternal-effect mutations of *Drosophila melanogaster* causing the production of double-abdomen embryos. *Genetics* **1122**, 802–22.

Nüsslein-Volhard, C. (1977) Genetic analysis of pattern formation in the embryo of *Drosophila melanogaster*. Characterisation of the maternal-effect mutant Bicaudal. *Roux's Arch.* **183**, 244–268.

Nüsslein-Volhard, C., and Wieschaus, E. (1980) Mutations affecting segment number and polarity in *Drosophila*. *Nature* **287**, 795–801.

Nüsslein-Volhard, C. Frohnhofer, H. G. and Lehmann, R. (1987) Determination of anteroposterior polarity in *Drosophila*. *Science, N.Y.* **238**, 1676–81.

Russell, M. A. (1985) Positional information in insect segments. *Devl. Biol.* **108**, 269–83.

Turing, A. M. (1952) The chemical basis of morphogenesis. *Phil. Trans. R. Soc. B* **237**, 37–72.

Van der Meer, J. M. (1984) Parameters influencing reversal of segment sequence in posterior egg fragments of *Callosobruchus* (Coleoptera). *Roux's Arch.* **193**, 339–356.

Winfree, A. W. (1984) *The Geometry of Biological Time*. Springer-Verlag, New York.

Turing prepatterns of the second kind simulated on supercomputers in three curvilinear coordinates and time

AXEL HUNDING

H. C. Ørsted Institute, Chemistry Laboratory III, University of Copenhagen, Universitetsparken 5, DK-2100 Ø Copenhagen, Denmark

INTRODUCTION

The spontaneous emergence and autoregulation of highly ordered patterns characterize living systems, but the physical-chemical processes involved are largely unknown. Turing (1952) demonstrated that autocatalytic biochemical control networks coupled to internal diffusion, but without any outside governing, could break up from the initial homogeneous state and form stable, well-defined, inhomogeneous concentration gradients and patterns. These prepatterns are related to the eigenfunctions of the Laplacian operator, and thus determined by the boundary conditions, i.e. the geometry of the system (cell, embryo) in which they arise. Turing's equations take the form

$$c_t = F(c) + D\Delta c \quad \text{in } \Omega \tag{1}$$

where vector c contains the concentrations of the biochemical reactants, $F(c)$ is the vector function determining the (enzyme-regulated) kinetics, and the last term adds Fickian diffusion. The components of c are assumed to be contained within the region Ω in question, and so they satisfy no-flux boundary conditions:

$$\frac{\partial c}{\partial n} = 0 \quad \text{on } \partial\Omega \tag{2}$$

Cell to Cell Signalling: From
Experiments to Theoretical Models
ISBN 0–12–287960–0

where $\partial/\partial n$ denotes the gradient along the outward normal to $\partial\Omega$. A particular example is the two-component biochemical control system suggested by Selkov in 1967:

$$\frac{\partial c_1}{\partial t} = v - \frac{k_1 c_1 c_2{}^\gamma}{1 + K c_2{}^\gamma} + D_1 \Delta c_1 \tag{3}$$

$$\frac{\partial c_2}{\partial t} = \frac{k_1 c_1 c_2{}^\gamma}{1 + K c_2{}^\gamma} - k_2 c_2 + D_2 \Delta c_2 \tag{4}$$

Here, component one is fed (homogeneously) from a source with rate v, and converted to component two with a rate displaying product activation with Hill constant γ greater than 1. Component two is in turn consumed by first-order kinetics.

Thus, the system is thermodynamically open, and a flow of free energy through the system keeps it far from chemical equilibrium. In such systems, symmetry breaking and spontaneous emergence of Turing structures is not in conflict with the second law of thermodynamics (Nicolis and Prigogine, 1977).

Bifurcations to eqn (1) may be sought by expanding c from the homogeneous stationary solution c_0: let $z = c - c_0$, we seek stationary solutions satisfying

$$L(z) + N(z) = 0 \tag{5}$$

where

$$L(z) = Mz + D\Delta z \tag{6}$$

and M is the Jacobian matrix $\partial F/\partial c$ evaluated at c_0; $N(z)$ is the remaining non-linear part from eqn (1), satisfying $N(0) = 0$. The problem

$$L(z) = 0 \tag{7}$$

has the solution $z = 0$ by construction. As eigenfunctions to the Laplacian in Ω satisfy

$$\Delta \varphi_{nml} = - k^2{}_{nml} \, \varphi_{nml} \tag{8}$$

non-trivial solutions to eqn (7) may arise, when

$$|M - k^2{}_{nml} D| = 0 \tag{9}$$

This equation for the determinand specifies where bifurcations may occur to inhomogeneous solutions of geometry φ_{nml}. As M depends on rate constants, eqn (9) defines critical values of rate constants, where an exchange of stability may occur between c_0 and $z = \varphi_{nml}$. Indeed, the matrix $M - k^2_{nml}D$ has an eigenvalue with positive real part, when the bifurcation point is passed, and thus any component along φ_{nml} of small fluctuations to the homogeneous solution c_0 amplify beyond this point. Eventually, a new stable inhomogeneous solution c is formed (see Fig. 1).

This stable solution may be constructed analytically using bifurcation theory.

A result from such bifurcation studies is that the patterns which emerge are largely independent of the chemistry details in eqn (1), but the possible patterns and their change-over into new patterns (secondary bifurcations) are defined by the boundary conditions, and the vanishing or non-vanishing of certain integrals over products of eigenfunctions for the region in question. Thus, it is meaningful to study bifurcations in reaction–diffusion systems with simple model systems like Selkov's (eqns (3)–(4)) as the results obtained should remain valid for a broad class of biochemical networks. This makes numerical studies on such model systems reasonable (Hunding, 1980).

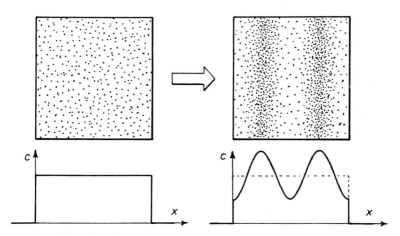

Fig. 1.   Creation of stable inhomogeneous solution.

NUMERICAL STUDY OF PREPATTERN FORMATION

Bifurcations to eqn (1) have been investigated by numerical solution of the coupled system of non-linear PDEs with no flux boundary conditions. The method of lines was used, and the resulting large system of coupled ODEs were solved with a stiff code (Gears), as the discretization of the Laplacian in three-dimensional curvilinear coordinates yields condition numbers in excess of 40,000.

The Jacobian of the ODE-system is a sparse, banded matrix, and thus the corrector step was solved by sparse matrix iteration (successive overrelaxation).

Complete chessboard ordering of meshpoints throughout 3D space was used. Hereby the RBSOR method becomes vectorizable and thus particularly well-suited for implementation on modern supercomputers, as large blocks of the iteration matrix, of dimension 1024 or more, may be processed in parallel. The complete solution of eqn (1) from the homogeneous state to the final stable Turing structure, thus obtains at remarkably sustained rate: 160–195 MFLOPS (CRAY X-MP) or 200–240 MFLOPS (AMDAHL VP 1100).

The development of the stiff code made simulation of 3D Turing structures possible. With the further improvement of the vectorized code, such studies have been made cheap, colloquially 'a Turing structure for a dollar'. Our 3D simulations have thus made it possible to compare theoretically obtained patterns with actual biological experiments.

TURING STRUCTURES AND BIOLOGICAL SELF-CONTROL OF SPATIAL DEVELOPMENT

Prepatterns arising from reaction–diffusion processes in an embryo may govern the activation of certain genes, which would then form a basis for reliable well-controlled cell differentiation (Wolpert and Stein, 1984).

The 3D simulations performed so far have been largely centred upon simple gradients and bipolar patterns within single cells. Indeed, the eigenfunctions φ_{nml} (eqn (8)) for a sphere include the functions

$$\varphi_{2ml} = j_2(\kappa_{21} r/R) \ Y_{2m}(\theta,\varphi) \quad m\varepsilon[-2,2] \tag{10}$$

where κ_{21} is the first extremum of the spherical Bessel function j_2, and Y_{nm} are the spherical harmonics. Analytical bifurcation theory shows that selection rules govern the pattern formation from the homogeneous distribution. Only Y_{20} emerges, the other functions are forbidden in this transition. This is confirmed in our numerical simulations. The pattern $j_2 Y_{20}$ is bipolar, yielding high concentrations at two spontaneously formed poles, and a small concentration ring in the equator region.

Such a pattern should be an ideal platform for spatial organization during cellular division processes (mitosis and cytokinesis), as it organizes the sphere into two identical halves. Consequently it is called the 'mitosis' pattern. Indeed, former theories on mitosis based on centrioles as the prime organizers are now abandoned, as these organelles seem to be unessential for the process; for example, laser destruction of one centriole before separation to the poles has no effect on the formation of the bipolar spindle. Also plant cells have no centrioles.

An indication of the presence of Turing structures in single cells was found by the author in a search for rudiments of a pure prepattern-governed chromosome separation process among primitive cells, a process which could be an early evolutionary state of mitosis. In the radiolarian *Aulacantha scolymantha*, nuclear division takes place by a much more haphazard process than the well-organized mitosis seen in animal cells.

The more than 1000 presumably identical chromosomes are crudely segregated to the two halves of the nucleus. In this process, however, the chromosomes confine themselves to highly symmetrical regions within the nucleus. The reasons for this are hard to explain by conventional spindle forces. It was shown, however (Hunding, 1981), that these highly symmetrical regions were strongly related to the geometry present in the Turing structures, which were found numerically for the dividing nucleus. Notably, a highly symmetrical, saddle-shaped structure, repeatedly photographed in *Aulacantha scolymantha*, is identified with the neutral regions of the pattern $j_1 Y_{10} + j_2 Y_{22}$, which emerges in numerical simulations.

So far, Turing's original equations (eqn (1)) have been discussed. There is some evidence, that one Turing structure could affect the parameters in a second system, thereby giving a Turing system of the second kind:

$$c_t = F(k(\underline{r}), c) + \operatorname{div} D(\underline{r}) \operatorname{grad} c \qquad (11)$$

Here, rate constants k and diffusion constants D are functions of position, $\underline{r}$. An example of this is the determination of cleavage plane orientation in early blastulas. In the egg, an axis known as the animal–vegetal axis is present. Suppose this is defined by a Turing structure, which sets up a simple gradient from one pole to the other, and that this first system defines position-dependent parameters in a second system.

This system was shown (Hunding, 1987) to give rise to bipolar mitosis prepatterns, which eventually are strictly perpendicular to the animal–vegetal axis, in accordance with the experimental finding, that the spindle orientates this way, and the first division plane is thus strictly through the animal–vegetal poles. The same equations account for the division plane orientation in the developing blastula, notably the B2, B4 stage. It is argued that precise spatial governing of division planes relative to a pre-established axis is a phenomenon which must require active governing, and that bifurcations in Turing systems of the second kind may provide such precise spatial orientation. This is then an important argument in favour of the idea that Turing structures are active in cellular development. The establishing of the upright $j_2 Y_{20}$ bipolar prepattern, which turns 90° to a new stable orientation is shown in Fig. 2.

Turing systems of the second kind may also explain several features of morphogenesis in insect embryos, notably the set-up of stripe prepatterns in *Drosophila* (Slack, 1984; Ingham, 1988). The *hunchback* and *Krüppel* genes

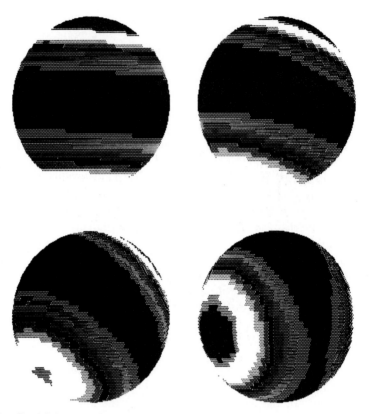

Fig. 2.   Establishment of the upright $j_2 Y_{20}$ bipolar prepattern, which turns 90° to a new stable orientation.

are read out in zones which suggest the presence of a biperiod prepattern. Such a pattern is demonstrably easily formed based on Turing's mechanism.

The biperiod prepattern, possibly in conjunction with double period patterns linked to the above so-called gap genes, may then create a Turing system of the second kind which yields directly a prepattern with seven stable stripes in the middle of the embryo. The transition to this state has several features in common with observed prepatterns in the embryo. A 3D-simulation is shown in Fig. 3. Thus, stabilization of stripes (and size regulation as well) (Hunding and Sorensen, 1988) obtains in Turing systems of the second kind.

In conclusion, Turing structures may participate in governing a number of biological events, from cell division to embryo organization. With current hardware, and software as described, the important phenomena of spatial self-organization in biosystems may now be studied quantitatively.

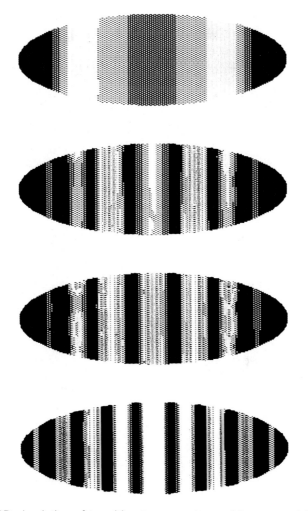

Fig. 3.   3D-simulation of transition to a prepattern with seven stable stripes.

REFERENCES

Hunding, A. (1980) Dissipative structures in reaction–diffusion systems: Numerical determination of bifurcations in the sphere. *J. Chem. Phys.* **72**, 5241–8.

Hunding, A. (1981) Possible prepatterns governing mitosis. *J. theor. Biol.* **89**, 353–85.

Hunding, A. (1982) Spontaneous biological pattern formation in the three-dimensional sphere. Prepatterns in mitosis and cytokinesis. In *Evolution of Order and Chaos in Physics, Chemistry and Biology* (ed. H. Haken), Vol. 17, pp. 100–11. Springer series in Synergetics.

Hunding, A. (1987) Bifurcations in Turing systems of the second kind may explain blastula cleavage plane orientation. *J. Math. Biol.* **25**, 109–21.

Hunding, A. and Sorensen, P. G. (1988) Size adaptation of Turing prepatterns. *J. Math. Biol.* **26**, 8827–39.

Ingham, P. W. (1988) The molecular genetics of embryonic pattern formation in *Drosophila. Nature* **335**, 25–34.

Nicolis, G. and Prigogine, I. (1977) *Self-organization in Nonequilibrium Systems.* Wiley, New York.

Slack, J. (1984) A Rosetta stone for pattern formation in animals? *Nature* **310**, 364–5.

Turing, A. M. (1952) The chemical basis of morphogenesis. *Phil. Trans. R. Soc. Lond.* B **237**, 37–72.

Wolpert, L. and Stein, W. D. (1984) Positional information and pattern formation. In *Pattern Formation* (eds G. M. Malacinski and S. V. Bryant). Macmillan, London.

Modelling plant growth and architecture

PH. DE REFFYE,[a] R. LECOUSTRE,[a] C. EDELIN[b] AND P. DINOUARD[a]

[a]*Laboratoire de Modélisation, CIRAD, Montpellier, France*
[b]*Labaratoire de Botanique Tropicale, Université des Sciences et Techniques du Languedoc, Montpellier, France*

The purpose of the present work is to present a model for plant development, based on measured characteristics of particular trees. When implemented on a computer and studied with the help of advanced graphical methods, this model provides a faithful representation of the growth and architecture of a large variety of plants.

THE BASIC FEATURES OF PLANT ARCHITECTURE

The growth of a plant is the result of the evolution of meristems, which are specific cellular tissues within the bud. At any given time, a bud can either die (abort), and then does not produce anything, or give birth to a flower, to an inflorescence (and then the bud dies), or to an internode.

The leaf axis is the basic structure of the plant architecture. It is the result of the activity of the bud situated at its tip, called the apical bud, and is made of a series of internodes. An internode is a part of a woody stem at the tip of which one can find one or several leaves. Internodes can be quite short (i.e. 1 mm), as is the case for the fir tree, or they can measure up to several tens of centimetres, as is the case for the bamboo. Between two internodes there is a node bearing leaves and buds; each node bears at least one leaf (the symptom of existence of the node). At the base of each leaf one finds an axillary, or lateral bud (see Fig. 1).

An axillary stem can either grow immediately (sylleptic ramification) or after some delay (proleptic ramification). The growth of the leaf axes results from the development of the apical buds. The growth unit is a central notion

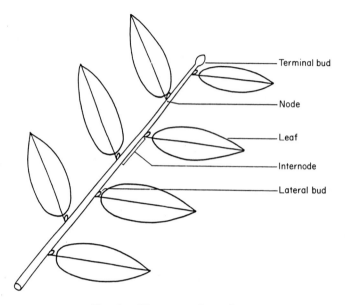

Fig. 1.   The vegetative axis.

for plant architecture: it consists of a sequence of internodes and nodes produced, usually in a short period of time, by the apical bud of the previous node. We shall distinguish between two cases: short growth units with few internodes (sometimes only one), and long growth units made of numerous internodes, each one usually being short.

Another important notion is the order of an axis. The order 1 axis of a plant is a sequence of growth units, each born from the apical bud of the previous unit, and the first one having grown out from the seed of the plant (Fig. 2). An order i axis ($i > 1$) is a sequence of growth units, the first internode being born from an axillary bud on an order $i-1$ axis called the bearing axis.

In the absence of traumas, for each variety of each species and each order, known functions control the relative positions of lateral buds – and consequently of the leaves – of any node with respect to the lateral buds of the previous one; this phenomenon, called phyllotaxy, can be of spiralled or distic nature.

Within a growth unit the ramification process can be:

(a) continuous, when each node of an axis is the starting point of an axis of higher order;
(b) rhythmic, when some nodes (but not all of them) are the starting point of an axis of higher order; or
(c) diffuse, when the nodes giving birth to an axis of higher order are located at random.

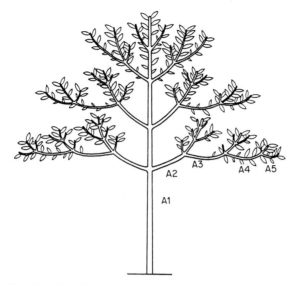

Fig. 2. Numbering of successive axes, A_i, of order i.

These types of ramifications depend on the order of the axis for a given variety and species. A monopod is a ramified system that includes a single number 1 axis and a finite number of axes of higher order; if there are k orders, such plant is called an order k monopod. A sympod is a system built by many replications of an order 1 or i axis module.

Another important parameter is the geometrical aspect of an axis with respect to the bearing axis. When the trend of the bearing axis is close to horizontal, its development is said to be plagiotropic; when vertical, it is said to be orthotropic. An orthotropic axis is generally associated with a spiralled phyllotaxy whereas plagiotropy is associated with distic phyllotaxy.

All these functional laws give rise to a relatively small number of plant architecture models (Hallé and Oldemann, 1970; Edelin, 1977, 1984; Hallé *et al.*, 1978; Hallé, 1979), some of which are shown in Fig. 3. Each architectural type is named after the author who first described it.

IMPLEMENTING THE MODEL

The development of a plant is the combination of three fundamental processes:

(1) *Growth*: an axis grows longer when its apical meristem forms and growth units develop.
(2) *Mortality*: axes die and come apart when meristems cease to operate.
(3) *Branching out*: lateral axes develop.

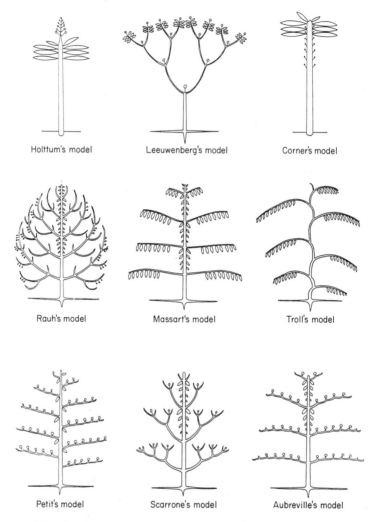

Fig. 3.   Some architectural models of plant development.

Our model for plant development is based on quantitative field observations carried out on homogeneous plant populations (clones, varieties). As to the modelling hypotheses, we assume the existence of an internal clock within each meristem, that combines probabilities of growth, mortality and branching at each time. All clocks are synchronized between themselves; rhythm ratios among clocks are invariable, and so are the probabilities of the various events. Plant architecture is thus the result of the succession of its periods of activity, independently of resting or slowing-down phases.

Although the growth of a plant depends on the interaction of genetic and

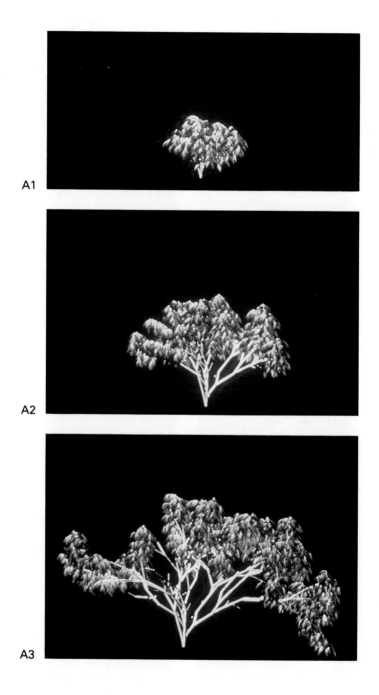

A1

A2

A3

Fig. 5. Graphical representations of some trees, obtained with the AMAP program for plant development described in the text. A: three growth stages of the lychee tree, B: cecropia, C: rubber tree, D: bamboo, E: cotton tree.

B

C

D

E

environmental factors, the modelling of its architecture from experimental data through stochastic processes is based on quite simple principles, whose consequences can easily be checked. Most of the well-known probability functions (binomial, Poisson, exponential, etc.) are indeed generated from relatively simple hypotheses. For instance, Poisson's distribution is reached by observing how the random occurrence of an event is realized within a given period of time, when the probability of occurrence is constant. In view of the stability of events during the linear stage of plant growth, this kind of function arises naturally, regardless of the complexity of the underlying cellular mechanisms.

THE MODELLING OF GROWTH

Only the instantaneous result of growth can be observed: branches are the trajectories of meristems, whose evolution can be retraced thanks to faithful pointers such as leaf scars and axial branches. Vegetative axes are formed by a succession of internodes resulting from the working of the apical meristems; such is the basis of our modelling.

Observing identical axes of the same age, one notes that the number of internodes varies according to a characteristic pattern; the latter changes with time, owing to the fact that the lengthening of internodes may cease to occur everywhere in a synchronized fashion. Each meristem is regularly submitted to 'growth tests', the result of which is either the success or the failure of the lengthening event. Probability laws may be used to model this process. The total amount of growth tests undergone by the meristem is called the dimension of the branch; the actual size of the branch, measured by the number of internodes, is thus equal to its dimension, or lower.

The growth of any population of identical vegetative axes is described by a simple random process in which a probability b_i of internode lengthening is associated with each growth test i. For each step of the population growth, there is a mean and a standard deviation of the instantaneous distribution of the amount of internodes. The relationship between mean and standard deviation therefore characterizes the growth process.

In the particular case of linear growth, the lengthening probability is constant, the relationship between mean and standard deviation is linear, and the slope yields a measure of the probability of meristem lengthening.

A tree top is formed by a primary axis and several secondary axes. The analysis of a mixture of homogeneous tops, proceeding from the plant summit downwards, provides a way to measure directly the various parameters that characterize the development of a particular plant. For T tops having followed the same growth patterns made of rank 1 and rank 2 branches, if lengthening probability P of rank 1 branches is constant, rank 2 branches have a lengthening probability b that is also constant. If rank 2 meristems undergo wN tests when rank 1 meristems undergo N tests, w represents the rhythm ratio parameter.

The tops can be described by the three parameters, P, b and w, and by the variable N. To determine P, b and w, three relations are necessary; these are obtained by watching on the plant all branches that start at the same distance of K internodes from the top, and measuring in internodes the mean and standard deviation of the size of these branches.

APPLICATION OF THE MODEL TO PARTICULAR PLANTS

Meristem lengthening

The growth of a variety of cotton tree was studied by Cognee and de Reffye (see de Reffye, 1976, 1979; de Reffye *et al.*, 1986). The parameters associated with the growth probability laws for this tree were determined by the analysis of the relationship mean/standard deviation, of the histogram of the distribution of internodes per stem at the end of growth, and by the analysis of the tops. The three methods give an estimate of P of about 0.8, with a total amount of 25 lengthening tests. The tops method yields the values $b = 0.88$ and $w = 0.27$.

The same methods have been used to calculate accurately the growth parameters for plants such as the lychee tree, cecropia, goupia and bamboo. For all these plants, lengthening probabilities remain constant; they vary from 0.54 (lychee) to 0.98 (cecropia). Rhythm ratios range from 0.27 (cotton tree) to 2 (goupia). In the case of the coffee tree, axillary buds possess a variable lengthening probability that depends on the amount of growth tests undergone by the meristem, i.e. its dimension.

Meristem mortality

The growth of vegetative axes may end because of a random accident, in which case a survival probability C can be allocated to each growth unit. The accident may also lead to the end of any growth, i.e. to the stop of the apical meristem inner clock: growth is then terminated and the axis eventually comes apart. An axis dimension therefore depends at the same time on the amount of lengthening tests, on the lengthening probability, and on the meristem viability at any developmental stage.

For the cotton tree, there is no outstanding growth accident, and all the stems stop growing at the end of the 25th lengthening test. For the coffee tree, meristems die before having reached their maximal dimension. Meristem viability for the lychee tree is constant.

Meristem branching

Axillary meristems develop either immediately (sylleptic pattern), or after a delay (proleptic pattern). We will only consider the former case. Generally, the branching probabilities of axillary buds of the same rank are not independent. In the case of the coffee tree, for instance, there are two axillary meristems per node; they give birth to 0, 1 or 2 branches. The distribution is not binomial, and a conditional probability coupling has to be considered to account for the observed distribution.

Let a be the branching probability of secondary axes and r the conditional probability for a second axillary bud to develop once a first one has done so. At any node on a coffee tree trunk the branching coupling probability between axillary meristems is found to be as high as 95%. Branching coupling is also found between successive nodes. In that way, branched out and barren areas are both found on vegetative axes.

In the case of the cecropia, a tree from Guyana, the layers of branches are distributed along the 400 plus internodes of the adult trunk. The probability of development of a layer is evaluated by dividing the amount of layers by the amount of available nodes. This probability can also be estimated from the amount of internodes between two successive layers, which varies according to a Pascal's function. Both estimates of a yield a value of about 0.05. The amount of axillary branches also follows a Pascal's distribution. If there is about a 5% chance for a branch developing from a trunk node, through coupling there is a 70% chance that the following node will produce a branch. Similar coupling coefficients have been found for the coffee tree and the lychee.

COMPUTER SIMULATION OF TREE DEVELOPMENT

On the basis of this model, we have developed a computer program, named AMAP, that can simulate and picture any plant architecture (de Reffye *et al.*, 1986). This application possesses two complementary aspects, namely, proper simulation and computer graphics.

Simulating plant architecture

The basic idea is to view the plant as the result of the activity of the meristems that have built it. Simulating plant growth thus reduces to simulating the activity of its meristems. Any viable meristem gives birth to a sequence of internodes, divided by nodes. The latter represent the basic elements in the simulation; their evolution is determined as a function of the three basic processes of growth, ramification and death.

An algorithm, called growth engine, governs the evolution of each node by means of a Monte Carlo method based on the probabilities determined by specific botanical laws fitted to statistical functions. Each test results in the occurrence of one of the three basic processes. To illustrate the results obtained with the model, two examples of random simulations of growth are shown in Fig. 4 for the lychee tree.

The order in which the nodes are processed is important in that it influences the outcome of the simulations and bears on the type of application of the program. Thus a first growth engine has initially been developed, which manages the node processing order on the basis 'first in, first out'. Such a simulation, which follows a prefixed order, greatly simplifies the management of the nodes on the computer. Although a plant built this way is botanically correct at the end of the simulation, its growth does not follow the actual pattern in time. Besides, this approach fails to provide any knowledge of the spatial environment of each node in the course of time.

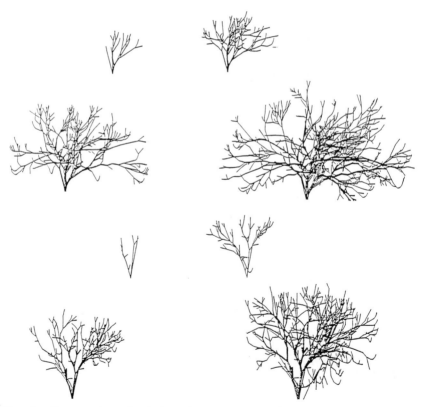

Fig. 4. Two random simulations for the growth of the lychee tree. Only the trajectories of the meristems are indicated.

This is why a new engine growth has been developed in a second stage. Here, the nodes are managed concurrently thanks to a list of events to come, that is continuously updated by the computer. As the engine follows an unfixed order, it is possible to obtain at any time during the simulation a complete topological knowledge of the whole plant. One can also process possible collisions with external obstacles, take into account the effect of shade on growth, systematically seek solar light, or simulate the competition between several plants. The parallel engine growth thus enables us to take into account the physical environment of the plant.

Graphical simulation of plant development

During the simulation, when completing a growth test, a node gives birth to an internode or to a vegetative organ (leaf, flower, etc.) whose geometrical characteristics can be determined so that a complete description of the plant is obtained. The nature of the component (internode, leaf, flower, etc.), its branching rank, its size, its position in space can all be taken into account for the graphical description, and faithful pictures are produced once each type of component is given a shape defined by graphical primitives specific of the plant (de Reffye *et al.*, 1986; Jaeger, 1987). Some examples of the pictures generated by the program are shown in Fig. 5.

CONCLUDING REMARKS

The present approach to plant architecture combines knowledge gained in botany, applied mathematics and computer graphics. It relies on a strong experimental basis, as it incorporates the data gathered by botanists and agronomists for each particular plant. The mathematical and botanical approaches complement each other, the mathematical model quantifying the botanical one. The advent of high-resolution computer graphics allows the production of a faithful, three-dimensional representation of the calculated plant.

Besides its interest for the process of plant development itself, the program foreshadows multiple applications in agronomy and forestry: interactions between plants and their environment (effect of fertilizers, pruning, density, etc.), calculation of the radiation transfer of any plant canopy, evaluation of damage by plant pests. The faithfulness of the simulated plants also allows consideration of applications of the model to the landscaping of urban environment.

REFERENCES

Edelin, C. (1977) Images de l'architecture des conifères. Thèse de l'Université des Sciences et Techniques du Languedoc, Académie de Montpellier.

Edelin, C. (1984) L'architecture monopodiale: l'exemple de quelques arbres d'Asie tropicale. Thèse de Doctorat d'Etat, Université des Sciences et Techniques du Languedoc, Académie de Montpellier.

Hallé, F. (1979) Les modèles architecturaux chez les arbres tropicaux: une approche graphique. In *Elaboration et Justification des Modèles. Applications en Biologie*, Vol. 2, pp. 537–50. Maloine, Paris.

Hallé, F. and Oldemann, R. A. A. (1970) *Essai sur l'Architecture et la Dynamique de Croissance des Arbres Tropicaux*. Masson, Paris.

Hallé, F., Oldemann, R. A. A. and Tomlinson, P. B. (1978) *Tropical Trees and Forests: An Architectural Analysis*. Springer-Verlag, Berlin.

Jaeger, M. (1987) Représentation et simulation de croissance des végétaux. Thèse de Doctorat es Sciences n° 328, Université Louis Pasteur, Strasbourg.

de Reffye, Ph. (1976) Modélisation et simulation de la verse du Caféier à l'aide de la résistance des matériaux. *Café-Cacao-Thé* **XX**, No. 4.

de Reffye, Ph. (1979) Modélisation de l'architecture des arbres tropicaux par des processus stochastiques. Thèse de Doctorat es Sciences n° 2193, Université de Paris-Sud.

de Reffye, Ph., Edelin, C., Jaeger, M. and Cabart, C. (1986) Simulation de l'architecture des végétaux. Actes du Colloque International "L'Arbre", Montpellier, 1985. *Naturalia Monspeliensia*, Special Issue pp. 223–45.

Part 3
Cell to cell signals in immunology

The broken idiotypic mirror hypothesis

J. URBAIN, F. ANDRIS, M. BRAIT, D. DE WIT,
M. KAUFMAN,[a] F. MERTENS AND F. WILLEMS

*Département de Biologie Moléculaire, Université Libre de
Bruxelles, Rue des Chevaux 67, B-1640 Rhode St Genèse,
Brussels, Belgium*
*[a]Service de Chimie Physique, Université Libre de Bruxelles,
Campus Plaine U.L.B., C.P.231, Boulevard du Triomphe, B-1050
Brussels, Belgium*

INTRODUCTION

Before a general theory of immunoregulation can be put forward, the
research worker is faced with at least three major problems to solve. The first
is understanding the mechanisms allowing self–non-self discrimination. A
self-antigen (S) is one that is continuously present from the beginning and
exposed to the immature immune system, whereas a foreign antigen is
present only transiently and is usually absent when the immune system
matures. Many persons term self-antigens all antigenic structures that are
encoded by the individual's germline. However, no one knows, for example,
if a ribosomal protein is considered a self-antigen by the immune system. Are
some internal cellular constituents processed and continuously exposed on
the cell membrane, as suggested by recent work? (Claverie and Kourilsky,
1986; De Plaen *et al.*, 1988).

Until now, two major concepts have been used to explain natural
tolerance. One theory holds that immature B cells (or T cells) are especially
sensitive to negative signalling. Immature B lymphocytes are inactivated as
soon as they encounter self-antigens. (For a review, see Nossal, 1983.) This
'clonal anergy' of early B lymphocytes could be followed by activation of
suppressor circuits that will allow maintenance of the 'tolerant state' (Bruyns
et al., 1976).

In another theory, every B lymphocyte, whether immature or not, is

Cell to Cell Signalling: From
Experiments to Theoretical Models
ISBN 0–12–287960–0

paralysable or inducible. Transduction of one signal, induced by antigen or anti-idiotype, would lead to unresponsiveness, whereas the same signal associated with a second signal (cooperative help) would promote differentiation of the B lymphocyte (Bretscher and Cohn, 1970; Cohn, 1981, 1985, 1986).

The exquisite sensitivity of immature B lymphocytes to negative signalling is well documented (Bretscher and Cohn, 1970). Some facts support the occurrence of suppressor circuits.

Whatever the correct theory, the network theory seems to stand in sharp contradiction to the self–non-self discrimination phenomenon. The idiotype network rests on the assumption (which is demonstrated experimentally) that complementary partners (idiotypes and anti-idiotypes) coexist within the repertoire of a single individual (Cazenave, 1977; Urbain *et al.*, 1977, 1980, 1981; Tasiaux *et al.*, 1978; Wikler *et al.*, 1979, 1980; Bona *et al.*, 1981; Jerne, 1984; Slaoui *et al.*, 1984; Vaz *et al.*, 1984; Wikler and Urbain, 1984; Kearney and Vakil, 1986; Urbain, 1986; Uytdehaag and Osterhaus, 1986). Therefore, as complementary lymphocytes emerge, they will cancel each other by virtue of self-recognition, leading to the paradox of the empty repertoire: diversity disappears as soon as it appears. Cohn has repeatedly stressed that no network theory is possible because there is no way to distinguish a self-epitope from a self-idiotope.

One goal of this chapter is to suggest that a network theory is possible on the basis of the self–non-self discrimination phenomenon.

The second problem of immunoregulation is the selection of actual and available repertoires. Before antigen arrival, the presence of so-called 'natural immunoglobulins' must be explained. Do these natural immunoglobulins represent the constitutive synthesis of the total repertoire (Vaz *et al.*, 1984; Dighiero *et al.*, 1985; Sourouson *et al.*, 1988)? Are they synthesized by a special subset of B lymphocytes (Lyl subset)? These natural immunoglobulins are claimed to be polyspecific, autoreactive and highly connected. Here we have a tautological problem, because being polyspecific, these immunoglobulins are autoreactive and highly connected (Vaz *et al.*, 1984; Dighiero *et al.*, 1985; Sourouson *et al.*, 1988). The problem of the actual repertoire of 'internally activated lymphocytes' is one of the most fundamental questions (Vaz *et al.*, 1984). As soon as the immune system is confronted with one antigen, the problem arises of the selection of available repertoires from a larger potential repertoire. In classical immunological thinking, it is often stated that this problem does not exist, because after establishment of the self–non-self discrimination, the rest is due to clonal selection of the anti-non-self repertoire. On the basis of experimental data, we will argue that the selection of available repertoires is connected to both self–non-self discrimination and idiotypic circuitry.

The third problem of immunoregulation, which will only be touched here, is the regulation of ongoing responses or the dynamics of an immune response (understanding the increase and decrease of binding affinity, the isotype switching, the appearance of memory cells, the induction of suppressor circuits, and the like) (see Kaufman *et al.*, 1985 for discussion). Before proposing a new network theory designed to explain self–non-self discrimination and the selection of available repertoires, we will describe some old and new findings that support the concept that idiotypic games are used by the immune system in the selection of available repertoires.

For years we have been investigating the selection of available repertoires and idiotypic choices using a tool called the 'idiotypic cascade' (Moser *et al.*, 1983; Francotte and Urbain, 1984; Gurish and Nisonoff, 1984; Manser *et al.*, 1984; Meek *et al.*, 1984; Roth *et al.*, 1985; Jeske *et al.*, 1986; Rathbun *et al.*, 1988). In the initial experiments, we start with an idiotype 'à la Oudin' (a private idiotype) (Oudin and Michel, 1969a,b). Then, conventional anti-idiotypic antibodies are obtained (Ab_2). These purified Ab_2 are injected into normal naive recipients who, as a result, develop an Ab_3 response. The major part of Ab_3 antibodies share idiotypic specificities with Ab_1, but they do not bind the antigen (subset id^+, ag^-). A small subset of Ab_3 (Ab_1') is strikingly similar to Ab_1: they share idiotopes and bind antigen.

If the animals that have been immunized with Ab_2 are subsequently challenged with antigen, the subset of Ab_1' antibodies is largely expanded. Using the idiotypic cascade to probe potential repertoires, we demonstrated that it is possible to induce antibacterial or antiviral antibodies in rabbits or mice that have never seen either the bacteria or the virus.

A particularly striking experiment used a rabbit Ab_2 directed against a private rabbit idiotype (anti-tobacco mosaic virus) (TMV). BALB/c mice dendritic cells were then pulsed *in vitro* with the rabbit Ab_2 and these pulsed dendritic cells were injected into syngeneic mice that had never seen the virus, which synthesized Ab_3 antibodies including anti-TMV antibodies. The treatment led to a striking priming effect after the first virus injection, the anti-idiotype-treated mice synthesized 300 µg (average) of specific anti-TMV antibodies per millilitre (as compared with 10 µg in the controls) (Francotte and Urbain, 1984).

The results of the idiotypic cascade can be explained in two ways that are not mutually exclusive: (1) The whole set of data can be interpreted in terms of double clonal selection: selection by anti-idiotypic antibodies followed by antigenic selection. (2) Conversely the induction of third-order antibodies (Ab_3) by Ab_2 could suppress the internal auto-Ab_2 and relieve the complementary Ab_1 from suppression. In other words, the idiotypic cascade could interfere with the inner life of the immune system. This second interpretation not only is an idiotypic dream, but also is supported by some experimental data.

For example, we induced Ab_3 antibodies in female rabbits by injecting them with an Ab_2 directed against a private anti-*Micrococcus luteus* idiotype. These female rabbits were crossed with naive males. Two months after birth, offspring, mothers and fathers were immunized with *Micrococcus luteus*. All mothers responded by the synthesis of large amounts of AB_1' antibodies (id^+, ag^+) sharing idiotypic specificities with the starting private Ab_1. The fathers produced unrelated anti-*Micrococcus* idiotypes. The surprising result was that half the offspring were synthesizing the same idiotype as were the mothers. These animals have assimilated 'the learning' of the mother by being exposed to maternal Ab_3 immunoglobulins. In this system, the interpretation cannot be simply clonal selection, because Ab_3 (id^+ ag^-, id^+ ag^+, id^- ag^-) cannot select Ab_1.

The reverse experiment (crossing of Ab_3 fathers with naive mothers) produced expected results: no idiotype crossreaction was seen between the fathers, mothers and offspring (Wikler *et al.*, 1980). Similar results have been obtained by others (Nossal, 1983; Jeske *et al.*, 1986).

PRIVATE AND RECURRENT IDIOTYPES IN THE ARSONATE SYSTEM

The immune response Ars–KLH (arsonate coupled to hemocyanin) in mice from the A/J strain is characterized by a major recurrent idiotype whose code name is CRI_A. (For a review see Tasiaux *et al.*, 1978.) The molecular basis for this recurrent idiotype has been unravelled. During the primary response, the anti-arsonate antibodies are characterized by the expression of a V-gene segment, VH id CR11, which can be associated with different D segments. As the response proceeds, the selection of a major canonical combination is observed: the recurrent idiotype, which is mainly associated with the heavy chain, is made up of the VH id CR11, the DF1 16.1 and the JH2 segments. (For details, see Wikler *et al.*, 1980; Siekewitz *et al.*, 1982; Gurish and Nisonoff, 1984; Rathbun *et al.*, 1988.) Most other strains of mice, which belong to other IgH haplotypes, do not express this idiotype when immunized with the same antigen. Using a panel of monoclonal anti-idiotypic antibodies and a panel of hybridoma, which are somatic variants of the germline combination, we were able to distinguish at least five idiotopes in the CRI_A molecules (Urbain, 1986). For simplicity, we will not discuss the problem associated with the light chain. BALB/c mice immunized with Ars–KLH usually do not express appreciable amounts of this idiotype and do not harbour in their genome the VH id CR11 segment. F1 mice (A/J × BALB/c) express in a recurrent fashion the CRI_A idiotype as well as the congenic mice CAL2O. So at first sight the situation seems to be very simple. Either the relevant genetic segments are present and the idiotype is expressed, or some genetic elements are absent and the idiotype is absent (Rathbun *et al.*, 1988). Recurrent expression of the CRI_A idiotype could be due to clonal selection of the antigen of relevant precursors able 'to sustain the generation of useful somatic variants'.

However, the phenomenon of idiotypic recurrence is apparently more complex, because the idiotypic dominance of the CRI_A idiotype is lost when irradiated A/J mice are repopulated with syngeneic bonemarrow or spleen cells. This disappearance is also observed in long-term radiation chimeras or in partially shielded irradiated A/J mice. Detailed analysis of the response in immunized BALB/c mice (Willems *et al.*, in preparation) revealed that some mice (a small percentage) synthesize significant amounts of a strongly crossreactive idiotype, which was designated CRI_A-like. Four of five idiotopes were shared between CRI_A and CRI_A-like idiotypes (Moser *et al.*, 1983; Meek *et al.*, 1984; Jeske *et al.*, 1986). Furthermore, when BALB/c mice are pretreated with a polyclonal Ab_2 directed against the CRI_A idiotype and immunized with Ars–KLH, all BALB/c mice synthesized large amounts of anti-arsonate antibodies characterized by the CRI_A-like idiotypic specificities. Thus, the CRI_A-like idiotype is expressed as an idiotype 'à la Oudin' in BALB/c mice immunized with antigen alone and as a recurrent idiotype, as a result of the idiotypic cascade.

Investigation of the molecular basis of the CRI_A-like idiotype 23 showed that the used VH segment was very different from the VH segment of the CRI_A: the VH segment from CRI_A-like molecules belongs to the VH IX family. However, the D segment and the light chains are nearly identical in CRI_A and CRI_A-like idiotypes (Meek *et al.*, 1984).

We next turned our attention to recombinant in-bred strains of mice AXC1, AXC2 and so forth. We will discuss only the data obtained from AXC1 mice, which harbour part of the VH locus of BALB/c mice and the CH locus of A/J mice. These AXC1 mice do not contain the VH id CR11 gene segment used in the CRI_A idiotype. When these mice are immunized with Ars–KLH alone, the CRI_A-like idiotype, which is expressed as an idiotype 'à la Oudin' in BALB/c mice, is now expressed in a recurrent fashion. This can be deduced from both serological analysis and sequence analysis (using hybridomas from AXC1 mice). However, and this is the most significant fact suggesting that the selection, when F_1 mice obtained from the cross between AXC1 and BALB/c mice are immunized with the same antigen (Ars–KLH) expression of the recurrent idiotype disappears. Results obtained in the backcross experiments clearly show that expression of this idiotype is under the control of genes that are unlinked to the IgH and IgK loci (Mertens *et al.*, unpublished).

THE BROKEN MIRROR HYPOTHESIS

Let us consider the potential repertoire of one individual towards given epitope E. Epitope E is able to bind to Ig receptors of several idiotypic clones, designated Ab_{1a}, Ab_{1b}, Ab_{1c}, ... Ab_{1i}, ... Ab_{1n}. According to the network assumption (Jerne, 1974), and as a result of diversity, the repertoire of the same individual will contain complementary (or anti-idiotypic) clones that we shall call Ab_{2a}, ... Ab_{2i}, ... Ab_{2n} (Urbain *et al.*, 1980, 1981).

With respect to antigens, Ab_1 is considered the idiotype and Ab_2 the anti-idiotype. However, in the absence of antigen and for reasons of symmetry, we could say that Ab_1 is the anti-idiotype of Ab_2 which is the idiotype of antigen X. Complementary partners (Ab_{1a} and Ab_{2a}, Ab_{1i} and Ab_{2i} and the like can be viewed as antagonistic. Ab_1 suppresses Ab_2, and Ab_2 suppresses Ab_1. The number of non-immunoglobulin self-antigens (somatic self) is large; therefore, a significant fraction of the potential immune repertoire should be endowed with antiself specificity. Hence some Ab_1 and some Ab_2 are often anti-S (anti-insulin, antihaemoglobin, etc.) in addition to their idiotypic properties.

Let us suppose, for example, that Ab_{2i} is antiself. The clones bearing antiself Ab_{2i} receptors will be inactivated. As a result, the weight of suppression on Ab_{1i} will be less than that on Ab_{1a}, ... Ab_{1n}. If antigen E is introduced, lymphocytes bearing Ab_{1i} idiotypic receptors will escape suppression more easily than will clones Ab_{1b}, and the like, still counteracted by those clones bearing Ab_{2a}, Ab_{2b}, ... receptors (Fig. 1). Therefore, the idiotype Ab_{1i} will tend to be a recurrent idiotype in one in-bred strain of mice or a dominant idiotype in one individual from one out-bred species; if the S antigen is polymorphic, within the species. In the beginning, everything is suppressed by virtue of symmetry. When the symmetry is broken by self–non-self discrimination, some clones escape suppression and determine the available repertoire. As such, the model explains why it is so difficult to induce syngeneic autoanti-idiotypic antibodies (being antiself) or the results of the idiotypic cascade (the induction of Ab_3 antibodies 'knocks' down the complementary Ab_2 and creates 'blind spots' in the Ab_2 repertoire). However, in this state, the model does not consider the distinction between self–

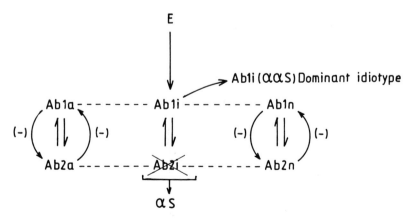

Fig. 1.   Escape from suppression (α S: antiself, α α S: anti-antiself).

epitopes (somatic self) and self-idiotopes (immunological self). Furthermore, the network is represented in an oversimplified way by describing a collection of n pairs of complementary partners. There is no reason why one Ab$_1$ should be connected to a single Ab$_2$ and so on.

Let us imagine a developing immature immune system (Fig. 2). It is important to realize that at this stage there is no endogenous synthesis of idiotypes (or anti-idiotypes). With the exception of idiotopes present on maternal immunoglobulins or on immunoglobulins introduced by the research worker, the endogenous idiotypes are present only on membrane immunoglobulin receptors. We assume that the functional inactivation of early B lymphocytes stems from the encounter with self-antigens that have been picked up by special presenting cells that we call self-presenting cells (SPC). SPC could be immature normal antigen-presenting cells (APC) or a special kind of cell whose main function is tolerance induction. Therefore, the immune system will not become tolerant to most idiotopes (except those of maternal immunoglobulins). These SPC can migrate to the thymus and

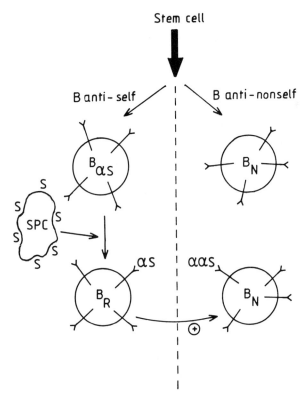

Fig. 2.   Immature immune system. B$_R$, long-lived regulatory B lymphocyte; B$_N$, normal complementary partner.

present to T cells the total array of self-antigens that is accessible to B lymphocytes.

Therefore, the library of developing B lymphocytes can be subdivided into two subsets. One subset does not meet self-antigens and will give rise to conventional B cells (B_N). The other subset, as a result of SPC meeting, will be diverted from the normal differentiation pathway (B_N). They will not die and will become long-lived lymphocytes unable to transform into normal antibody-producing cells (clonal anergy) but endowed with regulatory properties. We will call them B_R (B regulatory lymphocytes).

The main function of these B_R is to deliver positive signals of amplification to normal B lymphocytes, bearing receptors anti-idiotypic to B_R. Idiotypes, which will be favoured in an immune response, are anti-antiself antibodies. Ab_{1i} will be favoured because Ab_{2i} is antiself. As the immune system matures, a first repertoire (antiself) is established. This repertoire is long-lived and will shape, by positive loops, the available repertoires to non-self-antigens. As such, the model has very precise implications: (a) 'Tolerant' B lymphocytes are long-lived. Eventually they can be rescued and pushed to antibody secretion by LPS. (b) The induction of 'anergy' in an antiself–lymphocyte is not due solely to the encounter of the lymphocyte with the self-antigen. This self-antigen must be on the surface of the SPC. (c) The somatic self and the immunological self are treated the same way. Therefore, maternal idiotypes (or idiotypes introduced by the research worker at some critical period) should have dramatic effects in the shaping of available repertoires.

It is known that immature B lymphocytes are specially sensitive to tolerance induction, but the meeting of the B lymphocyte with the antigen is not sufficient to induce tolerance. For example, arsonate coupled to human immunoglobulin is an excellent tolerogen in neonatal mice but arsonate coupled to mouse gammaglobulin is a very poor tolerogen in mice (De Wit *et al.*, in preparation).

Let us consider the case in which an immature immune system is exposed to maternal idiotypes. These maternal idiotypes will be treated as self-antigens (id S). The complementary lymphocytes bearing anti-id S (or Ab_2) will be exposed to id S presented by SPC. As a result, B lymphocytes bearing Ab_2 receptors will be transformed into B regulatory lymphocytes, favouring the emergence of complementary B_N lymphocytes bearing anti-Ab_2 or Ab_1 receptors. Therefore, the young animal will have a tendency to produce the same idiotype as the mother, which makes sense in evolutionary terms. This is precisely the result we have shown in the experimental data just presented (Wikler *et al.*, 1980; Bordenave *et al.*, 1981; Stein and Söderström, 1984).

The properties of regulatory B lymphocytes described herein look similar to properties exhibited by Ly1 B lymphocytes (Sher *et al.*, 1987; Hayakawa and Hardy, 1988) (long-lived, anti-self-repertoires, regulatory properties and so forth). However, an alternative view is that these Ly1 B lymphocytes represent a special lineage of B lymphocytes, expressing a special repertoire of great protective value for the individual.

This network model is very precise, is testable, and does not avoid the problem of self–non-self discrimination. Recent data from Dorf and his group (Sher *et al.*, 1987) strongly support the idea of the existence of regulatory B lymphocytes helping complementary partners.

Acknowledgements: The authors are greatly indebted to L. Neirinckx and A. Tassin for typing the manuscript. This work was funded by contracts A.R.C. and CEE ST2J-0385-C (GDF).

REFERENCES

Bona, C., Hebër-Katz, H. and Paul, W. E. (1981) Idiotype-antiidiotype regulation. I. Immunization with an levan-binding myeloma protein leads to the appearance of autoanti-(anti-Id) antibodies and to activation of silent j clones. *J. exp. Med.* **153**, 851–967.

Bordenave, G., Babinet, C. and Michel, M. (1981) Idiotypic relationships between allotypically different rabbit antibodies. *Molec. Immunol.* **24**, 177–85.

Bretscher, P. and Cohn, M. (1970) A theory of non-self discrimination. *Science, N.Y.* **169**, 1042–9.

Bruyns, C., Urbain-Vansanten, Planard, C., De Vos-Cloetens, C. and Urbain, J. (1976) Ontogeny of mouse B lymphocytes and inactivation of antigen of early B lymphocytes. *Proc. natn. Acad. Sci. U.S.A.* **73**, 2462.

Cazenave, P. A. (1977) Idiotypic–antiidiotypic regulation of antibody synthesis in rabbits. *Proc. natn. Acad. Sci. U.S.A.* **74**, 5122–5.

Claverie, J.-M. and Kourilsky, P. (1986) The peptidic self model: a reassessment of the role of the major histocompatibility complex molecules in the restriction of the T cell response. *Ann. Inst. Pasteur* **137D**, 425–42.

Cohn, M. (1981) Conversation with Niels K. Jerne. *Cell. Immunol.* **61**, 425–36.

Cohn, M. (1985) Diversity in the immune system: preconceived idea or ideas preconceided? *Biochimie* **67**, 9–27.

Cohn, M. (1986) The concept of functional idiotype network for immune regulation mocks all and comforts none. *Ann. Immunol. (Inst. Pasteur)* **137C**, 57–100.

De Plaen, E., Lurquin, C., Van Pel, A., Mariame, B., Szikora, J. P., Wölfel, T., Sibille, C., Chomez, P. and Boon, T. (1988) Immunogenic (tum) variants of mouse tumor P815: cloning of the gene of tum-antigen P91A and identification of the tum-mutation. *Proc. natn. Acad. Sci. U.S.A.* **85**, 2274–8.

Dighiero, G., Lymberi, P., Holmberg, D., Lundquist, I., Countinho, A. and Avrameas, S. (1985) High frequency of natural autoantibodies in normal newborn-mice. *J. Immunol.* **134**, 765.

Francotte, M. and Urbain, J. (1984) Induction of anti-tobacco mosaic virus antibodies in mice by rabbit antiidiotypic antibodies. *J. exp. Med.* **160**, 1485–94.

Gurish, M. F. and Nisonoff, A. (1984) Structural properties and genetic control of an idiotype associated with antibodies to the *p*-azophenylarsonate hapten. In *Idiotypy in Biology and Medicine* (eds H. Köhler, J. Urbain and P. A. Cazenave), pp. 63–88. Academic Press, New York.

Hayakawa, K. and Hardy, R. (1988) Normal autoimmune and malignant CD5 + B cells: the LyB lineage. *A. Rev. Immunol.* **6**, 197.

Jerne, N. K. (1974) Towards a network theory of the immune system. *Ann. Immunol. (Inst. Pasteur)* **125C**, 373–89.

Jerne, N. K. (1984) Idiotypic networks and other preconceived ideas. *Immunol. Rev.* **79**, 1–29.

Jeske, D., Milner, E., Leo, O., Moser, M., Marvel, J., Urbain, J. and Capra, J. D. (1986) Molecular mapping of idiotopes of antiarsonate antibodies. *J. Immunol.* **136**, 2568.

Kaufman, M., Thomas, R. and Urbain, J. (1985) Towards a logical analysis of the immune network. *J. theor. Biol.* **114**, 527.

Kearney, J. F. and Vakil, M. (1986) Functional idiotype networks during B cell ontogeny. *Ann. Immunol. (Inst. Pasteur)* **137C**, 25–30.

Leo, O., Slaoui, M., Marvel, J., Milner, E., Hiernaux, J., Moser, M., Capra, D. and Urbain, J. (1985) Idiotypic analysis of polyclonal and monoclonal anti-p-azophenylarsonate antibodies of Balb/c mice expressing the major cross-reactive idiotype of the A/J strain. *J. Immunol.* **134**, 1734.

Manser, T., Huang, S-J and Gefter, M. (1984) The influence of clonal selection on the expression of immunoglobulin variable region genes. *Science, N.Y.* **226**, 1283.

Meek, K., Jeske, D., Slaoui, M., Leo, O., Urbain, J. and Capra, D. (1984) Complete amino acid sequence of heavy chain variable regions derived from two monoclonal anti-p-azophenylarsonate antibodies of Balb/c mice expressing the major cross-reactive idiotype of the A/J strain. *J. exp. Med.* **160**, 1070–86.

Moser, M., Leo, O., Hiernaux, J. and Urbain, J. (1983) Idiotypic manipulation in mice: Balb/c can express the crossreactive idiotype of A/J mice. *Proc. natn. Acad. Sci. U.S.A.* **80**, 4474–9.

Nossal, G. (1983) Cellular mechanisms of immunologic tolerance. *A. Rev. Immunol.* **1**, 33.

Oudin, J. and Michel, M. (1969a) Idiotypy of rabbit antibodies. I. Comparison of idiotypy of antibodies against other bacteria in the same rabbits or of antibodies against *Salmonella typhi* in various rabbits. *J. exp. Med.* **130**, 595.

Oudin, J. and Michel, M. (1969b) Idiotypy of rabbit antibodies. II. Comparison of idiotypy of various kinds of antibodies formed in the same rabbits against *Salmonella typhi*. *J. exp. Med.* **130**, 619.

Pollock, B. and Kearney, J. F. (1984) Identification and characterization of an apparent germline set of auto-antiidioptypic regulatory B lymphocytes. *J. Immunol.* **132**, 114–9.

Rathbun, G., Sanz, I., Meek, K., Tucker, P. and Capra, D. (1988) The molecular genetics of the arsonate idiotypic system of A/J mice. *Adv. Immunol.* **42**, 95.

Roth, C., Rocca-Serra, J., Somme, G., Fougereau, M. and Theze, J. (1985) The gene repertoire of the GAT immune response: comparison of the VK and D regions used by anti-GAT antibodies and monoclonal antibodies produced after antiidiotypic immunization. *Proc. natn. Acad. Sci. U.S.A.* **82**, 4788–92.

Sher, D., Dorf, M., Gibson, M. and Sedman, C. (1987) Lyl B helper cells in autoimmune viable motheaten mice. *J. Immunol.* **139**, 1811.

Siekewitz, M., Gefter, M., Brodeur, P., Riblet, R. and Marshak-Rothstein, A. (1982) The genetic basis of antibody production: the dominant anti-arsonate idiotype response of the strain-A mouse. *Eur. J. Immunol.* **12**, 1023.

Slaoui, M., Leo, O., Marvel, J., Moser, M., Hiernaux J. and Urbain, J. (1984) Idiotypic analysis of potential and available repertoires in the arsonate system. *J. exp. Med.* **160**, 1–11.

Sourouson, M., White-Scharf, E., Schartz, J., Gefter, M. and Schwartz, R. (1988) Preferential autoantibody reactivity of the preimmune B cell repertoire in normal mice. *J. Immunol.* **140**, 4173–9.

Stein, K. and Söderström, J. (1984) Neonatal administration of idiotype or antiidiotype primes for protection against *Escherischia coli* K13 infection in mice. *J. exp. Med.* **160**, 1001–11.

Tasiaux, N., Leuwenkroon, R., Bruyns, C. and Urbain, J. (1978) Possible occurrence and meaning of lymphocytes v bearing autoantiidiotypic receptors during the immune response. *Eur. J. Immunol.* **8**, 464–8.

Urbain, J. (1986) Idiotypic networks: a noisy background or a breakthrough in immunological thinking: the broken mirror hypothesis. *Ann. Immunol. (Inst. Pasteur)* **137C**, 57–100.

Urbain, J., Wikler, M., Franssen, J-D. and Collignon, C. (1977) Idiotypic regulation of the immune system by the induction of antibodies against antiidiotypic antibodies. *Proc. natn. Acad. Sci. U.S.A.* **74**, 5126–30.

Urbain, J., Cazenave, P. A., Wikler, M., Franssen, J-D., Mariamé, B. and Leo, O. (1980) Idiotypic induction and immune networks. In *Immunology 80* (eds J. Dausset and M. Fougereau). Academic Press, New York.

Urbain, J., Wuilmart, C. and Cazenave, P. A. (1981) Idiotypic networks in immune regulation. *Contemp. Topics Molec. Immunol.* **8**, 113.

Uytdehaag, F. and Osterhaus, A. (1986) Vaccines from monoclonal antiidiotypic antibody. Poliovirus infection as a model. *Curr. Topics Microbiol. Immunol.* **119**, 31.

Vaz, N. M., Martinez, C. and Couthino, A. (1984) The uniqueness and boundaries of the idiotypic self. In *Idiotypy in Biology and Medicine* (eds H. Köhler, J. Urbain and P. A. Cazenave), pp. 43–59. Academic Press, New York.

Wikler, M. and Urbain, J. (1984) Idiotypic manipulations of the rabbit immune response against *Micrococcus luteus*. In *Idiotypy in Biology and Medicine* (eds H. Köhler, J. Urbain and P. A. Cazenave), pp. 219–41. Academic Press, New York.

Wikler, M., Franssen, J-D., Collignon, C., Leo, O., Mariamé, B., Van de Walle, P., Degroote, D. and Urbain, J. (1979) Idiotypic regulation on the immune system. *J. exp. Med.* **150**, 184–95.

Wikler, M., Demeur, C., Dewasme, G. and Urbain, J. (1980) Immunoregulatory role of maternal idiotypes. Ontogeny of immune networks. *J. exp. Med.* **152**, 1024.

Immune networks: A topological view

ALAN S. PERELSON

Theoretical Division, Los Alamos National Laboratory, Los Alamos, NM 87545, USA

INTRODUCTION

The immune system is our primary defence against pathogenic organisms and as such has evolved strategies for recognizing antigens, i.e. foreign cells and molecules. The basic strategy, called clonal selection, is interesting from a theoretical viewpoint since it uses a pseudo-random process to perform specific pattern-recognition tasks. Further, the immune system is capable of learning and remembering patterns that it has recognized. One consequence of the pattern-recognition algorithm that the immune system uses is that certain novel components of the immune system such as antibody cannot be distinguished from antigen. Thus the immune system may recognize and respond to itself in a self-referential way. This leads to the idea, first presented by Jerne (1974), of the immune system as a network of interacting cells and molecules, normally looking inward, and being perturbed by exogenous antigens.

The basic defence cells of the immune system are a class of white blood cells known as lymphocytes. B lymphocytes are the cells that make antibody molecules. The other class of lymphocytes, T lymphocytes, are involved in cell–cell interactions. Some T cells are responsible for killing tumours or virally infected cells, whereas others secrete factors that promote the growth and differentiation of B cells and T cells.

Both B cells and T cells have receptor molecules on their surface that can recognize antigen. When antigen enters the body it encounters a large number of lymphocytes. If some of these cells have receptors that recognize the antigen these cells may become stimulated, begin proliferating and secreting molecules such as antibody that can lead to the elimination of the antigen. A single cell that recognizes the antigen can grow into a large clone

Cell to Cell Signalling: From
Experiments to Theoretical Models
ISBN 0–12–287960–0

of cells all of which are capable of fighting the antigen. The cells that grow into clones are selected by the antigen, hence the name clonal selection.

Although clonal selection operates at the level of both B cells and T cells, here I restrict the discussion to B cells and the antibodies they secrete. In broad outline a similar story applies to T cells.

The receptors on a B cell are made by a random genetic process during the maturation of a B cell in the bone marrow. The process involves choosing various immunoglobulin gene segments: V, D, J and C, from gene families, bringing these segments together and joining them in an error-prone way. During development each B cell constructs a particular variable-region gene and expresses it on its receptors. Many, perhaps most, B cells express receptors that are useless and detect nothing. Such cells die within a few days of leaving the bone marrow. However, on occasion a randomly made receptor is useful and detects something. The cells carrying this receptor are then amplified via clonal selection, and the receptor molecule, now known to be useful, is secreted as antibody. During the proliferation of the clone point mutations occur at unusually high frequency within the genes coding for the antibody molecule. Thus further refinements in specificity can occur during the course of an immune response.

For clonal selection to work as a means of generating protective antibodies the antibody repertoire needs to be complete, i.e. the population of B lymphocytes needs to be sufficiently large and diverse that essentially any antigen will be recognized by some lymphocyte in the population. Quantitative calculations by Perelson and Oster (1979), based on the idea of shape space, indicate that if antibodies are truly made at random the repertoire will be complete in all immune systems with 10^5 or more different elements. Interestingly, the smallest known immune system, that of a young tadpole, is estimated to have on the order of 10^5 different antibody types.

A consequence of the repertoire being complete is that the body recognizes its own antibodies as antigens. Antibodies are just protein molecules. A novel protein made by a stimulated B cell and secreted into the bloodstream should look to the rest of the immune system just the same as a novel molecule made by a bacterium or some other pathogen. If the antibody concentration gets high enough, the system recognizes it and responds to it by making an antibody that can bind to it. The unique portions of an antibody molecule that the immune system uses for recognition purposes are called idiotopes. The antibodies raised against other antibodies are called anti-idiotypic antibodies.

Because the immune system makes anti-idiotypic antibodies, one can imagine a cut on your finger, which allows bacteria to enter your body, stimulating an immune response which then cascades via antibodies being made against antibodies into an event involving all of the lymphocytes in your body. Thus there are two extreme views of an immune response. In classical clonal selection the antigen excites only those cells that can make antibody against the antigen. This, in general, is a rather small portion of the immune system; typically one in 10^5 cells respond to a given antigenic determinant. At the other extreme, the immune system is an idiotypic network and potentially the entire network is involved in the response to any antigen. How can we assess these two extreme views?

Networks, because of their large size and interconnected nature, are difficult to study experimentally. What may be *terra incognito* for experimentalists is *terra firma* for theorists. Network models have been vigorously pursued in many areas, particularly in neural science. In immunology, network models involving a small number of components were developed by Hoffmann (1975) and Richter (1975) soon after Jerne proposed the network idea. Later variations were developed by Hiernaux (1977), Seghers (1979) and others. In my view one of the substantial problems with these early models is that they focused on the developing dynamical equations for the network without properly considering the structure of the network. More recent models by Farmer *et al.* (1986) and De Boer (1988, in this volume) attempt to rectify this. Here I pursue the Farmer–Packard–Perelson model and examine the topological properties of the resulting network. I shall show that in some parameter regimes the model predicts a phase transition; on one side of the transition networks are rather sparse and responses should be dominated by single clones or a few interacting clones, as predicted by the clonal selection model. On the other side of the transition a highly interconnected network exists. The specific dynamic rules governing the growth of clones will determine how much of the network is involved in any particular response.

MODEL

Immunological recognition is governed by chemical interactions between molecules. For an antibody or B cell receptor to recognize an antigen it must bind it. Binding between antigen and antibodies generally involves short-range non-covalent interactions based on electrostatic charge, hydrogen binding, van der Waals interactions, etc. In order for the molecules to approach each other over an appreciable portion of their surfaces requires a complementarity in three-dimensional space of portions of the antigen and antibody. In some cases, the complementary regions may be planar, while in others they more closely resemble a bump and a groove. Both shape and charge distributions, as well as the existence in the appropriate complementary positions of chemical groups that can form hydrogen bonds and interact in other ways, are properties of antigens and antibodies that are important in determining the interactions between these molecules. Unfortunately, features such as the tertiary structure of molecules are known in only a few cases and tertiary structure is still very difficult to predict *à priori*. To mimic the required shape, charge and amino acid complementarity between molecules we have found it convenient to represent antibodies and antigens by binary strings (cf. Farmer *et al.*, 1986). Complementarity can then be defined by any of a number of rules, and the degree of complementarity can be quantified and used as a measure of the affinity of the antigen–antibody interaction. Figure 1 illustrates the simplest rule: two molecules are complementary if their binary string representations are complementary. In support of this rule there is evidence suggesting that if a peptide or protein is read from one strand of a double-stranded DNA molecule and a 'complementary peptide' is synthesized by reading from the complementary DNA strand, then the peptide and complementary peptide will bind specifically and with high affinity (Bost *et al.*, 1985a,b; Smith *et al.*, 1987; Shai *et al.*, 1987). Further, in the case of the hormone ACTH, antibodies against ACTH and antibodies against the complementary peptide seem to be an idiotypic–anti-idiotypic pair, leading to the speculation that idiotopes and anti-idiotopes may represent complementary sequences in the hypervariable regions of such immunoglobulin pairs (Smith *et al.*, 1987).

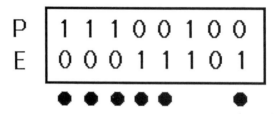

Fig. 1.   Complementarity can be assessed by aligning the antibody (or paratope string P) and antigen (or epitope string E) and then summing the number of positions at which a 1 is matched by 0. If this sum is greater than or equal to n_{good} we say the strings match and that the molecules they represent are complementary.

Other rules can also be used for determining complementarity. For example, since the strings represent molecules they need not be aligned when they interact. A match rule involving sequence shifting is discussed in Farmer *et al.* (1986). Molecules generally do not interact over their entire length, but rather interactions tend to be localized. Stadnyk (1987) uses a complementarity rule in which the number of adjacent complementary bits is important. Segel and Perelson (in this volume) represent antibody shape by a real number rather than a binary string. The complement of antibody of shape x is then an antibody of shape $-x$.

A graph can be used to represent the topology of an idiotypic network. A network graph can be generated by assigning each antigen and each antibody of different specificity to a node and then representing complementarity between two molecules as a link joining their nodes (Fig. 2). By setting a threshold in the degree of complementarity required for the formation of a link different network graphs can be obtained, each graph representing interactions at a different minimal level of affinity.

We find there is some resemblance between our randomly generated networks and experimentally obtained networks (Perelson, 1988). However, since only a few networks have been mapped it may be premature to ascribe great significance to this resemblance.

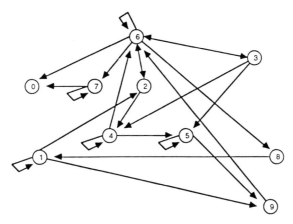

Fig. 2. A network graph generated using strings of length 8. Here a link has been drawn between two nodes if and only if at least 6 bits are complementary. See Perelson (1988) for further details.

PHASE TRANSITIONS IN IDIOTYPIC NETWORKS

When an antigen is injected into an animal the set of antibodies that the animal raises against the antigen are denoted Ab_1. These first-level antibodies are the usual antibodies of clonal selection theory. If the animal generates an anti-idiotypic response, then it will view Ab_1 antibodies as antigens and raise a set of antibodies, denoted Ab_2, that are to some degree complementary to members of the set Ab_1. Continuing in this way, Ab_i antibodies can lead to the generation of Ab_{i+1} antibodies. In experiments, Ab_1, Ab_2, Ab_3 and Ab_4 antibodies have been found (Bona and Pernis, 1984). In order to examine the structure of idiotypic networks, we have studied (Perelson, Packard and Stadnyk, in preparation) the topology of networks constructed using the complementarity match rule and the following algorithm: Construct a system containing N randomly generated antibodies each n bits long, and a single n bit antigen. Determine, according to the match rule, all antibodies that match the antigen and label these Ab_1. For each Ab_1 antibody then determine the matches with the remaining antibodies in the system. Any antibodies that match Ab_1 antibodies are placed in Ab_2. Continue in this manner, assigning an antibody to layer i if it matches an antibody in layer $i - 1$, and if it has not already been assigned to some previous layer. This latter condition ensures that each antibody is assigned to a unique layer and generates a network graph that is a rooted tree, the antigen being the root. Figure 3 is an example of a small network represented in this way.

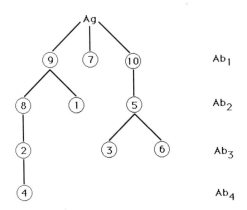

Fig. 3.   A network of ten antibodies drawn as a rooted tree.

There are various properties of networks represented as rooted trees that should be noted. First, all N antibodies in the system need not appear in the diagram. For example, if no antibody matches the antigen the diagram will contain only the antigen root. There is nothing special about the antigen, the diagram can 'die out' at any layer because of a lack of matches. Thus, beginning with any root all of the antibodies in the system need not be assigned to an idiotypic level. Second, the number of layers in the tree is variable. Some trees will have many layers, others few. Both the probability of the tree lacking some of the antibodies in the system and the number of layers in the tree depend on the probability that two strings match and hence on the match rule that we are using. For example, consider a system with 100 antibodies each represented by a 32-bit string (i.e. $N = 100$ and $n = 32$). Further, assume that the match rule is such that some threshold number of bits must be complementary for us to score two molecules as being complementary. Denote this threshold number n_{good}. If we choose $n_{good} = 32$, then we are essentially assured that the tree will be trivial, containing only the antigen. The probability of a match is $1/2^{32} \approx 2^{-10}$ and hence the system would need to contain 5×10^9 antibodies in order to expect even a single match.

Consider what happens as one varies n_{good} or equivalently the probability of a match. For small values of n_{good} all molecules in the system should match.

In fact, using the simple match rule of Fig. 1 we expect that on average half of the bits will match since 0 and 1 are chosen with equal probability. If all molecules match, then all of the antibodies in the system will be in layer 1. At values of n_{good} larger than $n/2$ there should be some matching, but not total matching. Thus one would expect trees with increasing numbers of layers. As n_{good} approaches n matches become rare, and both the number of layers and the number of antibodies in the tree will decrease. As reasoned above, when n_{good} reaches n we expect no antibodies in the tree if $N \ll 2^n$. Thus, we expect that a graph of the maximum layer reached in the tree versus n_{good} to approximate a smooth curve starting at 1, going through a maximum, and ending at 0. Surprisingly, we find via Monte Carlo simulations that although the curve has these general characteristics it seems to approach a curve with a singularity as N is increased. Curves for $N = 100$ and 500 are given in Fig. 4. The behaviour shown in the figure is typical for a system with a phase transition. There seems to be a critical value of n_{good}, called n_c, at which the number of layers in an idiotypic network rises very sharply. For the system with 500 antibodies an average of 29 layers are encountered at n_c. For systems with the diversity of the human immune system, i.e. containing 10^8 antibody types, I imagine that at the critical point hundreds or even thousands of layers may be present. Outside the phase-transition region the

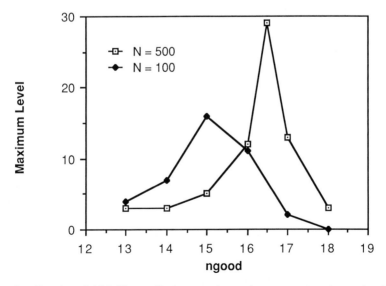

Fig. 4. Results of 100 Monte Carlo runs. In each run a network graph of the form shown in Fig. 3 was generated, and the maximum level reached recorded. The maximum of the runs is plotted for systems with $N = 100$ and $N = 500$ antibodies. The number of bits in each string, $n = 20$. In the graph for $N = 500$ a set of runs was done with $n = 32$ to mimic a connectance that could be attained when $n_{good} = 16.5$.

network may be minimal. Simulations with $N = 10^8$ are not feasible and thus it is important to understand how the phase-transition scales with system size.

I believe that a simple argument based on percolation theory (see Stauffer, 1985) on a Bethe lattice (or Cayley tree) can be used to explain the phase transition in idiotypic networks. A Bethe lattice is infinite so some modification of the following argument will be needed to make it rigorous for a finite system but the general idea should remain valid. In a Bethe lattice each node is connected to at most f other nodes. f is called the coordination number of the lattice. The lattice can be drawn in levels (or generations) in the same manner as an idiotypic network (see Perelson, 1980; Macken and Perelson, 1985). Let $E(i)$ be the expected number of nodes on the ith level, and let p be the probability that two nodes are connected. Then,

$$E(i + 1) = (f - 1)pE(i) \tag{1}$$

The factor $f - 1$ arises because each node on level i, $i \geqslant 1$, is connected to a node on level $i - 1$ and thus can be connected to at most $f - 1$ nodes on level $i + 1$. When

$$p > p_c = 1/(f - 1) \tag{2}$$

$E(i + 1) > E(i)$, and the lattice will grow without bound. When $p < p_c$, $E(i + 1) < E(i)$ and the graph will terminate after a finite number of levels. p_c is the threshold between these two qualitatively different behaviours and is called the critical percolation threshold.

For idiotypic networks each antibody can, in principle, be connected to all others. Hence $f = N - 1$ and $p_c = 1/(N - 2) \approx 1/N$. Further, from the match rule one can easily compute the probability that two n bit antibodies are complementary,

$$p = 2^{-n} \sum_{i = n_{good}}^{n} \binom{n}{i} \tag{3}$$

Evaluating eqn (3) for different values of n_{good} one finds $p = 0.02$, 0.006 and 0.001, when $n_{good} = 15$, 16 and 17, respectively. From eqn (2), $p_c = 0.01$, for $N = 100$. Hence one expects, n_c, the critical value of n_{good} to be approximately 15. This is what is found. Similarly, for $N = 500$, I predict n_c to be between 16 and 17. Figure 4 shows the validity of these two predictions.

What are the properties of the immune system on the two sides of the phase transition? In the 'pre-critical region', where $p < p_c$, i.e. $n_{good} > n_c$, the network is very sparse and composed of many unconnected components. Each antibody, on average, is connected to less than one antibody on the next level $(p(f - 1) < 1)$. Including the connection to the previous level each antibody is connected to less than two antibodies in the network. (De Boer, in this volume, p. 285 discusses an alternative criterion for the phase transition derived by Erdös and Rényi based on the average connectance of each node). In the pre-critical region no network *per se* exists. Rather there are many small discrete, non-interacting subnetworks. Each antigen is connected to a small number of antibodies and these in turn are connected to rather few other antibodies. With one antigen one cannot excite the entire immune system. In the 'post-critical region' $p > p_c$, i.e. $n_{good} < n_c$, the network is highly connected. In fact, there is a non-zero probability that all antibodies are part of a single network. As p is increased or n_{good} decreased the probability of a single global network increases. Concomitantly, the observed connectivity of the network increases. The number of antibodies at each level increases and in a finite system fewer and fewer levels are needed to account for all antibodies. Ultimately, when $p = 1$, all antibodies recognize the antigen and are on level 1.

Why is the existence of this phase transition interesting? Among immunologists the relevance of idiotypic networks to the functioning of the immune system is controversial. Some immunologists believe that idiotypic networks are an epiphenomenon and of no functional relevance, whereas other immunologists believe that idiotypic networks are the core of the immune system, accounting for all of the normal activity of the immune system in times of health and controlling the system in times of disease. The phase transition is a marker. On one side of it the network is so sparse that signals cannot propagate through many idiotypic levels. Some idiotypic–anti-idiotypic interactions will be present but the topology of the network will prevent a cascade of antibodies against antibodies from occurring. On the other side the network is highly connected and idiotypic interactions can lead to communication among all clones in the immune system. Signals can propagate very deeply into the idiotypic network and network interactions may dominate any response. Don't forget that from topology alone one cannot predict the dynamics of the immune response. It is the dynamics that will determine for any set of antibody and antigen concentrations and kinetic parameters whether or not deep penetration of the network will in fact occur.

From current estimates of parameters characterizing the immune system I believe that the immune system operates well into the phase-transition region where the network is highly connected. Using the percolation result, $p_c \approx N^{-1}$, we need only estimate N, the number of antibodies in the repertoire, and p the probability that two randomly chosen antibodies recognize each other. The size of the repertoire that is expressed in a given individual is not precisely known. A mouse, the experimental animal of choice in immunology, contains approximately 10^8 B lymphocytes. Average clone sizes are thought to be between 10 and 100 (Jerne, 1984), and thus the expressed repertoire would be between 10^6 and 10^7, values that agree with other estimates (Holmberg et al., 1986). Using these values for N, $p_c \leqslant 10^{-6}$.

In a variety of experiments the fraction of B cells that respond to an antigen has been measured (see Klinman and Press, 1975). Typical values are 10^{-5}. We can use this number as an estimate of the probability that a B cell responds to a random antigen or antibody. Thus $p \simeq 10^{-5}$ which is substantially larger than p_c. Consequently, I believe that under normal circumstances the immune system operates in the post-critical regime. As stated above, in this regime the immune network is highly connected and deep penetration is possible.

DISCUSSION

Topology is only one aspect of immune networks. It places bounds on what is possible. In the pre-critical region the network is very sparse and it is impossible for antigen or any internal perturbation of the network to spread throughout the immune system. In the post-critical region, the immune network is highly connected and it is possible for global excitation to occur. However, whether it does occur or not is now determined by the dynamical interactions between the elements of the immune system. In a dynamical model of idiotypic networks that I have developed with Lee Segel (Segel and Perelson, this volume p. 273), we have shown that the tuning of a single parameter can cause the behaviour of the model to switch from network-like, in which many clones and their anti-idiotypic clones are excited, to a model in which only a single clone that recognizes antigen is excited. In the real immune system T cells, antigen-presenting cells, various cytokines and growth factors must play an important, if not crucial, role in regulating the activity of the immune network. These controls seem to prevent global activation of the network, still leaving the question of whether they allow substantial network activity.

In some autoimmune diseases, such as systemic lupus erythematosus (Shoenfeld, 1988), one observes many different types of antiself antibodies. One might speculate that under such circumstances the normal controls on the immune system malfunction allowing us to see many of the interactions that are possible in the post-critical region. If many antibodies are raised against other antibodies some of these may crossreact with self components and contribute to the symptoms of autoimmune disease. Network interactions have been implicated in causing myasthenia gravis (Dwyer et al., 1986) as well as other autoimmune conditions.

CONCLUSIONS

Much remains to be done in order to understand the operation of the immune system. I believe that analytical and simulation models, such as the ones we have developed (Farmer *et al.*, 1986; Segel and Perelson, 1988), will be valuable tools. Having gained insights into the topology of immune networks we should be better able to develop dynamical models that confront the complexity underlying the immune system.

Acknowledgements: This work was performed under the auspices of the US Department of Energy. The simulations establishing the existence of a phase transition were done in collaboration with N. Packard and I. Stadnyk, during a visit to the Center for Complex Systems Research, University of Illinois, Urbana. I thank R. Bagley for help in generating the figures in this paper.

REFERENCES

Bona, C. A. and Pernis, B. (1984) Idiotypic networks. In *Fundamental Immunology* (ed. W. E. Paul), pp. 1372–5. Raven Press, New York.

Bost, K. L., Smith, E. M. and Blalock, J. E. (1985a) Similarity between the corticotropin (ACTH) receptor and a peptide encoded by an RNA that is complementary to ACTH mRNA. *Proc. natl. Acad. Sci. U.S.A.* **82**, 1372–5.

Bost, K. L., Smith, E. M. and Blalock, J. E. (1985b) Regions of complementarity between the messenger RNAs for epidermal growth factor, transferrin, interleukin-2 and their respective receptors. *Biochem. biophys. Res. Commun.* **128**, 1372–5.

De Boer, R. J. (1988) Symmetric idiotypic networks: connectance and switching, stability, and suppression. In *Theoretical Immunology, Part Two* (ed. A. S. Perelson), pp. 265–89. Addison-Wesley, Redwood City, CA.

Dwyer, D. S., Vakil, M. and Kearney, J. F. (1986) Idiotypic network connectivity and a possible cause of myasthenia gravis. *J. exp. Med.* **164**, 1310–18.

Farmer, J. D., Packard, N. H. and Perelson, A. S. (1986) The immune system, adaptation, and machine learning. *Physica* **22D**, 187–204.

Hiernaux, J. (1977) Some remarks on the stability of the idiotype network. *Immunochemistry* **14**, 733.

Hoffmann, G. W. (1975) A theory of regulation and self–nonself discrimination in an immune network. *Eur. J. Immunol.* **5**, 638–47.

Holmberg, D., Freitas, A. A., Portnoi, D., Jacquemart, F., Avrameas, S. and Coutinho, A. (1986) Antibody repertoires of normal BALB/c mice: B lymphocytes populations defined by state of activation. *Immunol. Rev.* **93**, 147–78.

Jerne, N. K. (1974) Towards a network theory of the immune system. *Ann. Immunol. (Inst. Pasteur)* **125C**, 373–89.

Jerne, N. K. (1984) Idiotypic networks and other preconceived ideas. *Immunol. Rev.* **79**, 5–24.

Klinman, N. R. and Press, J. L. (1975) The B cell specificity repertoire: its relationship to definable subpopulations. *Transplant. Rev.* **24**, 41–83.

Macken, C. A. and Perelson, A. S. (1985) *Branching Processes Applied to Cell Surface Aggregation Phenomena*. Lect. Notes in Biomath., Vol. 58, Springer-Verlag, New York.

Perelson, A. S. (1980) Mathematical immunology. In *Mathematical Models in Molecular and Cellular Biology* (ed. L. A. Segel), pp. 365–439. Cambridge University Press, Cambridge.

Perelson, A. S. (1988) Toward a realistic model of the immune system. In *Theoretical Immunology, Part Two* (ed. A. S. Perelson), pp. 377–401. Addison-Wesley, Redwood City, CA.

Perelson, A. S. and Oster, G. F. (1979) Theoretical studies of clonal selection: minimal antibody repertoire size and reliability of self- non-self discrimination. *J. theor. Biol.* **81**, 645–70.

Richter, P. H. (1975) A network theory of the immune system. *Eur. J. Immunol.* **5**, 350–4.

Segel, L. A. and Perelson, A. S. (1988) Computations in shape space: A new approach to immune network theory. In *Theoretical Immunology. Part Two* (ed. A. S. Perelson), pp. 321–43. Addison-Welsey, Redwood City, CA.

Seghers, M. (1979) A quantitative study of an idiotypic cyclic network. *J. theor. Biol.* **80**, 553–76.

Shai, Y., Flashner, M. and Chaiken, I. M. (1987) Anti-sense peptide recognition of sense peptides: Direct quantitative characterization with the ribonuclease S-peptide system using analytical high-performance affinity chromatography. *Biochemistry* **26**, 669–75.

Shoenfeld, Y. (1988) Analyses of the idiotypes and ligand binding characteristics of human monoclonal autoantibodies to DNA: Do we better understand systemic lupus erythematosus? In *Perspectives on Autoimmunity* (ed. I. R. Cohen), pp. 135–41. CRC Press, Boca Raton, FL.

Smith, L. R., Bost, K. L. and Blalock, J. E. (1987) Generation of idiotypic and anti-idiotypic antibodies by immunization with peptides encoded by complementary RNA: A possible molecular basis for the network theory. *J. Immunol.* **138**, 7–9.

Stauffer, D. (1985) *Introduction to Percolation Theory.* Taylor and Francis, Philadelphia.

Stadnyk, I. (1987) Schema recombination in pattern recognition problems. In *Genetic Algorithms and their Applications* (ed. J. J. Grefenstette). Lawrence Erlbaum Associates, Hillsdale, NJ.

Shape space analysis of immune networks

LEE A. SEGEL[a] AND ALAN S. PERELSON[b]

[a]Department of Applied Mathematics and Computer Science,
Weizmann Institute of Science, Rehovot 76100, Israel
[b]Theoretical Division, Los Alamos National Laboratory, Los
Alamos, NM 87545, USA

INTRODUCTION

In this chapter we outline our recent efforts to use the concept of (generalized) shape space as a basis for the study of the immune system. We begin with an explanation of the concept of shape space, which we then flesh out in a 'toy' model. Although exceedingly simple in comparison with the true complexities of the immune system, this model is already mathematically somewhat involved.

We summarize work already done that employs the 'toy' model to explore certain strategic issues concerning the immune system. These issues include (i) the trade-off between stability and sensitivity, (ii) the relative specificity of activation and suppression, and (iii) the behaviour of memory cells in a network environment. The chapter concludes with a sketch of our 'next generation model' wherein shape space ideas are still employed but the model explicitly includes lymphocytes, antibody and complexes of antibodies with each other and with receptors. Here only a B cell submodel is sketched, but the general ideas suffice to derive a model wherein both T and B cell populations are incorporated.

SHAPE SPACE

The immune system is a collection of molecules and cells. Most of the cells are characterized by a single key receptor molecule. The extent to which such receptors are occupied on a given cell is the major determinant of that cell's function.

Cell to Cell Signalling: From
Experiments to Theoretical Models
ISBN 0–12–287960–0

The basic idea of the shape space approach is to characterize the cells and molecules of the immune system by a 'generalized shape' variable x. The dimension of the shape space may be quite high, for generalized 'shape' not only includes the three-dimensional shape of molecules but also all other properties, such as charge, that affect the binding of one molecule to another. In any case the concentration of each cell or molecule of shape x can then be determined by an integro-differential equation that includes information about the interactions of molecules of different shapes.

For simplicity, in all our work so far we have restricted ourselves to a one-dimensional shape space. (To fix ideas, one can imagine that this simplification refers to a universe where only the electric charge x is important in determining binding.) Justification of this extreme simplification can be found in the hopes that (i) our conclusions can later be extended to higher dimensional shape spaces and that (ii) a few dimensions will eventually be found adequately to characterize real shapes, i.e. that eventually it will be found possible to approximate real shapes with a few well-chosen measurements.

Shape space was introduced by Edelstein and Rosen (1978) and Perelson and Oster (1979). Their considerations, while suggestive, were rather general as no dynamical equations were formulated. Our strategy has been first to formulate the simplest dynamical equations that captured at least some properties of immune system interactions. From these equations we attempted to draw some broad conclusions and directions for further research. The next step, which we have just begun, is to derive and analyse considerably more realistic equations. Here we present an outline of our progress to date.

FORMULATION OF THE 'TOY' MODEL

Let $b(x,t)$ denote the number per unit volume, at time t, of lymphocytes bearing receptors of shape x. We do not distinguish between B cells and T cells, and antibodies are not explicitly considered. (One can imagine that each B lymphocyte is associated with a constant number of antibodies.) The complement to shape x is considered to be shape $-x$, as is certainly a natural assumption if x represents electrical charge.

We regard cells as belonging to one of two classes, stimulated (probability α) and unstimulated (probability $1 - \alpha$).* Unstimulated cells die at rate d, while the bone marrow supplies new cells at a constant rate m. Stimulated x-cells reproduce at a rate r that is a function not only of the population level $b(x,t)$ of such cells but also, to account for non-specific factors, of the average

---

* Several of the succeeding paragraphs are very similar to a section in Segel and Perelson (1989b).

lymphocyte population size $B(t)$. We can represent antigens of shape x at time t by a source function in the $b(x,t)$ equation or with a separate population dynamic equation. However, antigen is neglected in the calculations to be described here. With all this, the basic equation of the model is

$$\frac{\partial b(x,t)}{\partial t} = m - db(x,t)\{1 - \alpha[b(x,t)]\} + r[B(t),b(x,t)] \, \alpha[b(x,t)] \, b(x,t) \tag{1}$$

If shape ranges between $-L$ and L, then the average population size B is given by

$$B(t) = \frac{1}{2L} \int_{-L}^{L} b(x,t) \, dx \tag{2}$$

To specify the function α, which depends implicitly on the distribution of clones in shape space $b(x,t)$, we assume that whether or not a cell is switched into an active state depends on a competition between activating and suppressing influences, A and S. We model activation and suppression as acting via two distinct receptor systems. For any fixed shape x, let $a(x,y)b(y,t)$ dy be the fraction of an x-cell's activating receptors that are bound by cells whose shape variable ranges between y and $y + dy$, where dy is a small number. Then

$$A(b) = \begin{array}{l} \text{fraction of } x\text{-cells'} \\ \text{activating receptors} \\ \text{bound when cell} \\ \text{distribution in shape space} \\ \text{of cells is given by } b(y,t) \end{array} = \int_{-L}^{L} a(x,y)b(y,t) \, dy \tag{3}$$

In essence $a(x,y)$ is the association constant for y binding to activating receptors on x-cells (assuming that a small fraction of receptors is bound). Analogously, for suppressive influences

$$S(b) = \int_{-L}^{L} s(x,y)b(y,t) \, dy \tag{4}$$

The probability of stimulation α is some function F of the amount of binding to the two classes of receptors:

$$\alpha = F(A,S) \tag{5a}$$

The function α increases (decreases) if more activator (suppressor) receptors are bound:

$$\partial F/\partial A > 0, \qquad \partial F/\partial S < 0 \tag{5b,c}$$

A simple possibility for F is the Michaelean expression

$$F(A,S) = \frac{A}{(p + qS) + A} \tag{6}$$

Interrupting the formulation temporarily, we acknowledge here that there is no universal agreement that active suppression, especially via 'suppressor' T cells, plays an important role in the normal immune system (see Möller, 1988; Janeway, 1988). One would expect the immune system to explode if only purely activating factors can be triggered by antigen. We have necessarily made a specific commitment to a mechanism by which a counterweight to activation is generated. We believe that it is likely that the conclusions that we shall draw will remain true even if 'suppression' is due to a lack of some activating factor, and that no specific 'suppressor receptors' in fact exist.

We now complete the formulation of our model. To that end, in order to assure that the fit of shape y to shape x decreases from a maximum when $y = -x$ (perfectly complementary shape), we assume Gaussian functions of $x-(-y)$ for the activation and suppression association factors a and s, with standard deviations σ_a and σ_s respectively. An important parameter is the ratio

$$\theta \equiv \sigma_s^2/\sigma_a^2 \tag{7}$$

For the proliferation rate r of activated cells we generally take

$$r(B,b) = r_0 e^{-\eta B^n} e^{-\lambda b} \qquad (r_0 = \text{constant}) \tag{8}$$

Numerical calculations are carried for a discrete version of eqn (1)

$$db_i(t)/dt = m - db_i(t)[1 - \alpha] + r[B(t),b_i(t)]\,\alpha \tag{9}$$

In discretizing the equations that supplement (9), integrals are replaced by sums. For example the activating contribution to the fraction $\alpha(A,S)$ of stimulated cells is (replacing (3))

$$A_i = \sum_{j=-N}^{N} a_{ij} b_j \tag{10a}$$

The interval $-L \leqslant x \leqslant L$ is broken into $2N$ sub-intervals of width Δ:

$$\Delta = L/N \tag{10b}$$

The lymphocyte population levels b_j are assigned values at the $2N + 1$ interval endpoints. To minimize 'edge effects' near $j = -N$ and $j = N$ we make the periodicity assumption $b(i) = b(i + 2N)$.

PREVIOUS RESULTS

We have shown (Segel and Perelson, 1988) that the governing system of equations possesses a uniform positive steady state $b_i \equiv \bar{b}$, $\bar{b}$ a constant. An approximate formula for $\bar{b}$, valid for the parameters employed here, is

$$\bar{b} \approx m/[d - q^{-1}(d + r_0)] \tag{11}$$

In this 'virgin' state all clones are equally represented. The virgin state is stable to small-amplitude perturbations providing $r < r_0^{\text{crit}}$, where r_0^{crit} is a certain function of the remaining parameters.

An immune system should be stable to small perturbations but not so stable that it will not respond to an antigenic challenge. We thus suggested (Segel and Perelson, 1988) that the parameters in our model should be set not too far into the stable domain of parameter space:

$$r_0 < r_0^{\text{crit}} \text{ but not } r_0 \ll r_0^{\text{crit}} \tag{12}$$

We showed that indeed if and only if (12) holds will our model immune system avoid massive response to insignificant doses of antigen but yet will respond appropriately to a significant antigenic challenge.

In Segel and Perelson (1988) we pointed out that our model possessed an unusual type of instability. As expected from numerous similar examples in biological modelling, instability could arise from short-range activation and long-range suppression (Meinhardt, 1982; Oster, 1988). However, there was also the possibility of instability when suppression is relatively short range, i.e. $\theta < 1$ in (7). Later (Segel and Perelson, 1989a) we carefully examined the cause of the possible instability. Our argument begins with the observation that the new instability stems from the fact that the origin $x = 0$ is not an arbitrary point but rather corresponds to self-matching shapes. (Antibodies of this type have been discovered by Kang and Kohler, 1986, and have been termed 'autobodies'.) Instabilities when inhibition is relatively short range come from anti-symmetric disturbances, i.e. from disturbances in which a relative surplus of cells that bear receptors of a given shape are associated with a relative deficit of cells bearing complementary receptors. Because of this, the roles of activation and suppression are in a sense interchanged.

Several theoreticians (e.g. De Boer, 1988; Hoffmann et al., 1988; Kaufman and Thomas, 1987) have demonstrated that immune memory can arise from models possessing more than one steady-state solution. Roughly speaking, the introduction of antigen drives the system to a new steady state. This state remains even after the antigen disappears, and thus constitutes a memory of the transient presence of the antigen in question.

By contrast, biologists have found evidence that antigenic challenge leads to the production of specific memory cells with relatively long lifetimes. The

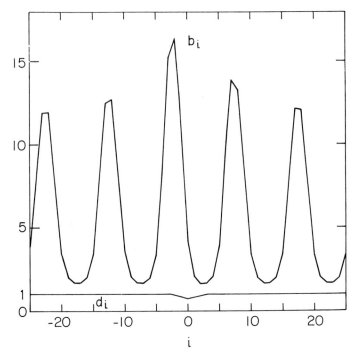

Fig. 1. Numerical solution of the discrete model (9). If $d_i \equiv 1$ then the uniform steady state $b_i = \bar{b} \approx 4.2$ is stable ($r_0^{crit} = 37.4$, $r_0 = 37.0$). Shown here is the highly non-uniform steady state that arises in the presence of a 'patch' of unstable shapes (memory cells, with smaller death rates d_i. When $d_i \equiv d < d^{crit} = 0.91$, the uniform state is unstable. Here $d_0 = 0.8$, $d_1 = d_{-1} = 0.87$, $d_2 = d_{-2} = 0.93$; for all other i, $d_i = 1$.) Parameters: $m = 1$, $\lambda = 0.06$, $\eta = 0$, $\sigma_a = 0.1$, $\sigma_s = 0.033333$, $q = 50$, $\Delta = 0.028$, $N = 25$, $p = 0.0001$.

continued presence of these cells is assumed to constitute the memory trace. We have examined how putative memory cells might alter the distribution of cell types in a crossreacting network. We find that if the network is barely stable then 'memory cells' cause extensive echoes throughout the network (Fig. 1). Changing parameters to render the network more stable remedies this defect. (Fig. 2. See Segel and Perelson, 1989b, for a considerably fuller treatment of how memory cells behave according to (9).) We are driven to the conclusion that a fairly restricted parameter setting must be made if our toy model is to behave properly with respect to the few aspects we have tested – stability, responsiveness and lack of overreaction to the presence of memory cells.

Our results to date lead us to conclude that shape space has proved itself as a suitable arena for the exploration of interesting questions concerning the operation of the immune system. The next step is to reconsider these and

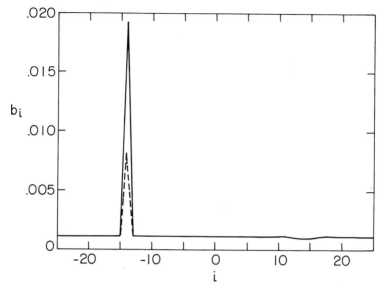

Fig. 2.   Demonstration that in contrast to their behaviour amidst the slightly stable background of Fig. 1, memory cells in a 'rather' stable background give rise to a minimal perturbation to the rest of the system. (Parameters as in Fig. 1 except $m = 0.01$, $\lambda = 0.5$.) Steady state when $d_{-14} = 1.3$; $d_i = 10$, $i \neq -14$ (solid line) (if $d_i \equiv 10$ then $\bar{b} = 0.001$, $r_0^{crit} = 297$) and when $d_{-14} = 2$ (dashed line). For reference, $d^{crit} = 1.22$. Note that if (11) is applied with the assumption $d_i \equiv d_{-14}$ then $\bar{b} = 0.007$ and $\bar{b} = 0.019$ respectively – i.e. the 'background' clones $i \neq -14$ scarcely influence the memory clone b_{-14}.

other questions with a better model. We have taken steps to formulate such a model. In the next section we report on progress in this direction.

B CELL SUBMODEL

In our next generation of models we explicitly consider cells and molecules, together with the chemical reactions involved in molecules binding to cellular receptors. Although this increases the tally of elements in our models we can reduce the number of differential equations needed by taking into account the fact that chemical reactions take place on a time scale (milliseconds to seconds, perhaps minutes) that is much shorter than the time scales that are characteristic of cellular changes (e.g. activation, proliferation – hours to days). We can provide formal justification of our simplifications by means of the method of multiple scales (see Lin and Segel, 1974, section 11.2, for an elementary exposition).

To examine how the simplification process goes, and to illustrate the type of modelling to be done, let us consider a submodel where only B cells and antibody are considered. (Antigen can easily be included and follows binding kinetics that are similar to antibody.) The basic chemical reactions are the binding of antibody to receptor sites (each B cell receptor has two independent sites)

$$A_f(y,t) + b(x,t)R_f(x,t) \underset{k_-(x,y)}{\overset{k_+(x,y)}{\rightleftharpoons}} R_b(x,y,t)b(x,t) \qquad (13a)$$

and the formation of antibody–antibody complexes:

$$A_f(y,t) + A_f(y',t) \underset{m_-(y,y')}{\overset{m_+(y,y')}{\rightleftharpoons}} C(y,y',t) \qquad (13b)$$

where $A(y,t)$ = total antibody concentration of shape y (at time t); $C(y,y',t)$ = antibody–antibody complex concentration; $b(x,t)$ = concentration of B cells bearing receptors of shape x; $R(x,t)$ = number of immunoglobulin (antibody) receptor sites on $b(x,t)$; $R_b(x,y,t)$ = number of x receptor sites per B cell that are complexed with antibody of shape y; and subscripts b and f are used to denote bound and free concentrations.

For the present we shall neglect receptor internalization and synthesis, so that the number of receptor sites per cell is constant, i.e.

$$R(x,t) = R = \text{constant}, \quad R_f(y,t) + \int R_b(x,y,t)dx = R \qquad (14a,b)$$

The integrals are taken over all possible shapes, i.e. $-\infty < x < \infty$. The key result of our simplification is that the total amount of antibody of shape y (whether free, bound in antibody–antibody complexes, or bound to receptor) is given by a function $F(y,t)$ that varies slowly with time:

$$A_f(y,t) + \int b(x,t)R_b(x,y,t)dx + \int C(y,y',t)dy' = F(y,t) \qquad (15)$$

F is known initially, at $t = 0$. It is augmented by antibody secretion from activated B cells and diminished by the destruction of antibody–antibody complex and the natural decay of serum antibody:

$$\partial F(y,t)/\partial t = s_A \left[\int R_b(y,x,t)dx \right] b(y,t) - \delta_C \int C(y,y',t)dy' - \delta_A A_f(y,t) \qquad (16)$$

Note that each 'y' B cell is assumed to secrete antibody at a rate s_A that depends on the fraction R_b of its receptor sites that are bound. Of course, there is also a dependence on the concentrations of factors secreted by T cells, such as interleukins. These growth and differentiation factors are regarded as being present at some suitable constant level in this B cell submodel.

The antibody secretion rate s_A is determined by an equation of the form

$$\partial s_A(y,t)/\partial t = Q_A \left[\int R_b(y,x,t)dx \right] - k_A s_A \qquad (17)$$

This equation has the property that the secretion rate ultimately reaches a steady value in a time scale determined by k_A. The function Q_A describes the dependence of the steady-state secretion rate on the extent of receptor binding. By using a differential equation to determine $s_A(y,t)$, we account for the delays involved in lymphocyte activation and the 'gearing up' for antibody secretion. (This generally takes of order 1–2 days.)

B cells proliferate at a rate r_B that depends on the number of receptor sites bound. There is an influx m_B from the bone marrow and a death rate d_B:

$$\partial b(x,t)/\partial t = r_B \left[\int R_b(x,y,t)dy, \int b(y,t)dy, b(x,t) \right] b(x,t) + m_B - d_B b(x,t)$$
(18)

Note that in a full model, interleukin concentrations would affect r_B, but here the concentration of such non-specific growth factors is assumed to be proportional to the total population size $\int b(y,t)dy$. The dependence of r_B on the specific clone size $b(x,t)$ models effects such as crowding in lymph nodes, spleen, etc.

Although the 'slowly varying' conservation law (15) is unconventional, a standard steady-state assumption can be used to calculate on a fast time scale the fraction of receptor on 'x' cells that are complexed with 'y' antibody:

$$\frac{R_b(x,y,t)}{R} = \frac{K_a(x,y)A_f(y,t)}{1 + \int K_a(x,y)A_f(y,t)dy}$$
(19)

Here the association constant K_a is defined as usual as

$$K_a(x,y) \equiv k_+(x,y)/k_-(x,y)$$
(20)

where k_+ and k_- are the rate constants corresponding to the kinetics of (13a). At lowest order the steady-state version of (13b) yields

$$C(y,y',t) = M_a(y,y')A_f(y,t)A_f(y',t)$$
(21)

where M_a is the association constant for antibody–antibody binding:

$$M_a(y,y') \equiv m_+(y,y')/m_-(y,y')$$
(22)

Since there is a lack of precise information on the difference in rate constants for reactions between antibodies and between antibodies and receptors, we take as a first approximation $M_a = K_a$. Equation (21) can be used to eliminate the complex from (15), which yields

$$A_f(y,t) + \int K_a(y,y')A_f(y,t)A_f(y',t)dy' + \int b(x,t)R_b(x,y,t)dx = F(y,t)$$
(23)

Thus our model consists of the five equations (16)–(19) and (23) for the B cell concentration b, the fraction of occupied B cell receptor sites R_b, the free antibody concentration A_f, the total antibody concentration F, and the antibody secretion rate s_A. The formulation is completed by suitable initial and boundary conditions. The former, in fact, are the conditions at the end of the fast transient; they can be ascertained as in the usual approach to quasi-steady-state assumptions (see, for example, Segel, 1988).

We hope to be able to report soon on the extent to which our earlier conclusions concerning the 'toy' model are borne out on more realistic models. Another high priority matter is to employ the more detailed models in attempts to discern to what extent the details of the biology affect the overall functioning of the immune system.

Acknowledgements: This work was performed under the auspices of the US Department of Energy and was partially supported by the US–Israel Binational Science Foundation (Grant no. 3777).

REFERENCES

De Boer, R. J. (1988) Symmetric idiotypic networks: connectance and switching, stability, and suppression. In *Theoretical Immunology, Part Two* (ed. A. S. Perelson), pp. 265–89. Addison-Wesley, Redwood City, CA.

Edelstein, L. and Rosen, R. (1978) Enzyme-substrate recognition. *J. theor. Biol.* **73**, 181–204.

Hoffmann, G. W., Kion, T. A., Forsyth, R. B., Soga, K. G. and Cooper-Willis, A. (1988) The *n*-dimensional network. In *Theoretical Immunology, Part Two* (ed. A. S. Perelson), pp. 291–319. Addison-Wesley, Redwood City, CA.

Kang, C.-Y. and Kohler, H. (1986) A novel chimeric antibody with circular network characteristics: autobody. *Ann. N.Y. Acad. Sci.* **475**, 114–22.

Kaufman, M. and Thomas, R. (1987) Model analysis of the bases of multistationarity in the humoral immune response. *J. theor. Biol.* **129**, 141–62.

Janeway, Jr., C. (1988) Do suppressor T cells exist? A reply. *Scand. J. Immunol.* **27**, 621–3.

Lin, C. C. and Segel, L. A. (1974) *Mathematics Applied to Deterministic Problems in the Natural Sciences*. Macmillan, New York. (Reprinted, 1988, by SIAM Publications, Philadelphia.)

Meinhardt, H. (1982) *Models of Biological Pattern Formation*. Academic Press, London.

Möller, G. (1988) Do suppressor T cells exist? *Scand. J. Immunol.* **27**, 247–50.

Oster, F. G. (1988) Lateral inhibition models of developmental processes. *Math. Biosci.* **90**, 265–86.

Perelson, A. S. and Oster, G. F. (1979) Theoretical studies of clonal selection: minimal antibody repertoire size and reliability of self–non-self discrimination. *J. theor. Biol.* **81**, 645–70.

Segel, L. A. (1988) On the validity of the steady state assumption of enzyme kinetics. *Bull. Math. Biol.* **50**, 579–93.

Segel, L. A. and Perelson, A. S. (1988) Computations in shape space: A new approach to immune network theory. In *Theoretical Immunology, Part Two* (ed. A. S. Perelson), pp. 321–43. Addison-Wesley, Redwood City, CA.

Segel, L. A. and Perelson, A. S. (1989a) A paradoxical instability caused by relatively short range inhibition. *SIAM J. Appl. Math.* (in press).

Segel, L. A. and Perelson, A. S. (1989b) Some reflections on memory in shape space. In *Theories of Immune Networks* (eds H. Atlan and I. R. Cohen). Springer-Verlag, Berlin (in press).

Information processing in immune systems: Clonal selection versus idiotypic network models

ROB J. DE BOER

Bioinformatics Group, University of Utrecht, Padualaan 8, 3584 CH Utrecht, The Netherlands

INTRODUCTION

Experimental immunology is a rapidly developing field in which an ever-increasing number of cell types, molecules and interactions between them are being described. Despite the availability of detailed experimental data, our understanding of the 'functioning' of the immune system remains very primitive. Simple issues, such as the immunity/memory phenomenon, and essential issues, such as self–non-self discrimination, are largely unresolved. Our approach to this problem is to view the immune system as a highly complex information-processing system. Information is provided in the form of (1) the (huge) repertoire of B and T cell receptors, (2) the repertoire of self-antigens, (3) MHC molecules, and (4) various foreign antigens that attempt to invade the system. This information is processed by interactions among the various cell types and molecules that constitute the immune system. As a result of the information-processing immune systems respond in a coordinated way both to pathogens and to self-antigens. In our 'bioinformatic' approach we attempt to pinpoint the interactions (i.e. the informatic processes) that account for the phenomena we are interested in. In our attempts to pinpoint the pivotal informatic processes in immunology, we previously discerned two levels of informatic organization: (1) profound networks of receptor–receptor (i.e. 'idiotypic') interactions, and (2) cellular and/or molecular interactions involved in the clonal (B and/or T cell) response to antigen. This chapter reviews our previous results. In combination, these results are counter-intuitive because information processing in the highly complex and seemingly powerful profound networks of idiotypic interactions turned out to be inferior to that of the simple autocatalytic interactions between helper T cells (Th) and the growth hormone (IL2) that these cells produce. Additionally it will be shown that Th may hinder the development of functional idiotypic interactions among B cells.

Networks of idiotypic interactions seem to be an inevitable property of immune systems. Jerne (1974, 1984) based his network theory on the 'completeness' axiom: if the repertoire of receptor molecules can collectively respond to virtually any antigen, i.e. if the repertoire is 'complete', receptors (i.e. idiotypes) should also respond to other complementary receptors (i.e. to anti-idiotypes). These networks of idiotypic interactions have often been compared to neural networks (see e.g. Jerne, 1974; Varela *et al.*, 1988). Neural networks, for which Hopfield equations provide the general paradigm system, are capable of complex computational processes (such as learning and memory) (Hopfield and Tank, 1986). It is thus frequently proposed that immune systems 'function' by means of similar profound network properties. We here analyse this proposal by means of a 'bottom up' approach: starting with basic assumptions about idiotypic interactions, and their effects on B cell proliferation, we develop 'fundamental' idiotypic network models. If immune systems do indeed function by means of idiotypic interactions, our models should eventually acquire regulatory (or computational) properties reminiscent of those described for neural network models. This is, however, not borne out by our analysis (De Boer, 1988; De Boer and Hogeweg, 1989b). In the next section we will review the problems that we have encountered with idiotypic network theory.

The 'clonal selection' level of the immune system, on the other hand, consists of the various cell types that are clonally activated by antigen (self or non-self). From the total repertoire antigen seems to 'select' the various clones that recognize it; this is 'clonal selection' theory (Burnet, 1959). These lymphocyte clones communicate via cell to cell contacts and/or via the production of signalling molecules (i.e. lymphokines). We have previously analysed complex 'networks' of such interactions among various clonally activated cell types. This analysis pinpointed helper T cell (Th) proliferation as the most crucial process in immunoregulation (De Boer *et al.*, 1985; De Boer and Hogeweg, 1987a,b). Therefore, in the next section we will concentrate on Th proliferation (i.e. on one clone).

In the final results section we present verbally the results obtained with a model that combines the features of the previous two sections. The model incorporates the Th interactions in the B cell idiotypic network model. Although the most crucial result of the Th model (i.e. the proliferation threshold) is omitted, the incorporation of Th cells hinders the development of functional idiotypic interactions. We conclude that idiotypic network theory is not inevitable, and we propose that more attention should be given to clonal selection systems.

A FUNDAMENTAL IDIOTYPIC NETWORK MODEL

For a detailed description of our symmetric network models we refer to De Boer and Hogeweg (1989a,b). In our models idiotypic interactions are symmetric. This follows from the idea of complementary idiotypes: if idiotype 'i' is a complement of 'j', 'j' is also complementary to 'i'. Hoffmann (1979, 1980) first proposed this simple and attractive theory. Note that symmetry occurs 'naturally' in the shape space model proposed by Segel (in this volume). In our model clones never recognize themselves (i.e. we set all A_{ii} to zero, see below).

In the models we consider N clones of B lymphocytes; each clone is regulated by three processes: (1) influx of newborn cells from the bonemarrow (S), (2) normal turnover of cells (D), and (3) proliferation (P). The rate of cell proliferation is governed by a growth function $G(X_i, \mathrm{Ag}_i, \alpha\mathrm{Id}_i)$, which depends on (1) the size of the clone (X_i), (2) the antigen (Ag_i), and (3) the total amount of anti-idiotype ($\alpha\mathrm{Id}_i$). The function $G(X_i, \mathrm{Ag}_i, \alpha\mathrm{Id}_i)$ is maximally one; proliferation then proceeds at a rate P per cell per day. The function (eqn (2)) is a log bell-shaped curve: too large anti-idiotype concentrations reduce the crosslinking and hence the rate of cell proliferation. For reasons of simplicity, antigen can only increase (up to a certain maximum) the rate of cell proliferation: the antigen dose–response curve is a simple saturation function. Antigen cannot grow and is eliminated by the clone that recognizes it. We thus propose the following model:

$$\alpha\mathrm{Id}_i = \sum_{j=1}^{N} A_{ij} X_j \qquad (1)$$

$$G(X_i, \mathrm{Ag}_i, \alpha\mathrm{Id}_i) = \frac{\mathrm{Ag}_i + \alpha\mathrm{Id}_i}{P_1 + F X_i + \mathrm{Ag}_i + \alpha\mathrm{Id}_i} \frac{P_2}{P_2 + \alpha\mathrm{Id}_i} \qquad (2)$$

$$X_i' = S_i - D X_i + P X_i G(X_i, \mathrm{Ag}_i, \alpha\mathrm{Id}_i) \qquad (3)$$

$$\mathrm{Ag}_i' = -\frac{K \mathrm{Ag}_i X_i}{K_1 + X_i} \qquad (4)$$

The parameter setting is: $S \approx 10$ cells day^{-1}, $D = 1$ day^{-1}, $P = 1.5$ day^{-1}, $P_1 = 10^3$, $P_2 = 10^6$, $F = 0.01$, $K = 1$, $K_1 = 10^5$. The virgin (i.e. unaffected) population density equals $S/D \approx 10$ cells. The influx is slightly different for each clone (to prevent settlement into unstable equilibria): S has a mean of 10 cells per day with a 10% standard deviation. Virgin populations are too small to evoke proliferation ($S/D \ll P_1$): idiotypic interactions are negligible in the virgin state. Large X_i populations ($F X_i \approx P_1$) cannot be stimulated by small antigen and/or anti-idiotype concentrations; this is a buffering effect (see De

Boer and Hogeweg, 1989a). Maximum proliferation proceeds at a rate $P - D = 0.5$ cells per cell per day (this corresponds to a doubling time of about 16 h). The affinity matrix (A_{ij}) is drawn randomly from a uniform distribution between zero and one. The low-D models have been analysed by GRIND (De Boer, 1983) which performs numerical 0-isocline analysis and numerical integration. High-D models are integrated by means of a variable step-size Runge–Kutta–Merson integrator implemented in NAG (1984).

Memory

Consider a network of two clones: X_1 and X_2, which see each other with maximum affinity (i.e. $A_{12} = 1$). In Fig. 1A this 2D network is analysed statically: the curved lines are the $X_{1'} = 0$ and the $X_{2'} = 0$ isoclines. The region in which X_1 expands is shaded; X_1 proliferates in the large shaded region situated at an intermediate X_2 population size. The isoclines intersect in three stable equilibria: the virgin state (V) and two immune states (I_1 and I_2). In the absence of antigen the system always remains in the virgin state; here idiotypic interactions are negligible. Figure 1B shows the same picture with a trajectory (i.e. the fat line) that was initiated by introducing antigen recognized by X_1 (i.e. Ag_1 was set to 10^4). In response to antigen, X_1 proliferates (the trajectory moves to the right), which, in turn, evokes the proliferation of X_2 (the trajectory moves upwards). As a consequence of the proliferation of X_1, antigen is rejected (not shown) and the system settles into the I_1 equilibrium. In this new state the system is immune to Ag_1: reintroduction of Ag_1 leads to rapid rejection because X_1 is already enlarged. X_2, however, is suppressed in the I_1 state: introduction of Ag_2 into this state never leads to rejection because X_2 fails to proliferate. Therefore, the previous exposure to Ag_1 is specifically remembered by the network.

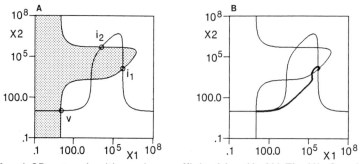

Fig. 1.   A 2D network with maximum affinity ($A_{12} = 1$). (A) The $X_1' = 0$ and $X_2' = 0$ isoclines define three stable equilibria: a virgin state (V) and two immune states (I_1 and I_2). The $X_1' > 0$ region is shaded. (B) The trajectory of a switch from the virgin state to an immune state (I_1) as it is evoked by an antigen dose of $Ag_1 = 10^4$ cells.

In both immune states both clones are enlarged, i.e. both clones pro-liferate. Thus the network maintains its immunity to Ag_1 by the mutual stimulation between X_1 and X_2. The reason why X_2 is nevertheless said to be 'suppressed' in the I_1 state is that it cannot proliferate any further. Any increase in X_1 decreases the rate of X_2 proliferation, whereas increasing X_2 increases X_1 proliferation. Antigenic stimulation of X_2 also fails to evoke additional proliferation and antigen cannot be rejected: X_2 is suppressed.

Absence of fading

Now consider a system with a third clone (X_3) that interacts with X_2. We again introduce Ag_1 and, as a consequence, X_1 and X_2 switch to the I_1 equilibrium. In the I_1 equilibrium the anti-idiotypic X_2 population maintains the proliferation of the large idiotypic X_1 population (note that $X_1 \gg X_2$). It is therefore to be expected that X_2 will also be able to initiate the proliferation of the (small, i.e. virgin) X_3 population. In our model we have implicitly (and quite reasonably) assumed that it should be easier to activate all cells of a small clone than of a large clone. Thus, on reasonable grounds, we expect X_2 to initiate the proliferation of X_3. X_3 as a consequence switches to an immune state that is comparable to that of X_1 in the I_1 state (not shown). This reasoning however can be continued: X_3 is expected to initiate the proliferation of the clone(s) to which it is connected, and so on. We conclude that the idiotypic activation signal fails to fade during its propaga-tion into the network: along the propagation pathways clones keep on switching to 'immune' or 'suppressed' states comparable to those described above.

Connectivity thresholds

The percolation of the idiotypic signal through the network depends not only on its extinction rate (i.e. on the above-mentioned fading), but also on the topology of the network. In highly connected networks the propagating signal will branch several times, and may hence become extinct on one branch but proceed along another. One can analyse the connectivity proper-ties of an idiotypic network by means of graph theory (Perelson, in this volume, analyses these properties by means of percolation theory and obtains similar results). A randomly connected, symmetric, idiotypic network corresponds to an (isotrophic) random undirected graph in which the idiotypic interactions are the edges (E) and in which the clones corre-spond to nodes (N). The connectivity properties of isotrophic random undirected graphs (in which E edges connect N nodes equiprobably) were analysed by Erdos and Renyi (1960), in Kauffman (1986). Several results of random infinite graphs (in Kauffman, 1986) are of interest for our (finite) random idiotypic networks. If $E \ll N$, only small isolated structures are found, in which any node is connected to a few others. Whenever the ratio of E/N exceeds $1/2$, a threshold is reached and most nodes are interconnected in one enormous structure. As E/N increases still further, more isolated nodes are incorporated into this very large connected structure. (In finite graphs these thresholds soften to sigmoids.) In our symmetric idiotypic networks, an average of one edge per two clones ($E/N = 1/2$) corresponds to an average of one idiotypic connection per clone.

The Erdos and Renyi theory thus predicts a phase transition: once the connectivity exceeds one idiotypic connection per clone, most clones suddenly become interconnected. Note that if the connectivity of the idiotypic network were to be lower, a large proportion of the clones would not be connected to the network at all (and would hence require other immunoregulatory mechanisms). In order to investigate whether immune systems function by means of network structures, we have to assume that the network connectivity usually exceeds one connection per clone. Thus most clones are interconnected in one large structure. However, if we combine this conclusion with our 'absence of fading' results, it follows that each perturbation of the network (by, for example, antigen) eventually affects all the clones. If most clones do become affected, the network becomes unresponsive to perturbation with novel antigens: the perturbation evoked by a new antigen is generally negligible in comparison to the effect of the vigorous network interactions. Moreover, the many clones that are involved in the immune response generate semi-chaotic oscillations of clones switching between the 'immune' and the 'suppressed' state. Antigen-specific memory/immunity hence becomes an erratic (i.e. oscillatory) phenomenon.

This 'extensive percolation' problem (that is apparently implicit in reasonable assumptions and theories) is illustrated in Fig. 2. We analyse a series of networks differing in size (i.e. N) and connectivity (i.e. the average number of connections per clone: NC). For each clone an average of NC affinity values (i.e. A_{ij} elements) were drawn from a uniform distribution between 0 and 1. A randomly chosen antigen (i.e. Ag_i, $1 \leqslant i \leqslant N$) was introduced into the virgin state of the N-dimensional system (i.e. all clones are sized around S_i/D). The antigen was removed after 25 days. The network is subsequently simulated (by numerical integration) until it settles into a stable equilibrium (this may take more than 3 years). In this equilibrium we score for each clone whether it is affected by the idiotypic cascade triggered by this antigen (i.e. we score the number of virgin clones). At the Erdos and Renyi threshold ($NC = 1$, i.e. the upper line in Fig. 2) most clones remain virgin following antigenic perturbation; this is irrespective of the network size. The random antigen apparently triggers a 'small isolated structure'. At a connectivity of $NC = 4$, i.e. above the threshold, nearly all clones are affected by the idiotypic cascade; this is also irrespective of the network size. The $NC = 2$ line is at an

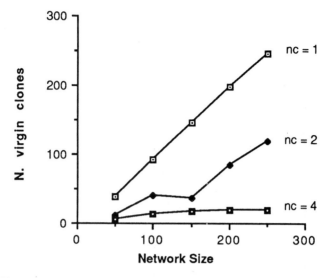

Fig. 2. The network properties of profound idiotypic networks. Each line in the figure represents a series of networks (each point is one network) varying in size (i.e. *N*). The various lines represent different connectivities in these networks (from top to bottom: *NC*=1, *NC*=2 and *NC*=4). If *NC*=1 most clones remain virgin, if *NC*=4 most clones are affected by the idiotypic cascade; this is largely independent of the size of the network.

intermediate level due to random fluctuations, i.e. its level depends on whether or not the random antigen triggers the largest connected structure.

No solutions

We have tried to overcome these unreasonable network properties in three ways (De Boer and Hogeweg, 1989b). Firstly, we used an affinity matrix which was no longer drawn from a random distribution, but which was based on complementary matches of idiotypes. Each clone was represented by a pattern of real numbers, and patterns were shifted to find the location of the optimum complementary match. (This resembles the complementary matching of bit-patterns used by Farmer *et al.*, 1986, or Perelson, in this volume.) Although these procedures generate a network topology that is different from those of random networks, we found a very similar phase transition around a connectivity of *NC* = 1. Secondly, we made inhibition more general than activation, i.e. we incorporated 'long-range inhibition' and 'short-range activation' (by using an affinity matrix for inhibition that required less optimum complementary matches). This is known to promote pattern formation (Meinhardt, in this volume), a difference in the range of inhibition and activation was also used in the shape space model (Segel, in this volume). In our model 'long-range inhibition' however failed to reduce the percolation: the idiotypic cascade involves a similar number of clones. Additionally, the network behaviour involves more (semi-chaotic) oscillations (that are initiated by the long-range interactions). In our models, long-range inhibition always implies long-range activation because clones that are inhibiting other clones will start to activate the other clones whenever they are inhibited themselves. Thirdly, we incorporated antibodies in the model. This also failed to reduce the percolation (even if we incorporated formation of antibody–antibody complexes). These additional results (De Boer and Hogeweg, 1989b) clearly support the robustness of our extensive percolation results. However, the interpretation does not become conclusive because it can always be argued that our model omits some essential element of idiotypic network theory. If this is indeed the case, we have at least demonstrated that the present idiotypic network theory is insufficient. It would be very important if this putative essential element were to be identified. However, we think the most likely explanation for our findings is that immune systems do not function by means of profound network structures. With regard to the classical idiotypic network theory we conclude that the properties of profound idiotypic network structures are unreasonable: very reasonable network models display extensive percolation of the idiotypic signal; this generates erratic immunity and a general unresponsiveness to subsequent antigenic perturbations.

HELPER T CELL PROLIFERATION

An alternative view about the functioning of immune systems is based on clonal selection theory (Burnet, 1959). We now therefore consider the cellular and/or molecular interactions among the cells constituting the 'clonal' response to one specific antigen. We have previously analysed complex models incorporating several of the various cell types that constitute the immune response to tumours (De Boer et al., 1985). We were able to show that, in these models, helper T cells (Th) played such a pivotal role that the immune response of the model could largely be characterized solely by the Th reactivity. Therefore, we here concentrate on Th proliferation, i.e. we reduce the clonal response of several cell types to that of one Th clone. All cells of this particular clone are maximally stimulated by antigen. These cells respond to this antigen in a peripheral (mouse) lymph node. For reasons of simplicity we consider a compartment within this lymph node in which all cells (of this clone) and IL2 molecules are mixed homogeneously. The compartment is seeded with cells of this clone by the thymus at a rate of S cells per day. IL2 molecules are produced locally, i.e. only by the cells in this volume.

There is a striking difference in the number of reasonable parameter estimates available for the Th proliferation model and for our simple model of B–B idiotypic interactions. The IL2 dose–response curves are known to: (1) be slightly sigmoid on a linear scale (Hooton et al., 1985), (2) reach 50% of their maximum around an IL2 concentration of 5 pM (Ashwell et al., 1986; Gullberg and Smith, 1986), and (3) be identical for proliferation and IL2-receptor binding (Robb et al., 1981). Our IL2 dose–response function $F(IL2)$ is a sigmoid saturation curve with 50% of its maximum at $IL2 = K_i$.

$$F(IL2) = \frac{IL2^U}{K_i^U + IL2^U}$$

We conceptually prefer to consider the number of IL2 molecules in our volume and not the IL2 concentration. Both IL2 production and IL2 absorption are measured in terms of molecules and not in concentrations. (In this model, however, IL2 molecules diffuse homogeneously over the compartment; thus the IL2 concentration is simply a constant fraction of our population of IL2 molecules.) One, maximally stimulated, helper T cell (Th) produces M molecules of IL2 per day. IL2 absorption depends on the presence of IL2 (according to $F(IL2)$): one cell maximally absorbs A molecules per day. IL2 decays or is removed at a rate R per molecule. The following differential equation is the formal equivalent of this verbal argument:

$$IL2' = M\,Th - A\,Th\,F(IL2) - R\,IL2 \tag{5}$$

The helper T cell population (Th) receives a constant influx (S) of precursor cells from outside the compartment (i.e. from the thymus). Th populations proliferate in response to IL2, at a maximum rate of P divisions per cell per day. The Th cells turn over at a rate D per cell per day:

$$Th' = S + P\,Th\,F(IL2) - D\,Th \tag{6}$$

For reasons of simplicity and clarity we deliberately kept this model very simple (i.e. minimal); similar phenomena, however, also occur in much more complex models incorporating antigen presentation, antigen growth, and cytotoxic T lymphocytes (De Boer and Hogeweg, 1987b).

Most parameters used in these models were determined experimentally by in vitro experiments. Due to discrepancies between the in vitro analysis and the in vivo process, and due to experimental difficulties, such parameters can only be estimated crudely. We nevertheless derive our parameters from the empirical literature because the only alternative is to choose them arbitrarily. Two parameters are entirely unknown: (1) the size of the homogeneous compartment, i.e. the rate of IL2 diffusion, and (2) the half-life (R) of IL2 in the lymph node. As a standard we choose a volume of 1 μl, and we assume a short half-life for IL2 ($R = 100$, i.e. 10 min). Both unknown parameters will be varied in our analysis. Note that because we assume that the IL2 molecules produced per helper T cell are diluted homogeneously over our standard volume, changing the size of the volume changes the IL2 concentration, i.e. changes the K_i parameter (the number of IL2 molecules required for 50% proliferation).

One helper T cell produces 4×10^7 IL2 molecules in 3 days (Vie and Miller, 1986); we round this to $M = 10^7$ molecules per cell per day. A lymphocyte absorbs 2000 IL2 molecules per hour (Gullberg and Smith, 1986); we round this to $A = 5 \times 10^4$ molecules per cell per day. The IL2 concentration at which proliferation reaches 50% of its maximum is 6.25 pM (Gullberg and Smith, 1986). However, if the experimental procedure circumvents the problem of IL2 consumption, this parameter varies around 4.4 pM (Ashwell *et al.*, 1986) (for an antigen-activated cytochrome-c-specific murine helper T cell clone). Even the latter might be an overestimate since T cells might have become refractory to IL2 during the assay (Ashwell *et al.*, 1986). For our standard volume of 1 μl we therefore round these figures to $K_i = 3 \times 10^6$ molecules (which roughly corresponds to an IL2 concentration of 5 pM). The slope of the IL2 dose–response curve is 1.45 (Hooten *et al.*, 1985); we round this to $U = 1.5$. Non-resting T cells are short-lived ($D = 1$) (Rocha, 1987); a maximum proliferation rate $P = 1.5$ then yields a 10-fold increase in 5 days (i.e. an average doubling time of 1 day). The influx of cells (S) from the thymus determines the virgin population size (in the absence of proliferation this is S/D). Influx corresponds to the antigenicity of the antigen recognized by the lymphocyte population (De Boer *et al.*, 1985).

Proliferation threshold

Figure 3A displays the Th$'$ = 0 and the IL2$'$ = 0 isoclines for various values of the influx (S) parameter. The region in which both IL2 and Th increase, i.e. the proliferation region, is shaded (for $S = 32$). If the influx of virgin cells is sufficiently small the isoclines intersect and form an attractor that is located at a low Th and IL2 population density. In this equilibrium Th cells fail to proliferate, i.e. the Th population is tolerant. Moreover, if the isoclines intersect, the model has two response modes: (1) 'proliferation' in the region where both Th and IL2 are large, and (2) 'tolerance' in the region where they are small. Because a primary immune response starts at a small Th clone size and a low IL2 concentration, stimulation of the system with antigen is not expected to evoke an immune response, i.e. to evoke Th proliferation: the system remains 'trapped' in the tolerance region. For the present parameters this tolerance region (and attractor) exists whenever $S < 10$ cells per day. If the influx of virgin precursor cells from the thymus is larger the isoclines do

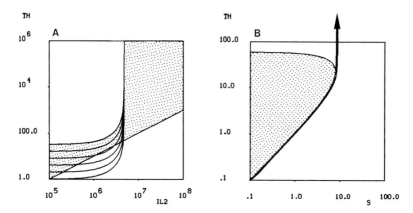

Fig. 3. The proliferation threshold. (A) The Th'=0 and IL2'=0 isoclines for various values of S (i.e. S=1, 2, 4, 8, 16, 32). The straight diagonal IL2'=0 isocline is independent of the S (influx) parameter. The bent Th'=0 isocline moves upwards if S increases. The region in which both Th and IL2 increase is shaded. The isoclines intersect (and form a 'tolerance' region and attractor) if $S<10$. (B) The Th'=0 isocline of the simplified model (no IL2 absorption and a quasi steady-state assumption for IL2). The proliferation threshold now shows as a catastrophe fold. The thick line is a trajectory corresponding to a slow increase in the S (influx) parameter. Proliferation is initiated if $S \approx 10$ cells per day.

not intersect, and the shaded (proliferation) region forms a continuum. Antigens that stimulate many ($S>10$) precursor cells, i.e. 'strong' antigens, therefore always evoke proliferation. We refer to this minimum Th population required to initiate proliferation as the 'proliferation threshold'.

Simplification

It can easily be shown (De Boer, 1989) that the form and location of the Th'=0 and IL2'=0 isoclines hardly change if the U (sigmoid) or A (absorption) parameters are varied (e.g. for $1 \leqslant U \leqslant 2$ and $10^3 \leqslant A \leqslant 10^6$). IL2 absorption plays no role in this model because one Th cell produces many more IL2 molecules than it can maximally absorb (i.e. $M \gg A$). Thus, we can omit absorption from the IL2 equation (eqn (5)) and still preserve our results. Moreover, because IL2 kinetics (production and turnover) proceed much faster than Th kinetics, we can make a quasi steady-state assumption for IL2 (trajectories do indeed run along the IL2'=0 isocline in Fig. 3A). Hence eqn (5) becomes:

$$IL2 = (M/R) \, Th \qquad (5a)$$

Figure 3B displays the simplified Th'=0 isocline (i.e. one based on eqns (5a) and (6)) as a function of the influx (S) of naive (virgin) precursor cells. The proliferation threshold now appears as a catastrophe fold. The thick line is a trajectory representing a slow increase in S: proliferation is initiated at $S \approx 10$. In this simplified model we can easily study the effect of the two unknown parameters (i.e. R and K_i). Elsewhere (De Boer, 1989) we have analysed this simplified $Th' = 0$ isocline as a function of the IL2 half-life (R) and the IL2 diffusion (K_i). The analysis demonstrates that the proliferation threshold always exists, but that its exact location (at the influx (S) axis) depends on the values of these parameters. The threshold located at $S \approx 10$ can therefore be treated as an example which is based on quite reasonable parameter estimates.

Tolerance

We conclude that Th populations that consist entirely of immuno-competent cells may fail to proliferate in the absence of any inhibitory signal. This intrinsic form of tolerance results from a paucity of IL2 in circumstances where insufficient Th cells, all producing IL2, have accumulated. The location of the threshold depends on the (1) rate of IL2 diffusion, and (2) IL2 turnover. A similar proliferation threshold exists in a variety of models (De Boer and Hogeweg, 1987a,b; Kevrekidis et al., 1988). In the present simple model the accumulation of Th cells is determined solely by the simple influx parameter S. The results become much more interesting if the model is made to account explicitly for the dynamics of, for instance, virgin and/or memory precursor subpopulations of this Th clone (De Boer and Hogeweg, 1987a). Because the number of accumulated (memory) precursor cells differs for (1) (self) antigens that are present at the time the immune system develops, and for (2) (foreign) antigens that invade the system at the time the system has fully matured, self and non-self antigens can be discriminated by simple processes that turn out to be intrinsic to Th activation and proliferation (De Boer and Hogeweg, 1987a).

Memory accumulation

Th populations (eqn (6)) in fact consist of cells in different stages of maturation. We here discern 'fully activated' cells (Th) and 'resting' cells (Tm). Activated cells become resting cells whenever they are insufficiently activated by antigen; resting cells can be re-activated by antigen. Activation with antigen is incorporated in the model by means of an arbitrary saturation function: $G(Ag)$. For reasons of simplicity, we call the resting cell stage the 'memory' stage, i.e. Tm cells. We assume our 'memory' cells to be long-lived (Th cells are short-lived). We thus propose the following model:

$$IL2 = (M/R) \text{ Th } G_h(Ag) \tag{5b}$$

$$Th' = Tm \ G_m(Ag) - Th \ (1 - G_h(Ag)) + S \ G_h(Ag) + P \ Th \ F(IL2) - D_h \ Th \tag{6a}$$

$$Tm' = -Tm \ G_m(Ag) + Th \ (1 - G_h(Ag)) - D_m \ Tm \tag{7}$$

The functions $G_m(Ag)$ and $G_h(Ag)$ specify the rates of antigenic (re)activation of Tm and Th respectively. Note that IL2 production, and the influx of activated Th cells (S), now also depend on antigenic stimulation (according to $G_h(Ag)$). D_h and D_m are the rates of Th and Tm turnover respectively (i.e. $D_h \gg D_m$). $F(IL2)$ is the above-mentioned IL2 dose–response function. We have assumed that the reactivation of memory cells requires a higher antigen concentration or higher antigen affinity than does the (re)activation of Th cells (i.e. $G_m(Ag) \ll G_h(Ag)$). However, recent empirical data (Cerrotini and MacDonald, 1989) suggest the reverse).

Self–non-self discrimination

Assume that all antigens (self or non-self) fail to initiate proliferation due to the proliferation threshold (i.e. assume that $S < 10$). This is a controversial assumption, but one which will give rise to interesting results. Self-antigens are present before the immune system develops (i.e. when $S = 0$, no influx of lymphocytes). As soon as cells recognizing this self-antigen (Ag) start emerging from the thymus they are immediately activated (i.e. $G_h(Ag)$ is saturated) and these cells persist as activated Th cells (cells are fully reactivated, i.e. no Tm are generated). Because $S < 10$ (as was assumed above), the Th population cannot proliferate: the self-antigen is tolerated. Now consider a foreign antigen and a Th clone that matches this antigen with maximum affinity. Due to the large repertoire of self-antigens and the multispecifity of lymphocyte receptors, this particular clone is expected to crossreact with some of the self-antigens. If this interaction with the self-antigens is low (if it is not low, we are again dealing with a self-reactive clone) this clone accumulates Tm cells (provided the foreign antigen remains absent). Whenever these cells emerge from the thymus they are poorly (re)activated by the crossreactive self-antigens (i.e. $G_m(Ag)$ and $G_h(Ag)$ are small); hence activated Th cells generally revert to the Tm stage. Because Tm cells are long-lived these cells persist: short-lived Th cells are replaced by long-lived Tm cells. As a consequence the total clone size (i.e. Th + Tm) increases markedly. Due to this increase the helper population will cross the proliferation threshold. Thus, whenever the foreign antigen is introduced (which saturizes $G_m(Ag)$ and $G_h(Ag)$) the accumulated memory cells are reactivated and proliferation is initiated. We conclude that these clonal Th models remain tolerant to self-antigens and respond to non-self-antigens.

Self–non-self discrimination can thus be generated without any suppressive mechanism.

In comparison to the above-analysed idiotypic network models that give percolation problems, our clonal Th proliferation models thus turn out to be very powerful. Moreover, the memory phenomenon that was generated by idiotypic interactions in the 2D network of Fig. 1 can also be accounted for by the Tm cells in the helper proliferation model. Following antigen rejection, the enlarged Th population is no longer restimulated by antigen (i.e. $G_h(Ag) = 0$) and all cells revert to the long-lived Tm stage. The Tm cells persist and account for memory if the system is restimulated with the same antigen. In conclusion, our comparison of idiotypic network models with Th proliferation models demonstrates that the 'clonal selection' systems are to be preferred.

IDIOTYPIC NETWORKS WITH Th CELLS

For immunological reasons we decided to combine our B cell idiotypic network model with our Th proliferation model. Classical B cells can only be stimulated and only proliferate and produce antibodies if they are 'helped' concomitantly by activated helper T cells. These Th cells provide growth and maturation factors that B cells require for propagation through the cell cycle. Moreover, it can be hypothesized that the incorporation of Th cells is one of the putative extensions of idiotypic network theory that might possibly reduce the extensive percolation. Th cells can be incorporated in two different ways: (1) as Th1d cells (Janeway, 1988), and (2) as classical MHC-restricted Th cells. The MHC-restricted cells give the most interesting results (De Boer and Hogeweg, 1989c); we will here discuss these results briefly. For reasons of minimalization, we have ignored the proliferation threshold in these Th–B idiotypic interactions (i.e. we have omitted the autocatalytic IL2 dynamics).

The MHC-restricted interaction between a Th and a B cell implies the processing of the antigens that the B cell presents to the Th cells. To put it briefly, B cells recognize the native antigenic protein, bind it (via the surface immunoglobulin (sIg) receptor), internalize it, and digest it into peptidic fragments. These fragments are presented to the Th cells in conjunction with class II MHC molecules. The Th receptor (TcR) recognizes this conjugate, i.e. it matches complementarily to MHC plus peptidic fragment. See Chesnut and Grey (1986) for a review of these (debated!) issues. For the present discussion it is important to establish that B and T cells 'see' antigen differently. If we now consider an idiotypic interaction between a B cell and a Th cell (i.e. a sIg–TcR interaction), we must assume that the B cell's sIg can directly match the T cell's TcR, but that the TcR can only match the sIg after it is processed and presented in conjunction with class II MHC molecules.

This interaction is therefore asymmetric: if the sIg matches complementarily to the TcR it is very unlikely that the same TcR matches the processed and conjugated form of this sIg. Thus the attractive symmetry assumption, which seems so natural in idiotypic B–B interactions, seems invalid for the idiotypic T–B interaction.

Now consider a Th–B idiotypic network (see Fig. 4). B_1 recognizes antigen (Ag), B_2 is anti-idiotypic to B_1 (this is a symmetric interaction). T_2 recognizes the processed idiotype on the surface of B_1, i.e. T_2 is anti-idiotypic to B_1. Because T_2 is closely connected to B_1, T_2 provides the 'help' that B_1 requires for proliferation. This is an asymmetric interaction: B_1 cannot recognize T_2, but T_2 recognizes and helps B_1. T_3 is similarly anti-idiotypic to B_2. If the system of Fig. 4 is perturbed with antigen, both B_1 and an antigen-specific Th cell (i.e. T_1) are activated. (This T_1 recognizes this antigen in a processed form as presented by an antigen-presenting cell.) It is generally assumed in immunology that the antigen brings B_1 and T_1 together (possible on the surface of antigen-presenting cells). Hence, the local production of lympho-kines by T_1 provides help for B_1 which results in B_1 proliferation. As a consequence, T_2 commences proliferation (being stimulated by the anti-idiotypic B_1 population). B_1 is thus further helped by an expanding T_2 population. Additionally, the anti-idiotypic B cells (i.e. B_2) are activated (i.e. receptors are crosslinked) by the expansion of B_1. However, observing the scheme of Fig. 4 (or the behaviour of our models, De Boer and Hogeweg, 1989c) we infer that B_2 cannot proliferate because these cells are not being helped. T_3, which is fully capable of helping B_2, is not activated by the idiotypic cascade (it would only become activated if B_2 were to proliferate). Thus the idiotypic signal stops at the level of T_2, i.e. the cascade involves only two clones. Because this is not really a network but is merely a Th–B cell cooperation system, we conclude that Th interactions help us to escape from

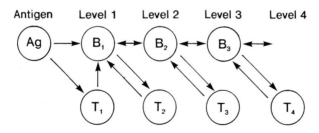

Fig. 4. A scheme of an idiotypic Th–B cell network. B_1 and an antigen specific Th cell (i.e. T_1) recognize antigen (Ag): they belong to level one. The B_1 idiotype is recognized by a B cell of the second level (i.e. B_2) and by a Th cell (T_2) that recognizes the (processed) B_1 idiotype. T_2 provides the growth and maturation factors that enable B_1 to respond to antigen (Ag) or to idiotype (B_2). B_2 is helped by T_3 (etc.). The figure shows that the anti-idiotypic response of B_2 is never initiated, because B_2's helper cells (T_3) are not activated by the idiotypic cascade.

the seemingly inevitable idiotypic network theory. Note, moreover, that (due to the asymmetry) this 2D Th–B network cannot account for memory: if antigen is rejected the sIg of B_1 no longer receives any antigenic or idiotypic signal. B_1 ceases to proliferate and reverts to the virgin state. As a consequence T_2 is no longer activated, and also reverts to the virgin state.

DISCUSSION

The most general conclusions of this overview of our previous (and present) work are: (1) that clonal immune system models are more powerful than idiotypic network models, i.e. even simple clonal Th proliferation models account for essential properties of immune systems, whereas reasonable idiotypic networks seem to imply unreasonable percolation problems, and (2) that functional idiotypic networks are not inevitable, but simply fail to develop if idiotypic B–B interactions are helper dependent (as most other immune reactions are). The second conclusion is very important for the first one, because it seems to allow us to concentrate on clonal selection models and to ignore idiotypic interactions. Further work on clonal models seems a worthwhile line to pursue. For instance, we think that the antigen presentation processes will prove to be of essential importance in immunoregulation. Switches from IgM to IgG production, from aspecific antigen presentation by monocytes to specific presentation by B cells, and from low-affinity to high-affinity immunoglobulins by somatic mutation, seem powerful ingredients for the clonal selection models that we plan to develop in the future.

Regulatory effects of profound idiotypic network structures, however, cannot be ruled out completely. Numerous idiotypic interactions have recently been described in early (developing) immune systems. It has been demonstrated that in the early repertoire of young mice, B cell clones producing IgM molecules have numerous idiotypic interactions with other B cell clones (Holmberg et al., 1986; Vakil and Kearny, 1986). The use of monoclonal antibodies has led to the description of multi-level (i.e. profound) network structures (Vakil and Kearny, 1986). These idiotypic interactions seem to play a role in the selection of the T and B cell repertoires because experimental manipulation of these early idiotypic interactions selects (positively or negatively) specific immune reactions during mature life (Vakil and Kearny, 1986). Other recent findings suggest that these early IgM networks may in fact be constituted by a special cell lineage: the $Ly1^+$ or $CD5^+$ B cells (Hayakawa, 1984). These cells occur predominantly during early life, are multispecific, and autoaggressive (Hayakawa, 1984). It seems quite likely that most of the IgM idiotypic interactions are helper independent (Th are expected to force these B cells to switch to the production of IgG). From our own results, we would indeed expect profound network structures for such multispecific and helper-independent B cells (even in the absence of antigenic triggering (De Boer, 1988; De Boer and Hogeweg, 1989b)). Analysing the possible role of the early IgM networks in repertoire selection and, possibly, in self–non-self discrimination thus seems another topic for further study.

With regard to the mature immune systems comprised of classical B and MHC restricted Th cells, i.e. the systems which display the phenomenon of antigen-specific memory and which also discriminate self from non-self, we however propose to concentrate on immune system models of clonal responses to antigen. The analogy with neural networks, which suggests that immune systems are 'cognitive' (Varela *et al.*, 1988) and 'know' about self-antigens and 'memorize' antigens has probably been pursued too far.

Acknowledgements: This work has been done in close collaboration with Dr Pauline Hogeweg. I am grateful to Ms S. M. McNab for linguistic advice.

REFERENCES

Ashwell, J. D., Robb, R. J. and Malek, T. R. (1986) Proliferation of T lymphocytes in response to IL2 varies with their state of activation. *J. Immunol.* **137**, 2572–8.

Burnet, F. (1959) *The Clonal Selection Theory of Acquired Immunity.* Cambridge University Press, Cambridge.

Cerottini, J.-C. and MacDonald, H. R. (1989) The cellular basis of T-cell memory. *Ann. Rev. Immunol.* **7**, 77–89.

Chesnut, R. W. and Grey, H. M. (1986) Antigen presentation by B cells and its significance for T–B interactions. *Adv. Immunol.* **39**, 51–94.

De Boer, R. J. (1983) *GRIND: Great Integrator Differential Equations.* Bioinformatics Group, University of Utrecht, The Netherlands.

De Boer, R. J. (1988) Symmetric idiotypic networks: connectance and switching, stability, and suppression. In *Theoretical Immunology, Part Two* (ed. A. S. Perelson), pp. 265–89. SFI Studies in the Science of Complexity, Vol. III. Addison-Wesley, CA.

De Boer, R. J. (1989) Clonal selection versus idiotypic network models of the immune system: a bioinformatic approach. Thesis, Bioinformatics Group, University of Utrecht, The Netherlands.

De Boer, R. J. and Hogeweg, P. (1987a) Immunological discrimination between self and non-self by precursor depletion and memory accumulation. *J. theor. Biol.* **124**, 343–69.

De Boer, R. J. and Hogeweg, P. (1987b) Self–nonself discrimination due to immunological nonlinearities: the analysis of a series of models by numerical methods. *IMA J. of Math. Appl. Med. Biol.* **4**, 1–32.

De Boer, R. J. and Hogeweg, P. (1989a) Memory but no suppression in low-dimensional symmetric idiotypic networks. *Bull. Math. Biol.* **51**, 223–46.

De Boer, R. J. and Hogeweg, P. (1989b) Unreasonable implications of reasonable idiotypic network assumptions. *Bull. Math. Biol.* **51**, 381–408.

De Boer, R. J. and Hogeweg, P. (1989c) Idiotypic interactions incorporating T–B cell co-operation. The conditions for percolation. *J. Theor. Biol.* (in press).

De Boer, R. J., Hogeweg, P., Dullens, H. F. J., De Weger, R. A. and Den Otter, W.

(1985) Macrophage T lymphocyte interactions in the anti-tumor immune response: a mathematical model. *J. Immunol.* **134**, 2748–58.

Erdos, P. and Renyi, A. (1960) On the random graphs. Publ. No. 5, Math. Inst. Hung. Acad. Sci.

Farmer, J. D., Packard, N. H. and Perelson, A. S. (1986) The immune system, adaptation, and machine learning. *Physica* **22D**, 187–204.

Gullberg, M. and Smith, K. A. (1986) Regulation of T cell autocrine growth. T4$^+$ cells become refractory to interleukin 2. *J. exp. Med.* **163**, 270–84.

Hayakawa, K., Hardy, R., Honda, M., Herzenberg, L. A., Steinberg, A. D. and Herzenberg, L. A. (1984) Ly-1 B cells: functionally distinct lymphocytes that secrete IgM autoantibodies. *Proc. natn. Acad. Sci. U.S.A.* **81**, 2494–8.

Hoffmann, G. W. (1979) A mathematical model of the stable states of a network theory of self-regulation. In *Systems Theory in Immunology* (eds C. Bruni, G. Doria, G. Koch and R. Strom), Vol. 32, pp. 239–57. Lecture Notes in Biomathematics. Springer-Verlag, Berlin.

Hoffmann, G. W. (1980) On network theory and H-2 restriction. In *Contemporary Topics in Immunobiology* (ed. N. L. Warner), Vol. 11, pp. 185–226. Plenum Press, New York.

Holmberg, D., Wennerstrom, G., Andrade, L. and Coutinho, A. (1986) The high idiotypic connectivity of 'natural' newborn antibodies is not found in the adult mitogen-reactive B cell repertoires. *Eur. J. Immunol.* **16**, 82–7.

Hopfield, J. J. and Tank, D. W. (1986) Computing with neural circuits: a model. *Science, N.Y.* **233**, 625–33.

Hooton, J. W., Gibbs, C. and Paetkau, V. (1985) Interaction of interleukin-2 with cells: quantitative analysis of effects. *J. Immunol.* **135**, 2464–73.

Janeway, C. A. Jr. (1988) Varieties of the idiotype-specific helper T cells – A commentary. *J. molec. Cell. Immunol.* **2**, 265–7.

Jerne, N. K. (1974) Towards a network theory of the immune system. *Ann. Immunol. (Inst. Pasteur)* **125C**, 373–89.

Jerne, N. K. (1984) Idiotypic networks and other preconceived ideas. *Immunol. Rev.* **79**, 5–24.

Kauffman, S. A. (1986) Autocatalytic sets of proteins. *J. theor. Biol.* **119**, 1–24.

Kevrekidis, I. G., Zecha, A. D. and Perelson, A. S. (1988) Modeling dynamical aspects of the immune response: I. T cell proliferation and the effect of IL-2. In *Theoretical Immunology, Part One* (ed. A. S. Perelson), pp. 167–97. SFI Studies in the Science of Complexity, Vol. II. Addison-Wesley, CA.

NAG (1984) Numerical Algorithms Group. Mayfield House, 256 Banbury Road, Oxford OX2 7DE, UK.

Robb, R. J., Munck, A. and Smith, K. A. (1981) T cell growth factor receptors: quantitation, specificity, and biological relevance. *J. exp. Med.* **154**, 1455–74.

Rocha, B. B. (1987) Population kinetics of precursors of IL2-producing peripheral T lymphocytes: evidence for short life expectancy, continuous renewal, and post-thymic expansion. *J. Immunol.* **139**, 365–72.

Vakil, M. and Kearny, J. F. (1986) Functional characterization of monoclonal auto-anti-idiotype antibodies isolated from the early B cell repertoire of BALB/c mice. *Eur. J. Immunol.* **16**, 1151–8.

Varela, F. J., Coutinho, A., Dupire, B. and Vaz, N. N. (1988) Cognitive networks: immune, neural, and otherwise. In *Theoretical Immunology, Part Two* (ed. A. S. Perelson), pp. 359–75. SFI Studies in the Science of Complexity, Vol. III. Addison-Wesley.

Vie, H. and Miller, R. A. (1986) Estimation by limiting dilution analysis of human IL2 secreting T cells: detection of IL2 produced by single lymphokine-secreting T cells. *J. Immunol.* **136**, 3292–7.

A model for the dynamics of the immune response based on idiotypic regulation

M. KAUFMAN[a] AND J. URBAIN[b]

[a]Université Libre de Bruxelles, Service de Chimie Physique, Campus Plaine U.L.B., C.P.231, Boulevard du Triomphe, B-1050 Brussels, Belgium
[b]Université Libre de Bruxelles, Départment de Biologie Moléculaire, Rue des Chevaux 67, B-1640 Rhode St Genèse, Brussels, Belgium

INTRODUCTION

Regulation of an immune response has mainly been investigated either in terms of network interactions between idiotypic and anti-idiotypic antibodies (Jerne, 1974; Richter, 1975; Urbain *et al.*, 1981; Gunther and Hoffmann, 1982; Urbain, 1986; De Boer, 1988), or in terms of circuits of antigen-specific cells with defined regulatory functions (Bretscher and Cohn, 1970; Grossman, 1984; Cohn, 1985). Both types of regulations have been experimentally demonstrated to occur, but their relative importance is not yet clear. It is most improbable, however, that antigen-specific and receptor-specific regulation operate independently of one another, and experimental data exist supporting the idea of an interplay between these two regulatory modes. In this framework we have developed a mathematical description of the humoral immune response which attempts to clarify some of the mechanisms modulating the intensity of an ongoing response. Our main purpose has been to bring out some significant feature that might explain how the magnitude of an antigenic stimulus determines the kind of response induced.

Antibody-mediated immune responses involve complex interactions between two large classes of lymphocytes, the B and T lymphocytes. B lymphocytes respond to antigen by proliferating and differentiating to plasma cells, the terminal antibody-producing cells. T lymphocytes function

Cell to Cell Signalling: From
Experiments to Theoretical Models
ISBN 0-12-287960-0

as regulatory cells and can be classified into T helper cells (T_H) which cooperate with the B cells to bring about an effective response, and T suppressor cells (T_S) which exert inhibitory effects thereby contributing to the decay in time of the response. The outcome of an immune response should depend crucially on the balance between these two opposite effects.

In many situations B and T cells reacting against some antigen have been shown to share similar idiotypic specificity (Slaoui *et al.*, 1986). Cell–cell interaction can therefore occur through antigen uptake by one cell and recognition of that antigen – in general in a processed form – by a regulatory cell, or through recognition between idiotypic and anti-idiotypic cell-surface receptors.

In our approach we consider a limited number of cell types, treating collectively all the cells belonging to each category, and focus on a small set of interactions which we believe to be especially relevant in generating a response to a given antigen. The cells come into close contact via antigen or via receptor–receptor recognition, the stimulatory or inhibitory character of the exchanged signals being, however, determined by the nature of the interacting partners.

In a first simple version of the model (Kaufman *et al.*, 1985; Kaufman and Thomas, 1987), the idiotypic interactions were introduced in an implicit way,

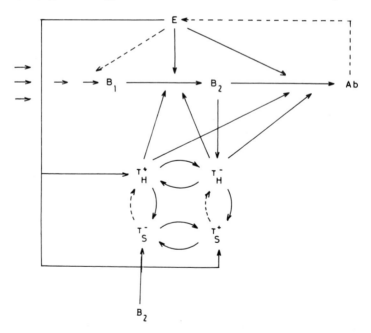

Fig. 1. Schematic interaction diagram: E, antigen; Ab, antibody; B_1, virgin B cells; B_2, more mature B cells; T_H, helper T cells; T_S, suppressor T cells. T^+ and T^- refer to antigen-specific and receptor-specific T cells respectively. Full lines indicate positive interactions or differentiation of the B cells. Broken lines correspond to negative interactions.

under the form of an antigen-independent autocatalytic component for the regulatory T_S and T_H cells. Moreover, we did not take into account the effect of antigen on the induction of the suppressor population. Here we present an extension of the model which involves explicitly a pair of antigen-specific and a pair of anti-idiotypic helper and suppressor T lymphocytes (Fig. 1). This extended version leads, for symmetric conditions, to results similar to those obtained with the previous model, indicating that the observed behaviour is not strongly dependent on the detailed structure of the regulatory loops. It offers, however, more possibilities and could therefore account for a greater variety of experimental observations.

LYMPHOCYTE INTERACTIONS AND KINETIC EQUATIONS

Our cellular network comprises two parts: the B cell differentiation pathway and a regulatory helper–suppressor T cell circuit. It can be summarized as follows (Fig. 1):

(1) Early immature B lymphocytes, which are steadily produced by the bone marrow, are highly sensitive to inactivation. At this stage, there is a negative effect of exposure to the antigen corresponding to their specificity (Nossal, 1983). The cells which escape this inactivation reach the 'virgin' B_1 stage. Antigen stimulates the B_1 cells having receptors for it, to proliferate and to differentiate to the terminal stage of production of antibody of the same specificity as the B cells. These different steps require B–T_H cooperation: helper T cells that recognize the antigen associated with the B cells or the idiotype on the immuno-globulin-surface receptors are activated to secrete lymphokines which promote B cell growth and differentiation. The antibody that is produced neutralizes the antigen, thus impairing its further inducing effect.

(2) The T helper cells that collaborate with the B cells result from the stimulation of precursor cells that were selected by the antigen or by receptor–receptor interactions, and similarly for the T suppressor cells that interact with the T_H cells expressing complementary receptors (Green et al., 1983; Dorf and Benacerraf, 1985). Suppression is here considered to act mainly on the development of the T_H population, rather than on the B cell functions. Idiotypic T–T interactions can also occur within each cellular subset (Sim and Augustin, 1983; Sim et al., 1986). We consider that they have a stimulatory influence (Hoffmann, 1980) and describe these positive regulations by cross-activation between complementary cells.

We focus primarily on the qualitative behaviour of the system and adopt a rather global description of the model in terms of the following set of differential equations:

$$dx_1/dt = k_1 F_{08}^- - k_2 x_1 (x_3 + x_4) F_{18} - d_1 x_1 \tag{1}$$

$$dx_2/dt = k_2 x_1 (x_3 + x_4) F_{18} - d_2 x_2 \tag{2}$$

$$dx_3/dt = k_h F_{38}^- F_{36}^- + k_3 F_{34} - d_3 x_3 \tag{3}$$

$$dx_4/dt = k_h F_{42}^- F_{45}^- + k_3 F_{43} - d_4 x_4 \tag{4}$$

$$dx_5/dt = k_s F_{58} + k_4 F_{54} + k_5 F_{56} - d_5 x_5 \tag{5}$$

$$dx_6/dt = k_s F_{62} + k_4 F_{63} + k_5 F_{65} - d_6 x_6 \tag{6}$$

$$dx_7/dt = k_7 x_2 F_{28} x_3 + k_7 x_2 x_4 - k_6 x_7 x_8 - d_7 x_7 \tag{7}$$

$$dx_8/dt = - k_6 x_7 x_8 - d_8 x_8 \tag{8}$$

where the x_i values, $i = 1$–8, are respectively the concentration of components B_1, B_2, T_H^+, T_H^-, T_S^+, T_S^-, Ab and E. The existence of constant levels of precursor cells is included in the corresponding kinetic constants. The detailed mechanisms of antigen presentation and lymphocyte interactions are not described. These stimulatory or inhibitory processes are represented globally by the Hill functions:

$$F_{ij} = \frac{x_j^n}{\theta_{ij}^n + x_j^n}, \quad F_{ij}^- = 1 - F_{ij} \tag{9}$$

with θ_{ij} a threshold parameter for the regulation of the ith variable by component j, and the Hill number n characterizing the steepness of the functions.

Proliferation of the B_2 cells is here assumed to be balanced by their differentiation to antibody-secreting cells. F_{08}^- and F_{18} describe, respectively, the blocking of the influx of virgin B_1 cells and their stimulation by the antigen. Antigen bridging is represented by F_{28}. We treat the concentration of T lymphokines acting on the B cells as being proportional to the concentration of the cells that produce them.

We centre our presentation on the case of a 'symmetric' T cell core, i.e. similar T^+/T^- and T^-/T^+ interactions are characterized by identical parameters (see appendix).

STEADY-STATE AND DYNAMICAL BEHAVIOUR

In the absence of antigen, the B and T pathways are uncoupled. The steady-state behaviour is then essentially determined by eqns (3)–(6) for the helper and suppressor T lymphocytes. They describe the coupling between two positive feedback loops which are known to generate multistationarity. The number of steady states is a function of the number of positive loops, and of the nature (positive or negative) and strength of the constraints between and on the loops (Thomas and Richelle, 1988; Thomas and d'Ari, 1989).

To analyse the T cell core we have combined the continuous description with a generalized logical formalization which involves logical parameters related to the kinetic constants of the continuous description (Snoussi, 1989). Although being essentially qualitative, this prior logical treatment predicts the range of parameter values to be used in the differential equations in order to observe a given behaviour. There may be up to four stable steady states. A typical situation that we shall consider here, corresponds to the coexistence of three stable states. In addition to the presence of virgin B_1 cells and no B_2 cells or antibody, these stable steady states are characterized by:

(1) the absence of T cells (virgin state, V);
(2) the presence of T suppressor cells (suppressed state, S); and
(3) a balance between T helper and T suppressor cells (memory state, M).

For the symmetric conditions considered, it can easily be shown that at these steady states: $x_3 = x_4$ and $x_5 = x_6$.

In the following we show that these stable regime states determine the outcome of an immunization as a function of the magnitude of the antigenic challenge and prior exposure to the antigen.

Primary responses

Figures 2a and c illustrate typical primary dose–response curves that are obtained with eqns (1)–(8), when given a single antigen injection starting from the virgin state (i.e. from a naïve system which has not yet been exposed to the antigen). The maximal quantity of unbound antibody is plotted versus the magnitude of the stimulus, for two sets of values of the activation rate constants k_h and k_s. The parameter values are given in the appendix.

Very low antigen doses lead to imperceptible primary responses and the system returns to the virgin state, V. Slightly higher doses lead, without the appearance of antibody to the induction of the suppressed state, S. Increasingly large doses of antigen lead to more and more effective immune responses, together with the establishment of the memory state, M. For high amounts of antigen, the memory state is still achieved, but the quantity of unbound antibody drops due to a saturation effect: whatever the magnitude of the challenge, the same *total* quantity of antibody is produced because rapid inactivation of the immature B cells becomes important in this region.

The two primary curves differ by their height, but also by the fact that in the conditions of Fig. 2c, the suppressed state is induced for a much broader range of antigen concentrations. The consequences are examined below.

Secondary responses

Extensive experimental studies have shown that prior exposure of an animal to antigen modifies its subsequent response to the same antigen. It has been

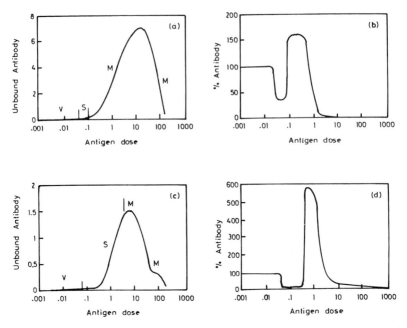

Fig. 2. (a) and (c) Primary dose–response curves. The maximum concentration of unbound antibody that is recorded after a single antigen injection is plotted versus the antigen dose used for stimulation, on a semilogarithmic scale. (a) $k_h=0.25$, $k_s=0.125$; (c) $k_h=k_s=0.1$. The letters V (virgin), S (suppressed) and M (memory) give the steady state in which the system settles when the antigen has decayed. (b) and (d) Secondary response curves. The effect of pretreatment of the system with different antigen doses (see text) is reflected in the secondary response to an optimal antigenic stimulation, x_8^*. The maximum antibody concentration observed after this optimal challenge is expressed relative to an untreated system, and plotted versus the antigen dose for pretreatment. Here, pretreatment is applied during 100 time units, followed by a rest period of 35 time units. (b) $k_h=0.25$, $k_s=0.125$ and $x_8^*=20$; (c) $k_h=k_s=0.1$ and $x_8^*=6$.

established that low or high primary doses can lead to a reduced secondary response, while medium doses increase the sensitivity to a next challenge (Mitchison, 1969). An example of experimental data on rats, taken from Shellam and Nossal (1968), is given in Fig. 4b.

Figures 2b and d summarize the dynamic behaviour obtained for our system after pretreatment. Starting from the virgin state V, one maintains a given antigen level in the system during some time interval, leaving thereafter the system to itself for a period. Then, the antibody response to a challenge which evokes an optimal response in an untreated system is recorded. The maximum concentration of antibody measured after the secondary injection is expressed in percentage of the maximum that is observed for an untreated system, and plotted as a function of the antigen dose used for pretreatment. Similar results are also obtained upon a single primary antigen injection as described above, with a shift of the curves toward higher antigen concentrations.

The model accounts for both low- and high-dose paralysis, with an intermediate region of increased response. Enhanced immunity is brought about by the establishment of the memory state during the primary encounter with the antigen. In low-dose paralysis, pretreatment activates preferentially the suppressor T cells which inhibit the development of the helper population needed for the differentiation of antibody-secreting cells. This suppression can only be achieved starting from a virgin system. Once the memory state settles, the suppressed state can no longer be induced. High-dose paralysis results from the strong inhibition of immature B cells: no antibody can be produced for a period after pretreatment. Inactivation of immature B lymphocytes has been described here in a very schematic way. It probably involves in itself a whole cellular circuitry that might include some type of T_s cells (Nossal, 1983). Recovery from high-dose paralysis occurs spontaneously and gradually. It reflects the generation time of new immunocompetent B lymphocytes in the absence of residual antigen. Departure from the suppressed state, on the contrary, has to be induced by conditions which favour the generation of T helper cells.

In Fig. 2b high-dose paralysis corresponds to a stronger inhibition than low-dose paralysis, which is furthermore observed for a narrow antigen concentration range. As shown on Fig. 3a suppression is lifted after a secondary optimal antigenic challenge. Figures 2c and d correspond to a lower activation rate of the T helper cells. This could, for example, result from a poor antigen presentation, a situation which is known experimentally

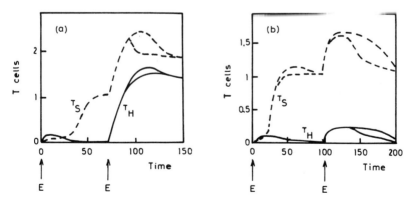

Fig. 3. Time evolution of the concentrations of T_H^+, T_H^-, T_S^+ and T_S^- cells upon two successive antigenic stimulations. (a) $k_h=0.25$, $k_s=0.125$. The first stimulation, with a low antigen dose ($x_8=0.055$), brings the system to the S state. After a second optimal stimulation ($x_8^*=20$), the M state is reached. (b) $k_h=k_s=0.1$. A first low antigen dose ($x_8=0.07$) induces the S state and the system remains in this suppressed state after a second optimal injection ($x_8^*=15$).

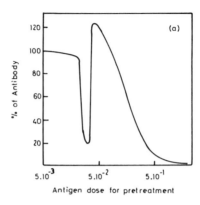

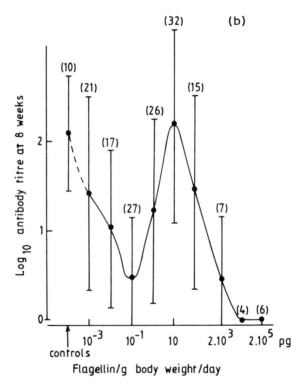

Fig. 4. (a) Dose–response curve after pretreatment obtained with the simpler version of the model involving a two-variable T cell core. (Redrawn from Kaufman and Thomas, 1987.) (b) Experimental illustration of a secondary dose–response curve showing two zones of paralysis separated by an immunity zone. (Redrawn from Shellam and Nossal, 1968.) Induction of paralysis in rats is obtained by daily injections of various doses of flagellin for 2 weeks followed by 10 μg of flagellin twice weekly until 10 weeks age. The number of animals in each group is listed in parentheses.

to favour the induction of paralysis. It leads here to a low primary response, however, once the memory state has been reached, the response to a next challenge is greatly enhanced. Low-dose paralysis is stronger and more easily achieved than for the first set of parameter values. After the secondary challenge the system settles back in the suppressed state (Fig. 3b) and recovery from paralysis requires particular manipulations such as, for example, the injection of a certain amount of already activated T_H cells.

The results presented here for the extended model are in good qualitative agreement with those obtained previously (Kaufman et al., 1985; Kaufman and Thomas, 1987) (Fig. 4a). The important point seems to be the positive character of the feedback loops implying the T_H or T_S cells, rather than the details of their structure and they could possibly consist of several steps. In the simpler version, idiotypic and anti-idiotypic T_H and T_S cells are lumped together to form a two-variable core which can easily be analysed by classical phase-plane techniques. The various situations described here can then be understood in terms of the size of the basins of attraction of the steady states as a function of some key parameters and the antigen level (Kaufman and Thomas, 1987; Kaufman, in preparation).

A critique of our model has been raised by Hoffmann (1987). First of all, the basic premises of the clonal selection theory are incorporated in the model: monospecificity of lymphocytes and triggering of lymphocytes pre-committed before antigen arrival. B and T lymphocytes interacting in our cellular network are selected and activated through antigen- or idiotype-specific mechanisms. Once triggered, the B cells proliferate and differentiate into monospecific antibody-producing cells. This basic formulation of the immune response is known to be substantially correct. Clonal expansion of the T cells, on the other hand, has not been included in the equations. Introduction in our two-variable-core model of additional terms describing antigen-dependent T cell proliferation does not lead, for reasonable parameter values, to *qualitative* changes (Kaufman, 1988). This is in agreement with experimental work by Melcher and Eichmann (1986) or Mitchison and Petersson (1983) suggesting that clonal proliferation of the T cells plays a minor role in the regulation of antibody responses.

Secondly, unlike the predictions of Hoffmann, and as surmised by a previous logical analysis of several alternative T cell regulatory cores (Kaufman, unpublished), our analysis remains valid when the T helper and T suppressor compartments explicitly consist of complementary pairs of anti-gen-specific and anti-idiotypic classes of cells, with direct interaction between the antigen and idiotypic T_S cells. As already stressed above, our results are essentially linked to the presence of positive feedback loops, an aspect which enters in the framework of symmetric network interactions put forward by Hoffmann himself (1980). Furthermore, interaction of antigen with both classes of T cells increases the variety of situations thereby leading to an enrichment of the dynamic properties of the model.

CONCLUDING REMARKS

Our approach is based on the concept that the immune system displays regulatory pathways in the absence of foreign antigen. This leads to an interpretation of the various modes of the humoral response in terms of transitions between multiple steady states. These switches are determined by the strength of the antigenic stimulus and the immunogenicity of the antigen. A previous encounter with the antigen can therefore lead to a state of enhanced response or to a state of paralysis.

The results account for a number of typical features of induction of paralysis to protein antigens, both as regards the type of cells that are engaged, as well as the conditions that have to be met to establish the tolerant states. At low sub-immunogenic doses, only the T cells are involved in the induction and maintainance of paralysis, while at high supraimmunogenic doses anergy of the B cells is the most important factor. In addition to the dosage requirements, the ability to induce paralysis, the extent and the duration of this non-responsiveness depend on the immunogenic capacity of the antigen and the regimen of administration. In accordance with the experiments, low-dose paralysis is more difficult to bring about with a strong immunogen than with a weakly immunogenic protein, which may furthermore lead to a prolonged state of unresponsiveness. Brief, prolonged or intermittent exposure, on the other hand, affects the response differently as a function of dose and immunogenicity.

Memory of the past antigenic history has here been studied mainly at the level of the regulatory T cells. One may consider, in addition, that during a successful immunization, a fraction of the stimulated B cells differentiate into B memory cells. Preliminary calculations (Kaufman, unpublished) have shown that upon successive challenges, higher and higher levels of antibody are then being produced. Moreover, high-dose paralysis can no longer be induced after an efficient response, in good agreement with the data.

Our description does not include a full characterization of all the cells and processes which are involved in the regulation of an humoral immune response. In spite of severe simplifications it gives, however, a reasonable view of several important experimental observations.

Acknowledgements: This work is supported in part by the Belgian programme on interuniversity attraction poles and in part by the CEE 'Actions de Stimulation' no. ST2J0385-C (GDF).

REFERENCES

Bretscher, P. and Cohn, M. (1970) A theory of non-self discrimination. *Science, N.Y.* **169**, 1042–9.

Cohn, M. (1985) In *Neural Modulation of Immunity* (eds R. Guillemin *et al.*), pp. 3–23. Raven Press, New York.

De Boer, R. J. (1988) Symmetric idiotypic networks: Connectance and switching, stability, and suppression. In *Theoretical Immunology, Part Two* (ed. A. S. Perelson), pp. 265–89. Addison-Wesley, CA.

Dorf, M. E. and Benacerraf, B. (1984) Suppressor cells and immunoregulation. *A. Rev. Immunol.* **2**, 127–57.

Green, D. R., Flood, P. M. and Gershon, R. K. (1983) Immunoregulatory T cell pathways. *A. Rev. Immunol.* **1**, 439–63.

Grossman, Z. (1984) Recognition of self and regulation of specificity at the level of cell populations. *Immunol. Rev.* **79**, 119–38.

Gunther, N. and Hoffmann, G. W. (1982) Qualitative dynamics of a network model of regulation of the immune system: A rationale for the IgM to IgG switch. *J. theor. Biol.* **94**, 815–55.

Hoffmann, G. W. (1980) On network theory and H-2 restriction. *Contemp. Topics Immunobiol.* **11**, 185–226.

Hoffmann, G. W. (1987) A critique of the Kaufman–Urbain–Thomas immune system network theory. *J. theor. Biol.* **129**, 355–7.

Jerne, N. K. (1974) Towards a network theory of the immune system. *Ann. Immunol. (Inst. Pasteur)* **125C**, 373–89.

Kaufman, M. (1988) Role of multistability in an immune response model: a combined discrete and continuous approach. In *Theoretical Immunology, Part One* (ed. A. S. Perelson), pp. 199–222. Addison-Wesley, CA.

Kaufman, M. and Thomas, R. (1987) Model analysis of the basis of multistationarity in the humoral immune response. *J. Theor. Biol.* **129**, 141–62.

Kaufman, M., Urbain, J. and Thomas, R. (1985) Towards a logical analysis of the immune response. *J. theor. Biol.* **114**, 527–61.

Melchers, I. and Eichmann, K. (1986) T cell regulation without clonal selection by antigen? Polygamous suppressor cells and monogamous helper cells at high frequencies. In *Paradoxes in Immunology* (eds G. W. Hoffmann, J. G. Levy and G. T. Nepom), pp. 137–48. CRC Press, Boca Raton, Florida.

Mitchison, N. A. (1969) In *Regulation and the Antibody Response* (ed. B. Cinader). Thomas, Springfield, Illinois.

Mitchison, N. A. and Petersson, S. (1983) Does clonal selection occur among T cells? *Ann. Immunol. (Inst. Pasteur)* **134D**, 37–45.

Nossal, G. (1983) Cellular mechanisms of immunological tolerance. *A. Rev. Immunol.* **1**, 33–62.

Richter, P. H. (1975) A network theory of the immune system. *Eur. J. Immunol.* **5**, 350.

Shellam, G. R. and Nossal, G. J. V. (1968) Mechanisms of induction of immunological tolerance. *Immunology* **14**, 273–84.

Sim, G. K. and Augustin, A. A. (1983) Internal images of major histocompatibility complex antigens on T-cell receptors and their role in the generation of T-helper cell repertoire. In *Immune Networks* (eds C. A. Bona and H. Kohler). *Ann. N.Y. Acad. Sci.* **418**, 272–81.

Sim, G. K., MacNeil, I. A. and Augustin, A. A. (1986) T helper cell receptors: idiotypes and repertoire. *Immunol. Rev.* **90**, 49–72.

Slaoui, M., Urbain, J., Lowy, A., Monroe, J., Benacerraf, B. and Green, M. I. (1986) Anti-idiotypic treatment of BALB/c mice induce CRI_A bearing suppressor cells with altered Igh restricted function. *J. Immunol.* **136**, 1968–73.

Snoussi, E. H. (1989) Qualitative dynamics of piece-wise differential equations: a discrete mapping approach. *Dynamics and Stability of Systems* (in press).

Thomas, R. and d'Ari, R. (1989) In *Biological Feedbacks*. CRC Press, Boca Raton, Florida (in press).

Thomas, R. and Richelle, J. (1988) Positive feedback loops and multistationarity. *Discrete Appl. Math.* **19**, 381–96.

Urbain, J. (1986) Idiotypic networks: a noisy background or a breakthrough in immunological thinking: the broken mirror hypothesis. *Ann. Immunol. (Inst. Pasteur)* **137C**, 57–100.

Urbain, J., Wuilmart, C. and Cazenave, P. A. (1981) Idiotypic regulation in immune networks. *Contemp. Topics molec. Immunol.* **8**, 113–48.

APPENDIX

The numerical simulations were performed with the following parameter values:

(a) B differentiation pathway

$k_1 = 1, k_2 = d_1 = d_2 = 0.1, k_7 = 0.4, k_6 = d_7 = 0.2, d_8 = 0.05$

$\theta_{08} = \theta_{28} = 0.1, \theta_{18} = 1$

(b) T cell regulatory core

$k_3 = 0.42, k_4 = 0.21, k_5 = 0.23, d_i = 0.2$ for $i = 3\text{--}6$.

$\theta_{38} = \theta_{58} = \theta_{42} = \theta_{62} = 0.1$

$\theta_{36} = \theta_{45} = \theta_{hs}, \theta_{34} = \theta_{43} = \theta_{hh}, \theta_{54} = \theta_{63} = \theta_{sh},$

$\theta_{56} = \theta_{65} = \theta_{ss}$ with $\theta_{sh}/\theta_{hh} = 0.95, \theta_{ss}/\theta_{hs} = 0.35$

The concentrations of T_H^+ (T_H^-) and T_S^+ (T_S^-) cells are reduced to the threshold values θ_{hh} and θ_{hs}, respectively.

$n = 2$ for all the Hill functions, and some non-linear Hill functions may be replaced by simple linear functions (Kaufman and Thomas, 1987) without qualitative modifications. Similar results are obtained for a broad range of parameter values.

Evolution of tumours attacked by immune cytotoxic cells: The immune response dilemma

R. LEFEVER,[a] J. HIERNAUX[a] AND P. MEYERS[b]

[a]Service de Chimie Physique, Université Libre de Bruxelles, Campus Plaine C.P.231, B-1050 Brussels, Belgium
[b]Laboratoire de Physiologie Animale, Université Libre de Bruxelles, 67, rue des Chevaux, B-1640 Brussels, Belgium

INTRODUCTION

Freely evolving tumours frequently exhibit multiphasic growth patterns. Figure 1, for example, shows a tumour which evolves in three distinct phases: (i) an initial progressive phase; (ii) a regressive phase during which the tumour's size decreases; and (iii) a second progressive phase which finally kills the host. Striking examples of multiphasic patterns are also observed in tumours which evolve towards a dormant state (Wheelock, 1981), or in tumours the cellular populations of which oscillate in the course of time. In the first case, as in Fig. 1, the initial progressive phase is interrupted by a regressive phase; subsequently, however, the resumption of the progressive process does not take place. Instead the tumour stays in a so-called dormant state where a reduced, approximately constant, tumour cell population survives. It is to be noted that dormancy is a property of the tumour rather than a property of the cells; the term should not be misunderstood as meaning that in a dormant tumour the cells no longer replicate. In the case of oscillatory tumours, numerous progressive and regressive phases alternate. An example in humans is granulocytic leukaemia, where oscillations of circulating white blood cell counts with a periodicity of 70 days have been reported in an untreated patient for several years (Gatti et al., 1973).

These examples of multiphasic growth patterns indicate that in many tumours cancer cells are not only continuously produced but are also continuously killed. In the absence of treatment, the unperturbed evolution

Cell to Cell Signalling: From
Experiments to Theoretical Models
ISBN 0–12–287960–0

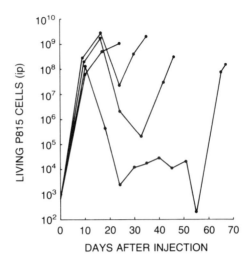

Fig. 1.   Patterns of tumour growth. Four DBA/2 mice were injected i.p. with 600 living P815 cells (P815 is a mastocytoma of DBA/2 origin). At various times thereafter, samples of the i.p. fluids were collected and diluted in culture medium that contained 0.4% Bactoagar. After a 3–4 day incubation at 37°C, the P815 colonies that contained more than six cells were counted. The number of living i.p. P815 was estimated, taking into account the cloning efficiency of *in vitro* growing P815, which was ±0.3. Peritoneal exudate cells of control animals do not form colonies under these conditions. Each curve represents an individual mouse and ends on the last sampling before the death of the animal.

of many tumours is the outcome of a competition between the progression mechanisms of malignant proliferation, i.e. in the first place the mechanism of replication of the cancerous cells and the systemic and the environmental factors which favour this replication, and the regression mechanisms, i.e. all the mechanisms that participate in cancerous cells' destruction and involve the host homeostatic defences. Furthermore, the variety of multiphasic patterns, illustrated by the above examples, suggests that there exists a variety of tumoural states, which differ by their dynamical properties. These differences manifest themselves at the cellular population level and cannot therefore be apprehended solely in terms of individual cell properties. In this perspective, multiphasic patterns reflect transitions phenomena taking place at the population rather than cellular level and which are due to shifts in the balance between progressive and regressive mechanisms.

It is generally admitted that the main natural defence against tumours is the immune system. Different immuno-dependent anti-tumour killer cells have indeed been identified: cytotoxic T lymphocytes (CTL) (Gorczynski, 1974), activated macrophages (Adams and Snyderman, 1979), natural killer (NK) cells (Heberman and Ortaldo, 1981) and antibody-dependent K cells (Adams *et al.*, 1984). These so-called effector cells kill cancer cells in cytolytic cycles which are part of a complex network of cellular and metabolic interactions. These interactions may modulate the production and the properties of effector cells as well as the malignant cell replication mechanism. They may also transform the malignant cells so that the latter lose their tumoural antigens and become unrecognizable by the effector cells. A relationship between tumour regression and the production of immune effector cells or between immunosuppression and tumour progression is supported by several experimental findings (see, for example, Broder *et al.*, 1978; Greene, 1980; North, 1984). In one experimental system, there is a correlation between the onset of immunosuppression and the size of the tumour (Bear, 1986). Tumour progression has also been related experimentally to the loss of key tumour-associated antigens in the stimulation of CTL (Uyttenhove *et al.*, 1983). These facts clearly illustrate the interdependence of neoplastic growth and of the anti-tumour immune response.

The relation between tumour dynamics and the immune system is, however, far from being an intuitively simple one. Not only, as mentioned above, do distinct multiphasic growth patterns occur, but furthermore a series of paradoxical situations is known where tumours escape an ongoing anti-tumoural immune response in quite unexpected manners: (i) immunostimulation of tumour growth has been described, i.e. a situation where immune effector cells seem to stimulate neoplastic proliferation (Prehn, 1977); (ii) dormancy which has already been mentioned and where a small number of tumour cells displaying immunogenic surface antigens can survive in a host which has destroyed a large quantity of progressive tumour cells (Wheelock, 1981; Hiernaux *et al.*, 1986); (iii) concomitant immunity where a second implant of the original tumour is rejected, whereas the progressive growth of the initial implant persists (Tuttle *et al.*, 1983). These paradoxical observations challenge the idea that tumour regression results from an active immune response.

We shall develop here the point of view that:

(i) In general the immune system's response to the presence of immunogenic tumoural cells is essentially aimed at the production of effector cells which recognize the cancer cells and kill them in a cell-mediated cytolytic cycle in which the processes of cancer cell recognition and lysis are kinetically independent.

(ii) As a result of the kinetic independence between recognition and lytic processes, the immune response inevitably meets, in the case of some strongly immunogenic tumours, a situation which we call the immune response dilemma. The feature characteristic of this situation is that it corresponds to dynamical conditions where any further increase in the recognition ability, i.e. specificity, of the effector cells, or any further increase in their number can only lead to a stabilization of the tumoural state.

(iii) Multiphasic patterns, as well as the paradoxical behaviours which seemingly challenge the notion that tumour regression results from an active immune response, are various aspects of the dynamics associated with the immune response dilemma. Two principal parameters govern this dynamics. First, a switching parameter β in terms of which a transition point can be defined which separates the situations for which tumour progression is certain, from the situations in which tumour regression becomes possible or certain. In other words, it defines sufficient conditions for progression and necessary conditions for regression. The second parameter playing a central role in tumour's dynamics is the avidity parameter κ. It expresses whether in the cytolytic cycle it is the recognition or the lytic process which is rate limiting. Depending upon the values of β and κ, tumours classify into three categories as far as the possibilities of immune rejection are concerned.

THE CYTOLYTIC REACTION PATHWAY AND THE SWITCHING PARAMETER β

In the following we focus on the cellular reactions between cancerous (target) cells and killer (effector) cells. We analyse how the conditions for growth or regression arise from the competition between target cells multiplication and the anti-target cell cytolytic reactions mediated by effector cells. More precisely, we consider the cellular reaction scheme (Hiernaux *et al.*, 1986; Meyers *et al.*, unpublished)

$$C \xrightarrow{\quad \lambda \quad} 2C \tag{1}$$

$$E_0 \xrightarrow[C]{\quad nb_0 \quad} E_1 \xrightarrow[C]{\quad (n-1)b_1 \quad} E_2 \xrightarrow[C]{\quad (n-2)b_2 \quad} \cdots \xrightarrow[C]{\quad b_n \quad} E_n$$

$$\downarrow{\scriptstyle l_1} \qquad\qquad \downarrow{\scriptstyle 2l_2} \qquad\qquad\qquad \downarrow{\scriptstyle nl_n} \tag{2}$$

$$E_0 + D \qquad\quad E_1 + D \qquad\qquad\quad E_{n\text{-}1} + D$$

in which λ is the cancer cell's (C) replication constant, b_i and l_i $(i = 0, \ldots, n)$, are the effector cells (E_i) binding and lysis constants. The cytolytic cycle schematized in (2) involves two kinds of processes which, in short, we call binding and lysis. This terminology, however, recovers two complex and kinetically independent sets of reactions. More precisely:

(i) Binding refers to the sequence of events associated with target recognition and the binding of the effector's receptors with the target's antigens. Effector cells, such as CTL, bear a large number of receptors (of the order of 10,000) recognizing the antigens present on the surface of tumoural cells. Therefore they easily form multicellular conjugates E_i in which a single CTL is simultaneously conjugated with i targets ($i = 1, \ldots, n$) (Zagury *et al.*, 1979). In practice, for steric reasons n is rarely larger than 3 or 4; exceptionally with very pure, homogeneous effector cell populations n may rise up to 10.

(ii) Lysis refers to the sequence of events associated with the killing of C and the recycling of E_i. It begins with the delivery of the lethal hit and ends with the dissociation of the effector/target complex E_i into a programmed to death target (D) and a functional effector/target complex E_{i-1} susceptible to participate anew in an identical cytolytic cycle.

The distinction between the recognition–binding process and the killing–recycling process is essential since it corresponds to processes relying on separate cellular compartments of the effectors: on one side, the surface membrane receptors which specifically recognize the tumoural antigens; on the other side, the cytoplasmic metabolic machinery responsible for the delivery of the lethal hit. The decomposition of the cytolytic process into these two steps is also experimentally supported by the fact that they are differently influenced by chemical factors, e.g. binding works with a variety of divalent cations while, without affecting the binding process, it is possible to suppress the lysis and hence the recycling by only diminishing the concentration in Ca^{2+}.

The balance between the proliferative step (1) and the cytolytic cycle (2) obviously depends on the parameters associated with these processes. The latter are not constant in time. During tumour development, the complex network of interactions which controls the functioning of the immune system and the properties of cancerous cells may change the value of any parameter entering (1–2). We shall not discuss here how or why these changes take place. This question relates to the organization and functioning of the immune system at the organismic level; it is still the object of much heated debates (see, for example, the chapters by Urbain and De Boer in this volume). Here we simply admit the existence of variations for the parameters λ, b_i and l_i; our aim is to investigate how these variations influence tumour dynamics and how they allow, in terms of a minimal set of dimensionless numbers, to apprehend the existence of a multiplicity of dynamical regimes and switching phenomena.

We first introduce a dimensionless parameter which describes in the reaction scheme (1–2) the balance between progression and regression. Consider, at time t and at point r, some tumour volume element which contains N cancer cells and E_T cytotoxic cells:

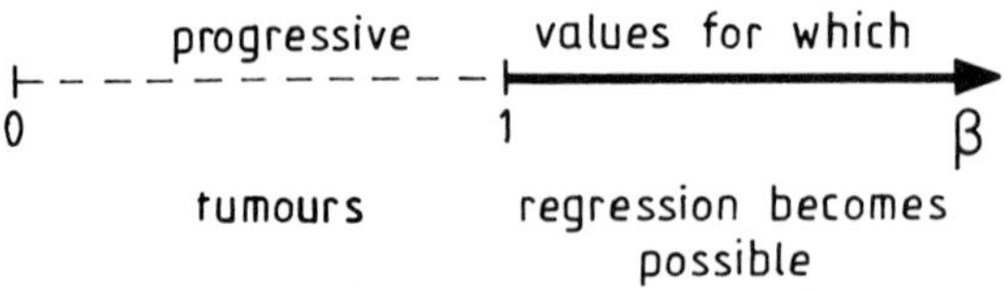

Fig. 2.   Switching parameter β. The values β∈(0,1) correspond to conditions where tumours are always progressive. A necessary (but not sufficient) condition for regression to take place is that β∈(1,∞).

$$E_{\mathrm{T}} = \sum_{i=0}^{n} E_i(r,t) \tag{3}$$

It is then intuitively clear that the dimensionless quantity

$$\beta \equiv \frac{nb_0 E_{\mathrm{T}}}{\lambda N} = \frac{\tau_{\mathrm{r}}}{\tau_{\mathrm{b}}} \tag{4}$$

plays the role of a switching parameter (Fig. 2) for the balance between progression and regression. If τ_{b}, the relaxation time of the binding process, is greater than τ_{r}, the relaxation time of the replication, i.e. if

$$\beta < 1 \tag{5}$$

the cytolysis rate is always smaller than the cancer cell proliferation rate. The occurrence of a progressive tumour is then certain: the normal tumourless state of the tissue is unstable with respect to any perturbations corresponding to the inoculation of however few cancer cells. The value $\beta = 1$ is a switching point: for tumour rejection to become possible, it clearly is necessary that binding be faster than replication, i.e. that

$$\beta > 1 \tag{6}$$

The tumourless state is then stable with respect to the inoculation or the 'spontaneous' appearance of a small number of cancer cells. Whether (6) is sufficient to insure the rejection of a fully developed tumoural state, as we show later, depends on whether or not the lytic process in (2), as compared to the binding, is rate limiting. In order to establish this result, we first present the kinetic equations governing the evolution of the cellular populations. This is done in the next section.

AN EVOLUTION EQUATION FOR THE STEADY-STATE GROWTH OF SOLID TUMOURS

The discussion which follows refers more particularly to the case of solid tumours. To start with, we consider the proliferative process (1). We derive an evolution equation governing the spatio-temporal behaviour of a cellular population which replicates in the absence of cytotoxic cells. For simplicity, we deal with one spatial coordinate r and admit that the size of the cells is constant and equal to a. The growth process can then be viewed as taking place on a one-dimensional lattice whose points are at a distance a from each other and occupied or not by a cell. The cells on the lattice do not move; the occupation of new lattice points by more cells can only occur via the replication of the cells located at neighbouring points. Accordingly, if $c(r,t)$ is the probability that at time t the lattice site at point r is occupied by a cancer cell, we may write that $c(r,t+\Delta t)$ is given by (we only sketch here the procedure for obtaining the desired evolution equation; a more detailed and general derivation encompassing a broader class of situations will be reported elsewhere (Lefever, unpublished)):

$$c(r,t+\Delta t) = c(r,t) + \frac{\lambda N \Delta t}{2}[1 - c(r,t)] \{c(r - a,t)[1 - c(r + a)]$$

$$+ [1 - c(r - a,t)]c(r + a,t) + 2c(r - a,t)c(r + a,t)\} + O[(\Delta t)^2] \qquad (7)$$

Taking the limit $\Delta t \to 0(\Delta t \ll (\lambda N)^{-1})$, expanding $c(r \pm a,t)$ in Taylor series around r and retaining terms up to $O(a^2)$, we straightforwardly obtain that the spatio-temporal evolution of the C cell population obeys the partial differential equation

$$\frac{\partial c}{\partial t} = \lambda N c (1 - c) + \frac{\lambda N a^2}{2}(1 - c) \left(\frac{\partial^2 c}{\partial r^2} \right) \qquad (8)$$

The meaning of the first (logistic) term in (8) is evident. It expresses the fact that at low cancer cell densities, i.e. for $c(r,t) \ll 1$, the population grows exponentially, while for $c(r,t) \to 1$, growth slows down to zero: the cells become completely close packed and their number per unit volume locally no longer changes. The second term in (8) accounts for the spreading of the tumour over space. In the region where $c(r,t) \to 1$, i.e. inside the tumour but far from its boundary, this term is negligible and the distribution of C is nearly uniform. On the other hand, outside the tumour, the distribution of C is (trivially) uniform; in fact equal to zero: $c(r,t) = 0$. Obviously this corresponds to the normal tumourless state of the tissue. As the simulations of Fig. 3 indicate, (8) admits heteroclinic solutions joining the tumour state

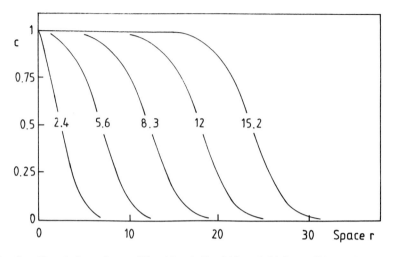

Fig. 3. Simulation of eqn (8) with $c(r,0)=\delta(r)$ as initial condition. The curves represent the spatial profile of the tumour at the times indicated (the curves are symmetrical with respect to the axis $r=0$). The growth of the tumour is modelled by a wave-like solution of (8), which after the initial transient regime, propagates over space at a constant velocity and without deformation.

$c = 1$ to the normal state $c = 0$. In the framework of our approach, tumour growth is thus modelled by a soliton type of solution propagating without deformation and at constant velocity over space; this propagation is mainly due to the replication of cells at the boundary of the tumour, i.e. in the region where $c(r,t)$ is different from zero or one and where $\partial c/\partial r$ is not vanishingly small.

Incorporating the cytolytic process (2) into the description yields the set of kinetic equations (in order to retain the essential features of the reaction scheme, it is sufficient to discuss the case $n=2$):

$$\frac{\partial c}{\partial t}=\lambda Nc(1-c)-2b_0E_Te_0c-b_1E_Te_1c+\frac{\lambda Na^2}{2}(1-c)\left(\frac{\partial^2 c}{\partial r^2}\right) \qquad (9)$$

$$\frac{\partial e_0}{\partial t}=-2b_0Ne_0c+l_1e_1+D\left(\frac{\partial^2 e_0}{\partial r^2}\right) \qquad (10)$$

$$\frac{\partial e_1}{\partial t}=2b_0Ne_0c-b_1e_1c+2l_2e_2-l_1e_1 \qquad (11)$$

$$\frac{\partial e_2}{\partial t}=b_1Ne_1c-2l_2e_2 \qquad (12)$$

which accounts for the competition between cancer cell proliferation and cell-mediated lysis. In writing (9)–(12), in order to avoid non-essential complications, we have neglected the presence of dead cancer cells on the lattice. The e_0, e_1, e_2 are the fractions of effector cells having bound 0, 1 or 2 target cells; they satisfy the conservation relation (cf. 3):

$$e_0 + e_1 + e_2 = 1 \tag{13}$$

Contrary to the tumoural cells, the free effectors E_0 are mobile. This motion is taken into account in (10) by the inclusion of a Fickian diffusion term.* Typically, the value of the 'diffusion' coefficient D, estimated from the rate at which the cells move over space, is of the order of

$$D \approx 10^{-9}\,\mathrm{cm^2\,s^{-1}} \tag{14}$$

This value has to be compared with that of the factor $a^2 \lambda N$ which in (9) multiplies the spatial derivative and which determines the rate of spatial progression of the tumour. Taking $a \approx 10^{-3}\,\mathrm{cm}$ and $\lambda N \approx 1\,\mathrm{day^{-1}}$ yields

$$a^2 \lambda N \approx 1.2 \times 10^{-11}\,\mathrm{cm^2\,s^{-1}} \tag{15}$$

which indicates that the time scale characterizing the spreading of the tumour is at least two orders of magnitude smaller than the one related with the infiltration of the tumour by the effector cells. Furthermore, while the time scale of (1) is $O(\mathrm{day^{-1}})$, that of (2) is $O(\mathrm{h^{-1}})$. This suggests that the effector cells quickly equilibrate in response to variations of $c(r,t)$, and therefore may be eliminated adiabatically in (9)–(12). Using this approximation, we obtain the simplified evolution equation:

$$\frac{\partial c}{\partial \tau} = c(1-c) - \beta \frac{(1+\kappa_0\rho c)c}{1+2\kappa_0 c + \kappa_0\kappa_1 c^2} + (1-c)\left(\frac{\partial^2 c}{\partial s^2}\right) \tag{16}$$

where time and space have been rescaled as

$$\tau = \lambda N t, \quad s = \frac{(\sqrt{2})r}{a} \tag{17}$$

and where

$$\kappa_0 \equiv \frac{b_0 N}{l_1}, \quad \kappa_1 \equiv \frac{b_1 N}{l_2}, \quad \rho \equiv \frac{b_1}{b_0} \tag{18}$$

Let us now examine the behaviours predicted on the basis of (16).

*In reality this may be a rather crude approximation. The effector cell motion generally is not simply random. A more refined treatment would take into account the possible existence of chemotactic effects.

THE AVIDITY PARAMETER κ AND THE IMMUNE RESPONSE DILEMMA

The second term of (16) represents the effect of cytolysis on the cancer cell population. Several authors have emphasized the mechanistic similarities existing between effector cell cytolytic cycles and enzymatic catalytic cycles (Thorn and Henney, 1976; Ziejlemaker *et al.*, 1977; Callewaert *et al.*, 1978; Garay and Lefever, 1978; Lefever and Garay, 1978). This led to models which postulate a Michaelian kinetics for the cytolytic process. Applied to *in vitro* cytotoxic assays, this approach has in several instances provided reasonably good fits of the experimental data. We examine here the predictions obtained in the framework of this assumption concerning the balance between progression and regression mechanisms.

The kinetics of the reaction scheme (2) is Michaelian when all binding constants b_i and all lytic constants l_i are equal:

$$b_i = b, \; l_i = l, \; \kappa_i = \frac{b_i N}{l_{i+1}} = \frac{bN}{l} \equiv \kappa, \text{ for all } i = 0, \ldots, n \qquad (19)$$

Indeed, the expression for the rate of cytolysis

$$V_{cyt} = \frac{\beta c}{1 + \kappa c} \qquad (20)$$

presents then the hyperbolic behaviour typical of Michaelian systems when plotted as a function of c (the targets c play the role of the substrate in enzymatic systems). Replacing the cytolytic term in (16) by (20), we obtain the evolution equation

$$\frac{\partial c}{\partial \tau} = c(1-c) - \frac{\beta c}{1 + \kappa c} + (1-c)\left(\frac{\partial^2 c}{\partial s^2}\right) \qquad (21)$$

which depends on two parameters only: the switching parameter β introduced earlier and the ratio

$$\kappa = \frac{bN}{l} \qquad (22)$$

whose value determines whether the binding ($\kappa < 1$) or the lysis ($\kappa > 1$) is rate limiting in the cytolytic cycle (2).

We shall call κ the avidity parameter. The reason for this is the following. The binding constant b is essentially a measure of the affinity or 'avidity' of the effector cells for the tumoural antigens: the greater b, the more promptly the effector cells recognize and bind the tumoural antigens, the greater also the selectivity of the effector cell receptors for these antigens. Once the presence of immunogenic tumoural antigens has triggered the development of an immune response, b is expected, at least initially, to increase in the course of time. Admitting that the value of the lysis constant l is characteristic of the nature of effector cells and relatively independent from their selectivity,* we expect the immune response to carry along an increase of the avidity κ (the lysis constant l stays essentially constant while the selectivity b and the number of effectors, E_T increase).

In summary, the dynamics of Michaelian effector/target systems is controlled by two dimensionless parameters: β and κ. The latter may be influenced in a variety of ways by the multitude of factors known to affect the kinetic coefficients (λ, E_T, b, l) entering their definition. At the onset of the immune response, however, we expect both β and κ to increase as a result of the increase of the product b and E_T.

The important point now is the following: though the origin of the increase of β and κ is the same, obviously the ranges $\Delta\beta$ and $\Delta\kappa$ over which they vary may notably differ, being in the first case proportional to $(\lambda N)^{-1}$ and in the second case to l^{-1}:

$$\Delta\beta \propto (\lambda N)^{-1}, \, \Delta\kappa \propto l^{-1} \tag{23}$$

From this simple observation, a straightforward prediction derives: tumours classify into three categories when subjected to an increasingly specific and/ or intense effector cell attack, i.e. corresponding to increasingly large values of b and/or E_T:

(i) Tumours for which rejection sets in as soon as the inequality $\beta \geqslant 1$ is satisfied. They are more easily rejected when the value of bE_T is large: the higher the avidity of the effectors, the more favourable the conditions for the elimination of the cancer cell population. These tumours are defined by the fact that the lysis and replication constants satisfy the inequality:

$$\mu \equiv \frac{\beta}{\kappa} = \frac{lr}{\lambda N} > \frac{1}{2} \tag{24}$$

with $r = \dfrac{E_T}{N}$ (effector/target ratio).

(ii) Tumours for which rejection is possible, but only if

$$\beta \geqslant \beta* \tag{25}$$

The value $\beta*$ depends on l, λ and r, and is necessarily greater than one.

---

*This is consistent with the well-established fact that binding and lysis are kinetically completely separate processes.

These tumours are defined by the inequalities

$$\frac{1}{2} > \mu = \frac{lr}{\lambda N} > \frac{1}{8} \tag{26}$$

(iii) Tumours which outrun and escape the immune response, whatever the value of bE_T, i.e. are progressive at all levels of the immune response and are not amenable to rejection by immunostimulation. These tumours are defined by the inequality

$$\frac{1}{8} > \mu = \frac{lr}{\lambda N} \tag{27}$$

The reasoning on which predictions i–iii is based is the following. Consider the situation inside the tumour at some point far away from the boundary so that the spatial term in (21) is negligible. As time goes on, $\partial c/\partial t \to 0$ and locally the cellular populations approach a steady state values c_s, e_{0s}, e_{1s} solution of the set of equations

$$0 = c_s(1 - c_s) - \frac{\beta c_s}{1 + \kappa c_s} \tag{28}$$

$$e_{0s} = \frac{1}{1 + \kappa c_s}, \ e_{1s} = 1 - e_{0s} \tag{29}$$

Clearly, if $\beta = \kappa = 0$, i.e. in the absence of immune response, the tumour grows freely and (21) reduces to (8). The solutions of (28)–(29) are then

$$c_s^0 = 0, \ e_{0s} = 1, \ e_{1s} = 0 \tag{30}$$

which corresponds to the normal (unstable) state, and

$$c_s^c = 1, \ e_{0s} = 1, \ e_{1s} = 0 \tag{31}$$

which corresponds to the tumoural (stable) state. Obviously, as the immune response unfolds and changes the values of β and κ, (30) always remains a solution of (28)–(29). The tumoural state (31), on the contrary, belongs to a branch of solutions which may or may not be physically acceptable depending on the values of β and κ, i.e. which may or may not correspond to real, non-negative solutions of (28). The (β,κ) values for which the solutions c_s^c are not physically acceptable clearly represent conditions sufficient for tumour rejection to be certain.

Typical curves giving c_s^c as a function of β for fixed values of κ are represented in Fig. 4. The distinguishing feature of these curves is the occurrence of an hysteresis loop for $\beta > 1$ when $\kappa > 1$. Therefore, two antagonistic effects, proceeding from the same cause, compete as the immune response develops. The immune system faces the immune response dilemma: to achieve the conditions for rejection it must increase the value of the switching parameter β (recall that a necessary condition for rejection is in any

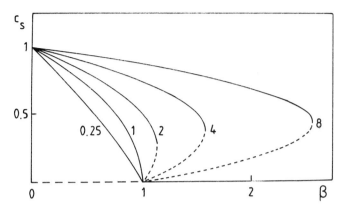

Fig. 4. Michaelian case. Solutions of (28) as a function of β and for the values of κ indicated. The parts of curves in broken line correspond to unstable states.

case $\beta > 1$), i.e. it must increase the value of E_T and/or b; in doing the latter, however, it necessarily increases the value of the avidity κ. This is counter-productive since it tends to stabilize the tumoural state. Especially, if

$$\kappa > 1 \tag{32}$$

the branch tumoural solutions c_s^c extends in the domain $\beta > 1$, up to the turning point value

$$\beta^* = \frac{(1+\kappa)^2}{4\kappa} \tag{33}$$

at which

$$c_s^c = c^* = \frac{1}{2}\left(1 - \frac{1}{\kappa}\right) \tag{34}$$

The sufficient condition for rejection to be certain is then,

$$\text{not simply } \beta > 1, \text{ but instead } \beta > \beta^* \tag{35}$$

In other words, as the immune response increases the values of b and E_T, β is running against increasing values of κ and β^*. Whether β can overtake β^* when κ is greater than 1, as summarized in Fig. 5, depends upon the ratio $\mu = lr/\lambda N$. One has indeed that

$$\frac{\beta}{\beta^*} = \left(\frac{8lr}{\lambda N}\right)\frac{\kappa^2}{(1+\kappa)^2} \tag{36}$$

Hence, as κ increases the ratio (36) tends asymptotically to $8lr/\lambda N$. Only if this value is greater than 1, i.e. if $\mu > 1/8$, does the possibility of immune rejection of the tumour exist; if $\mu < 1/8$, β^* always remains larger than β

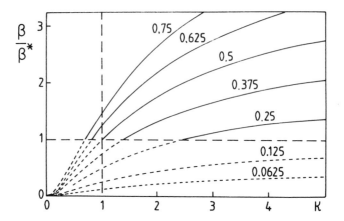

Fig. 5. Plot of β/β^* as a function of κ for the values of μ indicated. The values for which regression is certain have been drawn in full line. The curves crossing the line $\beta=1$ on the left hand side of the axis $\kappa=1$, reject the tumour as soon as β reaches the value 1; the curves crossing the line $\beta=1$ on the right hand side of the axis $\kappa=1$, reject the tumour for $\beta>\beta^*>1$.

which means that the tumour belongs to class iii above. The simplest way to deduce the condition defining class i, is to look at (22)–(23) and to express β in terms of κ and μ. One has $\beta=2\mu\kappa$; accordingly, if $\mu>1/2$, rejection will begin as soon as the condition $\beta\geqslant1$ is satisfied. Indeed, in that case, β increases more rapidly than κ so that β reaches 1 before κ ($\kappa=1$ is the value at which an hysteresis loop appears and $\beta=1$ for $\kappa=1/2\mu$). When $1/2>\mu>1/8$ (case ii), κ reaches 1 before β and the hysteresis loop appears. The turning point β^*, however, increases less rapidly than κ ($\beta^*\rightarrow\kappa/4$ for $\kappa\gg1$, cf. (33)) so that for

$$\kappa>\kappa^*=\frac{1+2\sqrt{2\mu}}{8\mu-1}\qquad(37)$$

β overtakes β^*.

These three situations are further illustrated by the plots of c_s^c as a function of β for different ratios $\mu=lr/\lambda N$ (Fig. 6). In case i (curves corresponding to $\mu\geqslant1/2$), c_s^c only exists for $\beta<1$. In case ii (curves corresponding to $1/2>\mu>1/8$), as $\mu\downarrow1/8$, β^* diverges towards infinity, and rejection rapidly becomes more and more difficult to achieve by the immune system (see (33) and (37)). For $\mu<1/8$, i.e. in case iii, the branch of solutions c_s^c splits into two separate curves. The upper one which contains at $\beta=0$, the point $c_s^c=1$ and is always stable; as usual it represents the tumoural state. The lower one which cuts the abscissa in $\beta=1$, represents a threshold state which is always unstable and is separated by a finite gap from the upper states. As a result, spontaneous rejection can no longer be achieved, even for $\beta\rightarrow\infty$.

The classification given above remains essentially valid in the more general

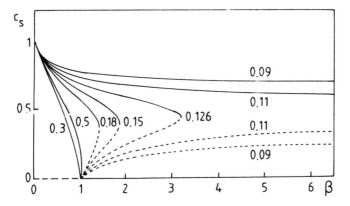

Fig. 6.   Michaelian case. Solutions of (28) as a function of β for the values of μ indicated.

non-Michaelian case. Some striking new features, however, arise in the plots of c_s^c. The two avidity parameters κ_0 and κ_1 may now have different values. In particular experimental results suggest that for some CTL the value of l_2 is markedly smaller than that of l_1. Figure 7 shows how this affects the conditions for rejection. Remarkably it reveals conditions where it appears, for $\beta > 1$, there exists an isolated branch of states which does not link up with the tumoural states found for $\beta < 1$. In other words, as the immune response develops and β increases, one successively passes from progressive growth to

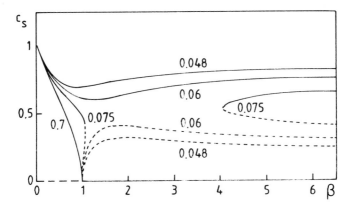

Fig. 7.   Solutions of (28) as a function of β for $\rho=1$, $\mu_1=l_1 r/\lambda N=0.7$ and the values of $\mu_2=l_2 r/\lambda N$ indicated. Remarkably, for some values of μ_2, the tumoural branch of states has two parts which do not link up: one part on the left hand side of the diagram which contains as a particular case for β=0, the solution corresponding to a freely growing tumour, i.e. $c=1$, and another isolated part on the right hand side, at large values of β, which indicates the existence of another domain in which tumour growth is possible (cf. the curve for $\mu_2=0.075$).

regression and then again to a situation where progressive growth may take place.

Another striking situation which may be encountered when the kinetics is non-Michaelian is represented in Fig. 8. In that case, stable, extremely small cancerous solutions may for $\beta < 1$ coexist with the usual fully developed tumoural states. The growth of the tumour may in such cases, when β increases, switch from a rapidly progressive phase to a regressive phase, and then again to a progressive but extremely slow growth phase.

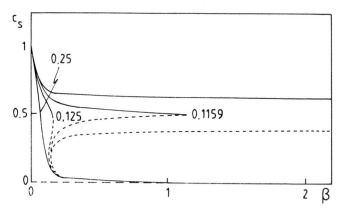

Fig. 8.   Solutions of (28) as a function of β for $\rho=320$, $\mu_1=0.25$ and the values of μ_2 indicated. The remarkable feature here is the possible existence of microcancer states. For example, for $\mu_2=0.1159$, there exists a range of values of $\beta<1$, where (28) admits beside the usual stable tumour state another stable (microcancer) tumoural state solution corresponding to very small cancer cell densities.

CONCLUDING REMARKS

A complete analysis of the properties of our model requires the detailed study of the various non-linear wave solutions to which the steady-state diagrams reported above give rise. This is beyond the scope of the present paper. We shall deal with this question elsewhere. Even so, from the results presented so far, a coherent picture already emerges which permits the interpretation of the essential aspects of cancerous growth curves such as those presented in the Fig. 1. Our approach also provides some new ideas concerning the dynamical conditions which lead to the paradoxical situations mentioned in the introduction, i.e. immunostimulation, concomitant immunity and dormancy. Let us conclude by reviewing some of the salient features which come out of the approach in this respect.

Progressive growth of a tumour may be observed either because the tumour is not (or not sufficiently) immunogenic which corresponds in our approach to situations where the values of the switching parameter β remain in the course of time less than 1, or it may correspond to one of the various manifestations of the immune response dilemma, e.g. the tumour is strongly immunogenic but the switching parameter cannot overtake the effects of the simultaneous rise of the avidity κ. The stabilization of the tumoural state which then takes place is a possible explanation for the apparently paradoxical phenomenon of immunostimulation described in some strongly immunogenic tumours. The results reported in Fig. 7 are in this respect particularly striking. Similarly, concomitant immunity finds a straightforward explanation: the primary tumour is not rejected as β increases because its evolution proceeds (see Fig. 6) on the upper branch of the case $\mu = 0.11$; the secondary inoculum is rejected because at the time of its implantation β has had the time to become greater than 1 and the normal state (30) is now stable and separated by a threshold from the tumoural state.

Multiphasic growth patterns such as those of Fig. 1 can be interpreted as follows. The initial growth phase corresponds to conditions where β is less than 1. Then, between days 10 and 20, partial regression is observed, suggesting that the necessary condition $\beta > 1$ is realized. If we admit that λN remains quasi-constant in the process, this means that the values of b and/or E_T have increased and that the avidity κ is still less than 1, or that $\kappa > 1$ but in such a way that $\beta > \beta^*$. When growth resumes this means that the system either reverses to $\beta < 1$ (or $\beta < \beta^*$), or that the value of β continues to increase and reaches a domain where a stable tumoural branch of states exists for $\beta > 1$ again. Experimentally, two explanations have been put forward to explain the resumption of progressive growth: (i) the loss of key tumour-associated antigens (Uyttenhove et al., 1983; Biddison and Palmer, 1977); (ii) suppressor T lymphocytes inhibit the production and the activity of the CTL. The former case would correspond here to a decrease of b; the second case would mainly give rise to a decrease of the value of E_T.

Dormancy is particularly interesting because it is likely that many human tumours pass through a dormant state. It is also particularly intriguing because it may arise at the end of a regression phase where all tumoural cells, but a very few, have been lysed. The results reported in Fig. 8 clearly indicate that such a condition arises in some domains of values for the parameters of the reaction scheme (1)–(2).

Acknowledgements: This work owes much to the continuous interest and support of Professors I. Prigogine and J. Urbain. This work is conducted pursuant to a contract with the National Foundation for Cancer Research.

REFERENCES

Adams, D. O. and Snyderman, R. (1979) Do macrophages destroy nascent tumors? *J. natl. Cancer Inst.* **62**, 1341–5.
Adams, D. O., Hall, T., Steplewski, Z. and Koprowski, H. (1984) Tumors undergoing rejection induced by monoclonal antibodies of Ig G$_{2a}$ isotype contain increased

numbers of macrophages activated by a distinctive form of antibody-dependent cytolysis. *Proc. natl. Acad. Sci. U.S.A.* **81**, 3506–10.

Bear, H. D. (1986) Tumor-specific suppressor T-cells which inhibit the *in vitro* generation of cytolytic T-cells from immune and early tumor-bearing host spleens. *Cancer Res.* **46**, 1805–12.

Biddison, W. E. and Palmer, J. C. (1977) Development of tumor cell resistance to syngeneic cell-mediated cytotoxicity during growth of ascitic mastocytoma P815-1/. *Proc. natl. Acad. Sci. U.S.A.* **74**, 288–91.

Bonmassar, E., Menconi, E., Goldin, A. and Cudkowicz, G. (1974) Escape of small numbers of allogenic lymphoma cells from immune surveillance. *J. natl. Cancer Inst.* **53**, 475–9.

Broder, S., Muul, L. and Waldmann, T. A. (1978) Suppressor cells in neoplastic disease. *J. natl. Cancer Inst.* **61**, 5–11.

Callewaert, D. M., Johnson, D. F. and Kearney, J. (1978) Spontaneous cytotoxicity of cultured human cell lines by normal peripheral blood lymphocytes. III. Kinetic parameters. *J. Immunol.* **121**, 710.

Garay, R. and Lefever, R. (1978) A kinetic approach to the immunology of cancer: stationary state properties of effector-target cell reactions. *J. theor. Biol.* **73**, 417–38.

Gatti, R. A., Robinson, W. A., Deinard, A. S., Nesbit, M., McCullough, J. J., Ballow, M. and Good, R. A. (1973) Cyclic leukocytosis in chronic myelogenous leukemia: new perspectives on pathogenesis and therapy. *Blood* **41**, 771–82.

Gorczynski, R. M. (1974) Evidence for *in vivo* protection against murine sarcoma virus induced tumors by T lymphocytes from immune animals. *J. Immunol.* **112**, 532–6.

Greene, M. I. (1980) The genetic and cellular basis of regulation of the immune response to tumor antigens. *Contemp. Topics Immunobiol.* **11**, 81–116.

Heberman, R. B. and Ortaldo, J. R. (1981) Natural killer cells: their role in defense against disease. *Science, N.Y.* **214**, 24–30.

Hiernaux, J. R. and Lefever, R. (1988) In *Theoretical Immunology, Part Two* (ed. A. S. Perelson), pp. 19–36. Addison-Wesley, CA.

Hiernaux, J. R., Lefever, R., Uyttenhove, C. and Boon, T. (1986) In *Paradoxes in Immunology* (eds G. W. Hoffmann, J. Levy and G. T. Nepom), pp. 95–106. CRC Press, Boca Raton, Florida.

Klein, E. (1972) Tumor immunology: escape mechanisms. *Ann. Inst. Pasteur (Paris)* **113**, 593–602.

Lefever, R. and Erneux, T. (1984) In *Nonlinear Electrodynamics in Biological Systems* (eds W. R. Adey and A. F. Lawrence), pp. 287–305. Plenum, New York.

Lefever, R. and Garay, R. (1978) Local description of immune tumor rejection. In *Biomathematics and Cell Kinetics* (ed. A.-J. Valleron). North-Holland, Amsterdam.

North, R. J. (1984) The murine antitumor immune response and its therapeutic manipulation. *Adv. Immunol.* **35**, 89–155.

Old, L. J., Boyse, E. A., Clarke, D. A. and Carswell, F. A. (1962) Antigenic properties of chemically induced tumors. *Ann. N.Y. Acad. Sci.* **101**, 80–106.

Prehn, R. T. (1977) Immunostimulation of the lymphodependent phase of neoplastic growth. *J. natl. Cancer Inst.* **59**, 1043–9.

Takei, F., Levy, J. G. and Kilburn, D. G. (1976) *In vitro* induction of cytotoxicity against syngeneic mastocytoma and its suppression by spleen and thymus cells from tumor-bearing mice. **116**, 288–93.

Thorn, R. M. and Henney, C. S. (1976) Kinetic analysis of target cell destruction by effector T cells. I. Delineation of parameters related to the frequency and lytic efficiency of killer cells. *J. Immunol.* **117**, 2213.

Tuttle, R. L., Knick, V. C., Stopford, C. R. and Wolberg, G. (1983) *In vivo* and *in vitro* anti-tumour activity expressed by cells of concomitantly immune mice. *Cancer Res.* **43**, 2600–5.
Uyttenhove, C., Maryanski, J. and Boon, T. (1983) Escape of mouse mastocytoma P815 after nearly complete rejection is due to antigen-loss variants rather than immunosuppression. *J. exp. Med.* **157**, 1040–52.
Wheelock, E. F. (1981) The tumor dormant state. *Adv. Cancer Res.* **34**, 107–40.
Zagury, D., Bernard, J., Jeannisson, P., Thierness, N. and Cerottini, J. C. (1979) Studies on the mechanism of T cell-mediated lysis at the single effector cell level. I. Kinetic analysis of lethal hits and target cell lysis in multicellular conjugates. *J. Immunol.* **123**, 1604–9.
Ziejlemaker, W. P., van Oers, R. H. J., de Goede, R. E. Y. and Schellekens, P. T. A. (1977) Cytotoxic activity of human lymphocytes: quantitative analysis of T cell and K cell cytotoxicity, revealing enzyme-like kinetics. *J. Immunol.* **119**, 1507.

Complex dynamical behaviour in the interaction between HIV and the immune system

ROY M. ANDERSON AND ROBERT M. MAY

Department of Pure and Applied Biology, Imperial College, London SW7 2BB, UK

INTRODUCTION

The basic reproductive rate, R_0, for most viral, bacterial and protozoan infections may be defined as the number of secondary infections produced, on average, by one infected individual in a population where essentially everyone is uninfected (Anderson and May, 1979). For the human immunodeficiency virus (HIV), which is the agent causing AIDS disease, R_0 for transmission within a defined risk-group can be estimated as the average number of new partners (homosexual, heterosexual, needle-sharing, etc.) that an infected individual acquires per unit time, multiplied by the average duration of infectiousness, multiplied by the probability that the infected individual will indeed transmit HIV infection to an uninfected partner (May and Anderson, 1987). There are currently uncertainties in each of these three components of R_0 for any group (Anderson and May, 1988). These uncertainties notwithstanding, it is clear that R_0 has exceeded unity among homosexual males and i.v. drug users in large cities in developed countries, and among heterosexuals in Central Africa, simply because self-sustaining HIV/AIDS epidemics have been observed in all these groups. But the current uncertainties mean, among other things, that it is not possible to say whether or not there will be a self-sustaining epidemic of HIV spread (albeit more slowly than the earlier epidemics among homosexual males and i.v. drug users) along purely heterosexual chains of transmission in developed countries.

These factual uncertainties, and their consequences for mathematical models of the transmission dynamics of HIV/AIDS, have been reviewed

Cell to Cell Signalling: From Experiments to Theoretical Models
ISBN 0–12–287960–0

Copyright © 1989 Academic Press Limited
All rights of reproduction in any form reserved

elsewhere (May and Anderson, 1988; Hyman and Stanley, 1988). The present paper focuses on one particular aspect of the larger picture, namely the patterns in the infectiousness of infected individuals, over the long and variable incubation time that elapses between infection with HIV and the clinical manifestation of full-blown AIDS disease. The simplest early models for the transmission dynamics of HIV/AIDS mostly followed the conventional epidemiological assumption that infected individuals exhibit some constant level of infectiousness from shortly after acquiring HIV until they moved out of the infected class (by developing AIDS and dying soon after, or possibly by moving to an asymptomatic category). But, as will be discussed in more detail in the next section, current indications are that patterns of infectiousness may vary significantly throughout the incubation period. We have emphasized elsewhere that such variability can introduce significant complications both into the dynamics of the epidemic, and into the interpretation and analysis of observed trends in seropositivity and AIDS incidence (May *et al.*, 1989). Our concern here is not with the transmission dynamics as such, but rather with taking some tentative first steps towards modelling the dynamics of the interaction between HIV and the immune system, with a view to gaining some insight into how an individual's infectiousness may fluctuate over the course of his or her infection.

In what follows, we first summarize some factual observations. We then present a deliberately oversimplified mathematical model for the interaction between HIV and the immune system, and analyse its properties. Of particular interest are the oscillatory or chaotic fluctuations in HIV abundance and T4 cell abundance that can result from the impact of a subsequent 'opportunistic' infection upon the activated HIV/immune system. We then present some possible modifications to our simple model, and discuss some of the overall implications.

We emphasize strongly that the present study is intended to do no more than indicate the complex dynamics – including the possibility of chaotic fluctuations in HIV abundance and in lymphocyte populations – that can arise in very simple models for the interaction between HIV and the immune system. Although our models are based on observed features of the interaction between HIV and the immune system, they do not capture all the observations in detail, nor do we believe they will survive as a better understanding of the interaction of HIV with the immune system develops. The simplest models nevertheless are useful, both in showing the kinds of complex dynamical behaviour that can easily arise in such systems and as a start toward constructing more detailed and more realistic models as our factual knowledge grows.

OBSERVATIONS ON HIV AND ANTIGENAEMIA IN INFECTED INDIVIDUALS

Clinical, immunological and epidemiological studies reveal great variability among patients in the rate at which individuals develop clinical manifestations of HIV-mediated disease, from the point of infection or seroconversion. The average incubation period of AIDS appears to be approximately 7–8 years (Medley *et al.*, 1988), but, over this period, marked fluctuations occur in blood serum titres of HIV antigen-specific antibodies and HIV antigen-aemia, both within and among individual patients. Despite this wide variability, there is a consistent general pattern of change in antigen concentration and antibody titres, as patients progress from seroconversion, through a period with no overt disease, to AIDS-related-complex (ARC), and finally to AIDS: two major peaks in antigenaemia are detectable, one 10–40 weeks after seroconversion, and the second during the manifestation of ARC and AIDS (for a review of the published data, see May *et al.*, 1989). Marked deviations from these average trends, however, occur in individual patients.

The severe immunosuppression typically generated by HIV infection is in part caused by a selective depletion of helper/inducer T lymphocytes (CD4+ cells) which express the major receptor for the virus (the CD4 molecule). Although a variety of immunological abnormalities accompany HIV infection and AIDS, many are thought to be related to the cytopathic action of the virus on CD4 lymphocytes. In general, there appears to be an ordered progression in immunological abnormalities from the point of seroconversion to the development of AIDS, with patients passing via states of no abnormalities, a depressed CD4/CD8 cell ratio, a depressed CD4/CD8 cell ratio and CD4 cell depletion, to a state of depressed CD4/CD8 cell ratio along with CD4 cell depletion and lymphopaenia. However, much variability is observed between patients in rates of progression, in the relationships between quantitative scores of lymphocyte abundances and ratios, and in associations between immunological abnormalities and symptoms of disease. Concomitant with this progression are changes in HIV antigenaemia and concentrations of antibodies specific to core and envelope HIV antigens. Detectable levels of HIV-specific antibodies (seroconversion) are present between 20 and 80 days after the point of infection (the average being around 60 days) and, after primary HIV infection, antibody titres to the core antigens P24 and P17 (*gag* proteins) then begin to decline as patients progress to ARC and AIDS.

It has been suggested that the concentrations of anti-P24 and anti-P17 antibodies are inversely correlated with antigenaemia, with a loss of reactivity to P17 being an earlier marker of disease progression. In contrast, antibodies to the envelope antigen P41 (*env* gene products) remain high following primary HIV infection and throughout the development of ARC and AIDS. Other studies suggest that a decline in antibody titres to the envelope protein (gp 110) and the reverse transcriptase enzyme (P51 and P65) are also predictive of progression to AIDS. It seems plausible to associate the apparent peaks in antigenaemia soon after infection, and as patients progress to ARC and AIDS, with the infectiousness of an infected patient to susceptible sexual partners.

In general, however, great variability is observed among patients, and during the incubation period within any one patient, in the relative concentrations of antibodies to the various HIV antigens. Not only do the durations and magnitudes of the early and late peaks in antigenaemia vary from individual to individual, but there are also indications of irregular (and brief) bursts of antigenaemia within the relatively long 'silent phase' between the early and late peaks (associated, respectively, with seroconversion and with the onset of ARC and AIDS). These data come from independent studies of three groups of infected patients attending clinics in London, which will be reported in detail elsewhere (Anderson et al., unpublished). At the US National Academy of Sciences Symposium on AIDS in April 1988, Fauci presented apparently similar data, with indications of irregular minor outbreaks throughout the 'silent period', which itself is of average duration 6 years or so (in the discussion of his paper, and in subsequent private communication, Fauci has emphasized the preliminary nature of these findings, and the schematic character of the graph he presented).

It is probably worth mentioning that highly irregular patterns in viraemia have also been observed in some other retrovirus infections. In particular, Payne et al. (1987) have studied the rectal temperature of horses infected with equine infectious anaemia virus (EIAV, a member of the lentivirus subfamily of retroviruses), and have documented irregular, brief bursts of higher temperature that are very similar to the tentative patterns in HIV antigenaemia just discussed. Although these data are vastly superior to the HIV data, their relevance is very indirect (in that EIAV is a quite different retrovirus, and the relation between rectal temperature and viral abundance is uncertain).

A SIMPLE MODEL

As a first step toward understanding the irregular nature of these tentatively observed patterns in antigenaemia, we consider the following model for the dynamics of the interaction between HIV and the various components of the human immune system that act either as a host for viral replication (as do CD4 lymphocytes) or as antagonists to viral population growth, spread and persistence (as do antibodies). Other equally simple but different models are, of course, possible, given our current ignorance. We give our reasons for choosing to work with the model defined below, but we cannot emphasize too strongly that we regard it as illustrative and not apodictic.

We begin by considering a population of free HIV virus at time t, $V(t)$, and a population of T lymphocytes (namely, those expressing the major receptor for HIV virus, the molecule CD4), which are subdivided into populations of immature non-activated cells, activated but uninfected cells, and infected cells, represented by $P(t)$, $X(t)$, and $Y(t)$, respectively. Immature cells are recruited (from production in the thymus) at a constant rate Λ, and are removed either by mortality at a per capita rate μ or by activation (as a result of contact with virus) at a net rate γPV, where γ is the probability of activation through contact. The population of activated but uninfected cells grows as a result of recruitment from the immature cells (at the rate γPV) and from proliferation in the activated state via clonal expansion (at a net per capita rate r, discounted for X-cell mortality); this population suffers losses as activated cells are infected with HIV (at a net rate βXV, where β is the probability of infection via contact). The population of infected cells, $Y(t)$, is assumed to be unable to proliferate, so that this population grows through recruitment from the activated but uninfected population (via infection by contact with the virus, at the rate βXV), and suffers deaths as a result of infection with HIV virus, at a per capita rate α.

The population of free virus (i.e. HIV outside infected cells), $V(t)$, is assumed to increase as a result of infected cells releasing large numbers of viral particles (λ per cell) when they die (at the rate αY); this produces free virus at the overall rate $\lambda \alpha Y$. The population is assumed to decrease as a result of mortality (at a per capita rate b), of absorption by activated cells (at a net rate $\delta(X+Y)V$), and of the activities of uninfected lymphocytes in stimulating antibody and cell-mediated viral destruction (at a net rate σXV).

These assumptions yield the following set of non-linear, differential equations for $P(t)$, $X(t)$, $Y(t)$ and $V(t)$:

$$dP/dt = \Lambda - \mu P - \gamma PV \tag{1}$$

$$dX/dt = \gamma PV + rX - \beta XV \tag{2}$$

$$dY/dt = \beta XV - \alpha Y \tag{3}$$

$$dV/dt = \lambda \alpha Y - bV - \delta(X+Y)V - \gamma XV \tag{4}$$

These assumptions embody a current opinion that the virus effectively infects only those mature cells that have been activated by previous contact with the virus (not all researchers hold this opinion: for a discussion, see Barnes, 1988). The model further assumes that the virus is cytopathic (*in vitro* studies suggest cell death 10–14 days after exposure to the virus), and that the virus is killed by the action of the immune system (which here is crudely represented by T cells only, although the loss term σXV is assumed roughly to represent T cell activation of B cells and associated antibody production, along with cell-mediated viral destruction).

Dynamical properties of the model

Despite the crudity and simplicity of these biological assumptions, the resulting model, eqns (1)–(4), is complex in form, and can exhibit a wide range of dynamical behaviour, depending on the values of the parameters.

In the absence of HIV ($V=0$), the system settles to an obvious stable state with no activated cells ($X=Y=0$) and the immature cells at the equilibrium value $P=\Lambda/\mu$.

When HIV is introduced, a phase-space analysis of the model reveals three qualitatively different patterns of dynamical behaviour. It is helpful first to define the net reproductive rate of the virus within the human host, R (which is not to be confused with the basic reproductive rate, R_0, for HIV infection in the human population, as discussed in the opening paragraphs). By neglecting the parameter b (the 'natural' viral mortality) in relation to other relevant rates, we find R to be:

$$R=\beta\lambda/(\delta+\sigma) \qquad (5)$$

If $R<1$, viral 'reproduction' is effectively below threshold, and the virus is unable to establish itself within the host; V and Y tend to zero, while the population of activated but uninfected cells increases definitely ($X\to\infty$), owing to clonal expansion, in this simplest model. If we crudely describe the action of T suppressor cells by the subtraction of a regulatory term, dX^2, in eqn (2), then the activated cells will obtain an equilibrium at a high population density, given by $X^*=r/d$. Whether or not T suppressor cells are included in the model, the individual host will be seropositive but uninfected ($X>0$ and $P=\Lambda/\mu$, but $V\to0$ and $Y\to0$).

If $R>1$, the virus can establish itself. Two qualitatively different possibilities can now be distinguished in the limit $b\to0$. If

$$r>\beta a\lambda(1-1/R)/\delta \qquad (6)$$

the virus can establish itself, but its effects are not sufficient to halt the growth of the population of activated cells (the rX term outruns the βXV term in eqn (2)). When the effects of suppressor cell regulation are included (dX^2 subtracted on the left hand side in eqn (2)), the system tends to settle to, or to oscillate around, a state with relatively high abundance of activated cells, both infected and uninfected. Conversely, when the inequality in eqn (6) is reversed (and $R>1$), the virus not only persists, but can regulate the activated but uninfected cell population; the system tends to exhibit an initial burst of high viraemia, and then to settle to, or oscillate around, a state where the abundance of activated T cells (although possibly high) is regulated by the virus (with the dX^2 suppressor cell term, if it is included, having negligible effect), even though viraemia may be low. In either the suppressor-cell regulated system (when eqn (6) pertains) or the virus-regulated system (when $r < \beta\alpha\lambda(1 - 1/R)/\delta$), the system tends to exhibit sustained oscillations (a 'stable limit cycle') when r and $\beta\alpha\lambda(1 - 1/R)/\delta$ are comparable in magnitude. When r is either much greater or much less than $\beta\alpha\lambda(1 - 1/R)/\delta$, the system tends to settle to a steady state, after the initial bout of viraemia.

More generally, the phase-space analysis is messier if the 'natural death rate' for free virus, b, is sufficiently large that bV exceeds, or is comparable in magnitude with, the lymphocyte-induced death rates $\delta(X + Y)V$ and σXV (as is the case for the numerical simulations in Figs 1–3). However, the broad division into three regimes – one where virus is eliminated, one with virus and activated T4 cells regulating each other, and one with virus present but suppressor cells primarily responsible for regulating antibody production – persists in this more complex system.

Effects of a subsequent infection

Suppose we now introduce a second infection – viral, bacterial or protozoan – which replicates within the host and which stimulates these same T cells to proliferate by clonal expansion. The dynamical behaviour produced by the introduction of such a subsequent infection upon the population of activated (but HIV-uninfected) T cells can be complicated, resulting in apparently chaotic fluctuations in HIV antigenaemia amongst other things.

We represent the population of such a subsequent, 'opportunistic' infectious agent by $I(t)$. The action of this population is described by adding an extra term to the right hand side of eqn (2) of the form kIX, to describe proliferation via contact:

$$dX/dt = \gamma PV + rX + kIX - \beta XV \tag{2a}$$

We also require an equation describing changes in $I(t)$:

$$dI/dt = cI - hIX \tag{7}$$

Here, c denotes the per capita reproductive rate of the opportunistic infection, and the term hIX represents the killing of the infectious agent by antibody or cell-mediated action (whose severity is proportional to the density of helper T cells).

In the absence of any HIV infection, the activated T cells always eliminate the opportunistic infection, after an initial bout of pathogen growth which stimulates proliferation of the lymphocytes. This is a standard response in models for the interaction between the immune system and 'ordinary' infections.

When the host individual has already experienced HIV infection, the

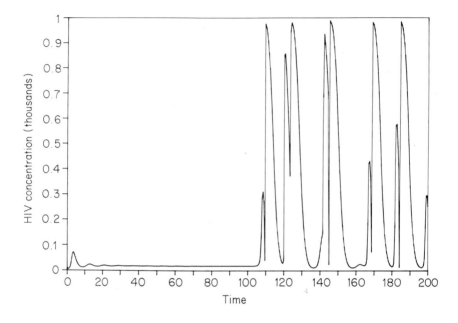

Fig. 1. This shows the concentration of free HIV virus in the blood as a function of time, as described by eqns (1)–(4) and (7). We see an initial peak in antigenaemia, which drops to a steady low level; at $t=100$ the opportunistic infection is introduced, and its interaction with the HIV-activated immune system produces apparently chaotic fluctuations in HIV antigenaemia thereafter. Specifically, this example is for the above system of equations, with the parameters having the broadly representative values $\Lambda=1$, $\mu=0.1$, $\gamma=0.01$, $r=1$, $d=0.001$, $\lambda a=10$, $b=1$, $\delta=0.01$, $\sigma=0.1$, $k=h=0.01$, $c=1$, $\beta=0.1$ and $a=2$ (all rates are in units of week^{-1} although, from the point of view of the qualitative dynamics of the system, the units may be taken to be arbitrary).

subsequent opportunistic infection can result either in oscillatory fluctuations (which can have large magnitudes) in the abundance of HIV and the opportunistic infectious agent, or in chaotic dynamical behaviour. What happens in this simple model is that the presence of HIV prevents the elimination of the opportunistic infection, and the interaction of the two agents with the immune system can trigger wide fluctuations (often of an erratic and apparently unpredictable or chaotic nature) in HIV abundance. Figures 1–3 show numerical simulations of the behaviour of the above model (with the suppressor-cell regulatory term, dX^2, subtracted from the right hand side of eqn (2) or (2a)), for the parameter choices specified in the captions. In Fig. 1, there is an initial peak in viraemia shortly after HIV infection; the system settles to low HIV levels until the opportunistic infection invades, whereupon apparently chaotic fluctuations in viraemia ensue. Figure 2 is similar, except here the initial HIV infection leads to small-amplitude oscillations in viraemia, before the opportunistic infection precipitates the system into large-amplitude, chaotic fluctuations. Figure 3 is again

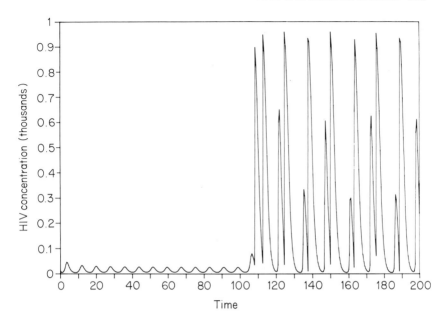

Fig. 2.   As for Fig. 1, except here the parameter β has the value β=0.03. The patterns in HIV antigenaemia are broadly similar to those in Fig. 1.

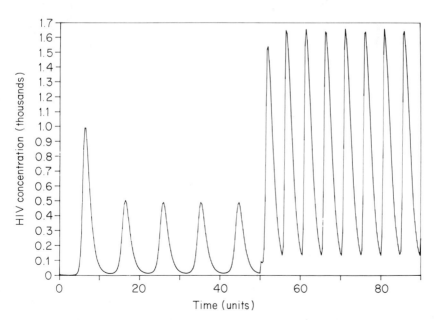

Fig. 3.   Again as in Fig. 1, except here the parameters β and α have the values β=0.03 and α=1. In this case, the opportunistic infection is introduced at *t*=50 (rather than *t*=100). As discussed in the text, with these parameter values there are initially marked oscillations in HIV antigenaemia, and the effect of the opportunistic infection is less dramatic than in Figs 1 and 2.

similar, except here the original oscillations in HIV viraemia are more pronounced, and the effects of the opportunistic infection are to produce oscillations that are more regular and more comparable with the initial fluctuations than is the case in Figs 1 and 2. Taken together, Figs 1–3 testify to the complex and diverse range of dynamical phenomena that can be exhibited by the simple set of eqns (1)–(4) and (7).

The general conclusion that we draw from this very simple model is that the dynamical tension created by the cytopathic influence of HIV on the lymphocyte population (which itself plays a central role in the immune defence of the host) can give rise to oscillatory or chaotic fluctuations in the abundances of HIV (free virus), infected and uninfected lymphocyte populations, and in populations of the opportunistic infectious agent.

SOME ALTERNATIVE MODELS

Many alternative versions of our basic model are equally reasonable, given present uncertainties about the detailed interactions between HIV and the immune system.

One major class of alternative models would omit the distinction between activated but uninfected (X) and infected (Y) populations of lymphocytes (Perelson, private communication). Our preliminary impression is that such models may be less liable to produce oscillatory or chaotic dynamics.

An important possible refinement of our basic system, eqns (1)–(4), is to allow for the possibility that the immune system is able to recognize and kill infected cells (our Y-population of T4 lymphocytes). Such an effect – whereby the immune response (proportional to X) is responsible for killing infected cells and virus – could be represented by subtracting a term κYX from the right hand side of eqn (3), to get

$$dY/dt = \beta XV - \alpha Y - \kappa YX \qquad (3a)$$

and at the same time having the killing of virus proportional to the direct and indirect effects of the activated lymphocyte population, as represented by $(\delta + \sigma)VX$, only,

$$dV/dt = \lambda \alpha Y - bV - (\delta + \sigma)VX \qquad (4a)$$

Preliminary numerical studies suggest this system is broadly similar to that studied above, with the important exception that one no longer tends to get a high proportion of infected T4 cells (which is an unrealistic feature of the numerical simulations shown in Figs 1–3, which are based on eqns (1)–(4)).

Another realistic refinement is to acknowledge the possible effects of the formation of syncytia, or groups of amalgamated cells. There is some evidence that such syncytial formation, caused possibly by the poor state of infected cells, may attract and bind-in the activated but uninfected cells, and thus remove them from functioning in the immune system. Such effects can be represented by substracting a term φXY from the right hand side of eqn (2). Again the indications from numerical simulations are that the dynamical possibilities are similar to those of the basic model, except that – especially if added to the refinements represented by eqns (3a) and (4a) – there tend to be significantly fewer infected cells (which accords better with observation).

These and other variations upon our basic model deserve exploration as the underlying immunological facts become better understood. As repeatedly emphasized, our primary aim has been to show the kinds of dynamical phenomena that can easily arise.

DISCUSSION

An alternative way of displaying the results in Fig. 1 (or any other such figure) is shown in Fig. 4a and b. These figures show the concentration of HIV, as free virus (plotted on the vertical axis), and the concentration of antibodies (plotted on the horizontal axis), as they would appear if samples were taken from the patient at unit time intervals (specifically, weekly) over the first 100 time units (weeks) following the original HIV infection (Fig. 4a), and over the second 100 time units following the acquisition of the subsequent opportunistic infection (Fig. 4b). These 'pseudo-data' are very like the preliminary data from three groups of patients in hospitals in London, which were referred to earlier (with details to be given in Anderson et al., 1989). That is, the patterns generated by numerical studies of the simple models described above are consistent with such tentative data as are available, and they support the conjecture that there may be oscillatory or chaotic dynamical interactions between HIV antigenaemia (which is assumed to reflect the concentration of free virus in the blood) and the immunological response of the patient (as measured by P24 antibody titres).

If we accept the model as a very crude caricature of the interaction between HIV and the immune system, it suggests that a marked rise in antigenaemia is triggered by the arrival, or activation, of an opportunistic infection that stimulates further proliferation of the T4 lymphocytes. The model suggests both that there will be wide-scale fluctuations in HIV abundance during this stage (suggestive of the progression of infected patients to ARC and AIDS), and that the host will be unable to control the population growth and persistence of the opportunistic agent. It follows that the wide variation observed in the incubation period of ARC/AIDS may reflect the timing of the arrival of an opportunistic infection that stimulates activated T4 cells to proliferate; those who develop AIDS rapidly following HIV infection tend to have high antigen concentrations throughout their shorter-than-average incubation periods. The model further suggests that the presence of an opportunistic infection at the time of infection by HIV would trigger high

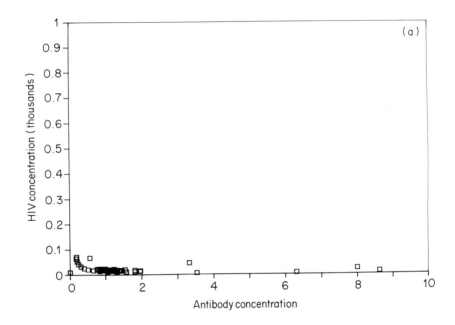

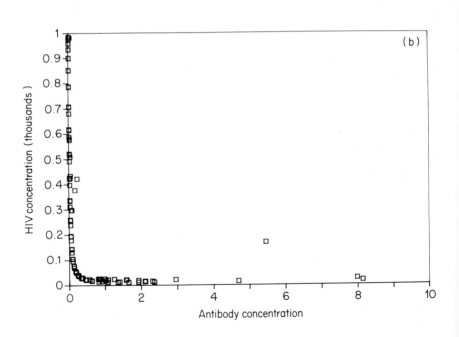

and widely fluctuating abundances of both pathogens. Interestingly, the model also suggests that there may be high prevalences of infected lymphocytes – but with wide fluctuations – during the phase in which the opportunistic infection is present (ARC/AIDS). Recent reports suggest that the very low percentage of infected cells in the peripheral blood (around 1 in 10^5) that are expressing virus at any given time may be due to the inability to detect infected cells that are not expressing virus. A growing proportion of cells appear to become infected, as patients progress to ARC and AIDS.

Some objections and alternatives to the above explanations should be noted. For one thing, if the onset of ARC and AIDS is associated with acquisition of some appropriate kind of second, opportunistic infection by an individual already infected with HIV, we should expect to find that incubation intervals are typically shorter among those groups (i.v. drug users, highly promiscuous homosexual men) where other studies have shown high incidence of other infections. No such marked correlation has, however, been noted. For another thing, other explanations are possible for sporadic upsurges in antigenaemia at unpredictable intervals over the 'silent phase' between seroconversion and ARC/AIDS. It could, for example, be that different strains or genotypes of HIV appear as the infection progresses (as is demonstrably the case, in a different context, for many protozoan infections); we might then see erratic bursts of antigenaemia as these various viral strains briefly escaped recognition by the immune system, and then were reduced by an appropriate immune response, one after another.

One qualitative feature of our models – which seems likely to be present in most models of this general kind – is the existence of a threshold criterion, eqn (5). This criterion determines whether or not HIV will persist in the infected host and is defined by a combination of the parameter values that control the dynamics of HIV and the immune system. One of the main parameters is β, the rate of infection of T4 cells. Note that eqn (5) suggests that it is not necessary totally to block infection of T4 cells to eliminate the virus ($\beta = 0$), but rather it suffices simply to reduce the value of β such that the right hand side of the equation is less than unity in value. Also note that enhancing the rate at which antibody (crudely described in the model by the

---

Fig. 4.   (a) This figure shows the concentration of free HIV virus (the vertical axis), plotted against the concentration of antibody cells (the horizontal axis), for a series of discreet points in time over the first phase (the first 100 time units) of the interaction between HIV and the immune system depicted in Fig. 1. (b) Similar to Fig. 4a, except here the relation between HIV concentration and antibody concentration is shown for the second phase (the second 100 time units) displayed in Fig. 1. That is, Fig. 4b contrasts with Fig. 4a in that it shows the HIV–antibody relations after the advent of a subsequent, opportunistic infection. As discussed in the text, Fig. 4a and b look like the preliminary results from data on three groups of patients in hospitals in London (Anderson et al., unpublished).

density of uninfected T4 cells) kills virus – the term σXV in eqn (4) – can also act to reduce the net reproductive rate of the virus to less than unity.

In conclusion, we stress that our deliberately oversimplified models are designed to make three main points.

First, our understanding of the immune system in general, and of its interaction with HIV in particular, would be much enhanced if more were known about the population biology and regulation of the immune system, regarded as a system of interacting populations of cells. The tendency to evoke increasingly complex explanations of the interactions between cells and factors, and to postulate more and more types of cells and factors, ignores the possibility that a few simple processes dominate observed dynamical behaviour. As shown here, very simple non-linear models can generate very complicated patterns of behaviour.

Second, our model suggests a possible interpretation for some of the relations among quantities that characterize the progression of disease in patients infected with HIV. The potential for highly oscillatory or chaotic dynamics may help explain the great variability from patient to patient, with a wide distribution in incubation times and substantial differences in transmission among infected and susceptible sexual partners (with infection resulting from a few sexual acts in certain partnerships, but not after very many acts in others). If non-linearities indeed produce chaotic dynamics in the abundance of free virus, then slightly different patient histories can have chaotically different histories in the course of infection and infectiousness.

Third, the formulation of models helps focus attention on precisely what needs to be measured to improve our understanding. Of special importance is the need for improved techniques for the detection both of free virus in blood, excretions and secretions, and of infected lymphocytes. We also need repeated and frequent (at short time intervals) assessments of changes in antigenaemia and immunological factors (such as antibodies, T4 cell abundances, measures of proliferation, etc.) throughout the long and variable incubation period of AIDS and during the course of the disease itself, in individual patients.

Finally, we end as we began by emphasizing that the models investigated here are oversimplified caricatures. They indicate the kinds of things that can be learned by systematic refinement of such an approach, but these simplest models do no more than that.

Acknowledgements: This work was supported in part by the UK Medical Research Council (RMA), and by the US National Science Foundation (RMM).

REFERENCES

Anderson, R. M. and May, R. M. (1979) Population biology of infectious diseases: part I. *Nature* **280**, 361–7.

Anderson, R. M. and May, R. M. (1988) Epidemiological parameters of HIV transmission. *Nature* **333**, 514–9.

Barnes, D. M. (1988) AIDS virus coat activates T cells. *Science, N.Y.* **242**, 515.

Hyman, J. M. and Stanley, E. A. (1988) A risk-based model for the spread of the AIDS virus. *Math. Biosci.* **90**, 415–73.

May, R. M. and Anderson, R. M. (1987) Transmission dynamics of HIV infection. *Nature* **326**, 137–42.

May, R. M. and Anderson, R. M. (1988) The transmission dynamics of human immunodeficiency virus (HIV). *Phil. Trans. R. Soc. B* **321**, 565–607.

May, R. M., Anderson, R. M. and Johnson, A. M. (1989) The influence of temporal variation in the infectiousness of infected individuals on the transmission dynamics of HIV. In *AIDS 1988: AAAS Symposium Papers* (ed. R. Kulstad), pp. 75–83. American Association for the Advancement of Science, Washington, D.C.

Medley, G. F., Anderson, R. M., Cox, D. R. and Billard, K. (1988) The distribution of the incubation period of AIDS. *Proc. R. Soc. B* **233**, 367–77.

Payne, S. L., Salinvoch, O., Nauman, S. M., Issel, C. J. and Montelaro, R. C. (1987) Course and extent of variation of equine infectious anemia virus during parallel persistent infections. *J. Virol.* **61**, 1266–70.

Part 4
Hormonal signalling: The reproductive system

The circhoral hypothalamic clock that governs the 28-day menstrual cycle

ERNST KNOBIL

Laboratory for Neuroendocrinology, The University of Texas Health Science Center, Medical School, P.O. Box 20708, Houston, Texas 77225, USA

Reproduction, in all vertebrates of both sexes studied to date, is the consequence of a complex cascade of neuroendocrine events initiated by the rhythmic activation of a central signal generator that, in mammals, appears to be situated in the mediobasal hypothalamus and is an intrinsic property of this structure. (See Knobil and Hotchkiss, 1988, for review.)

The existence of this signal generator was first evidenced by the observation that the pituitary gonadotropic hormones, in a variety of species including man, are secreted episodically (see Pohl and Knobil, 1982 and Plant, 1986 for review). In the ovariectomized monkey, wherein the phenomenon was first observed, rhythmic 'pulses' of the luteinizing hormone (LH), appear in the peripheral circulation with a period of approximately 1 h, hence the designation of this frequency as 'circhoral' (Dierschke *et al.*, 1970). It was surmised, at the time, that each pulse of LH released by the pituitary gland is the consequence of a bolus of luteinizing hormone releasing hormone (LHRH or GnRH, a 10 amino acid neuropeptide),* discharged from the nerve terminals of GnRH neurones in the median eminence into the pituitary portal circulation thence impinging on the gonadotrophs of the adenohypophysis. That GnRH is indeed the signal for the pulsatile release of LH has been clearly demonstrated by experiments performed in ovariectomized ewes by Clark and Cummings (1982). That the probable origin of these neuroendocrine signals is the region of the arcuate nucleus in the mediobasal hypothalamus, at least in the rhesus monkey, has been demonstrated by classical lesioning techniques (Plant *et al.*, 1978) and by *in vitro* superfusions

*Because the same neuropeptide appears to cause the release of both gonadotropic hormones it is commonly referred to as GnRH.

Cell to Cell Signalling: From
Experiments to Theoretical Models
ISBN 0–12–287960–0

of hypothalamic fragments (McKibbin and Belchetz, 1986). These observations and considerations gave rise to the notion of a GnRH pulse generator residing in the hypothalamus which controls LH secretion and, therefore, gonadal function. In the absence of the pulse generator, gonadotropic hormone secretion essentially ceases and gametogenesis and gonadal sex hormone secretion cannot occur.

In rhesus monkeys, each LH pulse and, by inference, each GnRH pulse is immediately preceded by a dramatic increase in electrical activity in the hypothalamus as detected by recording electrodes chronically implanted in the region of the arcuate nucleus (Wilson *et al.*, 1984 and Fig. 1). These electrical signals and the resultant LH pulses can be completely arrested or their frequency markedly reduced by barbiturates (Wilson *et al.*, 1984), a variety of α-adrenergic and dopaminergic blocking agents (Kaufman *et al.*, 1985) and by opiates (Kesner *et al.*, 1986).

In a more physiological context, progesterone and testosterone reduce the frequency of the pulse generator, actions apparently mediated by endo-

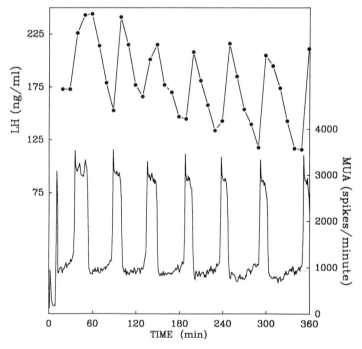

Fig. 1. The time courses of hypothalamic electrical activity and plasma LH concentrations in an ovariectomized rhesus monkey restrained in a primate chair (see Wilson *et al.*, 1984, for methodology employed). Note that the duration of the 'volleys' of electrical activity decreases with time. This is attributable to the 'stress' of restraint although, clinically, the animal seemed completely relaxed (Williams *et al.*, 1988).

genous opiates, at least in primates (Van Vugt *et al.*, 1984; Veldhuis *et al.*, 1984). In the physiological setting of the menstrual cycle, oestrogens do not seem to have such an effect (see Kesner *et al.*, 1987). More recently, it has been observed that the duration of pulse generator electrical activity is also under hormonal control (Williams *et al.*, 1988). Oestradiol foreshortens the duration of firing as does morphine and hypothalamic corticotropin-releasing factor.

The significance of the GnRH pulse generator in the control of reproductive function became apparent with the finding, in rhesus monkeys in whom the pulse generator had been destroyed by radiofrequency lesions, that normal LH secretion could only be re-established by the intermittent administration of GnRH at the physiological frequency of one injection per hour (see Knobil, 1980 for review). Higher frequencies or continuous infusion of the decapeptide desensitized the gonadotrophs and LH secretion fell to zero. Lower frequencies also interfered with normal gonadotropin secretion. Furthermore, completely normal, 28-day ovulatory menstrual cycles could be achieved in rhesus monkeys with hypothalamic lesions by the administration of GnRH at an unvarying frequency of one pulse per hour. These observations have been confirmed in women who were anovulatory because of GnRH pulse generator failure (see Filicori, in this volume).

In rhesus monkeys, as in women, relatively minor deviations from the physiological frequency of GnRH administration leads to derangements in gonadotropin secretion and, consequently, in ovarian function. In monkeys with hypothalamic lesions, for example, reducing the frequency of GnRH administration from one injection per hour to one every 90 min leads to a significant reduction in ovulation rate (Pohl *et al.*, 1983). It follows that, not only is the hypothalamic production of LHRH essential for normal ovarian function, it must also be delivered to its target cells within a relatively narrow window of frequencies.

How is this obligatory, circhoral signal from the hypothalamus translated into the 28-day ovarian cycle of women and rhesus monkeys? The central event in the ovarian cycle of all mammals is ovulation. This is triggered by an outpouring of gonadotropic hormones from the pituitary gland, the preovulatory gonadotropin surge. In many mammals, this is initiated by a complex and still incompletely understood interaction between rising oestrogen levels, the mediobasal hypothalamus and a circadian timing mechanism in the anterior hypothalamus. In some, the so-called reflex ovulators, the sensory input associated with copulation is also needed. In the higher primates, however, it appears that the only requirement of the central nervous system for the initiation of the preovulatory gonadotropin surge is the circhoral release of GnRH (see Knobil and Hotchkiss, 1988 for review and Filicori's chapter in this volume). During the follicular phase of the menstrual cycle, the mean frequency of LH and FSH pulses (and, therefore, of the GnRH pulse generator) does not change appreciably although a clear reduction in this frequency is observable at night. In response to this gonadotropic stimulation, the Graafian follicle grows and increases its output of oestradiol. When the plasma level of this steroid surpasses a certain threshold and remains above it for approximately 36 h a gonadotropin surge that eventuates in ovulation is initiated. The timing of ovulation is, therefore, determined by the ovary rather than by some cerebral mechanism as it appears to be in rodents (see Knobil, 1974, 1980 for review). Because this positive feedback action of oestradiol is fully demonstrable when the endogenous GnRH pulse generator is replaced by a programmed pump that delivers pulses of the neuropeptide at an unvarying amplitude and frequency, it may safely be concluded that the site of action of this steroid is at the level of the pituitary. In any case, it takes approximately 14 days for follicular oestradiol production to achieve the appropriate plasma level in response to the circhoral signals from the hypothalamo-hypophysial apparatus.

Following ovulation, the corpus luteum is formed and the remainder of the cycle is dominated by progesterone secreted by this new endocrine gland. Progesterone inhibits further follicular development by mechanisms that remain to be fully defined. It also prevents oestrogen-induced gonadotropin surges by an action at the level of the central nervous system (Pohl *et al.*, 1982). When the corpus luteum involutes, a new cycle is initiated, progesterone secretion falls and the progesterone block to further follicular development is removed. This decline in progesterone levels, incidentally, also leads to menstruation. While the mechanisms that cause the demise of the corpus luteum in the higher primates are unresolved, the fact remains that the functional lifespan of this gland, which is also dependent on the activity of the GnRH pulse generator and on the resultant pattern of LH secretion, albeit at a reduced frequency (*vide infra*), is also approximately 14 days. The sum of the two phases of the cycle, both obligatorily dependent on the proper functioning of the circhoral GnRH pulse generator, is 28 days.

As shown in the rhesus monkey and in women with non-functional GnRH pulse generators, replacement with exogenous GnRH at an unvarying frequency leads to luteal phases of normal duration (Knobil, 1980; Pohl *et al.*, 1983; Filicori, in this volume). During the luteal phase of the normal menstrual cycle, however, the frequency of LH pulses is significantly reduced but, because of the foregoing findings, the physiological significance of this phenomenon in the control of ovarian function is not clear. In any case, it does not appear to be involved in determining the functional lifespan of the corpus luteum (Hutchison and Zeleznik, 1985).

In summary, the menstrual cycle is timed by two ovarian clocks, each delimiting a period of 14 days. One is the Graafian follicle and its production of oestradiol, the other is the corpus luteum and its secretion of progesterone. Both are obligatorily dependent for their functioning on the proper operation of the hourly clock in the arcuate region of the hypothalamus, the GnRH pulse generator, that ensures the production and release of the gonadotropic hormones. The macrodynamics of gonadotropin delivery to the ovary, however, are dictated by the negative and positive feedback actions of the ovarian steroids on the hypothalamo-hypophysial system; oestradiol at the level of the pituitary gland and progesterone at the level of the hypothalamus.

REFERENCES

Clarke, I. J. and Cummins, J. T. (1982) The temporal relationship between gonado-tropic releasing hormone (GnRH) and luteinizing hormone (LH) secretion in ovariectomized ewes. *Endocrinology* **111**, 1737–9.

Dierschke, D. J., Bhattacharya, A. N., Atkinson, L. E. and Knobil, E. (1970) Circhoral oscillations of plasma LH in the ovariectomized monkey. *Endocrinology* **87**, 850–3.

Hutchison, J. S. and Zeleznik, A. J. (1985) The corpus luteum of the primate menstrual cycle is capable of recovering from a transient withdrawal of pituitary gonadotropin support. *Endocrinology* **117**, 1043–9.

Kaufman, J-M., Kesner, J. S., Wilson, R. C. and Knobil, E. (1985) Electrophysiologic manifestation of "LHRH Pulse Generator" activity in the rhesus monkey: influence of α-adrenergic and dopaminergic blocking agents. *Endocrinology* **116**, 1327–34.

Kesner, J. S., Kaufman, J-M., Wilson, R. C., Kuroda, G. and Knobil, E. (1986) The effect of morphine on the electrophysiological activity of the hypothalamic luteinizing hormone releasing hormone pulse generator in the rhesus monkey. *Neuroendocrinology* **43**, 686–8.

Kesner, J. S., Wilson, R. C., Kaufman, J-M., Hotchkiss, J., Chen, Y., Yamamoto, H., Pardo, R. R. and Knobil, E. (1987) Unexpected responses of the hypothalamic gonadotropin-releasing hormone "pulse generator" to physiological estradiol inputs in the absence of the ovary. *Proc. natl. Acad. Sci. U.S.A.* **84**, 8745–9.

Knobil, E. (1974) On the control of gonadotropin secretion in the rhesus monkey, *Recent Prog. Hormone Res.* **30**, 1–46.

Knobil, E. (1980) The neuroendocrine control of the menstrual cycle. *Recent Prog. Hormone Res.* **36**. 53–88.

Knobil, E. and Hotchkiss, J. (1988) In *The Physiology of Reproduction* (eds E. Knobil and J. D. Neill), pp. 1971–94. Raven Press, New York.

McKibbin, P. E. and Belchetz, P. E. (1986) Prolonged pulsatile release of gonadotropin-releasing hormone from the guinea pig hypothalamus *in vitro. Life Sci.* **38**, 2145–50.

Plant, T. M. (1986) Gonadal regulation of hypothalamic gonadotropin-releasing hormone release in primates. *Endocr. Rev.* **7**, 75–88.

Plant, T. M., Krey, L. C., Moossy, J., McCormack, J. T., Hess, D. L. and Knobil, E. (1978) The arcuate nucleus and the control of gonadotropin and prolactin secretion in the female rhesus monkey (*Macaca mulatta*). *Endocrinology* **102**, 52–62.

Pohl, C. R. and Knobil, E. (1982) The role of the central nervous system in the control of ovarian function in higher primates. *A. Rev. Physiol.* **44**, 583–93.

Pohl, C. R., Richardson, D. W., Marshall, G. and Knobil, E. (1982) Mode of action of progesterone in the blockade of gonadotropin surges in the rhesus monkey. *Endocrinology* **110**, 1454–5.

Pohl, C. R., Richardson, D. W., Hutchison, J. S., Germak, J. A. and Knobil, E.

(1983) Hypophysiotropic signal frequency and the functioning of the pituitary-ovarian system in the rhesus monkey. *Endocrinology* **112**, 2076–80.

Van Vugt, D. A., Lam, N. Y. and Ferin, M. (1984) Reduced frequency of pulsatile luteinizing hormone secretion in the luteal phase of the rhesus monkey. Involvement of endogenous opioids. *Endocrinology* **115**, 1095–101.

Veldhuis, J. D., Rogol, A. D., Samojlik, E. and Ertel, N. H. (1984) Role of endogenous opiates in the expression of negative feedback actions of estrogen and androgen on pulsatile properties of luteinizing secretion in man. *J. Clin. Invest.* **74**, 47–55.

Williams, C., Nishihara, M., Thalabard, J-C., Grosser, P. M., Hotchkiss, J. and Knobil, E. (1988) The hypothalamic GnRH pulse generator of the rhesus monkey: the duration of phasic electrical activity and its control. *Abstr. Proc. Soc. Neuroscience.*

Wilson, R. C., Kesner, J. S., Kaufman, J-M., Uemura, T., Akema, T. and Knobil, E. (1984) Central electrophysiologic correlates of pulsatile luteinizing hormone secretion in the rhesus monkey. *Neuroendocrinology* **39**, 256–60.

Cell to cell signalling through circulatory feedback: A mathematical model of the mechanism of follicle selection in the mammalian ovary

H. MICHAEL LACKER,[a] MITCHELL E. FEUER[a] AND ETHAN AKIN[b]

[a]Department of Biomathematical Sciences, The Mount Sinai Medical School, One Gustave L. Levy Place, New York, NY 10029, USA

[b]Department of Mathematics, The City College, New York, NY 10031, USA

INTRODUCTION

Hormonal feedback between the ovary and the hypothalamic–pituitary axis is essential in the regulation of ovarian follicle development. Each cell responding to a circulating hormone receives a signal that is the sum of the contributions of all cells that secrete it. The fact that each cell responds to a summed variable imposes an important symmetry on mathematical models that describe cell to cell interaction through hormonal feedback. This will be exemplified by the model of follicle selection proposed in this article. By the term follicle selection, we mean the physiological mechanism that determines the number of ovarian follicles that emerge with ovulatory maturity during each oestrous (menstrual) cycle. In humans, this number is typically one or two but, in general, each higher vertebrate has a range that is characteristic of the species or breed to which it belongs.

In all vertebrates there is a large dormant population of immature (primordial) follicles in both ovaries at birth. Each primordial follicle contains a single egg cell and a surrounding layer of supporting cells. At random times that approximate a Poisson process, primordial follicles in both ovaries will be triggered to start growing, that is, the supporting cells in these follicles will begin to divide. In the course of the next several days to several weeks (the length of time depends on the species) a follicle that is triggered to start growth will either fully mature and release its ovum into the oviduct or it will stop developing and begin to atrophy and eventually disappear from the ovary in a process called follicular atresia. In higher vertebrates, the overwhelming majority (>99% in humans) of follicles triggered in any given cycle undergo atresia and this process may begin at nearly any time or stage of development after a follicle has been activated to grow. Those few follicles in each cycle that escape atresia, develop in a coordinated fashion that culminates in the event of ovulation: the nearly simultaneous release of ova into the oviducts (fallopian tubes). Simultaneous here, means in a time that is short (minutes to hours) compared to the time of follicle development (days to weeks).

The number of follicles that attain full maturity in each ovulation cycle is a carefully regulated quantity. This is reflected by the relatively characteristic litter sizes of mammalian species and by the observation that litter size is unchanged after an ovary is removed. In fact, it has been repeatedly demonstrated in many mammalian species that the number of ovulatory follicles per cycle is nearly independent of the amount of ovarian tissue removed or the timing of its removal (see Jones, 1978 for a review). This conservation in ovulation number has come to be known as Lipschütz's Law of Follicular Constancy (Lipschütz, 1928) and any theory of follicle selection should be able to explain it.

WHAT IS THE EVIDENCE THAT CIRCULATORY FEEDBACK REGULATES FOLLICLE GROWTH AND OVULATION NUMBER?

Removing the pituitary (hypophysectomy) from an adult female results in the arrest of follicle growth at a relatively early stage of development. These immature follicles can be made to mature and ovulate by injecting pituitary extracts containing varying proportions of the two purified hormones FSH (follicle-stimulating hormone) and LH (luteinizing hormone). When these gonadotropic hormones are given in the correct time sequence, maturation to ovulation occurs even when the nerve supply to the ovary is severed and the ovary is transplanted to a different site (see Schwartz, 1974 for a review of the role of FSH and LH on follicle growth).

In intact animals, the variance in the ovulation number per cycle is significantly smaller than the mean ovulation number per cycle (Falconer *et al.*, 1961). Although artificial administration of FSH and LH to hypophysec-tomized animals can induce follicle development and ovulation there is, in general, a loss in the control of the number of follicles that ovulate. Not only is there a dosage-dependent change in the mean number of eggs that are released at the time of ovulation but the variance in the number of eggs released is much higher in the hypophysectomized animal even when the change in the mean ovulation number is taken into account (Falconer *et al.*, 1961). These facts suggest that developing follicles may not grow independently in response to pituitary secretion but instead, may interact by secreting one or more hormones that feed back on the pituitary. This hypothesis was first supported by the observation that significant elevations in both circulating FSH and LH occur following castration. In 1981 Zeleznik showed that the follicular hormone oestradiol, which is secreted into the circulation with increasing rates as a follicle matures, inhibits pituitary gonadotropin secretion with great sensitivity at a time when follicular development and selection is occurring. In some species, inhibin, a non-steroidal chemical isolated from follicular extracts, has also been shown to depress FSH secretion when injected into the circulation. Because the physiological role that inhibin plays in the regulation of follicle development and ovulation number is much less clear than that for oestradiol we will focus primarily on the oestradiol–gonadotropin negative feedback loop.

Interfering with this loop has significant effects upon follicle growth. This was dramatically demonstrated in the experiments of Biskind and Biskind (1944). In this experiment a small portion of ovary is transplanted under the splenic capsule of a castrated animal. Since the splenic venous effluent passes through the liver before reaching the hypothalamic–pituitary axis and since oestradiol is very effectively cleared from the circulation by the liver, circulating FSH and LH concentrations remain chronically elevated at near castrate concentrations in this preparation and continue to stimulate follicular growth in the transplanted tissue. The unrestrained growth produces large tumours of follicular origin in the spleen (Biskind and Biskind, 1944; Seager *et al.*, 1974). These tumours can be prevented if exogenous oestrogen is administered into the peripheral circulation or if a portacaval shunt is performed to allow oestradiol from the transplant to bypass the hepatic circulation (Seager *et al.*, 1974; Lee *et al.*, 1980). As might be expected, in both of these instances gonadotropin concentrations fall significantly below castrate levels. Moreover, in mammals with intact ovaries, physiological quantities of exogenous oestradiol can prevent ovulatory follicles from developing and anti-oestradiol antibodies injected into the circulation can, in monkeys, produce several mature follicles at a time when only one is ordinarily present (Zeleznik, 1981; Zeleznik *et al.*, 1985).

Paracrine versus endocrine interaction

Are endocrine interactions alone sufficient to account for ovulation control, or is local signalling between follicles exerted through chemical diffusion within the ovary (paracrine interaction) also important in regulating follicle development? Follicular interaction via a local diffusion mechanism should, in general, be stronger when two follicles are nearest neighbours than when the same two follicles are far apart or in different ovaries. On the other hand, the interaction between two follicles that occurs by hormonal feedback will depend on the contribution that each follicle makes to the total circulating hormone concentration, but not on the distance separating them. In other words, paracrine interactions are inherently spatially dependent while endocrine interactions are not.

Present evidence supports the hypothesis that spatially dependent interactions do not play a significant role in regulating the number of follicles that reach ovulatory maturity per cycle. This is based on the observation in multiple ovulators that the distribution of ova shed between left and right ovaries when conditioned on a given ovulation number per cycle satisfies binomial statistics (Danforth and de Aberle, 1928; Brambell, 1935; Brambell and Hall, 1936; Brambell and Rowlands, 1936; Falconer *et al.*, 1961; McLaren, 1963). This observation implies that the probability of an ovulatory follicle growing on any particular side is independent of the number of other ovulatory follicles growing on that side. While this is a natural consequence of a spatially independent circulatory feedback mechanism, it is not, in general, consistent with the assumption that local (paracrine) interactions are important in regulating ovulation number.

The general argument for this assertion can be understood by considering a specific example in which it is assumed that ovulation number per cycle is controlled by a single follicular secretion product exerting both endocrine and paracrine effects. More precisely, suppose that follicles secrete, in proportion to their maturity, a growth stimulating hormone E that diffuses and directly stimulates the growth of neighbouring follicles. Suppose also that E enters the bloodstream and inhibits in proportion to its circulating concentration the release of growth stimulating hormones from the pituitary. Thus, E acts directly as a local *paracrine stimulator* and indirectly through its circulatory feedback effect as a global *endocrine inhibitor*. Let us assume that as a consequence of these local and global interactions that the total ovulation number per cycle is regulated at a mean of eight and that an ovulation is as likely to occur on the right side as on the left ($p = 1/2$). The observation that the distribution of ova shed at the time of ovulation between sides satisfies binomial statistics implies that when we examine the number of occurrences where there are eight ovulations and where at least seven of these are in one ovary then half of the occurrences will be distributed 7 to 1 and half will be distributed 8 to 0.

This result is a natural consequence of a spatially independent circulatory feedback mechanism but it is not consistent with the presumed local stimulatory effect of E. If the local effect of E is to stimulate growth by diffusing to neighbouring follicles then we would expect a greater chance for the eighth ovulatory follicle to appear on the side where there are seven other large follicles growing and secreting E than on the side where there are none.

In other words, the presence of a significant paracrine effect in the regulation of follicle growth violates a fundamental assumption of the actual distribution that is observed in mammals; the assumption of independent Bernoulli trials. Paracrine stimulators will produce significant deviations from binomial statistics that favour the extremes of the distribution over the mean while local inhibitors will favour the mean over the extremes. Significant paracrine interactions are consistent with a binomial distribution only if the effects of local inhibitors were exactly balanced by the effects of local stimulators so that the total effect of the local factors could not be detected.

Although p is close to one-half, there is, in many mammals, a definite preference for ovulation to occur on a particular side. If the probability per unit time that a dormant follicle will start to grow is independent of side it is on, then a preference for one ovary can arise if one ovary has a somewhat larger pool of primordial follicles than the other. In this way the observation that p is not one-half can be explained without altering the assumption that a spatially independent interaction mechanism (circulatory feedback) regulates the growth of activated follicles. A binomial distribution would also be observed if pituitary secretion controlled follicle growth without any follicular feedback at all (local or circulatory). We have rejected this hypothesis for the following reasons: (1) it cannot explain the control in ovulation number per cycle that mammals actually achieve (the variance in ovulation number per cycle is much smaller than the mean ovulation number for that mammal); and (2) it is inconsistent with a demonstrated sensitive feedback loop involving follicular oestradiol and pituitary gonadotropins that has been shown to have significant effects on the regulation of follicle growth and ovulation number.

A MODEL OF THE REGULATION OF FOLLICLE MATURATION BY OESTRADIOL–GONADOTROPIN FEEDBACK

Let $h_1(t)$ and $h_2(t)$ represent the circulating concentrations of FSH and LH as a function of time, and let these hormones be removed from the circulation by a first-order process with rate constants γ_1 and γ_2 respectively. If we assume that these hormones are secreted into a blood volume V at rates σ_1 and σ_2 that depend upon the circulating concentration of oestradiol X then

$$V\frac{dh_1}{dt} = \sigma_1(X) - \gamma_1 h_1 \qquad (1)$$

$$V\frac{dh_2}{dt} = \sigma_2(X) - \gamma_2 h_2 \qquad (2)$$

If we measure the maturity of the ith follicle at time t by its oestradiol secretion rate into the circulation $s_i(t)$ and assume that the metabolic removal of oestradiol from the circulation is a first-order process with rate constant γ_3 then

$$V\frac{dX}{dt} = \sum_{j=1}^{N} s_j - \gamma_3 X \tag{3}$$

where the summation is over all follicles that are secreting oestradiol at time t. Assuming that the rate at which a follicle matures depends on both its maturity and the concentrations of FSH and LH in the circulation then

$$\frac{ds_i}{dt} = \bar{g}_i(s_i, h_1, h_2) \tag{4}$$

In most mammalian species, follicle maturation occurs over a period of days to weeks. In comparison the half-life of oestradiol, FSH and LH in the circulation and the response time of the pituitary–hypothalamic axis to circulating oestradiol are short (a few minutes to an hour or two depending on the species) (Baird *et al.*, 1969; Cargille *et al.*, 1969; Pedersen, 1970; Speroff and VandeWiele, 1971; Tsai and Yen, 1971; Tapper and Grant-Brown, 1975). This means that the dynamics of (4) will be slow compared to (1)–(3) and therefore the solutions of (1)–(3) will always be near equilibrium on the time scale set by (4). This approximation is equivalent to setting the left hand sides of (1)–(3) equal to zero which produces the following simplified dynamical system:

$$\frac{dx_i}{dt} = g_i(x_i, X), \quad i = 1, \dots, N \tag{5}$$

$$X = \sum_{j=1}^{N} x_j \tag{6}$$

where

$$x_i(t) = \frac{s_i(t)}{\gamma_3} \tag{7}$$

is the concentration that the ith follicle supports in the circulation at time t. Since $x_i(t)$ is directly proportional to $s_i(t)$, it is an equivalent measure of follicle maturity. The system (5)–(6) is connected to (1)–(4) by (7) and

$$g_i(x_i, X) = \bar{g}_i\left(\gamma_3 x_i, \frac{\sigma_1(X)}{\gamma_1}, \frac{\sigma_2(X)}{\gamma_2}\right) \tag{8}$$

Note that the effects of FSH and LH are still present in the system (5)–(6). Their effects have been represented implcitly by (8). Approximate solutions for $h_1(t)$ and $h_2(t)$ can be found by solving the system (5)–(6) to obtain $X(t)$ and then substituting into

$$h_1(t) = \frac{\sigma_1(X(t))}{\gamma_1} \tag{9}$$

$$h_2(t) = \frac{\sigma_2(X(t))}{\gamma_2} \qquad (10)$$

The function $g_i(x,X)$ in (5) determines the ith follicle's maturation rate for any given maturity x and circulating oestradiol concentration, X. Thus, any particular choice for $g_i(x,X)$ in the model represents a possible programme of follicle development. If it is reasonable to assume that follicles in a given individual or even a given mammalian breed or species inherit a similar plan of development then it seems natural to first consider an idealization where every model follicle of the system (5)–(6) inherits the same programme of development. More precisely, we will assume

$$g_i(x,X) = f(x,X) \text{ for all } i \qquad (11)$$

Substituting (11) into (5) yields the system

$$\frac{dx_i}{dt} = f(x_i, X), \qquad i = 1, \ldots, N \qquad (12)$$

$$X = \sum_{j=1}^{N} x_j \qquad (13)$$

In addition to assuming that all follicles share a common maturation programme, the system (12)–(13) has another important symmetry. The maturities $x_1, \ldots, x_N$ of the developing follicles are coupled to one another through the variable X and this variable is a symmetric function (the summation function) of these maturities. This symmetry reflects the idea of follicular interaction through circulatory feedback: Each follicle affects the maturation rate of every other follicle through its contribution to the circulating oestradiol concentration and the effect that this hormone has on gonadotropin release. At any given instant all model follicles receive the same circulating signal since there is only one set of values X, h_1 and h_2 for each t. Although all model follicles receive the *same* circulating signal at time t and obey the *same* maturation function they need not, in general, develop at the same rate if they have different maturities. This is because f depends on x as well as the circulating oestradiol concentration. Can an initially similar population of immature follicles each inheriting the same programme of growth and interacting through a common circulating hormone concentration account for follicle selection?

THEORY

The system (12)–(13) actually represents a class of models that becomes a specific model when a follicle maturation function $f(x,X)$ is chosen. Ideally this function would be determined directly from experiment but laboratory techniques are not yet available to explicitly measure this function in any species. We do not have a theory for analysing the solutions to (12)–(13) for arbitrary f, but it is possible to analyse the qualitative behaviour of (12) for a limited class of follicle maturation functions and demonstrate specific examples that regulate ovulation number (Akin and Lacker, 1984; Lacker and Akin, 1987). The class of models for which the theory can be analysed takes the form:

$$\frac{dx_i}{dt} = f(x_i, X), \quad i = 1, \dots, N \tag{14}$$

$$X = \sum_{j=1}^{N} x_j \tag{15}$$

$$f(x_i, X) = x_i \delta(X) \left[\rho(X) + \xi \left(\frac{x_i}{X} \right) \right] \tag{16}$$

where δ is any smooth, positive-valued function and ρ and ξ are arbitrary smooth functions defined on appropriate physiological domains. For δ and ρ the domain is $X > 0$ since negative values have no physical interpretation. The domain for ξ is the interval $[0,1]$ since the ratio x_i/X must physiologically lie in this interval. We will first study the case where the number of interacting follicles N is a fixed constant that does not change in time. The function

$$\varphi(x, X) = \delta(X) \left[\rho(X) + \xi \left(\frac{x}{X} \right) \right] \tag{17}$$

can be given a biological interpretation. For a given value of X

$$\tau_D(x, X) = \frac{\ln 2}{\varphi(x, X)} \tag{18}$$

represents the time it would take for a follicle with maturity x to double its oestradiol secretion rate. If, for example, the oestradiol secretion rate of a follicle were directly proportional to the number of oestradiol-secreting (granulosa) cells it contained, then φ would determine the doubling times of these cells as a function of follicle maturity (oestradiol production rate) and the circulating concentrations of FSH and LH (represented implicitly by X).

Since the δ function in (17) is assumed to be positive, a new time variable τ can be defined by

$$\tau(t) = \int_0^t \delta(X(s)) ds \text{ or } \frac{d\tau}{dt} = \delta(X(t)) \tag{19}$$

Scaling follicle maturity so that it is a fraction between 0 and 1 in the new time scale by letting

$$y_i(\tau) = \sqrt{\left(\frac{x_i(t(\tau))}{X(t(\tau))} \right)}, \quad i = 1, \ldots, N \tag{20}$$

allows the system (14)–(16) to be expressed in the following form

$$\left. \begin{aligned} \frac{dY}{d\tau} &= -\nabla_s V(Y), \quad Y \in S \\[2mm] V(Y) &= -\tfrac{1}{2} \sum_{i=1}^{N} \int^{y_i} s\xi(s^2)\,ds \\[2mm] S &= \left\{ Y : \sum_{i=1}^{N} y_i^2 = 1 \right\} \end{aligned} \right\} \tag{21}$$

$$\frac{d\bar{X}}{d\tau} = \bar{X}\left[\rho(\bar{X}) + \sum_{i=1}^{N} y_i^2 \xi(y_i^2) \right] \tag{22}$$

where $Y = (y_1, \ldots, y_N)$, $\bar{X}(\tau) = X(t(\tau))$. This change of coordinate system allows the dynamical system to be given a simple geometric interpretation. If we assign a coordinate axis to each of the N interacting follicles then the solution of (21) will trace a curve $Y(\tau)$ on the unit sphere S in N-space that represents the progression of the N-follicle system in the new maturity and time variables. The actual path taken on S depends upon the particular follicle maturation function f used in (16) and the initial choice of follicle maturities (the starting point $Y(0)$ on S). One can think of the function V in (21) as generating a relief map of hills, valleys and passes on the sphere S where $V(Y)$ defines the height above the surface S at Y. Note that V depends only on the ξ function of the follicle maturation programme of (16) and not on the δ or ρ functions.

According to (21) the solution curve $Y(\tau)$ obeys the following rule: at any point Y on S, $Y(\tau)$ will move on S along the direction of steepest descent of V and with speed equal to the slope of V in this direction. The notation $\nabla_s V(Y)$ represents the operation of taking the gradient of V on S; that is, taking the gradient of V in R^N and projecting it onto the tangent plane of S at Y. More precisely, $\nabla_s V = \nabla V - \langle \nabla V, Y \rangle Y$. Equilibria only occur at the critical points of V on S, that is, where $\nabla_s V(Y) = 0$. These points correspond to tops of hills (relative maxima), tops of passes (saddle points) or bottoms of valleys (relative minima). The tops of hills and passes, however, must be unstable since directions exist where the slightest local perturbation will allow descent to lower values of V. The only stable equilibria are the relative minima. Provided that Y does not start exactly at an unstable equilibrium of V, $Y(\tau)$ will end up trapped at the bottom of a well where no local path of further descent exist. Finding the eventual maturational state of the system of N interacting follicles, is therefore, equivalent to finding the local minima of V

on S and interpreting these local minima in terms of the original coordinate system. The components of $\nabla_s V(Y)$ are

$$-\tfrac{1}{2}y_i\left[\xi(y_i^2)-\sum_{i=1}^{N}y_j^2\xi(y_j^2)\right], \quad i=1,\ldots,N \qquad (23)$$

The components of $\nabla_s V$ will be 0 if either

$$y_i=0$$

or

$$\xi(y_i^2)=\sum_{j=1}^{N}y_j^2\xi(y_j^2)=\lambda(Y_e) \qquad (24)$$

Thus, every non-zero coordinate of an equilibrium, Y_e, must be assigned same number $\lambda(Y_e)$ by the function ξ. We will call an equilibrium with M positive coordinates (negative coordinates do not have physical meaning) an M-fold equilibrium. More precisely,

$$Y_e=\underbrace{(a_1a_2\ldots,a_M,}_{M}\cdot\underbrace{0,\ldots,0)}_{N-M} \qquad (25)$$

is an M-fold equilibrium if $a_i>0$ for $i=1,\ldots,M$ and

$$\lambda(Y_e)=\xi(a_1^2)=\xi(a_2^2)=\ldots=\xi(a_M^2) \qquad (26)$$

with

$$\sum_{i=1}^{M}a_i^2=1 \qquad (27)$$

Any Y whose coordinates are permutations of an M-fold equilibrium is also an M-fold equilibrium. The following stability theorem can be proved (Akin and Lacker, 1984):

Theorem 1

An M-fold equilibrium, Y_e, is stable (a local minimum of V on S) if and only if the common value $\lambda=\xi(a_i^2)>\xi(0)$ for all coordinates $a_i\neq0$, and either the derivative $\xi'(a_i^2)<0$ for all of the non-zero coordinates or $\xi'(a_i^2)\geqslant0$ for exactly one non-zero coordinate and

$$\sum_{i=1}^{M}[\xi'(a_i^2)]^{-1}>0 \qquad (28)$$

(By convention $1/0=+\infty$).

The most obvious way to satisfy (26) and obtain M-fold equilibria of (21) is make all M non-zero coordinates equal. For (25) this means that

$$a_1 = a_2 = \ldots = a_M \tag{29}$$

and (27) requires that

$$a_i = \sqrt{\left(\frac{1}{M}\right)}, \quad i = 1, \ldots, M \tag{30}$$

and that $\lambda = \xi(1/M)$. An M-fold equilibrium whose positive coordinates are equal will be called an M-fold symmetric equilibrium. Such equilibria will take the form

$$Y_e = \Bigg(\underbrace{\frac{1}{\sqrt{M}}, \frac{1}{\sqrt{M}} \cdots, \frac{1}{\sqrt{M}}}_{M}, \underbrace{0, \ldots, 0)}_{N-M} \Bigg), \quad M = 1, \ldots, N \tag{31}$$

or permutations of these coordinates.

AN EXAMPLE

Let us consider the physiological implications of this theorem by studying a specific example. Consider the class of follicle maturation functions (Lacker, 1981):

$$f(x,X) = x\{K - D(X - M_1 x)(X - M_2 x)\} \tag{32}$$

where K, D, M_1 and M_2 are positive real-valued parameters that are the same for all N interacting follicles. To be more specific, let us consider the case where

$$\begin{aligned}
K &= 4 \times 10^{-1} \text{ day}^{-1} \\
D &= 4 \times 10^{-3} \text{ (pg/ml)}^{-2} \text{ day}^{-1} \\
M_1 &= 3.85 \\
M_2 &= 15.15
\end{aligned} \tag{33}$$

Note that M_1 and M_2 are dimensionless parameters. The class of follicle maturation functions represented by (32) can be written in the form of (16) where

$$\delta(X) = X^2 \tag{34}$$

$$\rho(X) = \frac{K}{X^2} - D \tag{35}$$

$$\xi(p) = Dp(M_1 + M_2 - M_1 M_2 p), \quad p = \frac{x}{X} \tag{36}$$

If we define M_H as the harmonic mean of M_1 and M_2

$$\frac{1}{M_H} = \frac{1}{2}\left(\frac{1}{M_1} + \frac{1}{M_2}\right) = \frac{1}{6.14} \tag{37}$$

then the function ξ is a concave down parabola with a maximum of

$$\frac{1}{2}D\frac{M_1 + M_2}{M_H} = 6.2 \times 10^{-3} \tag{38}$$

at

$$p_{**} = \frac{1}{M_H} = \frac{1}{6.14} \tag{39}$$

The parabola intersects the x-axis at the origin and at $p_* = 2/M_H = 1/3.07$ (see Fig. 1).

$$\xi(0) = \xi\left(p_* = \frac{2}{M_H}\right) = 0 \tag{40}$$

Theorem 1 says that the symmetric M-fold equilibrium (31) will be stable if $\xi(1/M) > \xi(0) = 0$ and $\xi'(1/M) < 0$. Figure 1 shows that this will occur for all values of $1/M$ that lie in the interval (p_{**}, p_*), that is, for all integers M that satisfy

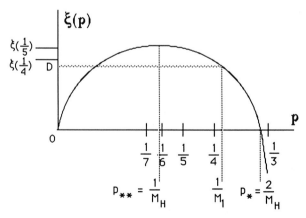

Fig. 1.   A graph of $\xi(p)$ for the follicle maturation function (32)–(33).

$$\tfrac{1}{2}M_H < M < M_H \tag{41}$$

For the parameter values in (33), the only integers in this interval are 4 and 5 and 6. Therefore the stable M-fold symmetric equilibria occur only for $M = 4$, 5 or 6.

Non-symmetric equilibria can be identified in the following way. Choose a horizontal line which intersects the graph of $\xi(p)$ at two points (Fig. 2). Let $(L, \xi(L))$ be the left intersection and $(H, \xi(H))$ be the right. Equation (26) requires that the line be horizontal, $\xi(L) = \xi(H)$. If there exist two integers M_L and M_H such that

$$LM_L + HM_H = 1 \tag{42}$$

then (25)–(27) will be satisfied and an M-fold equilibrium will exist with $M = M_L + M_H$. This equilibrium will have M_L coordinates with the value $\sqrt{L}$, and M_H coordinates equal to $\sqrt{H}$. The remaining $N-M$ coordinates will be 0

$$Y_e = (\overbrace{\sqrt{L}, \ldots, \sqrt{L}, }^{M_L} \overbrace{\sqrt{H}, \ldots, \sqrt{H}, }^{M_H} \underbrace{0, \ldots, 0}_{N-M}) \tag{43}$$
$$\underbrace{\phantom{\sqrt{L}, \ldots, \sqrt{L}, \sqrt{H}, \ldots, \sqrt{H}}}_{M}$$

The value of λ associated with this equilibrium will be $\xi(L) = \xi(H)$.

According to Theorem 1 all equilibria of the form (43) are unstable except those with M_L equal 0 or 1. We have already discussed the case when $M_L = 0$ since these are just the M-fold symmetric equilibria with $M_L = M$. When $M_L = 1$, we have

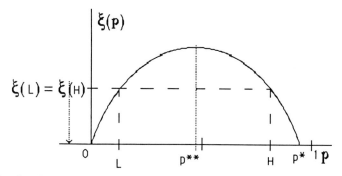

Fig. 2. The non-zero coordinates of a non-symmetrical equilibrium

$$Y_e = (\overbrace{\sqrt{L}, \ldots, \sqrt{L}, }^{M_L} \overbrace{\sqrt{H}, \ldots, \sqrt{H}'}^{M_H} \underbrace{0, \ldots, 0}_{N-M}) \quad \text{must satisfy } \xi(L) = \xi(H).$$
$$\underbrace{\phantom{\sqrt{L}, \ldots, \sqrt{L}, \sqrt{H}, \ldots, \sqrt{H}}}_{M}$$

$$Y_e = (\overbrace{\sqrt{L}, \sqrt{H}, \ldots, \sqrt{H}}^{M_H}, \underbrace{0, \ldots, 0}_{N-M}) \tag{44}$$

It can be shown that an equilibrium of this type will be stable if and only if $p^{**} > 1/2$ (Akin and Lacker, 1984; Lacker and Akin, 1987). Since this condition is not met for the parameter values in (33) (see Fig. 1), the only stable equilibria in this example will be of the form (31) with $M = 4$, 5 or 6. Defining the relative maturity of the ith follicle as the fraction of the total oestradiol that a follicle supports at any time i.e. $x_i/X = y_i^2$, then what we have shown so far is the following: If N follicles inherit the programme of circulatory feedback and growth given by (32) with parameters as in (33), then essentially, no matter where these follicles start on the unit sphere the follicle population will gradually split in such a way that 4, 5 or 6 follicles will approach the same relative maturity as $\tau \to \infty$ and all of the remaining follides will approach a relative maturity of zero. This result is independent of the size of the interacting follicle population, N, provided that $N \geqslant 6$.

One can think of the dynamical system as carving up the sphere S into regions. Each region is assigned a single number from the set $\{4, 5, 6\}$ depending on the value of the attracting M-fold symmetric equilibrium. To predict the actual oestradiol concentration that the ith follicle supports at time t, we need to solve for $X(t)$ and $t(\tau)$. This can be accomplished by substituting the solution of (22) into the equivalent of (19)

$$\frac{dt}{d\tau} = \frac{1}{\delta(\bar{X}(\tau))} \tag{45}$$

and integrating to obtain $t(\tau)$. The solution to (45) also implicitly determines $\tau(t)$. This inverse function exists since δ is defined to be positive valued. Substituting into the solution of (22) yields

$$X(t) = \bar{X}(\tau(t)) \tag{46}$$

Finally, (46) can be used with the solution of $y_i(\tau)$ to obtain follicle maturity expressed in terms of the oestradiol concentration that a follicle supports at time t,

$$x_i(t) = y_i^2(\tau(t))X(t), \quad i = 1, \ldots, N \tag{47}$$

This is related to the ith follicle's secretion rate, s_i, at time t by (7).

We can continue our qualitative analysis of the specific follicle maturation scheme described by (32)–(33) by using the relationships (45)–(47) between coordinate systems. Since $Y(\tau)$ approaches an equilibrium of type (31) with $M = 4$, 5 or 6, the summation term in (24) must satisfy

$$\sum_{j=1}^{N} y_j^2 \xi(y_j^2) \rightarrow \lambda(Y_e) = \xi(a_i^2), \text{ as } \tau \rightarrow \infty \tag{48}$$

where $\lambda(Y_e)$ can have the values

$$\xi\left(\frac{1}{4}\right) = 4.7 \times 10^{-3}, \xi\left(\frac{1}{5}\right) = 5.9 \times 10^{-3}, \text{ or } \xi\left(\frac{1}{6}\right) = 6.2 \times 10^{-3} \tag{49}$$

Thus, for τ sufficiently large (22) will approach

$$\frac{d\bar{X}}{d\tau} \approx \bar{X}[\rho(\bar{X}) + \lambda(Y_e)] \tag{50}$$

From (35) we have

$$\rho(\bar{X}) > -D, \text{ for all } \bar{X} \tag{51}$$

Therefore, for τ sufficiently large

$$\frac{d\bar{X}}{d\tau} > k(Y_e)\bar{X}, \ k(Y_e) = \lambda(Y_e) - D \tag{52}$$

Using the values given by (33) and (49), we see that the value of the constant $k(Y_e)$ will be positive for all three stable symmetric equilibria. The graph of the circulating oestradiol concentration as a function of τ will therefore lie above the function $\exp(k(Y_e)\tau)$ for sufficiently large τ. Hence, for the specific follicle maturation scheme described by (32)–(33), we have that the circulating oestradiol concentration, $X \rightarrow \infty$ as $\tau \rightarrow \infty$. This will occur essentially for any set of starting maturities of the N-follicle system independent of the size of N. From (47) we can now conclude that the 4, 5 or 6 follicles that approach a common non-zero relative maturity of $1/4$, $1/5$ or $1/6$ will also approach an absolute maturity that grows without bound as $\tau \rightarrow \infty$.

Of course, physiologically neither follicle maturity nor circulating oestradiol can grow without bound. Eventually X will become sufficiently large to trigger the gonadotropic surge mechanism that signals ovulation. At this time 4, 5 or 6 of our model follicles will have nearly the same maturity in the sense that the ratio of any pair approaches 1. All of the remaining follicles in the interacting population will be repressed towards zero relative maturity. The following argument shows that a follicle that approaches zero relative maturity also approaches an absolute maturity of zero, that is, undergoes atresia. Writing

$$\bar{X}(\tau) = \sum_{i=1}^{N} \bar{x}_i(\tau) \tag{53}$$

and substituting (53) into (22) yields

$$\frac{d\bar{x}_i}{d\tau}=\bar{x}_i\,[\rho(\bar{X})+\xi\,(y_i^2)\,] \qquad (54)$$

where follicle maturity is expressed as a function of τ. If $y_i\to0$ and $\tau\to\infty$, then $\xi(y_i^2)\to0$, $X\to\infty$, $\rho(X)\to-D$ and

$$\frac{d\bar{x}_i}{d\tau}\approx-D\bar{x}_i \qquad (55)$$

which decays exponentially to zero with rate constant D.

What happens to the time t as $\tau\to\infty$? Solving (45), we obtain

$$\lim_{\tau\to\infty}t(\tau)=\int_0^\infty\frac{d\tau}{\delta(\bar{X})}=\int_{\bar{X}(0)}^{\bar{X}(\infty)}\frac{1}{\delta(\bar{X})}\frac{d\tau}{d\bar{X}}d\bar{X} \qquad (56)$$

For $\tau>\tau_1$, where τ_1, is a sufficiently large number, we can use (52) to get

$$\lim_{\tau\to\infty}t(\tau)<\int_{X(0)}^{\tau_1}\frac{d\tau}{\delta(\bar{X})}+\frac{1}{k(Y_e)}\int_{X(\tau_1)}^\infty\frac{d\bar{X}}{\bar{X}\delta\,(\bar{X})} \qquad (57)$$

If $\delta\,(X)$ grows faster than X^ε as $X\to\infty$, where $\varepsilon>0$, then the right hand side of (57) will be bounded as $\tau\to\infty$. Equation (34) shows that this condition is met for the maturation function (32) so this means that X will 'blow-up' in a finite time T. The time T is interpreted as an idealized ovulation time. The value of T will depend on the initial distribution of follicle maturities $Y(0)$.

NUMERICAL SIMULATIONS

In the previous section we predicted that a system of N follicles maturing according to (32)–(33) and interacting through a common summed circulatory feedback signal, regulates ovulation number in the range of 4–6. We will test this theoretical prediction by solving the system (14)–(16) numerically with $f(x,X)$ given by (32)–(33). An example of the results are illustrated in the two graphs of Fig. 3. In each graph the number of interacting follicles $N=10$. The initial maturity of each follicle is chosen independently from a uniform probability distribution over the interval [0, 1]. In each example the follicle population gradually splits into ovulatory and atretic groups. Although there are no explicit thresholds built into any of the equations, there appears to be an initial maturity that separates ovulatory and atretic groups. This separating maturity automatically adjusts itself so that the number of follicles that emerge as ovulatory is in the predicted range of 4–6. The domains of attraction associated with the stable ovulation number 6 are very small since

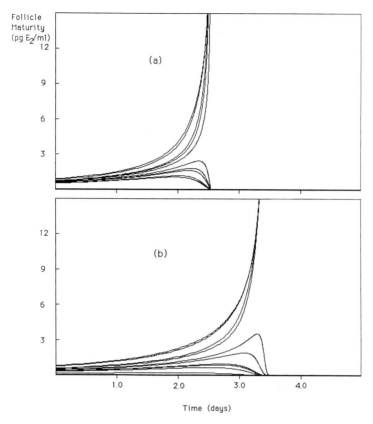

Fig. 3.   Two numerical solutions of the system (14)–(16) with $f(x,X)$ given by (32)–(33). In both graphs the number of interacting follicles $N=10$. The initial maturity of each follicle is chosen independently from a uniform probability distribution over the interval [0, 1]. In (a) five follicles gradually emerge as ovulatory while in (b) the ovulation number is four.

this ovulation number was not observed in over 1000 trials with initial conditions chosen as in Fig. 3. This is not too surprising since $p = 1/6$ lies very close to edge of the stability interval (p_{**}, p_*), (see Fig. 1).

Similar but not identical follicles

A feature of the numerical solutions shown in Fig. 3 is that follicle maturation curves do not cross. Since the maturation curves are continuous functions of time, in order for one follicle to overtake another there must exit a crossover time t_c when both have the same maturity, x_c. Let $X_c = X(t_c)$ be the circulating oestradiol at this time. Since both follicles inherit the same maturation function, $f(x,X)$, their maturation rates as well as their maturities are identical at t_c. It follows from the uniqueness theorem for the initial value problem of ordinary differential equations (Pontryagin, 1962) that the two follicles will have the same maturation curve for all times t on which the solution of the system of differential equations exists, that is, from $[0,T]$. Thus, if two follicles do not have the same initial maturity, their maturation curves will never meet.

The assumption that all follicles of an interacting follicle population inherit exactly the same maturation function is crucial to the preceding argument. Biologically it is realistic to relax this restriction of the model and consider the case of a system of N follicles that obey similar but not identical maturation schemes. If follicles i and j have the same maturity x_c at time t_c,

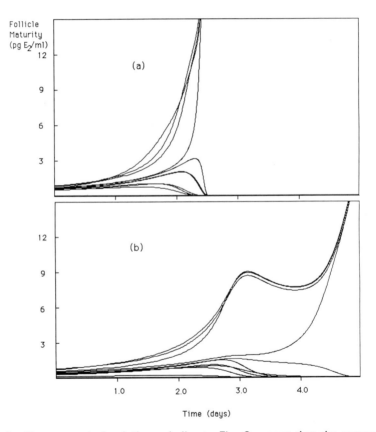

Fig. 4.   Two numerical solutions similar to Fig. 3 except that the parameter values of the maturation function (32) are independently and normally distributed among the 10 follicles so that each follicle obeys a slightly different maturation function. The mean value of the normal distributions for M_1, M_2, K and D are given by (33). The standard deviation of M_1 and $M_2=0.10$ and the standard deviation of K and D are 0.12 and 0.12×10^{-2}, respectively. In both (a) and (b) the initial maturities are the same as those in the corresponding graphs of Fig. 3.

and $f_i(x_c,X_c) \neq f_j(x_c,X_c)$ then the maturation rates of i and j will be different at t_c and the maturities will cross on any open interval containing t_c.

The assertion that an initially smaller follicle can overtake a larger follicle when the members of the interacting follicle population inherit similar but not identical maturation functions is shown in Fig. 4. Each follicle in this figure is assigned a maturation function of the form (32), but the parameter values M_1, M_2, K and D are now independently and normally distributed among the members of the follicle population with mean values that are identical to (33).

Random follicle entry

Thus far we have considered the behaviour of a system of interacting follicles where the number of follicles is a fixed size N. Physiologically the size of the interacting population changes in time as new follicles are activated at different times from the reserve pool. Figure 5 shows the results of activating follicles at random times that are generated by a Poisson process whose mean activation rate is 12 follicles per day. Each follicle is assumed to support a circulating oestradiol concentration of 2 pg/ml at the time of activation. At any time the oestradiol concentration X in the circulation is the sum of the contributions made by the activated follicle population. Although all follicles

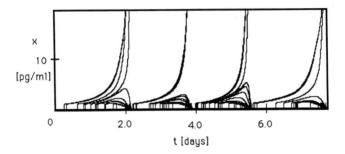

Ovulation Time (Days)	Ovulation Number
2.145	5
3.905	4
5.575	4
7.785	4

Fig. 5. Cycles of follicle maturation with controlled ovulation number result when follicles are activated at random times. Activation times are generated by a Poisson process with mean rate 12 follicles per day. Each activated follicle is given the same initial maturity of 2 pg/ml oestradiol and obeys the same maturation function (32)–(33). The circulating oestradiol $X(t)$ is the sum of the maturities of all activated follicles at time t. When $X=200$ pg/ml the emergent ovulatory population is removed, simulating ovulation.

obey the same maturation function (32) with parameters given by (33) and interact through the common feedback signal X, 4 or 5 follicles repeatedly emerge from the interacting population as ovulatory. When X reaches a sufficiently large value (arbitrarily chosen as 200 pg/ml in this case) these follicles are assumed to ovulate, that is, they are removed from the interacting population and their contribution to X is set to 0. Physiologically, this corresponds to triggering the ovulatory surge mechanism when the circulating oestradiol concentration is sufficiently high. Despite the fact that the activation process in Fig. 5 is chaotic, a temporal order emerges where the maturation curves of ovulatory follicles becomes synchronized and the non-ovulatory follicles become fully repressed by the time of ovulation.

Lipschütz's Law of Follicular Constancy

Figure 6 simulates the physiological consequences of suddenly removing an ovary during a period of follicular maturation. The parameters are the same as in Fig. 5. Activated follicles are randomly assigned with equal probability to one of two sets (ovaries). At $t = 1$ day, those follicles assigned to one set are removed producing a sudden drop in the circulating oestradiol concentration. The mean activation rate is also reduced by a factor of two after $t = 1$ day. In Fig. 6 the two largest follicles were in the ovary that was suddenly removed. These follicles would have been in the ovulatory group had they not been removed. Nevertheless, the ovulation number is maintained at four in accordance with Lipschutz's Law of Follicular Constancy (see Introduction). Two follicles that would have become atretic are automatically

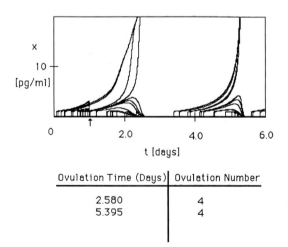

Ovulation Time (Days)	Ovulation Number
2.580	4
5.395	4

Fig. 6.   Simulation of single ovary removal at $t=1$ (arrow). Ovulation number is conserved.

recruited by the circulatory feedback interaction mechanism into the ovulatory group.

Anovulation

Another phenomenon arises when follicles are allowed to activate at random times from a reserve pool and interact according to the follicle maturation scheme (32)–(33). In addition to the cyclic emergence of a well-controlled number of ovulatory follicles, transitions from cyclic to anovulatory behaviour occasionally spontaneously occur (see Fig. 7). During such an anovulatory state several follicles get 'stuck' at an intermediate level of maturity (Lacker and Akin, 1987; Lacker et al., 1987a).

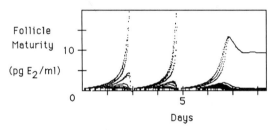

Fig. 7.   Spontaneous transition from cyclic to steady anovulatory behaviour. The parameters are the same as Fig. 5.

Similar anovulatory states have been described in some members of some species including humans (Yahia and Taymor, 1970) and are believed, in some cases, to have deleterious effects on oestrogen-sensitive tissues in the breast and uterus (Speroff et al., 1973). The assertion that continual random injection of follicles into interacting population induces the anovulatory state is supported by the numerical experiment shown in Fig. 8a and b. In Fig. 8a an anovulatory state is achieved in which three follicles are 'stuck' at an intermediate maturity. Figure 8b is exactly the same simulation as Fig. 8a except that at $t=3$ days (arrow) the random entry of follicles is stopped. From this point on, the number of interacting follicles is a fixed number N. An additional follicle joins the three with intermediate maturity and all four follicles ovulate in accordance with the theoretical predictions given earlier. This result supports the idea that rate of follicle injection in Figs 7 and 8a is sufficient to increase the oestradiol concentration to a level that represses the remaining follicle population. In principle, anovulatory states produced by this mechanism will be transient since follicle activation is stochastic and the suppressing effects that it may produce will fluctuate in time.

The theory also predicts other mechanisms that can produce anovulatory states similar to that shown in Fig. 7. For example, if we construct the

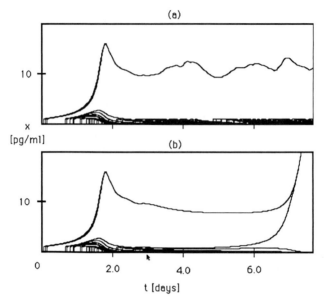

Fig. 8.   (a) An anovulatory state is produced under the same conditions as Fig. 7. In (b) the activation times and all other parameters are identical to (a) except that at *t*=3 (arrow) activation of new follicles stops. The anovulatory state becomes unstable and four follicles eventually ovulate.

maturation function $f(x,X)$ in (16) using the δ and ξ functions defined in (34) and (36) but define

$$\rho(X) = \frac{K}{X^2} - D^*$$
(58)

where D^* is given a value that lies in the interval $(\xi(1/4), \xi(1/5))$ then in addition to the behaviour depicted in Fig. 7 an anovulatory state can be produced in which four follicles will become 'stuck' at some intermediate maturity (Lacker and Akin, 1987). This arises because (50) now has a unique stable equilibrium when Y_e is 4-fold symmetric.

Variations on a theme

The theory described earlier can be used to construct follicle maturation functions that regulate ovulation number in different ranges. For example, the stable range of ovulation numbers can be adjusted by moving the location of both the maximum, p_{**}, and the *x*-intercept, p_*, of $\xi(p)$ in Fig. 1. Changing the time scale or maturity scale can be accomplished by altering the parameter values of K and D in (33) or by choosing new functions for δ, ρ or ξ.

The follicle maturation function

$$f(x,X) = AxX \left[\frac{K}{X} - d + a \left(\frac{x}{X} \right)^2 + b \left(\frac{x}{X} \right) \right] \qquad (59)$$

which can be written in the form of (16) with

$$\delta(X) = AX, \; \rho(X) = \frac{K}{X} - r, \; \xi(p) = ap^2 + bp + (r - d)$$

$A = 0.0034 \; (\text{pg/ml} - \text{days})^{-1}, \; K = 100 \; \text{pg/ml}, \; r = 5$

$$a = \begin{cases} -9.0625, & \text{for } x \leqslant p_c, \\ -12.5, & \text{for } x > p_c, \end{cases} \quad b = \begin{cases} 14.5, & \text{for } x \leqslant p_c \\ 20.0, & \text{for } x > p_c \end{cases}$$

$$d = \begin{cases} 5.0, \text{ for } x \leqslant p_c \\ 7.2, \text{ for } x > p_c \end{cases} \qquad (60)$$

$$p_c = 0.8$$

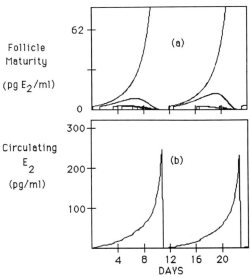

Fig. 9.   Follicles are activated by a Poisson process at a mean rate of 1 per day. At the time of activation each follicle supports a circulating oestradiol concentration of 2.5 pg/ml. All follicles obey the maturation function (59) with parameters given by (60). (a) Shows the individual follicle maturation curves while (b) is the graph of the circulating oestradiol concentration X which at any time is the sum of the y-values of the curves in (a). When the circulating value of $X = 250$ pg/ml is reached, the emergent dominant follicle is removed from the interacting follicle population (ovulation).

regulates ovulation number at 1. The parameters a, b and d, have been chosen so that $f(x,X)$ is both continuous and smooth. Figure 9 shows simulated follicle maturation and corresponding circulating oestradiol curves. Numerical simulations (Lacker *et al.*, 1987b) using this follicle maturation function have been shown to mimic several effects that exogenous oestradiol produces on follicle development in the primate menstrual cycle (Clark *et al.*, 1981; Dierschke *et al.*, 1985; Zeleznik, 1981).

Why do so many follicles die?

We have demonstrated that the range of possible ovulation numbers depends to a large extent on simple graphical properties of the function $\xi(p)$. We have also shown that the range of possible ovulation numbers for any given follicle maturation function of the form (16) is essentially independent of the size of the interacting follicle population, N. It is this fact that allows the number of ovulatory follicles per cycle to be virtually independent of the amount of ovarian tissue removed or the timing of its removal (Lipschutz's Law of Follicular Constancy).

Although the size of the interacting follicle population does not change the range of stable ovulation numbers, it does produce subtle effects on the frequency of occurrence of these ovulation numbers. This is shown in Fig. 10. This figure summarizes the results of many trials similar to Fig. 3 in which the number of interacting follicles is either 1000 (top), 100 (middle), or 30 (bottom). All follicles obey the same maturation function whose form is given by (32). The parameter values of M_1 and M_2 have been chosen so that the predicted range of ovulation numbers per cycle is 7 to 12. The graphs on the left show the frequency of occurrence of these ovulation numbers and the graphs on the right show the spread of ovulation times around the mean time for that follicle population size, N. (The mean time increases with N.) There are two effects associated with decreasing N: (1) There is a shift in the distribution of ovulation numbers towards integers in the higher end of the stable range and (2) there is a wider dispersion in ovulation times. In terms of the theory given earlier, these effects are the result of changes in the size and geometry of the capturing regions associated with stable symmetric equilibria as the dimension of the unit sphere is decreased.

As a mammal ages there is a gradual decline in the number of follicles that are activated in each cycle. In humans, this is associated with an increase in the dizigotic twinning rate and an increase in the spread of ovulation times (McArthur, 1954; Parkes, 1976; Vollman, 1977; Korenman *et al.*, 1978). Both of these observations are consistent with the model results described above. These results suggest an important physiological role for the large number of follicles that are activated but which atrophy and disappear in each cycle. They help the hormonal feedback mechanism to keep the

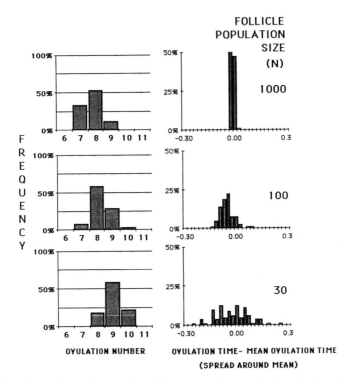

Fig. 10. The effect of interacting follicle population size on the distribution of ovulation times and numbers. Larger numbers of interacting follicles improve control of ovulation time and favour smaller ovulation numbers. Each graph represents the results of 80 cycles. Every follicle obeys the maturation function (32) with $M_1=6.1$, $M_2=5000.0$, $K=1.0$ and $D=1.0$. Initial maturities are chosen at random from a uniform distribution in the interval $[0, 10^{-5}]$. Statistics (mean-$\pm$standard deviation): $N=1000$ (top), ovulation number$=7.79\pm0.65$, ovulation time$=4.37\pm0.01$; $N=100$ (middle), ovulation number$=8.28\pm0.67$, ovulation time$=5.55\pm0.04$; $N=30$ (bottom), ovulation number$=9.04\pm0.65$, ovulation time$=6.33\pm0.10$.

ovulation number down and tighten control on the timing of maturation. These functions may represent an evolutionary adaptation of a mechanism that was originally designed to release a large number of eggs into the external environment. The model suggests that this mechanism was not discarded in the mammal but adapted to a new (and in this case opposite) role; that of helping a superimposed layer of control mechanisms to reduce the ovulation number per cycle and thus allow the mammal to cope with the additional demands of embryonic development in a confined, internal and more controlled environment.

Acknowledgements: This work has been partially supported by the Marcus and Bertha Coler Philanthropic Fund of the Jewish Communal Fund of New York.

REFERENCES

Akin, E. and Lacker, H. M. (1984) Ovulation control: The right number or nothing. *J. Math. Biol.* **20**, 113–32.

Biskind, M. S. and Biskind, G. S. (1944) Development of tumors in the rat ovary after transplantation into the spleen. *Proc. Soc. Exp. Biol. Med.* **55**, 176–9.

Baird, D. T., Horton, R., Longcope, D. and Tait, J. F. (1969) Steroid dynamics under steady state conditions. *Rec. Prog. Horm. Res.* **25**, 611–64.

Brambell, F. W. (1935) Reproduction in the common shrew. I. The estrous cycle of the female. *Phil. Trans. R. Soc. Lond. B* **225**, 1.

Brambell, F. W. and Hall, K. (1936) Reproduction in the lesser shrew. *Proc. Zool. Soc. Lond.* **106**, 957.

Brambell, F. W. and Rowlands, I. W. (1936) Reproduction of the bank vole I. The estrous cycle of the female, *Phil. Trans. R. Soc. Lond. B* **226**, 71.

Cargille, C. M., Ross, G. T. and Yoshimi, T. (1969) Daily variation in plasma follicle stimulating hormone, luteinizing hormone, and progesterone in the normal menstrual cycle. *J. clin. Endocr. Metab.* **29**, 12–9.

Clark, J. R., Dierschke, D. J. and Wolf, R. C. (1981) Hormonal regulation of ovarian folliculogenesis in rhesus monkeys. III. Atresia of the preovulatory follicle induced by exogenous steroids and subsequent follicular development. *Biol. Reprod.* **25**, 332–41.

Danforth, C. H. and de Aberle, S. B. (1928) The functional interrelation of the ovaries as indicated by the distribution of fetuses in mouse uteri. *Am. J. Anat.* **41**, 65.

Dierschke, D. J., Hutz, R. J. and Wolf, R. C. (1985) Induced follicular atresia in rhesus monkeys: Strength-duration relationships of the estrogen stimulus. *Endocrinology* **117**, 1397–403.

Falconer, D. S., Edwards, R. G., Fowler, R. E. and Roberts, R. C. (1961) Analysis of differences in the number of eggs shed by the two ovaries of mice during natural estrous and after superovulation, *J. Reprod. Fert.* **2**, 418–37.

Goodman, A. L., Nixon, W. E., Johnson, D. K. and Hodgen, G. D. (1977) Regulation of folliculogenesis in the cycling rhesus monkey: Selection of the dominant follicle. *Endocrinology*, **100**, 155–61.

Greenwald, G. S. (1961) Quantitative study of follicular development in the ovary of the intact or unilaterally ovariectomized hamster. *J. Reprod. Fert.* **2**, 351–61.

Jones, R. E. (1978) Control of follicular selection. In *The Vertebrate Ovary* (ed. R. E. Jones), Chapter 22, pp. 763–86. Plenum, New York.

Karsch, F. J., Weick, R. F., Hotchkiss, J., Dierschke, D. J. and Knobil, E. (1973) An analysis of the negative feedback control of gonadotropin secretion utilizing chronic implantation of ovarian steroids in ovariectomized rhesus monkeys. *Endocrinology* **93**, 478–86.

Korenman, S. G., Sherman, B. M. and Korenman, J. C. (1978) Reproductive hormone function: The perimenopausal period and beyond. *Clin. Endocr. Metab.* **7**, 625–43.

Lacker, H. M. (1981) Regulation of ovulation number in mammals: A follicle interaction law that controls maturation. *Biophys. J.* **35**, 433–54.

Lacker, H. M. and Akin, E. (1987) How do the ovaries count?, *Math. Biosci.* **90**, 305–32.

Lacker, H. M. and Peskin, C. S. (1981) Control of ovulation number in a model of ovarian follicular maturation. In *Lectures On Mathematics in the Life Sciences* (ed. S. Childress), Vol. 14, pp. 21–58. The American Mathematical Society, Providence.

Lacker, H. M., Beers, W. H., Meuli, L. E. and Akin, E. (1987a) A theory of follicle selection: I. Hypotheses and examples, *Biol. Reprod.* **37**, 570–80.

Lacker, H. M., Beers, W. H., Meuli, L. E. and Akin, E. (1987b) A theory of follicle selection: II. Computer simulation of estradiol administration in the primate. *Biol. Reprod.* **37**, 581–8.

Lee, S., Chandler, J. G., Broelsch, C. E., Yoichi, E., Condon, J. K., Charters, A. C., Yen, S. S. C. and Orloff, M. J. (1980) The effect of hepatic interposition on ovary-pituitary interaction. *J. Microsurgery* **1**, 440–6.

Lipschütz, A. (1928) New developments in ovarian dynamics and the law of follicular constancy. *Br. J. exp. Biol.* **5**, 283–91.

McLaren, A. (1963) The distribution of eggs and embryos between sides in the mouse. *J. Endocr.* **27**, 157.

McArthur, N. (1954) Statistics of twin births in Italy. *Ann. Eugen.* **17**, 249.

Meuli, L. E., Lacker, H. M. and Thau, R. B. (1987) Experimental evidence supporting a mathematical theory of the physiological mechanism regulating follicle development and ovulation number. *Biol. Reprod.* **37**, 589–94.

Parkes, A. S. (1976) *Patterns of Sexuality and Reproduction*. Oxford University Press, Oxford.

Pedersen, T. (1970) Follicle kinetics in the ovary of the cyclic mouse. *Acta Endocr.* **64**, 304–23.

Pontryagin, L. S. (1962) *Ordinary Differential Equations* (translated by L. Kacinskas and W. B. Counts), pp. 18–25. Addison-Wesley, Palo Alto.

Ross, G. T. and Lipsett, M. B. (1978) Hormonal correlates of normal and abnormal follicle growth after puberty in humans and other primates. *Clin. Endocr. Metab.* **7**, 561–75.

Seager, S. E., Boyns, A. R. and Blumgart, L. H. (1974) Histological and hormonal studies in rats bearing ovarian implants in the spleen. Effects of portacaval anastomosis. *Eur. J. Cancer* **10**, 35–40.

Schwartz, N. B. (1974) The role of FSH and LH and of their antibodies on follicle growth and on ovulation. *Biol. Reprod.* **10**, 236.

Speroff, L. and VandeWiele, R. L. (1971) Regulation of the human menstrual cycle. *Am. J. Obstet. Gynecol.* **109**, 234.

Speroff, L. R., Glass, H. and Kase, N. G. (1973) *Clinical Gynecologic Endocrinology and Infertility*. Williams & Wilkins, Baltimore.

Tapper, C. M. and Grant-Brown, K. (1975) The secretion and metabolic clearance rates of estradiol in the rat. *J. Endocr.* **64**, 215–27.

Tsai, C. C. and Yen, S. S. C. (1971) Acute effects of intravenous infusion of 17b-estradiol on gonadotropin release in pre- and post-menopausal women. *J. clin. Endocr. Metab.* **32**, 766–71.

Vollman, R. R. (1977) *The Menstrual Cycle*. W. B. Saunders, Philadelphia.

Yahia, C. and Taymor, M. L. (1970) Variants of the polycystic ovary syndrome. In *Meigs and Sturgis Progress in Gynecology* (eds S. H. Sturgis and M. L. Taymor), Vol. 5, pp. 163–71. Grune & Stratton, New York.

Zeleznik, A. I. (1981) Premature elevation of systemic estradiol reduces serum levels of follicle stimulating hormone and lengthens the follicular phase of the menstrual cycle in rhesus monkeys. *Endocrinology* **109**, 352–5.

Zeleznik, A. J., Hutchison, J. S. and Schuler, H. M. (1985) Interference with the gonadotropin-suppressing actions of estradiol in macaques overrides the selection of a single preovulatory follicle. *Endocrinology*, **117**, 991–9.

The emergence of the dominant ovarian follicle in primates: A random driven event?

J. C. THALABARD,[a] G. THOMAS[b] AND M. METIVIER[c]

[a]INSERM U292 and Laboratory for Neuroendocrinology
UTMSH, Houston, Texas 77225, USA
[b]Department of Toxicology, Fernand Widal Hospital, 75010
Paris, France
[c]Department of Applied Mathematics, Ecole Polytechnique,
91160 Palaiseau, France

INTRODUCTION

In adult female primates, the main feature of the reproductive function is the monthly maturation of one, and generally only one, ovarian follicle, followed by the release of the mature ovum into the fallopian tract. This periodic phenomenon is a major limiting event of the whole reproductive process. The hormonal mechanisms underlying this cyclic production essentially result from the dynamic interactions between the ovaries, the gonadotropin-secreting cells of the pituitary, and the hypothalamus (Knobil, 1980). Although the molecular and biochemical mechanisms at play at the ovarian level throughout the menstrual cycle have been extensively investigated, both *in vivo* and *in vitro*, a comprehensive yet simple description of these mechanisms is still lacking.

It is generally accepted that, at birth, primate ovaries comprise a non-renewable pool of oocytes, which are arrested in the prophase of their first meiotic division (Greenwald and Terranova, 1988). This stockpile is then progressively depleted during the reproductive lifespan, as primordial folli-cles continuously leave the quiescent pool to enter a growth process. This growth process is under the control of the gonadotropins, particularly the follicle-stimulating hormone (FSH) (Gougeon and Lefèvre, 1983; Hodgen and Di Zerega, 1983; Gougeon, 1986; Greenwald and Terranova, 1988).

The lifespan of a growing follicle overlaps several menstrual cycles

(Gougeon and Lefèvre, 1983; Gougeon, 1986). At the end of any normal menstrual cycle, the 10–20 largest and most mature follicles are triggered into the growth process by the rise of plasma FSH concentration. During the follicular phase of the next menstrual cycle, the triggered follicles participate into an interacting process, which ends up with the selection of one, and generally only one, follicle: the so-called dominant follicle. Concomitantly, the other follicles of the interacting pool undergo stagnation or atresia (Hodgen and Di Zerega, 1983).

   We propose a model which describes the complex phenomena of follicular maturation and dominance as a competitive system involving two opposite feedback mechanisms. This competitive system is described by a set of stochastic differential equations which govern the emergence of the dominant follicle.

THE MODEL

General assumptions

Our model is schematically illustrated in Fig. 1. The selection process is the result of the interactions between a limited number (N) of follicles. At each time point t, each follicle within this pool is almost equally able to respond to the gonadotropin stimuli (LH and FSH plasma concentrations). Here 'almost' implies that, although the underlying deterministic growth process is common to all follicles, random fluctuations will affect individual responses.

   During the follicular phase, the leading as well as limiting factor is the

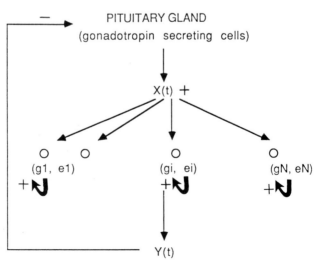

Fig. 1.   Schematic description of the model (see text).

specific action of FSH on the granulosa cells. FSH is responsible for both the increase in size of the granulosa cell pool, and the release of oestradiol into the bloodstream. The growth rate of the follicles is a non-decreasing function of FSH concentration. FSH is responsible for both the initiation of preovulatory follicular growth and the selection of a single preovulatory follicle. Furthermore there exists a threshold level above which plasma FSH concentration must rise in order to initiate the growth process (Zeleznik, 1981, 1988; Zeleznik et al., 1985). The magnitude of the follicular response to FSH at a given time depends on both the size of the follicle and the concentration of oestradiol at that particular time, suggesting a positive feedback mechanism at the ovarian level.

Circulating oestradiol exerts a negative feedback on FSH secretion by the pituitary. Taking into account the differences in time scale between FSH plasma kinetics and follicular growth, ultradian fluctuations of plasma FSH concentration are assumed negligible and only daily mean values are considered.

The evolution of any individual follicle is the net result of the FSH-driven growth process and the independent continuous atretic process.

Mathematical formulation

We denote by $1, 2, \ldots, i, \ldots, N$ the N interacting follicles. Let $g_i(t)$ and $e_i(t)$ represent the size and oestradiol secretion, respectively, of the ith follicle at time t.

$x(t)$ and $y(t)$ are the plasma concentrations at time t of FSH and oestradiol, respectively. Plasma FSH and oestradiol are assumed to follow time-invariant mono-exponential elimination kinetics, as characterized by their respective volumes of distribution V_x and V_y, and their respective decay constants k_x and k_y. In general form, the model is specified by the following equations:

$$dx/dt = f'(y)dy/dt - (k_x/V_x)x \tag{1}$$

$$dg_i/dt = v(\zeta, g_i, x) \tag{2}$$

$$e_i = s(g_i) \tag{3}$$

$$dy/dt = (\Sigma e_i)/V_y - (k_y/V_y)y \quad (i = 1, \ldots, N) \tag{4}$$

where f is a non-increasing function of y, with derivative $f'(y)$; ζ is a random variable, which models the (random) sensitivity of the growth rate of a given follicle at a given time, in response to the FSH stimulus. For fixed ζ, v is a non-decreasing function of both g_i and x; s is a non-decreasing function of g_i, relating oestradiol secretion to follicular size.

Illustration

Simulations were carried out with the following specific functions:

$$f(y) = (d-a)(y/c)^b/[1+(y/c)^b] + a$$

The four-parameter logistic function was chosen in order to avoid singularities and boundary problems, and to allow some physiological interpretation of the parameters. However, a variety of functions were found to yield essentially identical qualitative results.

$$v(\zeta, g_i, x) = k_1(x/u(g_i, x) - 1) - k_2$$

with

$$u(g_i, x) = k_3 \exp[-(k_4 g_i^\alpha + \zeta)]$$
$$\zeta \text{ normal, } N(0, \sigma^2)$$

$$s(g_i) = k_s g_i^\beta$$

The simulations were carried out on an IBM PS/2, using Euler's method of integration and the internal random number generator. The parameters were set at the following values: $k_x = 0.025$ min^{-1}; $k_y = 0.004$ min^{-1}; $V_x = 3000$ ml; $V_y = 3000$ ml; $k_1 = 0.25$; $k_2 = 0.1$; $k_3 = 5$; $k_4 = 0.1$; $k_5 = 0.5$; $\sigma = 0.1$; $\beta = 1$; $\alpha = 4$.

The initial FSH concentration was set at 4.0. (The initial oestradiol concentration was calculated by inverting f.) All follicles were assumed to have an initial size of 1.5. Ovulation was considered to take place when one of the follicles reached a size of 20 at least.

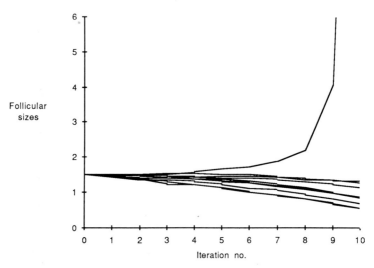

Fig. 2.   Growth curves of ovarian follicles simulated as described in the text.

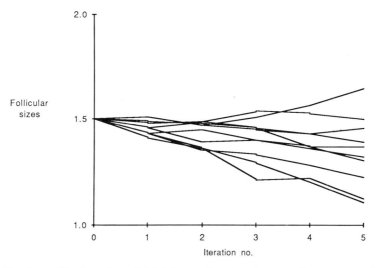

Fig. 3. Magnification of the initial times, showing the growth curve crossings.

Figure 2 illustrates a typical interacting process, for $a=7$, $b=2$, $c=25$ and $d=0.5$. It is noteworthy that the process ends up with the selection of a single dominant follicle. Another important feature of the model is the possibility of growth curve crossings, which are readily seen on Figure 3. In particular, the follicle that is eventually selected is not the fastest growing at early times.

Figure 4 illustrates the threshold effect of follicle size on the deterministic growth rate, as a function of the FSH concentration.

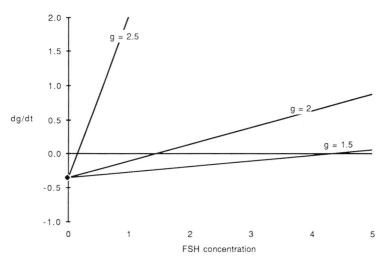

Fig. 4.. Deterministic dependence of follicular growth rate (dg/dt) on follicular size (g) and FSH concentration. Note the threshold effect, accounting for the atretic process.

The critical parameters of the model are obviously those governing the growth rate, i.e. k_1, k_2, k_3 and α. However, the structural properties of the model have yet to be investigated.

DISCUSSION

Several mathematical models that have been proposed to describe the menstrual cycle were concerned with the fluctuations of plasma hormone levels only (Bogumil *et al.*, 1972). The follicular dominance phenomenon is a critical step in the course of an ovulatory cycle. Starting from the existence of a leading follicle surrounded by a pool of non-individualized follicles, Thomson *et al.*, (1969) could not describe the selection process. More recently, this issue has been addressed specifically from the mathematical point of view (Lacker, 1981; Lacker *et al.*, 1987a,b; Meuli *et al.*, 1987). Lacker's model offers a great flexibility and is able to simulate a wide range of physiological and pharmacological situations. However, his model implies that follicular growth curves will never cross, so that the selected follicle will invariably be the largest one initially. This result is at variance with data from ultrasonic follow-ups of normal or stimulated cycles, which provide evidence for a great variability of growth curve profiles, together with the occurrence of curve crossings during the early period of follicular development.

In our model, the assumption of unequal initial follicle sizes is unnecessary, the emergence of a single dominant follicle after an initial period of apparent instability being wholly accounted for by the random differences between individual follicle responses to the FSH stimulus. Whether this apparent randomness merely reflects our ignorance of the follicular microenvironment i.e. the actual local FSH concentrations (Hodgen and Di Zerega, 1983; Greenwald and Terranova, 1988), or is an intrinsic property of the system is still an open question. Nevertheless, our results emphasize the usefulness of the stochastic approach to biological modelling at the macroscopic level. Another critical feature of our model is the non-linearity of the follicular response as a function of FSH concentration, suggesting some striking similarities with competitive species models in ecology (May, 1976).

As a final word of caution we should be reminded that, however powerful as exploratory and pedagogical tools, mathematical models cannot pretend at replacing the complexity of biological phenomena.

REFERENCES

Bogumil, R. J., Ferin, M., Rootenberg, J., Speroff, L. and Van de Wiele, R. L. (1972) Mathematical studies of the human menstrual cycle. I. Formulation of a mathematical model. *J. clin. Endocr. Metab.* **35**, 126–43.

Gougeon, A. (1986) Dynamics of follicular growth in the human: a model from preliminary results. *Human Reproduction* **1**, 81–7.

Gougeon, A. and Lefèvre, J. (1983) Evolution of the diameters of the largest healthy and atretic follicles during the human menstrual cycle. *J. Reproduction* **69**, 497–502.

Greenwald, G. S. and Terranova, P. F. (1988) Follicular selection and its control. In *The Physiology of Reproduction* (ed. E. Knobil and J. Neill), pp. 387–445. Raven Press, New York.

Hodgen, G. and Di Zerega (1983) The ovarian triad. *Rec. Prog. Horm. Res.* **39**, 1–73.

Knobil, E. (1980) The neuroendocrine control of the menstrual cycle. *Rec. Prog. Horm. Res.* **36**, 53–88.

Lacker, H. M. (1981) Regulation of ovulation number in mammals: a follicle interaction law that controls maturation. *Biophys. J.* **35**, 433–55.

Lacker, H. M., Berrs, W. H., Meuli, L. E. and Akin, E. (1987a) A theory of follicular selection: I. Hypothesis and examples. *Biol. Reprod.* **37**, 581–8.

Lacker, H. M., Berrs, W. H., Meuli, L. E. and Akin, E. (1987b) A theory of follicular selection: II. Computer simulation of estradiol administration in the primate. *Biol. Reprod.* **37**, 581–8.

May, R. M. (1976) Simple mathematical models with very complicated dynamics. *Nature* **261**, 459–67.

Meuli, L. E., Lacker, H. M. and Thau, R. B. (1987) Experimental evidence supporting a mathematical theory of the physiological mechanism regulating follicle development and ovulation number. *Biol. Reprod.* **37**, 589–94.

Thomson, H. E., Horgan, J. D. and Delfs, E. (1969) A simplified mathematical model and simulations of the hypophysis-ovarian endocrine control system. *Biophys. J.* **9**, 278–91.

Zeleznik, A. J. (1981) Premature elevation of systematic estradiol reduces serum levels of follicle stimulating hormone and lengthens the follicular phase of the menstrual cycle in rhesus monkeys. *Endocrinology* **109**, 352–5.

Zeleznik, A. J. (1988) Selection of the dominant follicle. Abstract S-22 World Congress of Endocrinology, Kyoto, Japan.

Zeleznik, A. J., Hutchinson, J. S. and Schuler, H. M. (1985) Interferences with the gonadotropin-suppressing actions of estradiol in macaques overrides the selection of a single preovulatory follicle. *Endocrinology* **117**, 991–9.

The critical role of signal quality: Lessons from pulsatile GnRH pathophysiology and clinical applications

MARCO FILICORI

Center for Chronobiology of Reproduction, University of Bologna, Italy

INTRODUCTION

Human reproduction is regulated by the action of several hormones produced by the hypothalamic–pituitary–gonadal axis. The hypothalamus sec-
retes gonadotropin-releasing hormone (GnRH) into the pituitary portal
circulation. GnRH stimulates the release of pituitary gonadotropins (luteinizing hormone, LH; follicle-simulating hormone, FSH) into the peripheral
circulation. The action of gonadotropins is essential for gonadal steroid
production and for normal gamete maturation. In the male, stimulation by
gonadotropins results in spermatozoa maturation and production of testosterone; in the female the gonadotropins are responsible for oestradiol and
progesterone secretion as well as for oocyte and follicular maturation.

The existence of gonadotropic hormones has been known for several
decades. Since the late 1960s it has been possible to measure LH and FSH
with specific radioimmunoassays. The introduction of this new analytical
methodology permitted the further characterization of the role of gonadotropins and the identification of the characteristic chronobiological secretion of
LH. When several consecutive blood samples, drawn at a few minute
intervals, were analysed in monkeys (Dierschke *et al.*, 1970) and in the
human (Nankin and Troen, 1971) it became evident that LH levels fluctuate
and rapidly change within a few minutes. Further studies in men (Spratt *et
al.*, 1988) and women (Yen et al, 1972; Santen and Bardin, 1973) established
more precisely the pattern of pulsatile gonadotropin secretion in normal
subjects. These, and other early studies, clarified that episodic LH secretion
changes dramatically across the normal menstrual cycle. LH pulses are

Cell to Cell Signalling: From
Experiments to Theoretical Models
ISBN 0–12–287960–0

relatively frequent and have a small amplitude in the follicular phase of the cycle; increase in amplitude and frequency around the pre-ovulatory mid-cycle LH surge; and are characterized by low frequency and high amplitude in the luteal phase of the cycle. In men, episodic LH secretion is more stable and pulses occur at a mean interval of one pulse every 2 h in a normal subject.

In spite of these detailed descriptions of pulsatile LH secretion in human males and females the physiologic relevance of this chronobiological pattern escaped most investigators until the end of 1970s. The classical physiological studies of Knobil's group (Knobil, 1980) established that the action of hypothalamic GnRH is strictly dependent upon a pulsatile pattern of pituitary stimulation. These studies (for further details see Knobil, in this volume) utilized rhesus monkeys with stereotaxic hypothalamic lesions that destroyed the arcuate nucleus and eliminated endogenous GnRH secretion.

When exogenous GnRH was administered to these animals it was found that only a pulsatile type of administration (one pulse every 60 min) was capable of re-establishing gonadotropin secretion. On the other hand, the continuous administration of GnRH paradoxically suppressed LH and FSH levels in blood. These important studies prompted a resurgence of investigations on episodic gonadotropin secretion in normal and abnormal subjects and permitted the clinical application of exogenous pulsatile GnRH to stimulate gonadal function and of GnRH analogues to suppress pituitary and reproductive functions in several disorders (Filicori and Flamigni, 1988).

Gonadotropin-releasing hormone had been available for clinical testing since 1971. However, it was soon evident that the administration of a single daily dose of GnRH was not sufficient to consistently stimulate gonadal function and to induce ovulation or pubertal development. Also, the use of superactive GnRH analogues, i.e. compounds that are synthetized with changes in their amino acid sequence rendering these substances several times more potent than native GnRH, did not appear to improve pituitary and gonadal stimulation in anovulatory patients. The progress in the understanding of GnRH and gonadotropin physiology permitted the application of these compounds in a rational fashion and the achievement of excellent clinical results in several disorders.

Nowadays, pulsatile GnRH is utilized to stimulate ovulation in women and puberty in males, while GnRH analogues are employed to suppress gonadal function in several disorders such as endometriosis, uterine leiomyoma, precocious puberty, prostatic cancer and polycystic ovarian disease (Filicori and Flamigni, 1988).

PULSATILE GONADOTROPIN SECRETION IN THE NORMAL MENSTRUAL CYCLE

To assess more precisely the pattern of gonadotropin secretion in the normal menstrual cycle we accrued a large normative database in 36 normal women of 18–35 years of age (Filicori *et al.*, 1986). All of these women had one blood sample drawn daily for one complete menstrual cycle and during this month they were admitted to the Clinical Research Center of the Massachusetts General Hospital for a period of 24 h of blood samples drawn at 10 min intervals. A total of 51 pulsation studies were obtained and analysed with two different mathematical analytical procedures. This study confirmed that LH peaks in the follicular phase of the normal menstrual cycle occur at a frequency of almost 1 pulse per h. Thus, the frequency of LH pulses in normal women appears to be similar to that found in rhesus monkeys (Knobil, 1980). Furthermore, an impressive slowing of LH pulses during the hours of sleep was evident in women studied in the early follicular phase of the cycle (Fig. 1). In most subjects in the early follicular phase LH peaks

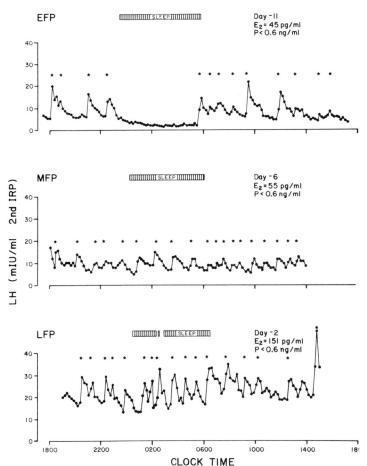

Fig. 1. Pulsatile LH secretion in the early (EFP), mid- (MFP) and late follicular phase (LFP) of the normal menstrual cycle. Notice the remarkable cessation of pulsatile LH secretion during sleep in the EFP. (Reproduced from Filicori *et al.*, 1986.)

appeared to stop at the beginning of sleep to resume only when patients woke up in the morning. When patients interrupted sleep during the night the occurrence of isolated LH peaks was recorded. Thus, it appears that a modulatory substance blocks the hypothalamic GnRH pulse generator during sleep; however, this feature is limited to the early follicular phase of the cycle. The recent studies by Rossmanith and Yen (1987) suggest that endogenous opiates may be the compounds responsible for such modulation of gonadotropin secretion at night. At the end of the follicular phase, when the pre-ovulatory LH surge begins, the amplitude of LH pulses increases dramatically and LH levels increase accordingly. Other groups (Djahan-

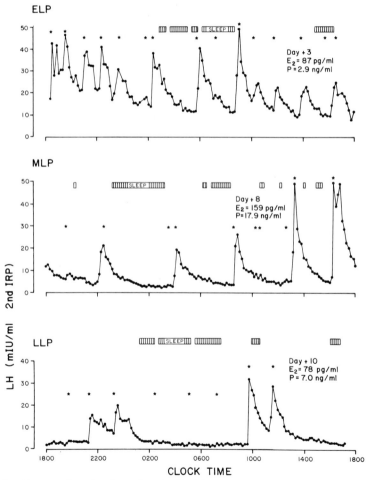

Fig. 2. Progressive slowing of pulsatile LF secretion across the luteal phase: typical secretory patterns in the early (ELP), mid- (MLP), and late luteal phase (LLP) of the normal menstrual cycle. (Reproduced from Filicori et al., 1986.)

bakhch et al., 1984) have suggested that the frequency of LH pulses may also increase at the time of the mid-cycle surge.

After ovulation, progesterone levels begin to increase in normal cycles. At the same time the frequency of LH pulses decreases so that a progressive slowing of LH pulse frequency is clearly evident across the normal luteal phase (Fig. 2). As few as one pulse of LH in 24 h may be present in normal subjects in the late luteal phase of the cycle. Recent studies in rhesus monkeys have indicated that this slowing of LH pulses is not incompatible with normal corpus luteum function and that the corpus luteum may be recovered by pulsatile LH secretion even after 24–36 h of hypothalamic pituitary quiescence (Hutchison and Zeleznik, 1985). Another phenomenon that emerges in the course of the normal luteal phase in humans is the appearance of episodic progesterone secretion from the corpus luteum. We have shown (Filicori et al., 1984) that no progesterone pulses are present in the early luteal phase, suggesting that at this stage the corpus luteum is independent of gonadotropin stimulation. However, in the mid- and late luteal phase clear progesterone pulses emerge and an impressive correspondence exists between LH and progesterone pulses (Fig. 3).

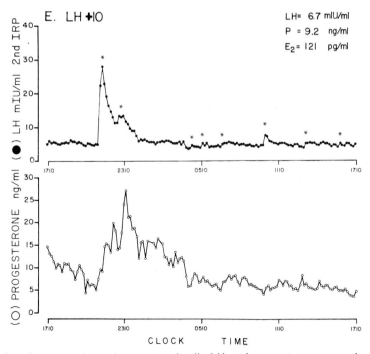

Fig. 3. Correspondence between pulsatile LH and progesterone secretion in a normal subject studied in the late luteal phase of the normal menstrual cycle. (Reproduced from Filicori et al., 1984.)

The pulsatile signal appears to reverberate across the reproductive axis. Pulses of hypothalamic GnRH can be measured in experimental animals (Levine *et al.*, 1982) and in particular human experimental conditions (Antunes *et al.*, 1978). An excellent correspondence exists between GnRH peaks measured at the hypothalamic level and LH pulses determined in the peripheral blood circulation. Thus, the excellent correspondence existing between LH and progesterone pulses in the mid-luteal phase of the menstrual cycle (Filicori *et al.*, 1984) and the relationship between LH and testosterone pulses identified in hypogonadal men (Spratt *et al.*, 1988) and normal men treated with synthetic androgens (Vigersky *et al.*, 1976), demonstrate that the hypothalamic signal can be recognized by the gonad after signal transduction at the pituitary level. Finally, recent studies by the group of Wilson *et al.* (1984) have shown that in rhesus monkeys an excellent relationship also exists between multi-unit electrical activity in the hypothalamus and LH pulses in the peripheral circulation.

THE ROLE OF DERANGED PULSATILE LH SECRETION IN REPRODUCTION

Although a pulsatile pattern is clearly essential for proper GnRH function, the role of LH pulse amplitude and frequency modulation across the menstrual cycle is unclear. As indicated in the previous section a fast LH pulse frequency is characteristic of the early part of the menstrual cycle (follicular phase) while a slow frequency is present following ovulation. The amplitude of LH pulses appears to be strictly regulated, as evidenced by the narrow standard error present in the estimation of this parameter in the follicular phase of normal menstrual cycles (Filicori *et al.*, 1986). The study of pulsatile gonadotropin secretion in several ovulatory disorders indicates that the alteration in LH pulse pattern ranges from an absolute absence of LH peaks, as in primary hypogonadotropic amenorrhoea, to decrements in LH amplitude and pulse frequency, and to pathological increments of both frequency and amplitude of LH peaks (Fig. 4). Thus, it appears that most ovulatory disorders are associated with deranged pulsatile gonadotropin secretion. However, the exact implications of these derangements are poorly understood.

We chose to study patients with polycystic ovarian disease (PCOD) in order to clarify the role of an excessive pituitary stimulation upon the gonad (Filicori *et al.*, 1988a). Polycystic ovarian disease is usually associated with oligomenorrhoea or secondary amenorrhoea and deranged ovulation. We selected six patients with PCOD and an anovulatory condition. These patients had an abnormal pulsatile LH secretion as determined with blood samples drawn every 10 min for 12 consecutive hours; they were given exogenous pulsatile GnRH at a physiological dose and frequency (5 μg every

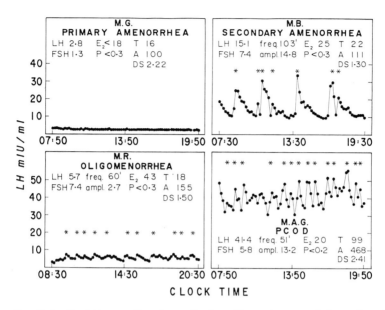

Fig. 4. Pulsatile LH secretion in patients with various ovulatory disorders. Serum samples were drawn every 10 min for 12 h. The specific disorders, the mean concentrations of reproductive hormones, and the pulse parameters are indicated in each panel.

60 min via the Zyklomat Ferring pump) in order to stimulate ovulation and achieve conception. However, the pattern of gonadotropin and gonadal steroid secretion in these patients was severely altered. These patients showed an excessive LH response to exogenous GnRH in the early follicular phase of the treatment cycles; furthermore, oestradiol and progesterone levels in the luteal phase were reduced, suggesting inadequate corpus luteum function. We then treated these patients with a GnRH analogue to block gonadotropin secretion and to induce a pharmacologic condition of hypogonadotropic hypogonadism. GnRH analogue administration was chosen to abolish endogenous gonadotropin secretion and to transform these patients into hypogonadotropic patients and thus isolate the specific role of deranged gonadotropin secretion in the genesis of PCOD. Immediately after GnRH analogue suppression treatment with pulsatile GnRH was resumed at the same dose and frequency as previously. In the post-GnRH analogue cycles PCOD patients achieved more physiological gonadotropin and gonadal levels in the follicular phase; ovulation was obtained in all cycles and corpus luteum secretion of oestradiol and progesterone appeared to be much more normal (Fig. 5). Furthermore, four of these patients conceived as compared to only one in the pre-GnRH analogue cycles. The pattern of gonadotropin secretion was particularly remarkable in these patients, showing a physio-

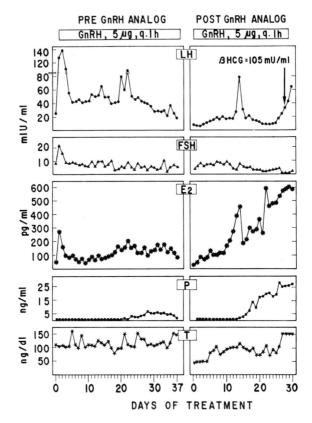

Fig. 5.   Pulsatile GnRH administration in a PCOD patient. The figure shows the response to the same regimen of pulsatile GnRH (5 µg every 60 min) before (pre-analogue panels) and after (post-analogue panels) pituitary suppression with a GnRH analogue. Notice conception and the improvement of endocrine response in the post-GnRH analogue cycle. (Reproduced from Filicori *et al.*, 1988a.)

logical increments of the FSH to LH ratio in the early follicular phase of the cycle and normal LH concentration throughout the follicular phase.

This study indicates that abnormal gonadotropin secretion is a critical parameter in the deranged ovulatory function of PCOD patients. Furthermore, these data suggest that the correction of dysfunctional pulsatile LH secretion in PCOD patients may result in a more physiologic ovulatory pattern. Thus, these data further emphasize the critical role of normal pulsatile gonadotropin secretion in the genesis of a physiologic menstrual cycle.

THE ROLE OF LH PULSE FREQUENCY IN THE FOLLICULAR PHASE OF THE MENSTRUAL CYCLE

We previously demonstrated that the interval of LH pulses in the follicular phase of normal menstrual cycles is approximately circhoral. Patients with hypothalamic amenorrhoea usually present an average LH intrapulse interval of 120–180 min (Reame et al., 1985). Thus, it appears that modest slowings of LH frequency may be sufficient to cause deranged ovulatory and menstrual function. In order to test this concept we selected 32 patients with primary hypogonadotropic amenorrhoea (Filicori et al., 1988b). As shown in Fig. 4 these patients are characterized by virtual absence of endogenous GnRH and LH pulses. Thus, just like Knobil's monkeys, these patients represent a unique model to test the effect of exogenous administration of different regimens of pulsatile GnRH without interference from endogenous GnRH secretion. In these patients we administered pulsatile GnRH at 60 or 120 min intervals. We maintained dose and frequency of GnRH administration stable across the menstrual cycle and we drew daily blood samples for the determination of gonadotropins and gonadal steroids. Our study showed that the ovulatory rate was more elevated in patients receiving pulsatile GnRH at 1 h intervals and that the mid-cycle LH surge appeared to be more elevated and more physiologic than in patients receiving a slower GnRH regimen. Thus, pulsatile LH frequency by itself appears to be a critical parameter in the development of physiological menstrual cycles and in the regulation of ovulation.

CONCLUSIONS

The data accrued in the last two decades on the physiologic signal pattern required for the regulation of the reproductive axis indicate that the quality of the signal received at each level of the hypothalamic–pituitary–gonadal axis is extremely important for normal reproductive function. The hypothalamic pulse-generating system and its modulation is sophisticated and complex. An extremely precise message appears to be essential for the appropriate regulation of this specific type of cell to cell signalling system. Despite normal integrity of all the organs participating in the reproductive axis the signal arising from them maybe altered and thus affect the function of endocrine glands.

Thus, the human reproductive axis may be compared to a computer; just as in computers, we can distinguish between hardware and software. In the reproductive axis the hardware is represented by the organs and glands producing different hormones; the software is composed of the chronobiological signal arising from these cells. An abnormal software (i.e. a deranged pulsatile secretion of GnRH and gonadotropins) is by itself capable of severely impairing reproductive function. Low body weight, heavy exercise and stress in general affect the pattern of pulsatile LH secretion, thus preventing normal ovulation. It is likely that in our ancestors this mechanism would block reproductive function at times of severe stress and malnutrition. Thus, reproductive competency could be limited to periods of good nutrition and lack of environmental dangers. Finally, the history of the discovery of the chronobiological regulation of the reproductive axis is a fascinating example of how relevant physiological information can be rapidly transformed in useful clinical and therapeutical applications.

Acknowledgements: Ms Silvia Arsento provided excellent secretarial assistance in the preparation of this manuscript.

REFERENCES

Antunes, J. L., Carmel, P. W., Housepian, E. M. and Ferin, M. (1978) Luteinizing hormone. Releasing hormone in human pituitary blood. *J. Neurosurg.* **49**, 382–6.

Dierschke, D. J., Bhattacharya, A. N., Atkinson, L. E. and Knobil, E. (1970) Circhoral oscillations of plasma LH levels in the variectomized rhesus monkey. *Endocrinology* **87**, 850–3.

Djahanbakhch, O., Warner, P., McNeilly, A. S. and Baird, D. T. (1984) Pulsatile release of LH and estradiol during the periovulatory period in women. *Clin. Endocr.* **20**, 579–89.

Filicori, M. and Flamigni, C. (1988) GnRH agonists and antagonists. Current clinical status. *Drugs* **35**, 63–82.

Filicori, M., Butler, J. P. and Crowley, W. F., Jr (1984) Neuroendocrine regulation of the corpus luteum in the human. Evidence for pulsatile progesterone secretion. *J. clin. Invest.* **73**, 1638–47.

Filicori, M., Santoro, N., Merriam, G. R. and Crowley, W. F., Jr (1986) Characterization of the physiologic pattern of episodic gonadotropin secretion throughout the human menstrual cycle. *J. clin. Endocr. Metab.* **62**, 1136–44.

Filicori, M., Campaniello, E., Michelacci, L., Pareschi, A., Ferrari, P., Bolelli, G. and Flamigni, C. (1988a) Gonadotropin-releasing hormone (GnRH) analog suppression renders polycystic ovarian disease more susceptible to ovulation induction with pulsatile GnRH. *J. clin. Endocr. Metab.* **66**, 327–33.

Filicori, M., Ferrari, P., Michelacci, L., Pareschi, A., Campaniello, E., Meriggiola, C., Ucci, N., Valdiserri, A. and Flamigni, C. (1988b) The physiologic role of gonadotropin-releasing hormone (GnRH) pulse frequency in the human menstrual cycle. *Endocrinology* **122** (Suppl.) 240, (Abst. 878).

Hutchison, J. S. and Zeleznik, A. J. (1985) The corpus luteum of the primate menstrual cycle is capable of recovering from a transient withdrawal of pituitary gonadotropin support. *Endocrinology* **117**, 1043–9.

Knobil, E. (1980) The neuroendocrine control of the menstrual cycle. *Rec. Prog. Horm. Res.* **36**, 53–88.

Levine, J. E., Pau, K. F., Ramirez, V. D and Jackson, G. L. (1982) Simultaneous measurement of luteinizing hormone-releasing hormone and luteinizing hormone release in the unanesthetized, ovariectomized sheep. *Endocrinology* **111**, 1449–55.

Nankin, H. R. and Troen, P. (1971) Repetitive luteinizing hormone elevations in serum of normal men. *J. clin. Endocr. Metab.* **33,** 558–60.

Reame, N. E., Sauder, S. E., Case, G. D., Kelch, R. P. and Marshall, J. C. (1985) Pulsatile gonadotropin secretion in women with hypothalamic amenorrhea: evidence that reduced frequency of gonadotropin-releasing hormone secretion is the mechanism of persistent anovulation. *J. clin. Endocr. Metab.* **61,** 851–8.

Rossmanith, W. G. and Yen, S. S. C. (1987) Sleep-associated decrease in luteinizing hormone pulse frequency during the early follicular phase of the menstrual cycle: evidence for an opioidergic mechanism. *J. clin. Endocr. Metab.* **65,** 715–8.

Santen, R. J. and Bardin, C. W. (1973) Episodic luteinizing hormone secretion in man. Pulse analysis, clinical interpretation, physiologic mechanism. *J. clin. Invest.* **52,** 2617–28.

Spratt, D. I., O'Dea, L. ST. L., Schoenfeld, D., Butler, J., Rao, P. N. and Crowley, W. F., Jr (1988) Neuroendocrine–gonadal axis in men: frequent sampling of LH, FSH, and testosterone. *Am. J. Physiol.* **254,** E658–66.

Vigersky, R. A., Easley, R. B. and Loriaux, D. L. (1976) Effect of fluoxymesterone on the pituitary-gonadal axis: the role of testosterone–estradiol binding globulin. *J. clin. Endocr. Metab.* **43,** 1–9.

Wilson, R. C., Kesner, J. S., Kaufman, J. M., Uemura, T., Akema, T. and Knobil, E. (1984) Central electrophysiologic correlates of pulsatile luteinizing hormone secretion in the rhesus monkey. *Neuroendocrinology* **39,** 256–60.

Yen, S. S. C., Tsai, C. C., Nattolin, F., Vandenberg, G. A. and Ajabor, L. (1972) Pulsatile patterns of gonadotropin release in subjects with and without ovarian function *J. clin. Endocr.* **34,** 671–5.

Episodic hormone secretion in the regulation and pathophysiology of the reproductive system in man

T. O. F. WAGNER, G. WENZEL, C. DETTE, G. DAEHNE,
A. GOEHRING, O. VOSMANN, I. MESSERSCHMIDT,
J. BRUNS, R. WÜNSCH AND A. VON ZUR MÜHLEN

*Abteilung Klinische Endokrinologie, Medizinische Hochschule,
Hannover, Konstanty-Gutschow-Strasse 8, D-3000 Hannover
61, FRG*

INTRODUCTION

When blood samples are withdrawn at regular intervals (minutes) and serial hormone determinations are plotted over time, discrete secretory events called pulses become discernible. This so-called pulsatile secretion of hormones, which has been found in many different endocrine systems of almost any species, is the result of a complex interaction of neurocrine and endocrine mechanisms.

The reproductive system has been extensively analysed with respect to episodic hormone secretion and its underlying mechanisms of regulation and pathophysiology. Thus, the reproductive system can be understood as a rather well-studied model of hypothalamic–pituitary gland interactions. In such a system a releasing factor released from hypothalamic neurones stimulates the pituitary cell secretion of glandotropic hormones (e.g. luteinizing hormone, LH, and follicle-stimulating hormone, FSH), the circulation then transports this signal to the responsive glands (ovary or testis) and the respective hormones are released into the bloodstream. The glandular product (testosterone, oestradiol, etc.) has a negative feedback action at the hypothalamic and pituitary level (Urban *et al.*, 1988).

Ever since this episodic nature of hormone secretion was first observed many speculations as to its physiological importance and pathophysiological relevance have been undertaken. It seems to be necessary, though, to clearly differentiate the episodic nature of secretion at the hypothalamic level from that at the pituitary level. In the following we will try to summarize some of the data available on the physiologic regulation of gonadotropin secretion in the normal male and the sparse information available on the importance of episodic secretion in the pathophysiology of disorders of the male reproductive function.

METHODS AVAILABLE FOR THE STUDY OF EPISODIC HORMONE SECRETION

Since the introduction of radioimmunoassay (RIA) into the study of endocrinology many different techniques of analysing series of data for the presence of pulses or bursts of secretion have been proposed. The different techniques have been summarized recently (Merriam, 1987; Urban *et al.*, 1988), therefore, only a few aspects which can be considered prerequisites for the analysis of hormonal data shall be addressed. At a basic level it is important to have hormone measurements of sufficient specificity and sensitivity. In the last few years most groups interested in episodic hormone secretion have concentrated on the perfection of computer programs for pulse identification and only very little information is available on the optimation of assay methods for these studies. As most peak identification methods rely on precision as a measure of noise to define the level that has to be surpassed by a significant signal (threshold method), assay accuracy is of major importance. If there is any doubt about the correlation of RIA and bioassay all series have to be evaluated by a bioassay analysis (Dufau and Veldhuis, 1987). When bioassay figures for LH determinations in healthy males are compared to RIA results, a higher peak amplitude (0.5- to 11-fold) is reproducibly found in the bioassay. Moreover, a dissociation of 'bio-peaks' from 'immuno-peaks' occurred in approximately 30% (Dufau and Veldhuis, 1987). Our own analysis (Fig. 1) of LH determinations with bio- (B) and immunoassay (I) indicates a significant difference in the B/I-ratio when oligospermic males (slow pulsing oligospermia, Wagner *et al.*, 1985) are compared to healthy individuals (Wenzel *et al.*, 1988). If no homologous bioassay is available it may be impossible to measure 'right' hormone levels. Because of the large heterogeneity of LH (micro-heterogeneity), bioassay evaluations of all LH determinations by RIA are of crucial importance. The data shown in Figs 1 and 2 illustrate that differences between normal and abnormal states may not be discernible by RIA and only become evident by analysis with bioassay.

The sampling interval has been clearly shown to be of great importance for the results obtained with most computer programs (Veldhuis *et al.*, 1984). Although most investigators agree that LH in the normal male is secreted

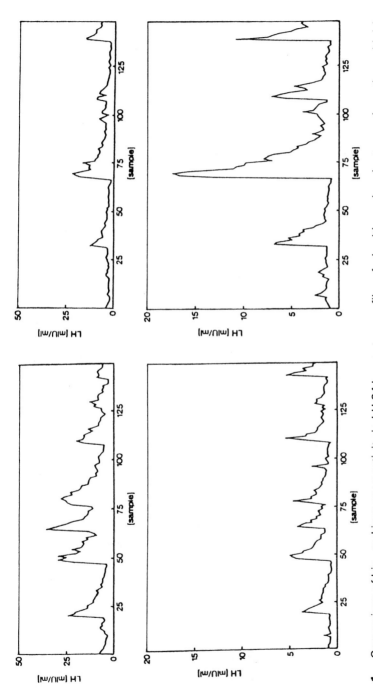

Fig. 1. Comparison of bio- and immunoreactivity in LH 24 h secretory profiles of a healthy male volunteer and a patient with 'slow pulsing oligospermia syndrome'. Left: Healthy volunteer; top, bioassay; bottom, immunoassay. Right: 'Slow pulsing oligospermia'; top, bioassay; bottom, immunoassay. Methods: Blood samples were taken every 10 min over a 24 h period. All samples were analysed in one assay (IRMA or mouse Leydig cell *in vitro* bioassay; standard MRC 68/40; mean of duplicate determinations are shown). Coefficient of variation (cv) for the IRMA was below 6% and below 8% for the bioassay, respectively.

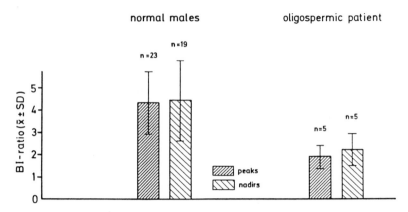

Fig. 2. Bio- to immunoratio of LH during 24 h secretory profiles in healthy male volunteers and a patient with 'slow pulsing oligospermia syndrome'. Methods: for details see Fig. 1.

with approximately 10 pulses per 24 h when 'normal' sampling intervals are used (5–20 min), the picture looks very different with higher sampling frequencies. When samples are taken at 1 min intervals over a 24 h time period 90–120 pulses are detected. Still, the overall view of the data is in good agreement with the normal sampling interval but at closer inspection so called 'micro-pulses', which are identified as significant pulses by most of the computer programs, become visible (Wagner *et al.*, 1987). So far it is not clear whether these micro-pulses are of any biological relevance or significance. One recent study did find a circhoral secretory pattern of LH and testosterone (Veldhuis *et al.*, 1989). In this study the secretory pattern of testosterone as a result of the stimulation pattern by LH could be used as a control for the analysis of the LH secretory pattern. If episodic hormone secretion in pathophysiological states is the study object it is necessary to prove beforehand that the so-called normal secretory profile is reproducibly 'normal'. Only very few data on the reproducibility of the episodic hormone secretory pattern are available.

To better characterize our normative database we studied four volunteers at three different occasions with the 24 h secretory protocol. All volunteers came back for another study 4 weeks and 8 weeks after the first blood sampling procedure over 24 h (Fig. 3). We found the large inter-individual variation in secretory profile contrasted with a rather preserved intra-individual secretory pattern. In two of the volunteers the study was repeated after another 6 months and 12 months (data not shown) and still the secretory pattern stayed rather constant intra-individually (Wenzel *et al.*, 1987). The data available on the healthy male have been summarized recently (Urban *et al.*, 1988). A large cooperative analysis of almost all data gathered (more than 120 24-h profiles of healthy volunteers) has not yet been completed.

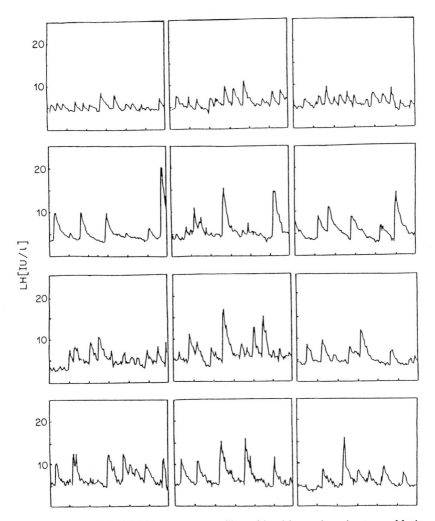

Fig. 3. Repeated LH 24 h secretory profiles of healthy male volunteers. Methods: Four healthy male volunteers (from top to bottom) were studied at three different occasions (from left to right). After a basal study (10 min sampling duration) a second and third sampling was performed after 4 and after 8 weeks. All samples of one individual were analysed in one assay (cv less than 6%) with an IRMA.

EXOGENOUS MODULATION

Infusions of testosterone and dihydrotestosterone at twice the daily production rate of testosterone in normal men significantly reduces LH pulse frequency and mean LH concentrations (Santen, 1975). As this approach may seem pharmacologic instead of physiologic the use of anti-oestrogens and anti-androgens may be preferred. Flutamide for example, an anti-androgen, increases LH pulse frequency and LH pulse amplitude (Urban *et al.*, 1987). Dopamine infusion reduces the mean LH level, but does not influence LH pulse frequency. The dopamine antagonist metoclopramide increased mean LH levels without influencing frequency as well. Therefore, both drugs can only exert their effect via changing the pulse amplitude or an underlying basal secretory rate (Boden *et al.*, 1972; Dette *et al.*, 1988).

ENDOGENOUS MODULATION

As most of the studies where exogenous drugs or hormones are used have the pharmacological drawback in common, several studies were done where endogenous modulation of the gonadotropin secretion was studied. As physical strain is well known to significantly change the gonadotropin secretory activity (amenorrhoea in marathon runners etc.) we studied the effect of near exhausting physical strain on the pulsatile secretion of gonadotropins (Messerschmidt *et al.*, 1988). A 4 hour treadmill test with 50% of the maximal oxygen consumption did not effect pulse frequency and/or amplitude. Even if the workload was increased to 4 h of 75% of oxygen consumption (VO_2 max) (continuous succession 10 min 75% VO_2 max, 10 min at 5% VO_2 max, total duration 4 h), the LH secretory pattern did not change. The Zeitgeber seems to be resistant to this short period of physical strain. Psychological strain over the same period of time was capable of wiping out all LH secretion in two out of five volunteers (unpublished data).

DISORDERS AS MODELS OF EPISODIC HORMONE SECRETION

Idiopathic hypothalamic hypogonadism is characterized by a completely abolished pulsatile release of LH-releasing hormone (LHRH) and consequently no pulsatile gonadotropin secretion. Only the pulsatile infusion of LHRH can reconstitute normal gonadotropin secretory patterns. Normal gonadal function can be reinstituted either by pulsatile infusion of LHRH or by once-daily injection of gonadotropins. This proves that at the hypothalamic–pituitary interaction a pulsatile signal is absolutely necessary for a normal function (Hoffman and Crowley, 1982). When gonadotropins are substituted a continuous or a slow release from a depot of injection is sufficiently effective to normalize spermatogenesis in the same percentage of patients as with pulsatile LHRH infusion. At the pituitary–gonad interaction the pulsatile signal does not seem to be of such vital importance (Schopohl *et al.*, 1987).

With oligospermia another pathophysiological condition could be distinguished where pulse frequency or amplitude disorders may be involved. In some of the patients with oligospermia and elevated FSH levels we were able to detect a decreased LH pulse frequency as compared to normal controls. When these patients were treated with a normal frequency pulsatile LHRH infusion, not only did FSH levels normalize, but the sperm count also ameliorated. We therefore called this 'the slow pulsing oligospermia syndrome' which may be the first syndrome described where pulse frequency disorder is of pivotal importance (Wagner et al., 1985).

CONCLUSIONS

In the reproductive system episodic hormone signalling has been shown to be an absolute necessity for the interaction of the hypothalamic neurones with the pituitary gland. Only episodic stimulation of the pituitary gonadotropins is capable of releasing sufficient amounts of bioactivity for normal reproductive function. Although, as a consequence of this episodic stimulation gonadotropins are secreted episodically as well under normal conditions, the importance of this episodic signal for the dependent glands (gonads) has not been proved. The pituitary thus seems to be a transmitter between the necessarily episodic neurocrine central compartment and the not necessarily episodic endocrine peripheral compartment. Whether indeed fractional changes of frequency or amplitude of the central signal are the major cause of disorders cannot be answered so far. In the reproductive system a complete lack of gonadotropin secretion has been shown to be due to a complete lack of LHRH stimulation. Whether the pulse frequency disorder named slow pulsing oligospermia syndrome is only an example of many types of these disorders has to be clarified. The reproductive system seems to be an ideal model to study the existence and cooperation of episodic and non-episodic compartments in a complex system.

REFERENCES

Boden, G., Lundy, L. E. and Owen, O. E. (1972) Influence of levodopa on serum levels of anterior pituitary hormones in man. *Neuroendocrinology* **10**, 309.

Dette, C., Daehne, G., Wenzel. G., Wagner, T. O. F. and von zur Mühlen, A. (1988) Short-term effects of dopamine agonists and an antagonist on episodic gonadotropin secretion in normal males and of patients Klinefelter's syndrome. *Acta Endocr.* **117**, 34.

Dufau, M. L. and Veldhuis, J. D. (1987) Pathophysiological relationships between the biological and immunological activities of lutenizing hormone. In *Balliere's Clinical Endocrinology and Metabolism* (ed. H. G. Burger), Vol. 1, pp. 153–65. W. B. Saunders, Philadelphia.

Hoffman, A. R., Crowley, Jr., W. F. (1982) Induction of puberty in men by long-term pulsatile administration of low-dose gonadotropin-releasing hormone. *New Engl. J. Med.* **307**, 1237.

Merriam, G. R. (1987) Techniques for pulse analysis: a survey. In *Episodic Hormone Secretion: From Basic Science to Clinical Application* (eds T. O. F. Wagner and M. Filicori), pp. 37–50. TM-Verlag, D-Hameln.

Messerschmidt, I., Vosmann, O., Wenzel, G., Wagner, T. O. F. and von zur Mühlen, A. (1988) Effect of a four hour near-exhausting physical strain on episodic LH and testosterone secretion in normal men. In *8th International Congress of Endocrinology–Abstracts*, p. 145. Japan Convention Services.

Santen, R. J. (1975) Is aromatization of testosterone to estradiol required for inhibition of luteinizing hormone secretion in men? *J. clin. Invest.* **65**, 1555.

Schopohl, J., von Zumbusch, R., Mehltretter, G., Eversmann, T. and von Werder, K. (1987) Male hypogonadotropic hypogonadism: treatment with GnRH or gonadotropins. *Acta Endocr.* **114**, 21.

Urban, R. J., Davis, M. R., Johnson, M. L. and Veldhuis, J. D. (1987) Operating characteristics of the human hypothalamo-pituitary axis: studies with a selective, potent non-steroidal antagonist of endogenous androgen action. In *Program of the 69th Annual Meeting of the Endocrine Society*, p. 201. Indianapolis.

Urban, R. J., Evans, W. S., Rogol, A. D., Kaiser, D. L., Johnson, M. L. and Veldhuis, J. D. (1988) Contemporary aspects of discrete peak-detection algorithms. I. The paradigm of the luteinizing hormone pulse signal in men. *Endocr. Rev.* **9**, 3–37.

Veldhuis, J. D., Evans, W. S., Rogol, A. D., Drake, C. R., Thorner, M. O., Merriam, G. R. and Johnson, M. L. (1984) Performance of LH pulse detection algorithms at rapid rates of venous sampling in humans. *Am. J. Physiol.* **247**, 554E.

Veldhuis, J. D., Evans, W. S., Urban, R. J., Rogol, A. D. and Johnson, M. L. (1989) Physiological attributes of the luteinizing hormone pulse signal in the human: cross-validation studies in men. *J. Androl.* (in press).

Wagner, T. O. F., Brabant, G. and von zur Mühlen, A. (1985) Slow pulsing oligospermia. In *Pulsatile LHRH Therapy of the Male* (ed. T. O. F. Wagner), pp. 111–7. TM-Verlag, D-Hameln.

Wagner, T. O. F., Goehring, A., Prechel, A., Krause, A. and von zur Muehlen, A. (1987) Prolonged high frequency plasma sampling technique using a miniaturized reverse plasmapheresis system in the evaluation of episodic gonadotropin secretion in the male. In *Episodic Hormone Secretion: From Basic Science to Clinical Application* (eds T. O. F. Wagner and M. Filicori), pp. 61–4. TM-Verlag, D-Hameln.

Wenzel, G., Wagner, T. O. F. and von zur Mühlen, A. (1987) Pulse frequency and pulse amplitude of LH are not sufficient to characterize the LH secretory pattern in the normal male. *Acta Endocr.* **114**, 131–2.

Wenzel, G., Hoppen, H. -O. and Wagner, T. O. F. (1988) Episodic LH secretion in slow pulsing oligospermia. In *8th International Congress of Endocrinology–Abstracts*, p. 142. Japan Convention Services.

Frequency coding in intercellular communication

ALBERT GOLDBETER AND YUE-XIAN LI

Faculté des Sciences, Université Libre de Bruxelles, Campus Plaine, C.P. 231, B-1050 Brussels, Belgium

INTRODUCTION

Intercellular communication often proceeds in a pulsatile, rhythmic manner. Thus, nerve cells transmit information in the form of trains of action potentials of varying frequency. Some species of cellular slime moulds, like *Dictyostelium discoideum*, communicate by pulses of cyclic-AMP secreted with a periodicity of 5–10 min. Moreover, an increasing number of hormones are found to be secreted in a pulsatile manner into the circulation. The physiological efficiency of all these pulsatile signals appears to be closely related to their frequency. We wish to examine here, from a theoretical point of view, the molecular bases that allow for the frequency coding of pulsatile signals in intercellular communication.

The prototype of periodic hormone signalling is that of the gonadotropin-releasing hormone, GnRH (also referred to as LHRH), secreted by the hypothalamus with a frequency of one 6-min pulse per h in the rhesus monkey and in humans (Dierschke *et al.*, 1970; Carmel *et al.*, 1976; Crowley *et al.*, 1985). The GnRH signal induces the release of the gonadotropin hormones LH (luteinizing hormone) and FSH (follicle-stimulating hormone) by target cells in the pituitary (Schally *et al.*, 1971). Experiments on rhesus monkeys rendered unable to release the GnRH signal autonomously have shown that the constant infusion of GnRH does not match the pulsatile signal delivered at the physiological frequency in eliciting the secretion of gonadotropin hormones by the pituitary (Belchetz *et al.*, 1978; Knobil, 1980; Wildt *et al.*, 1981). Moreover, the efficiency of the pulsatile signal closely depends on its frequency (Pohl *et al.*, 1983): thus, two or three pulses of GnRH per hour elicit a much lower secretion of LH and FSH. These

Cell to Cell Signalling: From
Experiments to Theoretical Models
ISBN 0–12–287960–0

Copyright © 1989 Academic Press Limited
All rights of reproduction in
any form reserved

observations have led to the establishment of effective therapeutic programmes in treatments of clinical disorders of GnRH secretion, in both male
and female (see the articles by Filicori, and by Wagner *et al.*, in this volume).

In *D. discoideum*, the amoebae respond chemotactically to cyclic-AMP
(cAMP) pulses secreted after starvation by cells behaving as aggregation
centres (Gerisch, 1987). Experiments in cell suspensions have shown that the
cAMP signals delivered after starvation at 5 min intervals also promote cell
differentiation; constant cAMP signals do not show this effect (Darmon *et
al.*, 1975; Gerisch *et al.*, 1975). Here also the frequency of stimulation is
essential, since pulses of cAMP delivered at 2 min intervals, much as constant
stimuli, fail to promote cell differentiation (Wurster, 1982).

Both in the case of GnRH in the reproductive system and in that of cAMP
in *D. discoideum*, the efficiency of the pulsatile, rhythmic signal can be
comprehended in terms of desensitization of the target cell. In both instances,
periodic signalling can be viewed as an optimal mode of intercellular
communication that allows for maximum responsiveness in cells that retain
the capability to adapt to constant stimuli (Goldbeter, 1987, 1988). The
desensitization that underlies the latter property can take multiple forms
such as receptor modification, as in *Dictyostelium* amoebae (see below),
downregulation by receptor internalization, or some other mechanism of
post-receptor desensitization relying, for example, on the synthesis of an
intracellular inhibitor of the cellular response triggered by the stimulus.

In order to investigate the efficiency of periodic stimuli in intercellular
communication, we shall consider a general model of a receptor which, upon
binding of a ligand that behaves as an agonist, elicits a cellular response; as a
result of receptor desensitization, this response adapts to constant stimuli.
We show that such a receptor system also adapts to periodic stimuli, i.e. the
response triggered by the rhythmic, pulsatile signal progressively reaches a
constant amplitude, and is characterized by the same periodicity. The
analysis of the model indicates (Li and Goldbeter, 1989) that desensitization
of the target cell provides a molecular basis for frequency coding of the
periodic signal. Moreover, there exists an optimal pattern of periodic
signalling that allows for maximal responsiveness of the target cell.

We apply the results to the periodic secretion of the GnRH hormone in the
reproductive system and to intercellular communication by cAMP pulses in
D. discoideum amoebae, before discussing the relevance of this analysis to
other examples of pulsatile hormone signalling and to other types of
frequency coding processes in biology.

A GENERAL MODEL BASED ON RECEPTOR
DESENSITIZATION FOR ADAPTATION TO CONSTANT
STIMULI

The model considered (Fig. 1) is that of a receptor which exists in two states,

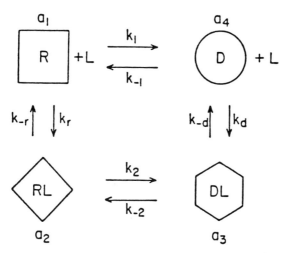

Fig. 1. Model of a receptor that exists in an active (R) and desensitized (D) state. Binding of ligand L by the two receptor states triggers a change in the activity A (eqn (1)) related to the cellular response. Each receptor state contributes to A with a weight factor a_i, called the activity coefficient. Desensitization results from the inequality $a_3, a_4 < a_1, a_2$.

active (R) and desensitized (D). Both states bind the ligand L, and thereby form the complexes $X \equiv RL$, $Y \equiv DL$. Each of the four receptor states contributes to the 'activity' A defined by eqn (1), with a certain weighting factor $a_i (i = 1, \ldots, 4)$, referred to as 'activity coefficient'. In order to keep the analysis as general as possible, we will not specify at this stage the nature of the activity coefficients or how the cellular response is related to the quantity A; possible examples are discussed below.

$$A(t) = a_1 R(t) + a_2 X(t) + a_3 Y(t) + a_4 D(t) \qquad (1)$$

This model has previously been analysed with respect to exact adaptation to constant stimuli (Segel et al., 1986), and the results were applied to the phenomenon as it occurs in bacterial chemotaxis and in cAMP secretion by cellular slime moulds (Knox et al., 1986). Basically, the question of how a sensory system displays exact adaptation to a step increase in stimulus amounts, in this model, to finding the conditions in which the activity increases and then returns to the same value at steady state, A_0, regardless of the magnitude of the stimulus.

For given values of the activity coefficients a_1 and a_4, the basal value A_0 of the activity in the absence of stimulus is given by eqn (2):

$$A_0 = a_1 R_0 + a_4 D_0 = R_0(a_1 + a_4 K^{-1}) = [R_T/(1 + K_1)](a_1 K_1 + a_4) \qquad (2)$$

where the equilibrium constant K_1 is the ratio of the resensitization and desensitization rate constants k_{-1} and k_1 (see Fig. 1) and R_T denotes the total receptor concentration. Exact adaptation will occur if, for all constant stimuli,

$$A(t) \rightarrow A_0 \text{ for } t \rightarrow \infty \qquad (3)$$

The steady-state concentrations of the four receptor species depend in a non-linear manner on the final level of the stimulus, L. Therefore, one may expect that at each level of stimulus, there will be a reshuffling of the receptor between the four species R, X, Y and D so that relation (1) will generally not produce, at steady state, the basal value A_0 defined by eqn (3). This is indeed what is found by analysing the dynamic behaviour of the model schematized in Fig. 1. Rather surprisingly, however, Segel *et al.* (1986) found that exact adaptation does occur, regardless of the stimulus level, when relations (4) hold for the activity coefficients of the X and Y complexes:

$$a_2 = \frac{a_1(k_2 + k_{-1}) - a_4(k_2 - k_1)}{k_{-1} + k_1}$$

$$\qquad (4)$$

$$a_3 = \frac{-a_1(k_{-2} - k_{-1}) + a_4(k_{-2} + k_1)}{k_{-1} + k_1}$$

When the transitions between the active and desensitized states of the

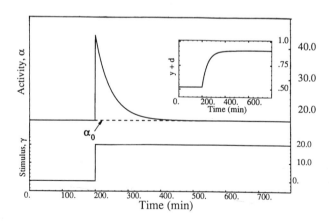

Fig. 2. Exact adaptation to a single step increase in stimulus. Shown is the change in the normalized activity α triggered by a step increase in ligand L from 0.01 to 20 (in K_R units). Parameter values are $a_1=20$, $a_4=1$, $k_1=0.004$ min^{-1}, $k_{-1}=k_2=0.02$ min^{-1}, $k_{-2}=0.002$ min^{-1}, $k_r=1$ μM^{-1} min^{-1}, $k_{-r}=25$ min^{-1}, $k_d=2$ μM^{-1} min^{-1}, $k_{-d}=1$ min^{-1}. The other activity coefficients are given the values prescribed by the exact adaptation conditions (4), namely, $a_2=85.1667$, $a_3=10$. *Inset:* the return of the activity to the basal state is accompanied by the evolution of the fraction of desensitized receptor to a higher steady-state value.

receptor are purely conformational, the cyclic nature of the receptor 'box' of Fig. 1 imposes on the equilibrium constants the detailed balance condition (5) where $K_2 = k_{-2}/k_2$, with $c = K_R/K_D$ and $K_R = k_{-r}/k_r$, $K_D = k_{-d}/k_d$:

$$K_1 = K_2 c \tag{5}$$

On the other hand, the particular condition (5) does not hold when the transitions between R and D and X and Y represent reversible covalent modification of the receptor (Segel et al., 1986; Waltz and Caplan, 1987), as in the case of the slime mould D. discoideum, where desensitization is accompanied by receptor phosphorylation (Devreotes and Sherring, 1985; Vaughan and Devreotes, 1988; see also Gundersen et al., in this volume). Figure 2 shows a typical response of the receptor system to a step increase in stimulus, when conditions (4) that produce exact adaptation are satisfied.

RESPONSE OF THE RECEPTOR MODEL TO PERIODIC STIMULATION

In order to address the efficiency of periodic signalling in intercellular communication, we consider the simplest form of periodic stimulus, namely that of a square wave in which the ligand is elevated from a basal value L_0 to a value L_1 during a time τ_1, before returning to the basal value during a time τ_0 (see bottom panel in Fig. 3). The period and amplitude of the stimulation are thus given by eqns (6) and (7), respectively, where γ denotes the ligand concentration L divided by the dissociation constant K_R of the complex formed by the ligand with the receptor in the R state:

$$T = \tau_0 + \tau_1 \tag{6}$$

$$\delta\gamma = \gamma_1 - \gamma_0 \tag{7}$$

To quantify the response of the receptor system, we consider the quantity α_T which is the activity $A(t)$ normalized by division through the total receptor concentration R_T, and integrated during the interval where it exceeds the basal activity $\alpha_0 = A_0/R_T$ (see the hatched area in the middle panel in Fig. 3). When we adopt the exact adaptation conditions (4) for the receptor system, the basal activity represents a natural reference state.

The theoretical analysis indicates (Li and Goldbeter, 1989) that after a transient phase, the periodic change in the normalized activity $\alpha(t) = A(t)/R_T$ elicited by the square-wave signal reaches a constant amplitude; at the same time, the level of modified receptor also reaches a constant, periodic variation (Fig. 3). The upper trace in Fig. 3 shows that the receptor undergoes desensitization during the time of stimulation, and initiates a phase of resensitization as soon as the stimulus is returned to a low level.

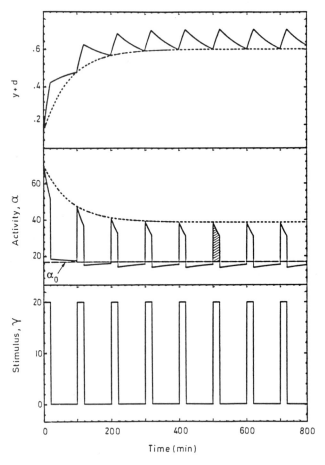

Fig. 3. Adaptation of the receptor system to periodic stimulation. The changes in activity (middle panel) and fraction of desensitized receptor (top) triggered by the square-wave stimulation (bottom) are indicated. Parameter values are as in Fig. 2; moreover, the pulse duration τ_1 and the pulse interval τ_0 are equal to 20 min and 80 min, respectively (redrawn from Li and Goldbeter, 1989, with permission from *Biophysical Journal*).

FREQUENCY CODING OF PULSATILE SIGNALS

How the integrated activity varies as a function of the period of the pulsatile signal is shown in Fig. 4, for three different values of the stimulus amplitude. The curves are established for a given value of the stimulus duration τ_1, for increasing values of the interval τ_0 that separates successive stimuli. As the latter interval progressively rises, there is a corresponding increase in the integrated activity generated by the receptor. The amplitude of the changes in activity depends on the amplitude of the periodic stimulation.

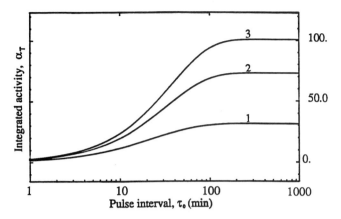

Fig. 4. Frequency coding based on receptor desensitization. Shown is the integrated activity (α_T) as a function of the interval τ_0 between two successive pulses of ligand; the duration of each pulse is fixed at the value $\tau_1 = 6$ min. The basal value of the ligand is $\gamma_0 = L_0/K_R = 0.001$. The curves are obtained for three different values of the stimulus amplitude $\delta\gamma = \gamma_1 - \gamma_0$: (1) 0.1, (2) 0.5, (3) 10. Parameter values are $K_1 = 10$, $K_2 = 0.1$, $c = K_1/K_2 = 100$, $k_1 = 0.00195$ min^{-1}, $k_2 = 0.645$ min^{-1}, $a_1 = 20$, $a_2 = 101$, $a_3 = 10$, $a_4 = 1$; other parameter values are as in Fig. 2.

That the response of the receptor system progressively rises with the duration of the pulse interval results from the fact that the time allowed for resensitization between successive stimuli increases, for a given value of the stimulus duration. The fraction of receptor in the active state thus becomes larger as the period of the signal increases; hence the increase in activity generated by the binding of the ligand. Desensitization thus provides the basis for a coding of periodic signals in terms of frequency.

OPTIMAL FREQUENCY OF PULSATILE SIGNALLING

As shown by Fig. 4, the integrated activity over a period will reach a saturating level once the interval between successive stimuli becomes sufficiently large. This is due to the fact that a further increase in the pulse interval will not allow any further increase in the fraction of active receptor, since the latter is already fully resensitized. The question arises, however, as to whether the integrated activity α_T provides a complete measure of the response of the receptor system. In particular, this measure does not convey any information as to the number of significant responses that the receptor system is able to produce in a given time interval. The latter measure is provided by the quantity α_R, defined as cellular responsiveness (Li and Goldbeter, 1989), according to eqn (8):

$$\alpha_R = \frac{\alpha_T}{\alpha_{Tstep}}\frac{\alpha_T}{T} \tag{8}$$

The first term in the definition of α_R provides a measure of the magnitude of the response; it scales the integrated activity α_T generated by one pulse of the periodic stimulus, with respect to the integrated activity generated by a single-step increase in the ligand concentration (the latter constant stimulus can be viewed as a periodic signal for which the pulse duration τ_1 becomes infinite). The second term in eqn (8) yields a measure of the number of responses that the receptor system generates in a given time interval.

In contrast with the integrated activity α_T which reaches a maximum, constant level as the period of stimulation increases (Fig. 4), there exists a pattern of periodic stimulation that produces a maximum value of the cellular responsiveness α_R. As demonstrated in Fig. 5, this optimal mode of periodic stimulation corresponds to a particular value of the pulse duration, τ_1^*, and of the pulse interval, τ_0^*. These optimal values depend on the kinetic constants that govern desensitization and resensitization of the receptor (Li and Goldbeter, 1989).

The results of Fig. 5 showing the existence of an optimal frequency of

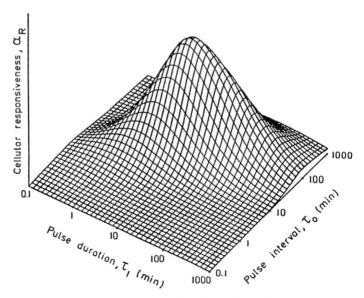

Fig. 5. Optimal pattern of periodic signalling. The cellular responsiveness α_R (eqn (8)) is calculated as a function of the pulse duration τ_1 and the pulse interval τ_0 of a square-wave stimulation similar to that shown in the bottom panel of Fig. 3. The pair of values of τ_1 and τ_0 that maximize cellular responsiveness corresponds to the optimal pattern of periodic stimulation. The activity coefficients are as in Fig. 2, while the values of the kinetic parameters are those in Fig. 4 (redrawn from Li and Goldbeter, 1989, with permission from *Biophysical Journal*).

stimulation can be generalized to the case of a more realistic periodic signal in which the ligand is added at regular intervals, and removed in a first-order process (Li and Goldbeter, 1989). There also, a particular pattern of periodic stimulation exists, that optimizes cellular responsiveness.

APPLICATION TO EXPERIMENTAL SYSTEMS

Owing to its generality, the above analysis applies to a large class of periodic signalling in intercellular communication. We shall focus here on its application to two distinct experimental systems and will show in each case how the optimal signalling frequency is determined by the characteristics of the desensitization and resensitization processes in target cells.

Pulsatile hormone signalling: the case of GnRH

The secretion of the gonadotropin hormones LH and FSH by the pituitary in response to pulses of GnRH released by the hypothalamus represents the prototype of periodic hormone signalling. The hormone GnRH is secreted as a 6 min pulse, approximately once per hour, in higher primates and man (Dierschke et al., 1970; see also the papers by Knobil and Filicori in this volume). A constant GnRH signal fails to elicit the release of LH and FSH (Belchetz et al., 1978), while a change in signalling frequency results in a reduced physiological effect (Knobil, 1980; Pohl et al., 1983).

The efficiency of periodic signalling is due to the fact that pituitary gonadotrope cells become desensitized under constant stimulation by GnRH (Smith and Vale, 1981; Adams et al., 1986; Conn et al., 1986, 1987). It is therefore interesting to apply the present analysis to the periodic stimulation of gonadotrope cells by GnRH. The results of Fig. 5 were obtained so as to yield the optimal pattern of periodic stimulation observed for GnRH, i.e. a 6 min pulse once per hour, corresponding to $\tau_1^* = 6$ min, $\tau_0^* = 54$ min. Owing to the absence of experimental data, these results were obtained for an *ad hoc* choice of values for the kinetic parameters characterizing receptor desensitization and resensitization.

In turn, the analysis suggests that the efficiency of the hourly 6 min signal implies the existence of a process of desensitization of the GnRH receptor, whose characteristic time would be of the order of 1–2 min, while resensitization upon removal of the ligand would take place with a characteristic time of 30 min. Experiments in cultures of pituitary cells corroborate the existence of an optimal pattern of periodic stimulation by GnRH, close to that observed *in vivo* (Liu and Jackson, 1984; McIntosh and McIntosh, 1986). Studies on pituitary cell cultures also indicate that the decrease in gonadotropin release upon stimulation by a sizeable signal of GnRH is biphasic: a fast decrease, in a few min, could correspond to the predicted fast desensitization process, while a second, much slower phase could be due to downregulation of the receptor by internalization (Naor et al., 1982; Zilberstein et al., 1983).

The time course analogy with desensitization in other sensory and hormonal systems such as bacterial chemotaxis (Koshland, 1979; Springer et al., 1979) and the β-adrenergic system (Lefkowitz and Caron, 1986), suggests that the desensitization predicted for the GnRH receptor could also originate from some covalent modification, e.g. reversible phosphorylation. In the case of GnRH, it is likely, however, that the desensitization of pituitary cells takes multiple forms. Besides receptor desensitization and downregulation, a process of post-receptor desensitization appears to be involved (Smith et al., 1983); this process differs from the depletion of intracellular LH and FSH pools (Jinnah and Conn, 1986).

The extension of the present analysis to the case in which the stimulus elicits the secretion of gonadotropin hormones, as well as the concomitant synthesis of an intracellular inhibitor of this response, shows that an optimal frequency of stimulation also exists in such conditions. Whereas in the present analysis this frequency is largely dictated by the resensitization time constant of the receptor, it is then primarily governed by the characteristic time of disappearance of the inhibitor (Y. X. Li and A. Goldbeter, in preparation).

The cAMP signalling system in *Dictyostelium* amoebae

In many respects, the mechanism of intercellular communication in cellular slime moulds resembles hormone signalling in higher, multicellular organisms. While cAMP is often used as a second messenger that is synthesized intracellularly in response to hormonal stimuli, in the slime mould *D. discoideum*, it behaves at the same time as first and second messenger (Konijn, 1972). Indeed, these amoebae synthesize cAMP in response to a cAMP stimulus (Roos et al., 1975; Shaffer, 1975); the latter extracellular signal controls both the aggregation and the differentiation of the amoebae after starvation (Devreotes, 1982; Gerisch, 1987).

The observation that pulses of cAMP promote cell differentiation when delivered with a periodicity of 5 min, in contrast with constant stimuli (Darmon et al., 1975; Gerisch et al., 1975) or with signals delivered with the higher frequency of one pulse every 2 min (Wurster, 1982) suggests the existence of an optimal frequency of periodic stimulation in the slime mould. As in the case of GnRH, the efficiency of periodic signalling is associated with the process of desensitization of target cells under constant stimulation. Here, desensitization appears to be primarily due to the reversible phosphorylation of the receptor, upon binding of extracellular cAMP (Klein et al., 1985; Devreotes and Sherring, 1985; Vaughan and Devreotes, 1988; see also Gundersen et al., in this volume).

When inserting into the model the values taken from the literature for the kinetic constants governing the reversible phosphorylation of the cAMP receptor in *D. discoideum*, one obtains, as in Fig. 5, an optimal pattern of periodic stimulation that is now characterized by the values $\tau_1{}^* = 3.8$ min and $\tau_0{}^* = 5$ min, for the pulse duration and the pulse interval (Li and Goldbeter, 1989). These values are close to those observed for the periodic signal of cAMP generated by cells which behave as aggregation centres after starvation (Devreotes, 1982; Gerisch, 1987).

A difference exists between the case of *D. discoideum* and that of periodic hormone signalling. While in the former case the process of receptor desensitization is part of the oscillatory mechanism (Martiel and Goldbeter, 1987; Goldbeter, 1988; see also Gundersen *et al.*, in this volume), in the case of GnRH the origin of the periodicity lies in a pulse generator, distinct from target cells, located in the hypothalamus (Wilson *et al.*, 1984; Lincoln et al., 1985; Knobil, 1987; see also Knobil, in this volume).

Because of its general structure, the model presented here allows us to apply the results about frequency coding to widely different systems such as those utilizing GnRH or cAMP pulses. Besides the values of the parameters governing receptor desensitization, what differs in each case is the link between the activity $A(t)$ and the actual cellular response. Thus, in the slime mould, the receptor-triggered changes in activity are somehow related to the intracellular synthesis of cAMP by adenylate cyclase (Segel *et al.*, 1986; Knox *et al.*, 1986; Barchilon and Segel, 1988), while in the case of GnRH the activity coefficients could represent ionic conductances measuring the influx of calcium ions upon stimulation by the hormone (Li and Goldbeter, 1989), since the elevation of intracellular calcium appears to be the initial process, elicited by GnRH binding to the receptor, that triggers LH and FSH release by pituitary cells (Conn *et al.*, 1986, 1987).

THE UBIQUITY OF FREQUENCY CODING IN BIOLOGICAL SYSTEMS

We have shown how the theoretical analysis of a model based on receptor desensitization allows one to comprehend the possible molecular bases of frequency coding in intercellular communication. This process, however, is not limited to cAMP signalling in the cellular slime moulds and to the periodic secretion of GnRH in the reproductive system. In fact, periodic signalling appears to be widespread in biological systems, and extends from intracellular signalling to communication between identical cells – as in the slime mould *D. discoideum*; distinct cells within an organism – as in hormone signalling; and multicellular organisms belonging to the same species.

Frequency coding in intracellular signalling

In a variety of cells, the stimulation by hormones or neurotransmitters triggers a train of cytosolic calcium oscillations (for a review, see Berridge and Galione, 1988, and Berridge and Cuthbertson in this volume). The frequency of these intracellular oscillations depends on the magnitude of the extracellular signal. This raises the possibility that the subsequent cellular response, e.g. protein phosphorylation, is governed by the frequency of calcium oscillations, rather than by the level of cytosolic calcium *per se* (Woods *et al.*, 1987; Berridge *et al.*, 1988; Jacob *et al.*, 1988; Dupont and Goldbeter, in this volume). There exists a close relation between Fig. 4 above, and the right panel of figure 3 in the chapter by Berridge in this volume. In both cases, the cellular response is modulated by the frequency of the signal. However, the modulation occurs in opposite directions: for hormone signalling, the cellular response rises and finally reaches a plateau as the period of the signal increases, i.e. as the frequency diminishes, while the inverse occurs for Ca^{2+} signalling. The difference arises from the type of mechanism responsible for the encoding of the pulsatile signal, namely, a putative protein phosphorylation for Ca^{2+} oscillations, and receptor desensitization for cAMP and GnRH pulses.

Frequency coding in intercellular communication

The pulsatile release of cAMP signals that control aggregation and differentiation in the slime mould *D. discoideum* represents the prototype of periodic signalling between unicellular organisms. As discussed above, the efficiency of the cAMP signal delivered at the frequency of one pulse every 5 min results from the kinetic characteristics of the reversible phosphorylation that underlies receptor desensitization. When two successive stimuli are separated by only 2 min, the receptor has not enough time to resensitize before the next pulse of cAMP. In contrast, the receptor has sufficient time to fully resensitize when successive stimuli are separated by 5 min. This explains the physiological efficiency of cAMP pulses delivered at 5 min intervals, in contrast to those applied at 2 min intervals (Wurster, 1982) or to constant stimuli (Darmon *et al.*, 1975; Gerisch *et al.*, 1975), since the latter give rise to a reduced response, owing to desensitization.

A similar explanation holds for the existence of an optimal frequency of stimulation in the case of the pulsatile release of hormones. The prototype of this type of periodic signalling is the pulsatile release of GnRH by the hypothalamus. Here, the optimal pattern of stimulation is a 6 min pulse delivered once an hour. The efficiency of this periodic signal results from the desensitization to GnRH of target cells in the pituitary. Although this desensitization can take multiple forms, such as receptor modification or downregulation, or may even arise from a post-receptor mechanism, the present theoretical analysis points to the existence of a fast desensitization process, taking place in a few minutes, while resensitization should proceed in some 30 min. These time courses are required to account for the observed physiological efficiency of the hourly 6 min signal.

The process of target cell desensitization thus provides a natural molecular basis for frequency coding of periodic stimuli in hormonal communication. In the words of Knobil (1981), 'the temporal patterns of hormonal stimulation may be of greater consequences to the effector system than are the levels of hormone in the blood'. This concept has had a number of important clinical applications in recent years, primarily in the treatment of disorders of the reproductive function, both in females (Santoro et al., 1986; Filicori, in this volume) and males (Wagner, 1985; Wagner et al., in this volume). Normal levels of gonadotropin hormones are restored in patients lacking the appropriate pulsatile secretion of GnRH when the latter hormone is supplied in a pulsatory manner, at the physiological frequency (Crowley et al., 1985). This treatment is successfully used for the induction of ovulation, permitting pregnancy, in women suffering from sterility owing to disorders of GnRH secretion (Leyendecker et al., 1980; Reid et al., 1981).

The number of hormones which are found to be secreted periodically into the circulation, and whose physiological effect is maximum when delivered in a pulsatory manner, is steadily increasing. Thus either one or both of these phenomena has been observed for insulin and glucagon (Matthews et al., 1983; Weigle et al., 1984; Komjati et al., 1986), growth hormone and growth hormone-releasing factor (Borges et al., 1984; Bassett and Gluckman, 1986), thyrotropin (Brabant et al., 1987), cortisol and adrenal corticotropin hormone (Van Cauter, 1987) as well as corticotropin-releasing hormone (Avgerinos et al., 1986). This list, which is by no means exhaustive, demonstrates that frequency coding is widely used in hormonal communication and might well represent the rule rather than the exception in this context. Such conclusion will likely prove to be of profound clinical significance.

Another type of intercellular communication where the information appears to be carried, at least in part, by frequency is that of signal transduction in the nervous system (Rapp, 1987; Lestienne and Strehler, 1987). An example related to neurosecretion is given by Cazalis et al. (1985). The cellular basis for the identification of temporal patterns of stimulation may differ here from desensitization, as it may involve the spatial summation of neuronal discharges, relying on the topographic distribution of synapses (Lestienne and Strehler, 1987).

A recent report suggests that frequency coding may extend to the control of developmental processes by growth factors. Thus, experiments in cell cultures indicate (Brewitt and Clark, 1988) that the lens tissue grows and becomes transparent only when subjected to periodic pulses of PDGF (platelet-derived growth factor); the lens becomes opaque when subjected to constant stimulation by this factor. Here also, as with hormones, or cAMP in *D. discoideum*, it is likely that receptor desensitization is the molecular process that allows for the efficiency of periodic versus constant stimuli. Other growth factors may thus turn out to be more effective when delivered in a pulsatory manner; the optimal interval between successive stimuli should largely be dictated by the characteristic time of receptor resensitization in target cells.

Frequency coding of communications between organisms

While unicellular organisms may communicate by pulsatile signals, as exemplified by the slime mould *D. discoideum*, higher organisms also appear to resort to periodic signalling. This is particularly the case for insects, which use frequency coding to communicate by pheromones (Bossert, 1969; Conner, 1985; Conner *et al.*, 1985) or bioluminescence (Lloyd, 1983). Periodic auditory stimuli are generated by a number of insects, e.g. the fruit fly *Drosophila* whose courtship song frequency is specific of the species (Kyriacou and Hall, 1986).

It appears, therefore, that frequency coding may well be one of the most important roles of biological rhythms. This type of communication, which can be more reliable than signalling based on amplitude (Rapp *et al.*, 1981), is used for the transduction of extracellular signals into intracellular oscillations of second messengers such as calcium, for insect communications, and in a wide variety of cell to cell signalling processes among which the periodic, pulsatile secretion of hormones represents a most important class.

Acknowledgments: This work was supported by the Belgian National Incentive Program for Fundamental Research in the Life Sciences, launched by the Science Policy Programming Services of the Prime Minister's Office (SPPS).

REFERENCES

Adams, T. E., Cumming, S. and Adams, B. M. (1986) Gonadotropin-releasing hormone (GnRH) receptor dynamics and gonadotrope responsiveness during and after continuous GnRH stimulation. *Biol. Reprod.* **35**, 881–9.

Avgerinos, P. C., Schurmeyer, T. H., Gold, P. W., Tornai, T. P., Loriaux, D. L., Sherins, R. J., Cutler, G. B. Jr. and Chrousos, G. P. (1986) Pulsatile administration of human corticotropin-releasing hormone in patients with secondary adrenal insufficiency: restoration of the normal cortisol secretory pattern. *J. clin. Endocr. Metab.* **62**, 816–21.

Barchilon, M. and Segel, L. A. (1988) Adaptation, oscillations and relay in a model for cAMP secretion in cellular slime moulds. *J. theor. Biol.* **133**, 437–46.

Bassett, N. S. and Gluckman, P. D. (1986) Pulsatile growth hormone secretion in the ovine fetus and neonatal lamb. *J. Endocr.* **109**, 307–12.

Belchetz, P. E., Plant, T. M., Nakai, Y., Keogh, E. J and Knobil, E. (1978) Hypophysial responses to continuous and intermittent delivery of hypothalamic -gonadotropin-releasing hormone. *Science, N.Y.* **202**, 631–3.

Berridge, M. J. and Galione, A. (1988) Cytosolic calcium oscillators. *FASEB J.* **2**, 3074–82.

Berridge, M. J. and Rapp, P. E. (1979) A comparative survey of the function, mechanism and control of cellular oscillations. *J. exp. Biol.* **81**, 217–79.

Berridge, M. J., Cobbold, P. H. and Cuthbertson, K. S. R. (1988) Spatial and temporal aspects of cell signalling. *Phil. Trans. R. Soc. Lond. B* **320**, 325–43.

Borges, J. L. C., Blizzard, R. M., Evans, W. S., Furlanetto, R., Rogol, A. D., Kaiser, D. L., Rivier, J., Vale, W. and Thorner, M. O. (1984) Stimulation of growth hormone (GH) and somatomedin C in idiopathic GH-deficient subjects by intermittent pulsatile administration of synthetic human pancreatic tumor GH-releasing factor. *J. clin. Endocr. Metab.* **59**, 1–6.

Bossert, W. H. (1969) Temporal patterning in olfactory communication. *J. theor. Biol.* **18**, 157–70.

Brabant, G., Brabant, A., Ranft, U., Ocran, K., Köhrle, J., Hesch, R. D. and von zur Mühlen, A. (1987) Circadian and pulsatile thyrotropin secretion in euthyroid man under the influence of thyroid hormone and glucocorticoid administration. *J. clin. Endocr. Metab.* **65**, 83–8.

Brewitt, B. and Clark, J. I. (1988) Growth and transparency in the lens, an epithelial tissue, stimulated by pulses of PDGF. *Science, N.Y.* **242**, 777–9.

Carmel, P. W., Araki, S. and Ferin, M. (1976) Pituitary stalk portal blood collection in rhesus monkeys: evidence of pulsatile release of gonadotropin-releasing hormone (GnRH). *Endocrinology* **99**, 243–8.

Cazalis, M., Dayanithi, G. and Nordmann, J. J. (1985) The role of patterned burst and interburst interval on the excitation-coupling mechanism in the isolated rat neural lobe. *J. Physiol., Lond.* **369**, 45–60.

Conn, P. M., Staley, D., Harris, C., Andrews, W. V., Gorospe, W. C., McArdle, C. A., Huckle, W. R. and Hanson, J. (1986) Mechanism of action of gonadotropin releasing hormone. *A. Rev. Physiol.* **48**, 495–513.

Conn, P. M., McArdle, C. A., Andrews, W. V. and Huckle, W. R. (1987) The molecular basis of gonadotropin-releasing hormone (GnRH) action in the pituitary gonadotrope. *Biol. Reprod.* **36**, 17–35.

Conner, W. E. (1985) Temporally patterned chemical communication: is it feasible? In *Perspectives in Ethology*, Vol. 6, *Mechanisms* (eds P. P. G. Bateson and P. H. Klopfer), pp. 287–301. Plenum Press, New York and London.

Conner, W. E., Webster, R. P. and Itagaki, H. (1985) Calling behavior in arctiid moths: the effects of temperature and wind speed on the rhythmic exposure of the sex attractant gland. *J. Insect Physiol.* **31**, 815–20.

Crowley, W. F., Jr., Filicori, M., Spratt, D. I. and Santoro, N. F. (1985) The physiology of gonadotropin-releasing hormone (GnRH) secretion in men and women. *Rec. Prog. Horm. Res.* **41**, 473–531.

Darmon, M., Brachet, P. and Pereira da Silva, L. H. (1975) Chemotactic signals induce cell differentiation in *Dictyostelium discoideum*. *Proc. natn. Acad. Sci. U.S.A.* **72**, 3163–6.

Devreotes, P. N. (1982) Chemotaxis. In *The Development of Dictyostelium discoideum* (ed. W. F. Loomis), pp. 117–68. Academic Press, New York.

Devreotes, P. N. and Sherring, J. A. (1985) Kinetics and concentration dependence of reversible cAMP-induced modification of the surface cAMP receptor in *Dictyostelium*. *J. biol. Chem.* **260**, 6378–84.

Dierschke, D. J., Bhattacharya, A. N., Atkinson, L. E. and Knobil, E. (1970) Circhoral oscillations of plasma LH levels in the ovariectomized rhesus monkey. *Endocrinology* **87**, 850–3.

Gerisch, G. (1987) Cyclic AMP and other signals controlling cell development and differentiation in *Dictyostelium. A. Rev. Biochem.* **56,** 853–79.

Gerisch, G., Fromm, H., Huesgen, A. and Wick, U. (1975) Control of cell contact sites by cAMP pulses in differentiating *Dictyostelium* cells. *Nature* **255,** 547–9.

Goldbeter, A. (1987) Periodic signaling and receptor desensitization: from cAMP oscillations in *Dictyostelium* cells to pulsatile patterns of hormone secretion. In *Temporal Disorder in Human Oscillatory Systems* (eds L. Rensing, U. an der Heiden and M. C. Mackey), pp. 15–23. Springer, Berlin.

Goldbeter, A. (1988) Periodic signaling as an optimal mode of intercellular communication. *News Physiol. Sci.* **3,** 103–5.

Jacob, R., Merritt, J. E., Hallam, T. J and Rink, T. J. (1988) Repetitive spikes in cytoplasmic calcium evoked by histamine in human endothelial cells. *Nature* **335,** 40–5.

Jinnah, H. A. and Conn, P. M. (1986) Gonadotropin-releasing hormone-mediated desensitization of cultured rat anterior pituitary cells can be uncoupled from luteinizing hormone release. *Endocrinology* **118,** 2599–604.

Karsch, F. J. (1987) Central actions of ovarian steroids in the feedback regulation of pulsatile secretion of luteinizing hormone. *A. Rev. Physiol.* **49,** 365–82.

Klein, C., Lubs-Haukeness, J. and Simons, S. (1985) cAMP induces a rapid and reversible modification of the chemotactic receptor in *Dictyostelium discoideum. J. Cell Biol.* **100,** 715–20.

Knobil, E. (1980) The neuroendocrine control of the menstrual cycle. *Rec. Prog. Horm. Res.* **36,** 53–88.

Knobil, E. (1981) Patterns of hormone signals and hormone action. *New Engl. J. Med.* **305,** 1582–3.

Knobil, E. (1987) A hypothalamic pulse generator governs mammalian reproduction. *News Physiol. Sci.* **2,** 42–3.

Knox, B. E., Devreotes, P. N., Goldbeter, A. and Segel, L. A. (1986) A molecular mechanism for sensory adaptation based on ligand-induced receptor modification. *Proc. natn. Acad. Sci. U.S.A.* **83,** 2345–9.

Komjati, M., Bratusch-Marrain, P. and Waldhausl, W. (1986) Superior efficacy of pulsatile versus continuous hormone exposure on hepatic glucose production *in vitro. Endocrinology* **118,** 312–9.

Konijn, T. M. (1972) Cyclic AMP as first messenger. *Adv. Cyclic Nucleot. Res.* **1,** 17–31.

Koshland, D. E., Jr (1979) A model regulatory system: bacterial chemotaxis. *Physiol. Rev.* **59,** 811–62.

Kyriacou, C. P. and Hall, J. C. (1986) Interspecific genetic control of courtship song production and reception in *Drosophila. Science, N.Y.* **232,** 494–7.

Lefkowitz, R. J. and Caron, M. G. (1986) Regulation of adrenergic receptor function by phosphorylation. *J. molec. Cell. Cardiol.* **18,** 885–95.

Lestienne, R. and Strehler, B. L. (1987) Time structure and stimulus dependence of precisely replicating patterns in monkey cortical neuronal spike patterns. *Brain Res.* **437,** 214–38.

Leyendecker, G. L., Wildt, L. and Hansmann, M. (1980) Pregnancies following intermittent (pulsatile) administration of GnRH by means of a portable pump ("Zyklomat"): a new approach to the treatment of infertility in hypothalamic amenorrhea. *J. clin. Endocr. Metab.* **51,** 1214–6.

Li, Y. X. and Goldbeter, A. (1989) Frequency specificity in intercellular communication: The influence of patterns of periodic signaling on target cell responsiveness. *Biophys. J.* **55,** 125–45.

Lincoln, D. W., Fraser, H. M., Lincoln, G. A., Martin, G. B. and McNeilly, A. S. (1985) Hypothalamic pulse generators. *Rec. Prog. Horm. Res.* **41,** 369–419.

Liu, T. C and Jackson, G. L. (1984) Long term superfusion of rat anterior pituitary cells: Effects of repeated pulses of gonadotropin-releasing hormone at different doses, durations and frequencies. *Endocrinology* **115**, 605–13.

Lloyd, J. E. (1983) Bioluminescence and communication in insects. *A. Rev. Entomol.* **28**, 131–60.

Martiel, J. L and Goldbeter, A. (1987) A model based on receptor desensitization for cyclic AMP signaling in *Dictyostelium* cells. *Biophys. J.* **52**, 807–28.

Matthews, D. R., Naylor, B. A., Jones, R. G., Ward, G. M. and Turner, R. C. (1983) Pulsatile insulin has greater hypoglycemic effect than continuous delivery. *Diabetes* **32**, 617–21.

McIntosh, J. E. A. and McIntosh, R. P. (1986) Varying the patterns and concentrations of gonadotropin-releasing hormone stimulation does not alter the ratio of LH and FSH release from perifused sheep pituitary cells. *J. Endocr.* **109**, 155–61.

Naor, Z., Katikineni, M., Loumaye, E., Vela, A. G., Dufau, M. L and Catt, K. J. (1982) Compartmentalization of luteinizing hormone pools: dynamics of gonadotropin releasing hormone action in superfused pituitary cells. *Molec. Cell. Endocr.* **27**, 213–20.

Pohl, C. R., Richardson, S. W., Hutchison, J. S., Germak, J. A and Knobil, E. (1983) Hypophysiotropic signal frequency and the functioning of the pituitary-ovarian system in the rhesus monkey. *Endocrinology* **112**, 2076–80.

Rapp, P. E. (1987) Why are so many biological systems periodic? *Prog. Neurobiol.* **29**, 261–73.

Rapp, P. E. and Berridge, M. J. (1977) Oscillations in calcium-cyclic AMP control loops form the basis of pacemaker activity and other high frequency biological rhythms. *J. theor. Biol.* **66**, 497–525.

Rapp, P. E., Mees, A. I. and Sparrow, C. T. (1981) Frequency encoded biochemical regulation is more accurate than amplitude dependent control. *J. theor. Biol.* **90**, 531–44.

Reid, R. L., Leopold, G. R and Yen, S. S. C. (1981) Induction of ovulation and pregnancy with pulsatile luteinizing hormone-releasing factor: dosage and mode of delivery. *Fert. Steril.* **36**, 553–9.

Roos, W., Nanjundiah, V., Malchow, D. and Gerisch, G. (1975) Amplification of cyclic-AMP signals in aggregating cells of *Dictyostelium discoideum. FEBS Lett.* **53**, 139–42.

Santoro, N., Filicori, M. and Crowley, W. F. Jr. (1986) Hypogonadotropic disorders in men and women: diagnosis and therapy with pulsatile gonadotropin-releasing hormone. *Endocr. Rev.* **7**, 11–23.

Schally, A. V., Arimura, A., Kastin, A. J., Matsuo, H., Baba, Y., Redding, T. W., Nair, R. M. G. and Debeljuk, L. (1971) Gonadotropin releasing hormone: one polypeptide regulates secretion of luteinizing and follicle-stimulating hormones. *Science, N.Y.* **173**, 1036–8.

Segel, L. A., Goldbeter, A., Devreotes, P. N. and Knox, B. E. (1986) A mechanism for exact sensory adaptation based on receptor modification. *J. theor. Biol.* **120**, 151–79.

Shaffer, B. M. (1975) Secretion of cyclic AMP induced by cyclic AMP in the cellular slime mould *Dictyostelium discoideum. Nature* **255**, 549–52.

Smith, M. A. and Vale, W. W. (1981) Desensitization to gonadotropin-releasing hormone observed in superfused pituitary cells on cytodex beads. *Endocrinology* **108**, 752–9.

Smith, M. A., Perrin, M. H. and Vale, W. W. (1983) Desensitization of cultured pituitary cells to gonadotropin-releasing hormone: evidence for a post-receptor mechanism. *Molec. Cell. Endocr.* **30**, 85–96.

Springer, M. S., Goy, M. F. and Adler, J. (1979) Protein methylation in behavioral control mechanisms and in signal transduction. *Nature* **280**, 279–84.

Van Cauter, E. (1987) Pulsatile ACTH secretion. In *Episodic Hormone Secretion:*

From Basic Science to Clinical Application (eds T. O. F. Wagner and M. Filicori), pp. 65–75. TM-Verlag, Hameln, FRG.

Vaughan, R. and Devreotes, P. N. (1988) Ligand-induced phosphorylation of the cAMP receptor from *Dictyostelium discoideum*. *J. biol. Chem.* **263**, 14538–43.

Wagner, T. O. F. (ed.) (1985) *Pulsatile LHRH Therapy of the Male*. TM-Verlag, Hameln, FRG.

Wagner, T. O. F. and Filicori, M. (eds) (1987) *Episodic Hormone Secretion: From Basic Science to Clinical Application*. TM-Verlag, Hameln, FRG.

Waltz, D. and Caplan, S. R. (1987) Consequence of detailed balance in a model for sensory adaptation based on ligand-induced receptor modification. *Proc. natn. Acad. Sci. U.S.A.* **84**, 6152–6.

Weigle, D. S., Koerker, D. J. and Goodner, C. J. (1984) Pulsatile glucagon delivery enhances glucose production by perfused rat hepatocytes. *Am. J. Physiol.* **247**, E564–E568.

Wildt, L., Haüsler, A., Marshall, G., Hutchison, J. S., Plant, T. M., Belchetz, P. E. and Knobil, E. (1981) Frequency and amplitude of gonadotropin releasing hormone stimulation and gonadotropin secretion in the rhesus monkey. *Endocrinology* **109**, 376–85.

Wilson, R. C., Kesner, J. S., Kaufman, J. M., Uemura, T., Akema, T. and Knobil, E. (1984) Central electrophysiologic correlates of pulsatile luteinizing hormone secretion in the rhesus monkey. *Neuroendocrinology* **39**, 256–60.

Woods, N. M., Cuthbertson, K. S. R. and Cobbold, P. H. (1987) Agonist-induced oscillations in cytoplasmic free calcium concentration in single rat hepatocytes. *Cell Calcium* **8**, 79–100.

Wurster, B. (1982) On induction of cell differentiation by cyclic AMP pulses in *Dictyostelium discoideum*. *Biophys. Struct. Mech.* **9**, 137–43.

Zilberstein, M., Zakut, H. and Naor, Z. (1983) Coincidence of down-regulation and desensitization in pituitary gonadotrophs stimulated by gonadotropin-releasing hormone. *Life Sci.* **32**, 663–9.

Part 5
Signal transduction based on calcium oscillations

Intracellular calcium oscillators

K. S. ROY CUTHBERTSON

Department of Human Anatomy and Cell Biology, University of Liverpool, P.O. Box 147, Liverpool L69 3BX, UK

INTRODUCTION

Oscillations of intracellular free calcium ('free Ca') have now been observed in a wide variety of cell types (Berridge *et al.*, 1988). We need to understand these oscillations in order to understand how calcium is mobilized in cells and also to understand the function of calcium signals in cells.

Experimental results on free calcium oscillations in two cell types will be discussed, together with theoretical models of the oscillator mechanisms. The first cell type, the rat hepatocyte, is popular with biochemists, and there is a substantial body of data on the biochemical basis for calcium signalling obtained from this cell type. This enables models of the calcium oscillator in hepatocytes to be based on the phosphatidylinositol signalling pathway with some confidence. The second cell type, the mouse oocyte, provides an unusually direct and special case of cell to cell signalling, namely fertilization. The calcium signal in this situation consists of a strongly structured episode of oscillations, with evidence of two distinct underlying oscillators. The proposed theoretical model for the oscillators in the mouse oocyte has biochemical features in common with the hepatocyte model, but the oscillator mechanism itself is quite different.

Oscillations may be fundamental to the operation of calcium second messenger systems, and may have an ancient evolutionary origin. It is therefore plausible that several different calcium oscillators may have evolved in parallel. There are apparent advantages to oscillatory modes of calcium signalling, especially pulsatile modes, and these may have encouraged the evolution of calcium oscillators.

Cell to Cell Signalling: From
Experiments to Theoretical Models
ISBN 0–12–287960–0

FREE CALCIUM OSCILLATIONS IN CELLS

Hepatocytes

Repetitive calcium transients

One of the functions of the liver is to store glucose as glycogen. The release of glucose from glycogen is under the control of both cyclic-AMP and calcium signals. Hepatocyte receptors that mobilize calcium do so through the PI-signalling system, that is they promote hydrolysis of phosphatidylinositol (4,5)bisphosphate (PtdIns(4,5)P$_2$), releasing inositol (1,4,5)trisphosphate (Ins(1,4,5)P$_2$) which in turn releases Ca^{2+} from internal stores (Burgess *et al.*, 1984). Woods *et al.* (1986) showed with aequorin measurements in single rat hepatocytes that this calcium mobilization occurs as repetitive transients (Fig. 1).

The period of the free Ca oscillations decreases from several minutes to less than 20 s with increasing agonist concentrations. The minimum period and the duration of each transient depend on the agonist type (Woods *et al.*, 1987a). Transient durations are 7 s for adrenalin acting at α_1 receptors, the α_1 noradrenalin analogue phenylephrine, and the purinergic agonist ADP; 10 s for vasopressin; 15 s for angiotensin II; and 27 s for ATP (all at 37°C). Curiously, both ADP and ATP are thought to act at P_{2Y} receptors, yet the kinetics of their calcium responses differ markedly. This indicates either a

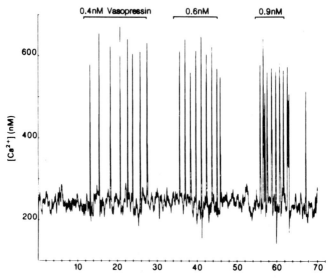

Fig. 1.   Repetitive free Ca transients in a hepatocyte responding to vasopressin at three increasing concentrations, showing a corresponding increase in transient frequency. (Reproduced with permission from *Nature*.)

novel subtyping of P_{2Y} receptors, or the intimate involvement of an agonist–receptor complex in the oscillator itself (Cobbold et al., 1988b).

The calcium oscillator model

The key property of the calcium oscillations is their pulsatile form. The calcium transients are 'all or none' like action potentials. This immediately implies that the oscillator mechanism is non-linear. The term 'repetitive calcium transients' is often used to describe this type of oscillation.

This model follows the basic elements of the now widely accepted scheme for the PI-signalling system (Fig. 2), in which agonist-bound receptors activate G-proteins (G_p) by stimulating the substitution of bound GDP by GTP. Activated G_ps in turn stimulate phospholipase C (PLC) to hydrolyse PtdIns(4,5)P_2 into Ins(1,4,5)P_3 and diacylglycerol (DG) (Cockcroft, 1987). Ins(1,4,5)P_3 releases calcium from the endoplasmic reticulum (ER), with three molecules required per release site (Meyer et al., 1988), and DG,

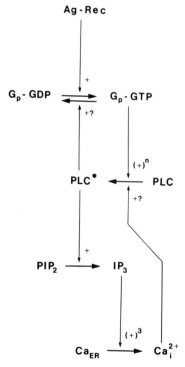

Fig. 2. Flow diagram of the PI-signalling cascade. Hypothetical pathways indicated by '?'. Ag: Agonist; Rec: receptor; Gp: G-protein; PLC: phospholipase C; PLC*: activated PLC; PIP$_2$: phosphatidylinositol (4,5)bisphosphate; IP$_3$: inositol (1,4,5)trisphosphate; Ca$_{ER}$: calcium stored in the endoplasmic reticulum; Ca$_i^{2+}$: intracellular free ionized calcium.

together with free Ca, activates protein kinase C (PKC) (Berridge, 1987). Additional features are proposed which allow the system to produce pulsatile oscillations. First, the step from G_p to PLC would be non-linear, requiring either multiple G_p to activate each PLC molecule, perhaps with cooperative binding of the G_p to PLC, or positive feedback, e.g. from free Ca, or both. The behaviour of G_p would depend on the receptor type, either because there were subtypes of G_p specific to each receptor, or because the G_p were effectively permanently bound to receptors, which would modify their properties. The effect of the receptors on G_p would be reciprocal to the modification of receptor affinities by G_p (Bojanic and Fain, 1986).

To complete the model a negative feedback or inactivation step is required.

The PKC feedback model. Phorbol esters reduce the frequency of free Ca transients (Woods *et al.*, 1987b). This observation led to the proposal that activation of the phorbol ester target PKC could provide the negative feedback required by the model (Woods *et al.*, 1987a; Cobbold *et al.*, 1988a). In this scheme the free Ca rises are switched off by negative feedback from PKC, which is activated by both DG and free Ca during the first few seconds of the transient. The PKC phosphorylates either activated G_p and thus inactivates it, or receptors to reduce their affinity, or both. This model has a problem in explaining how phorbol esters (which are presumed to be activating PKC) can reduce the frequency of the oscillations, and, indeed, inhibit them totally, without consistently reducing transient amplitude (Woods *et al.*, 1987b). It has to claim that the activation of PKC during each transient is maximally intense (i.e. saturates and is not further enhanced by phorbol esters) but brief, so that a recovery can occur between the transients. Phorbol esters continue to activate PKC during the interval between transients, thereby reducing the transient frequency.

G_p–PLC Interaction. An alternative source of negative feedback could come from the G_p–PLC interaction itself. If the inactivation process were contained within the G_p–PLC complex it would be tied closely to the activation process, and thus simplify the production of relatively brief transients. This could be achieved by reciprocal activation of the enzyme activities of G_p and PLC. Examples of the mutual activation of enzymes exist, such as between the components of tryptophan synthetase (Hatanaka *et al.*, 1962). In this scheme activated PLC would enhance the GTPase activity of the G_p bound to it. Conversion of GTP to GDP would inactivate G_p. There is a precedent for the enhancement of GTPase activity, namely the GAP protein, a factor in cells which can increase the GTPase activity of G_ps 1000-fold (Gibbs *et al.*, 1988). The model proposed here requires the hypothesis that PLC possesses GTPase-promoting activity when stimulated by G_ps.

The kinetics of the inactivation process, and hence the transient duration, could depend on G_p subtype. Many different G_ps are known, and there is some evidence of specificity between G_ps and receptors (Lochrie and Simon, 1988).

In hepatocytes oscillations can also be promoted by aluminium fluoride (Woods, Cuthbertson and Cobbold, unpublished observations) which bypasses receptors. Aluminium fluoride (AlF_4^-) is thought to activate GDP-bound G_ps by substituting for the gamma-phosphate of GTP. For the proposed model to account for the induction of oscillations by AlF_4^- it is necessary to invoke the cooperative binding of G_p to PLC which is already included in the model, and to make the additional assumption that there is a basal rate of activation of G_ps. Then the spontaneous accumulation of GTP-bound G_p and its subsequent degradation by activated PLC will allow oscillations to occur even though AlF_4^--stimulated G_p has no GTP to hydrolyse. This hypothesis predicts differences between AlF_4^- and agonist-stimulated oscillations. There are indeed differences, as AlF_4^--stimulated oscillations are considerably less regular than agonist-induced oscillations.

This mechanism could also work for other systems where G_ps are involved, e.g. adenylate cyclase (Gilman, 1984) and ion channels (Dunlap *et al.*, 1987). More generally still, perhaps GTP–GDP cycles underly a broad range of cyclic biological activity. For example, microtubule assembly/disassembly cycles, which can develop into overt oscillations *in vitro* under special conditions, involves GTP binding and hydrolysis (Melki *et al.*, 1988). The period of these oscillations depends on the rate of regeneration of GTP on tubulin, which is analogous to the control of calcium oscillation frequency by the rate of GTP binding to G_ps, as proposed here.

Evidence for PI oscillations. There are no direct measurements of parameters of the PI system in single cells which could indicate whether or not it can oscillate. However, the evidence in the hepatocyte that the shape of free Ca transients and other characteristics depend on the agonist type favour an oscillation mechanism that is close to the receptors (Woods *et al.*, 1987a). Unless $Ins(1,4,5)P_3$ production itself oscillates it is hard to see how the connection could be maintained. There would have to be at least one additional signalling pathway between receptors and the ER. The other arm of the PI-signalling system, the activation of PKC, can be ruled out because phorbol esters do not readily convert transients from one agonist-dependent type to another (Woods *et al.*, 1987b).

The same objection applies to an alternative model proposed by Meyer and Stryer (1988) in which ER-bound calcium, rather than activated G_p, is the 'sawtooth parameter' that determines the transient frequency. That model also fails to reproduce the invariance of the transient duration for a given agonist, which is observed to be independent of transient frequency.

Fertilization in mouse oocytes

Repetitive calcium transients

In the mouse oocyte the signal for the resumption of meiosis and thus the initiation of embryonic development is mediated by calcium. The calcium response to fertilization, as measured with aequorin, consists of repetitive transients (Cuthbertson and Cobbold, 1981, 1985). Free Ca transients rising to above 1 μM and lasting for about 1 min are repeated at ca. 20 min intervals for about 4 h (Fig. 3A). During this time the male and female pronuclei are formed and the second polar body is extruded. A consistent feature of the timing of the transients is that the first period is half the duration of the subsequent intervals.

Oocyte oscillator model

Shortly after the first elucidation of the PI-signalling system Berridge (1984) pointed out that it is highly sensitive to changes in energy metabolism. Directly or indirectly ATP concentrations affect the PI cycle at four points: production of PtdIns from phosphatidic acid and inositol (requiring CTP); production of $PtdIns(4,5)P_2$ from PtdIns (requiring two ATP-dependent phosphorylations); and activation of PLC by G_p (requiring GTP). It has been known for some time that the glycolytic pathway is inherently oscillatory (Hess, 1979). In cells where glycolysis has the potential to determine ATP concentrations significantly there is, therefore, the possibility that glycolytic oscillations may pace the PI system and the resulting free Ca transients. The multiple points of involvement of ATP in the PI cycle introduce considerable non-linearity into the coupling between ATP concentrations and Ca release, perhaps enabling glycolytic oscillations to trigger sharply rising free Ca transients.

This model can plausibly be applied to the repetitive free Ca transients in the mouse oocyte during fertilization. The periods and durations of these transients are substantially greater than in hepatocytes, for example, suggesting different underlying mechanisms. Glycolysis is essentially inactive in the unfertilized mouse oocyte, perhaps because there is insufficient fructose-6-phosphate (Barbehenn et al., 1978), the substrate for the key enzyme phosphofructokinase. Mouse oocytes have glycogen stores (Ozias and Stern, 1973) which could be released by the activation of glycogen phosphorylase-b by free Ca during the first transient, thereby activating glycolysis. The period of the glycolytic oscillator in extracts of mammalian muscle can be of the order of 10 min (Tornheim and Lowenstein, 1975) which matches the period of the fertilization transients.

A prediction of this model is that NADH concentrations should oscillate with the same frequency as the calcium transients. To test this, the autofluorescence of mouse oocytes was measured at NADH wavelengths during

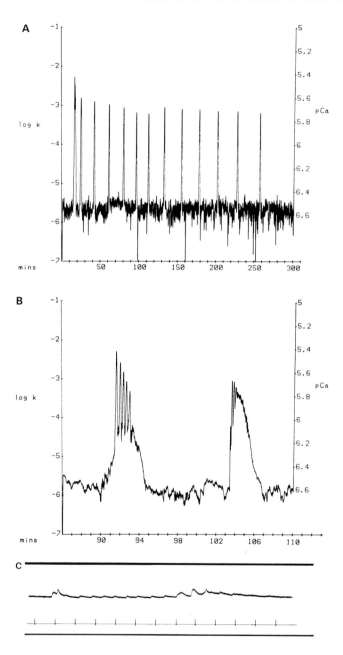

Fig. 3. (A) Repetitive free Ca transients in a mouse oocyte during fertilization. (B) The first two transients in a fertilization sequence showing superimposed fast oscillations. (C) NADH autofluorescence oscillations in a mouse oocyte during fertilization. The time marks are at 20 min intervals. (Fig. 3A and B reproduced with permission from *Nature*.)

fertilization. Oscillations were observed of similar period and pattern to the free Ca oscillations (Fig. 3C). In particular, the first period was about half the duration of later periods. The waveform, however, was sawtoothed rather than pulsatile. The last few NADH oscillations were larger, perhaps because of the cessation of calcium transients, which may feed back on glycolytic oscillations by reducing ATP concentrations. Simultaneous free Ca and NADH measurements will be required to test these proposed relationships.

ER oscillator model

A second, distinct kind of Ca oscillation occurs in the mouse oocyte during fertilization, and this probably requires a different explanation to the first kind. Superimposed on the long Ca transients are faster oscillations, with periods of 20 s on the first transient and 10 s subsequently (Fig. 3B). Similar oscillations can be induced by phorbol esters (Cuthbertson and Cobbold, 1985). These oscillations can give the appearance of varying the Ca concentration around an underlying Ca level, and the amplitude of these fluctuations is greater when the underlying free Ca level is higher. This suggests that the calcium release process may have a regenerative component.

Repetitive membrane hyperpolarizations occur in the hamster oocyte during fertilization (Igusa et al., 1983), which may be due to the activation of K^+ channels by regenerative releases of Ca^{2+} (Igusa and Miyazaki, 1983). The corresponding free Ca transients have been detected with Ca^{2+}-selective microelectrodes (Igusa and Miyazaki, 1986) and aequorin (Miyazaki et al., 1986). These may correspond to the faster of the two types of free Ca oscillation detected with aequorin.

Other examples of multiple oscillations occur in intestinal epithelial cells responding to histamine (Yada and Okada, 1984) and in hepatocytes responding to ATP (Cobbold et al., 1988b). An oscillator at the site of Ca release, i.e. the ER, is particularly plausible for such faster Ca oscillations superimposed on more slowly repeated transients.

There is now evidence that calcium release from the ER by a constant concentration of Ins(1,4,5)P$_3$ introduced into hepatocytes (Capiod et al., 1987) and Xenopus oocytes (Miledi et al., 1987) induce oscillatory membrane voltage or conductance changes which can be deduced to be due to calcium oscillations. Berridge has proposed a model for these oscillations (Berridge et al., 1988) in which the ER has both Ins(1,4,5)P$_3$-sensitive and Ins(1,4,5)P$_3$-insensitive regions. Ca uptake into the Ins(1,4,5)P$_3$-insensitive region that exceeds a threshold capacity triggers Ca release, setting up an oscillatory cycle.

An alternative possibility is that uptake and release of Ca by the ER are, for reasons of efficiency, constrained to be mutually exclusive, so that a futile cycle is avoided. The mechanism that controls the switch from uptake (which gives a set-point for free Ca at the usual 'resting level') to release (triggered by $Ins(1,4,5)P_3$, calcium-release-of-calcium, or other factors) might be suffi-ciently complex to oscillate. The oscillator might be confined to the ER membrane, and be based on as yet unknown properties of its components, such as the Ca channels themselves.

BIOLOGICAL ADVANTAGES

Why do calcium signals oscillate? The possible biological advantages can be discussed for three aspects of the signalling process: the control of the signal; the effects of the signal; and the demands on the cell made by producing the signal.

Control

An obvious requirement for the control of any signal is that it should be produced reliably. It is worth pausing here to consider what the signal is. This depends on the biological context. For fertilization the signal initiates embryonic development. This is a binary decision, or switch, or 'threshold phenomenon' (Campbell, 1983). However, even here the signal may have secondary aims, such as control and timing of polar body production (Cuthbertson and Cobbold, 1985) and early biochemical changes. In the hepatocyte there is also a switch, from glycogen synthesis to glycogenolysis. This switch needs to be sustainable yet reversible. The rate of glycogenolysis depends on the stimulus strength, so an analogue signal is also present. The control of an analogue signal carried by calcium ions presents special problems. A third type of signal may occur where several different processes need to be differentially controlled. A pulsatile calcium signal can act in all these modes. It can act as a sustainable switch, and also carry an analogue signal of wide dynamic range.

An analogue signal can be represented by the level of some carrier, or by the amplitude or frequency modulation of a carrier oscillation. One advan-tage of frequency encoding is resistance to distortion by noise (Rapp et al., 1981). An amplitude-modulated calcium signal would have to be exceptio-nally accurate because of non-linearity in the response of effector proteins (see next section). The problem of signal noise may be particularly acute for peptide hormones, which can act at sub-nanomolar concentration. Only a few agonists molecules may be presented to small cells with restricted extracellular spaces, adding Poisson noise and buffering by the receptors themselves to the noise from fluctuations in bloodflow through capillaries, variable pressures in a moving animal body, and so on. A signal transduction mechanism should also be resistant to noise in the intracellular parameters of the signal pathway.

The model mechanisms for producing pulsatile free Ca transients involving build-up of activated G_p act like an integrator of incoming stimuli (first messengers) between transients. These would act as smoothing filters for the incoming signal. They would also provide a good way of combining the information from several agonists at once. Most cells have receptors for more than one agonist. For example, hepatocytes have receptors for noradrenalin, adrenalin, ATP, ADP, vasopressin and angiotensin, all of which act through the calcium pathway.

Frequency modulation at the cellular level may interact with frequency modulation at the organ or whole body level, analogously to the way that pacemaking cells in the heart control the frequency of the heart beat and hence the circulation of the blood. Two possibilities worth considering are hormone release, which may be pulsatile, and the functioning of the central nervous system. As well as pacing oscillations at a higher level, intracellular oscillations might interact with frequency-modulated incoming signals, such as hormonal signals and signals in nerves.

Effects

Calcium-binding effector proteins tend to bind several calcium ions, and therefore act non-linearly with respect to free Ca. Calmodulin binds four calcium ions and aequorin three. Enzymes controlled through calmodulin will, at most free Ca concentrations, either be switched fully on or be fully off. Only in a very narrow band of free Ca concentrations will an intermediate proportion of enzyme molecules be activated. Calmodulin-activated enzymes may, however, be switched on at slightly different concentrations or with different kinetics, since calmodulin can be covalently bound to its target enzyme (e.g. phosphorylase kinase). An amplitude-modulated calcium signal would have to be very finely pitched to have any possibility of inducing intermediate levels of enzyme activity reliably. A calcium transient of sufficient magnitude and duration will, in contrast, unambiguously switch on all its target enzyme molecules. Intermediate levels of activation can be achieved, when time averaged, by modulation of the frequency of repetition of the transients. In a whole tissue averaging would also occur over the population of cells because of asynchrony of the transients (assuming they are not coupled).

The output of the system may be smoothed by slow kinetics downstream of the free Ca transients. There is though, no necessary reason in general why such smoothing would be required. Indeed, pulsing of metabolic or catabolic pathways may be advantageous, since it would allow temporal compartmentation of the reactions and consistent concentrations of substrates. This might simplify the optimizing of enzyme parameters.

In hepatocytes the duration of calcium transients depends on the agonist type (Woods *et al.*, 1987a). This raises the possibility of differential effects mediated by a family of effector proteins, each with its own set of kinetic parameters. Similarly, if amplitudes vary, larger transients could bring in a different range of effects with lower K_a values for free Ca. Differential effects according to pulse frequency are also theoretically possible, but this would run counter to the natural elegance of frequency encoding.

Signal production

A powerful advantage of an oscillating mechanism is that it is sustainable. At some point during each cycle the system may return to its starting point, and so continue indefinitely. For the calcium signal in particular, a finite calcium store is sufficient to maintain an indefinitely sustained signal if release is periodic. The calcium store can refill between transients. The ATP hydrolysed during the cycle can be replenished.

Specific examples of energy advantages for oscillatory systems have been claimed for cAMP signalling in slime moulds and for the glycolytic oscillator (see Rapp, 1987). An oscillatory PI-dependent calcium signalling system could also have energy advantages. The Ins(1,4,5)P$_3$ signal is energy intensive. Three phosphate groups must be supplied from ATP (one via CTP) for each Ins(1,4,5)P$_3$ molecule produced. Free Ca is removed with time constants of the order of seconds or less by ATP-dependent processes. If we assume that these energy costs are proportional to the integral of the product of time and free Ca, and that the output of the system is proportional to the product of time and the fourth power of free Ca (because of the non-linearity of calmodulin activation) then a pulsatile free Ca system uses less energy per unit of output than one requiring a sustained rise in free Ca.

SUMMARY

Three types of calcium oscillator model involving the PI-signalling system have been described here. The first involves feedback within the PI system, and is proposed as a model for the pulsatile oscillations that hormones induce in hepatocytes. This model consists of a linear component (the accumulation of activated G$_p$) and a non-linear component (the PI pathway plus the calcium-release system). The second model involves control of the PI system by another oscillator (based on the glycolytic oscillator but with feedback from calcium), and is proposed as an explanation of the calcium response during fertilization in mouse oocytes. The third model puts the site of the oscillator at the calcium-release mechanism of the ER rather than in the PI system itself. It is proposed to account for the faster calcium oscillations which may co-exist with oscillations of the other two types.

Pulsatile calcium signals have advantages in terms of sustainability, energy efficiency, resistence to noise and unambiguous signal transduction. They therefore provide a robust form of communication between cells.

REFERENCES

Barbehenn, E. K., Wales, R. G. and Lowry, O. H. (1978) Measurement of metabolites in single preimplantation embryos; a new means to study metabolic control in early embryos. *J. Embryol. exp. Morph.* **43**, 29–46.

Berridge, M. J. (1983) Rapid accumulation of inositol trisphosphate reveals that agonists hydrolyse polyphosphoinositides instead of phosphatidylinositol. *Biochem. J.* **212**, 849–58.

Berridge, M. J. (1984) Inositol trisphosphate and diacylglycerol as second messengers. *Biochem. J.* **220**, 345–60.

Berridge, M. J. (1987) Inositol trisphosphate and diacylglycerol: two interacting second messengers. *A. Rev. Biochem.* **56**, 159–93.

Berridge, M. J., Cobbold, P. H. and Cuthbertson, K. S. R. (1988) Spatial and temporal aspects of cell signalling. *Phil. Trans. R. Soc. Lond. B* **320**, 325–43.

Bojanic, D. and Fain, J. B. (1986) Guanine nucleotide regulation of [³H]vasopressin binding to liver plasma membranes and solubilized receptors. *Biochem. J.* **240**, 361–5.

Burgess, G. M., Godfrey, P. P., McKinney, J. S., Berridge, M. J., Irvine, R. F. and Putney, J. W. (1984) The second messenger linking receptor activation to internal calcium release in the liver. *Nature* **309**, 63–6.

Campbell, A. K. (1983) *Intracellular Calcium, its Universal Role as Regulator.* Wiley, New York.

Capiod, T., Field, A. C., Ogden, D. C. and Sandford, C. A. (1987) Internal perfusion of guinea-pig hepatocytes with buffered Ca²⁺ or inositol 1,4,5-trisphosphate mimics noradrenaline activation of K⁺ and Cl⁻ conductances. *FEBS Lett.* **217**, 247–52.

Cobbold, P., Cuthbertson, R., and Woods, N. (1988a) The generation of repetitive free calcium transients in a hormone-stimulated hepatocyte. In *Hormones and Cell Regulation. No. 12* (ed. J. Nunez, J. E. Dupont and E. Carafoli), pp. 135–46. INSERM/John Libbey Eurotext, Paris.

Cobbold, P. H., Woods, N., Wainwright, J. and Cuthbertson, K. S. R. (1988b) Single cell measurements in research on calcium-mobilising purinoceptors. *J. Receptor Res.* **8**, 481–91.

Cockcroft, S. (1987) Polyphosphoinositide phosphodiesterase: regulation by a novel guanine nucleotide binding protein, G_p. *Trends biochem. Sci.* **12**, 75–8.

Cuthbertson, K. S. R. and Cobbold, P. H. (1981) Free Ca²⁺ increases in exponential phases during mouse oocyte activation. *Nature* **294**, 754–7.

Cuthbertson, K. S. R. and Cobbold, P. H. (1985) Phorbol esters and sperm activate mouse oocytes by inducing sustained oscillations in cell Ca²⁺. *Nature* **316**, 541–2.

Dunlap, K., Holz, G. G. and Rane, S. G. (1987) G proteins as regulators of ion channel function. *Trends Neurosci.* **10**, 241–4.

Gibbs, J. B., Schaber, M. D., Allard, W. J., Sigal, I. S. and Scolnick, E. M. (1988) Purification of ras GTPase activating protein from bovine brain. *Proc. Natnl. Acad. Sci. U.S.A.* **85**, 5026–30.

Gilman, A. G. (1984) G Proteins and dual control of adenylate cyclase. *Cell* **36**, 577–9.

Hatanaka, M., White, E. A., Horibata, K. and Crawford, I. P. (1962) A study of the catalytic properties of *Escherichia coli* tryptophan synthetase, a two-component enzyme. *Arch. Biochem. Biophys.* **97**, 596–606.

Hess, B. (1979) The glycolytic oscillator. *J. exp. Biol.* **81**, 7–14.

Igusa, Y. and Miyazaki, S. (1983) Effects of altered extracellular and intracellular calcium concentration on hyperpolarizing responses of the hamster egg. *J. Physiol. Lond.* **340**, 611–32.

Igusa, Y. and Miyazaki, S. (1986) Periodic increase of cytoplasmic free calcium in fertilized hamster eggs measured with calcium-sensitive electrodes. *J. Physiol.* **377**, 193–205.

Igusa, Y., Miyazaki, S. and Yamashita, N. (1983) Periodic hyperpolarizing responses in hamster and mouse eggs fertilized with mouse sperm. *J. Physiol. Lond.* **340**, 633–47.

Lochrie, M. A. and Simon, M. I. (1988) G protein multiplicity in eukaryotic signal transduction systems. *Biochemistry* **27**, 4957–65.

Melki, R., Carlier, M.-F. and Pantaloni, D. (1988) Oscillations in microtubule polymerization: the rate of GTP regeneration on tubulin controls the period. *EMBO J.* **7**, 2653—9.

Meyer, T. and Stryer, L. (1988) Molecular model for receptor-stimulated calcium spiking. *Proc. natn. Acad. Sci. U.S.A.* **85**, 5051–5.

Meyer, T., Holowka, D. and Stryer, L. (1988) Highly cooperative opening of calcium channels by inositol 1,4,5-trisphosphate. *Science, N.Y.* **240**, 653–5.

Miledi, R., Parker, I. and Sumikawa, K. (1987) Oscillatory chloride current evoked by temperature jumps during muscarinic and serotonergic activation in *Xenopus* oocyte. *J. Physiol. Lond.* **383**, 213–29.

Miyazaki, S., Hashimoto, N., Yoshimoto, Y., Kishimoto, T., Igusa, Y. and Hiramoto, Y. (1986) Temporal and spatial dynamics of the periodic increase in intracellular free calcium at fertilization of golden hamster eggs. *Devl Biol.* **118**, 259–67.

Ozias, C. B. and Stern, S. (1973) Glycogen levels of preimplantation mouse embryos developing *in vitro*. *Biol. Reprod.* **8**, 467–72.

Rapp, P. E. (1987) Why are so many biological systems periodic? *Prog. Neurobiol.* **29**, 261–73.

Rapp, P. E. and Berridge, M. J. (1981) The control of transepithelial potential oscillations in the salivary gland of *Calliphora erythrocephala*. *J. exp. Biol.* **93**, 119–32.

Rapp, P. E., Mees, A. I. and Sparrow, C. T. (1981) Frequency dependent biochemical regulation is more accurate than amplitude dependent control. *J. theor. Biol.* **90**, 531–44.

Tornheim, K. and Lowenstein, J. M. (1975) The purine nucleotide cycle. Control of phosphofructokinase and glycolytic oscillations in muscle extracts. *J. biol. Chem.* **250**, 6304–14.

Woods, N. M., Cuthbertson, K. S. R. and Cobbold, P. H. (1986) Repetitive transient rises in cytoplasmic free calcium in hormone stimulated hepatocytes. *Nature* **319**, 600–2.

Woods, N. M., Cuthbertson, K. S. R. and Cobbold, P. H. (1987a) Agonist induced oscillations in cytoplasmic free calcium concentration in single rat hepatocytes. *Cell Calcium* **8**, 79–100.

Woods, N. M., Cuthbertson, K. S. R. and Cobbold, P. H. (1987b) Phorbol-ester-induced alterations of free calcium ion transients in single rat hepatocytes. *Biochem. J.* **246**, 619–23.

Yada, T. and Okada, Y. (1984) Electrical activity of an intestinal epithelial cell line: hyperpolarizing responses to intestinal secretagogues. *J. Membrane Biol.* **77**, 33–44.

Cell signalling through cytoplasmic calcium oscillations

MICHAEL J. BERRIDGE

Unit of Insect Neurophysiology and Pharmacology, Department of Zoology, Downing Street, Cambridge CB2 3EJ, UK

INTRODUCTION

Many cells display endogenous oscillations in intracellular calcium. There are two basic mechanisms for generating such oscillations, the oscillator may be located either within the plasma membrane or within the cytoplasm. Examples of the former are the sino-atrial node, the parabolic burster in *Aplysia* and the calcium spiking described in endocrine cells (Schlegel *et al.*, 1987). An interplay between membrane conductances results in the periodic opening of channels which gate calcium and it is the influx of external calcium which sets up the oscillations. Such membrane oscillators are very dependent upon external calcium to maintain oscillatory activity. This article will concentrate on cytosolic calcium oscillators which are most likely concentrated within the endoplasmic/sarcoplasmic reticulum (ER/SR). They are much less dependent on external calcium since the oscillation is the result of a cyclical uptake and release of calcium by the ER/SR which can continue for variable times even in the complete absence of external calcium. However, the oscillations are dependent on external calcium in as much as the flow of calcium across the plasma membrane can influence the oscillator, particularly its frequency. Oscillator frequency is sometimes found to vary with agonist concentration and it is this observation which has attracted considerable attention because it raises the possibility that second messengers might be frequency encoded.

DISTRIBUTION AND CHARACTERISTICS OF CYTOPLASMIC OSCILLATORS

The development of techniques to monitor the level of calcium in single cells has led to an ever-increasing list of cells displaying oscillatory activity (Table 1). The frequency of such oscillators varies considerably, ranging from periods of less than 1 s up to several minutes. In general, however, most periods lie within the 5–60 s range (Table 1). For any given cell, the period can vary depending upon external conditions with agonist concentration being the most important determinant. Of the stimuli summarized in Table 1, a large number are known to act through the phosphoinositide receptor mechanism. A characteristic feature of this signal transduction mechanism is

Table 1.   Summary of cells showing cytosolic calcium oscillations

Cell	Stimulus	Period (s)	Reference
Rat myocyte[d]	Caffeine	0.3–3	Kort et al. (1985)
Astrocytes[e]	TPA	0.3–10	MacVicar et al. (1987)
Parotid gland[a]	Carbachol	5	Gray (1989)
Lacrimal gland[e]	Acetylcholine	5–10	Evans and Marty (1986)
Gonadotropes[a]	GnRH	6	Shangold et al. (1988)
β-cells[a]	Carbamylcholine	12–25	Prentki et al. (1988)
Mouse oocytes[b]	TPA	17–35	Cuthbertson and Cobbold (1985)
Rat hepatocytes[b]	Vasopressin	18–240	Woods et al. (1986)
Macrophages[a]	Cell spreading	19–69	Kruskal and Maxfield (1987)
Xenopus oocytes[e]	Acetylcholine	20	Miledi et al. (1982)
HeLa cells[e]	Histamine	20–33	Sauve et al. (1987)
L cells[c]	—	20	Ueda et al. (1986)
Smooth muscle[a]	Phenylephrine or histamine	30–48	Ambler et al. (1988)
Fibroblasts (REF52)[a]	Gramicidin + vasopressin	35–100	Harootunian et al. (1989)
Endothelial cells[a]	Histamine	40–125	Jacob et al. (1988)
B lymphocytes[a]	Antigen	50–75	Wilson et al. (1987)
Hamster eggs[b]	Fertilization	55	Miyazaki et al. (1986)
Sympathetic neurones[a]	K+ depolarization and caffeine	60–120	Lipscombe et al. (1988)
Sympathetic ganglion[e]	Caffeine	ca. 120	Kuba and Nishi (1976)
Mouse oocyte[b]	Fertilization	600–1800	Cuthbertson et al. (1981); Cuthbertson and Cobbold (1985)

Calcium oscillations were measured either directly by using various indicators such as Fura-2[a], aequorin[b], Ca^{2+} electrode[c] or indirectly by using endogenous calcium-sensitive processes such as contraction[d] or calcium-activated potassium channels.[e]

that a membrane lipid, phosphatidylinositol(4,5)bisphosphate (PtdIns(4,5)P$_2$) is hydrolysed to give diacylglycerol (DG) and inositol-(1,4,5)trisphosphate (Ins(1,4,5)P$_3$) (Berridge and Irvine, 1984; Nishizuka 1984, 1988; Berridge, 1987). These two products then function as second messengers to initiate a bifurcating signal cascade. The DG remains within the plane of the membrane where it activates protein kinase C (PKC) (Nishizuka, 1984, 1988). The stimulatory effect of the natural messenger DG can be mimicked by phorbol esters which can influence the cytosolic oscillator both positively (Cuthbertson and Cobbold, 1985) and negatively (Woods *et al.*, 1987; Harootunian *et al.*, 1989). Through its ability to mobilize calcium from the ER/SR, the Ins(1,4,5)P$_3$ released from the membrane has a central role in controlling the cytosolic oscillator and thus features significantly in the models outlined below.

MODELS OF CYTOSOLIC OSCILLATORS

The models of cytosolic oscillators fall into two main groups (Fig. 1). There are those where the release of calcium is driven by the periodic generation of the calcium-mobilizing messenger Ins(1,4,5)P$_3$ and may thus be considered as

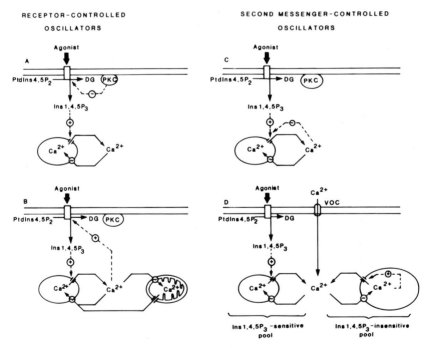

Fig. 1. Summary of receptor-controlled and second messenger-controlled models of cytoplasmic oscillators (see text for further details).

receptor-controlled oscillators. In this case, the ER/SR system is a passive component of the oscillator since it is being driven by an external signal and the models attempt to explain how the generation of Ins(1,4,5)P$_3$ might be periodic. The second group comprise second messenger-controlled models where the oscillator is thought to reside within the ER/SR itself with the process of calcium-induced calcium release (CICR) being an important feature in some cases.

Models based on Ins(1,4,5)P$_3$ pulsing

A characteristic feature of the bifurcating phosphoinositide-signalling system is that the DG/PKC limb of the pathway can feed back to inhibit the initial transduction event (Berridge, 1987). The existence of this negative feedback loop is the basis of a model (Fig. 1A) described in detail elsewhere in this volume (see Cuthbertson, p. 435).

Another model to explain the periodic formation of Ins(1,4,5)P$_3$ is based on a positive feedback loop whereby calcium acts to stimulate the enzyme phosphoinositidase which hydrolyses PtdIns(4,5)P$_2$ (Fig. 1B) (Meyer and Stryer, 1988). The idea is that the agonist begins to hydrolyse PtdIns(4,5)P$_2$ to form Ins(1,4,5)P$_3$ which by mobilizing some calcium will initiate the feedback loop whereby calcium activates more Ins(1,4,5)P$_3$ to result in the rapid formation of sufficient messengers to generate a full calcium signal. An important component of this model is the cooperativity of Ins(1,4,5)P$_3$ in releasing calcium from the ER/SR (Meyer et al., 1988). The production of Ins(1,4,5)P$_3$ is terminated as calcium is rapidly sequestered back into the internal pool. Meyer and Stryer (1988) have proposed that the mitochondria might contribute to this termination phase by helping to rapidly reduce the calcium concentration below a level where it activates the production of Ins(1,4,5)P$_3$.

The verification of these models based on oscillating levels of Ins(1,4,5)P$_3$ will have to await the development of techniques for measuring this second messenger in single cells.

Models based on the ER/SR having an endogenous oscillator

The two models in this class are characterized by having the oscillator within the ER/SR system where it is activated by the steady input of various signals such as Ins(1,4,5)P$_3$ or by calcium itself. In those cases where Ins(1,4,5)P$_3$ is the initiating signal, it is necessary to explain how a constant level of Ins(1,4,5)P$_3$ is translated into the periodic release of calcium. Somehow the effect of Ins(1,4,5)P$_3$ must pass through cycles of activation and inactivation. In one model, the action of Ins(1,4,5)P$_3$ is thought to be periodic by virtue of a negative feedback loop operated through calcium (Payne et al., 1988;

Prentki et al., 1988; Gray, 1989). Once Ins(1,4,5)P$_3$ is elevated it releases calcium to initiate the transient but as calcium builds up it feeds back to inhibit further release (Fig. 1C). An increase in calcium has already been shown to inhibit both the binding of Ins(1,4,5)P$_3$ to its receptor (Worley et al., 1987) as well as its ability to mobilize calcium (Suematsu et al., 1984; Thierry and Klee, 1986). As calcium is pumped back into the internal reservoir this inhibition is removed thus allowing Ins(1,4,5)P$_3$ to initiate another transient.

The second model attempts to explain oscillatory activity on the basis of the phenomenon of calcium-induced calcium release (CICR) by the ER/SR (Fig. 1D). In this case, the signal which initiates oscillatory activity is calcium itself. CICR was first described in skinned cardiac cells where it was thought to offer an explanation for excitation–contraction (E–C) coupling (Endo et al., 1970). The idea was that a small rise in calcium somehow triggers the release of calcium from the internal reservoirs. A critical aspect of this mechanism concerns the site of action of the trigger calcium. For the mechanism to function in E–C coupling it would seem that the calcium must act from the cytoplasmic side. However, there are indications that calcium can act from within the cisternal compartment, in this case it is the build-up of calcium inside the ER/SR cisternae which provides the trigger to release calcium (Fig. 1D). It is this mechanism which is most likely to play a role in inducing oscillatory activity. Skinned cardiac cells provide a beautiful demonstration of such an oscillator. Permeabilized heart cells continued to contract rhythmically when immersed in an appropriate Ca^{2+}/EGTA buffer (Fabiato and Fabiato, 1975). The SR exposed to the buffer takes up calcium until the concentration reaches some critical threshold whereupon calcium is spontaneously released back into the medium to stimulate a contraction. As calcium is pumped back into the SR the muscle relaxes but a new cycle begins as the level of calcium within the SR reaches the threshold to trigger another release.

This phenomenon of CICR could account for the oscillations described in hamster eggs following fertilization (Igusa and Miyazaki, 1986; Miyazaki, 1988), sympathetic ganglion cells (Kuba, 1980) and Xenopus oocytes (Berridge, 1988). CICR could also account for the oscillations described in cultured sympathetic neurones (Lipscombe et al., 1988) and in fibroblasts (Harootunian et al., 1989). In all of these examples, it seems that calcium oscillations are induced by an overloading of internal reservoirs resulting from an elevation of cytosolic calcium. This overload of calcium can be obtained either by an enhanced entry of calcium across the plasma membrane, most likely through the opening of voltage-operated calcium channels (VOCs), or by release of internal stores by Ins(1,4,5)P$_3$. In the latter case, the idea is that Ins(1,4,5)P$_3$ releases calcium from one pool which, upon being taken up by an Ins(1,4,5)P$_3$-insensitive pool, overloads this pool thus triggering it to release calcium back into the cytosol (Fig. 1D) (Berridge,

1988; Berridge *et al.*, 1988). There is considerable evidence for separate calcium pools in cells with the Ins(1,4,5)P$_3$-sensitive pool usually comprising approximately 30–50% of the total ER/SR pool. There already are suggestions from work on eggs (Busa *et al.*, 1985) and smooth muscle (Iino *et al.*, 1988) that the Ins(1,4,5)P$_3$-insensitive calcium pool may release calcium by CICR.

The key feature of this oscillator based on CICR is that frequency is very dependent upon two key parameters – the rate of calcium sequestration and the threshold level of the triggering process. Since it is the filling of the ER/SR pool which sets the stage for the next release, the inter-burst period and hence frequency will depend upon the rate at which the pool fills with calcium. One way of speeding up the rate of sequestration is to increase the intracellular level of calcium by raising the level of external calcium which was found to accelerate oscillation frequency in myocytes (Kort *et al.*, 1985), ventricular muscle (Allen *et al.*, 1984), fibroblasts (Harootunian *et al.*, 1989), hamster egg (Igusa and Miyazaki, 1983) and sympathetic ganglion cells (Kuba and Nishi, 1976). The rate of calcium entry across the plasma membrane can also be adjusted by varying membrane potential. Hyperpolarization, which enhances calcium entry by increasing the electrical gradient, was found to accelerate the oscillator in sympathetic ganglion cells (Kuba, 1980) and hamster eggs (Igusa and Miyazaki, 1983). The interval between calcium bursts can also be influenced by artificially introducing extra calcium into the cell. In the case of sympathetic ganglion cells, triggering an action potential will introduce a pulse of calcium which can speed up the appearance of the next spontaneous event (Kuba, 1980). A similar phenomenon has been described in medaka eggs where the iontophoretic injection of calcium immediately following a spontaneous calcium transient had no effect but became increasingly effective towards the expected time of the next spontaneous event (Igusa and Miyazaki, 1983). In summary, the interval between spontaneous events depends on the time it takes for the calcium concentration within the pool to reach the threshold level necessary to trigger the next discharge.

Another way of altering the frequency of the CICR oscillator is to alter the threshold so that release is triggered at different levels of cisternal calcium (Fig. 2). It has been suggested that caffeine may accelerate calcium oscillations by lowering the threshold for CICR (Berridge and Galione, 1988). Caffeine has been found to accelerate calcium oscillations in sympathetic ganglion cells (Kuba and Nishi, 1976), guinea-pig atrial cells (Glitsch and Pott, 1975) and rat myocytes (Kort *et al.*, 1985). If caffeine acts by lowering the threshold, then release will be triggered when the store is partly filled (Fig. 2) which should result in a smaller release of calcium which is exactly what is found in heart cells. Increasing the concentration of caffeine resulted in a progressive decrease in contraction amplitude (Glitsch and Pott, 1975; Kort *et al.*, 1985). There was no change in contraction amplitude when frequency

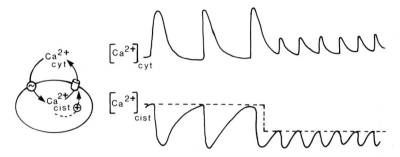

Fig. 2. Model of calcium-induced calcium release (CICR). An accumulation of calcium within the ER/SR cisterna (Ca^{2+}_{cist}) is thought to provide the trigger to stimulate the release of calcium into the cytoplasm (Ca^{2+}_{cyt}). Both the frequency and amplitude of the oscillations in Ca^{2+}_{cyt} can be altered by varying the threshold of CICR (dashed line).

was varied by altering the external concentration of calcium (Kort *et al.*, 1985).

FUNCTIONAL SIGNIFICANCE OF CALCIUM OSCILLATIONS

At the beginning I pointed to the existence of two separate mechanisms for generating calcium oscillations. The function of many membrane oscillators is clear, for example the oscillations in the sino-atrial node provide the pacemaker rhythm which drives the heart. Calcium spiking in neural and endocrine cells regulate secretory activity. But what might be the function of cytosolic calcium oscillators?

One possibility is that they provide an efficient mechanism of using calcium as a second messenger which overcomes the problem of overloading the cell with this ion. Cells at rest maintain a low intracellular level of calcium and any increase is immediately counteracted by potent homeostatic mechanisms which pump calcium out of the cell or return it to the internal pools. One way of using calcium as a messenger in the face of such a powerful homeostatic mechanism might be to elevate it for brief periods. By presenting calcium in the form of transient spikes the cell avoids becoming overloaded with calcium. One of the most interesting aspects of the oscillations in blowfly salivary gland (Rapp and Berridge, 1981), hepatocytes (Woods *et al.*, 1986), insulin-secreting pancreatic cells (Prentki *et al.*, 1988), *Xenopus* oocyte (Berridge, 1988) and endothelial cells (Jacob *et al.*, 1988) is that they are sensitive to hormone concentration. This relationship has led to the suggestion that the second messenger calcium might operate through a frequency-encoded rather than an amplitude-dependent mechanism (Fig. 3). The classical view of second messenger action is that they operate through an

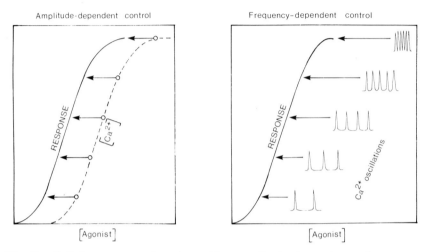

Fig. 3.   Cellular responses based on either amplitude-dependent or frequency-dependent control.

amplitude-dependent mechanism – changes in agonist concentration are translated into variations in the levels of some messengers such as calcium (Fig. 3). The alternative is that they are frequency dependent such that changes in agonist concentration are transformed into variations in the frequency of a constant-amplitude calcium transient. In the case of the insect salivary gland, the narrow 5-HT dose range which stimulated fluid secretion coincided exactly with that which induced the change in oscillator frequency (Rapp and Berridge, 1981). A similar correlation between calcium oscillation frequency and growth hormone secretion has been described in somatotropes (Holl *et al.*, 1988). The advantages of employing a frequency-encoded signalling system is discussed in detail elsewhere (Rapp, 1987).

CONCLUSION

Many cells display oscillations in intracellular calcium especially when stimulated with agonists which hydrolyse the inositol lipids. One of the messengers derived from the phosphoinositide receptor mechanism is $Ins(1,4,5)P_3$ which functions to mobilize calcium from internal stores. A number of models have been developed to explain how this calcium-mobilizing action of $Ins(1,4,5)P_3$ might become periodic. Some models consider that oscillations develop from a periodic synthesis of $Ins(1,4,5)P_3$ resulting from various feedback loops which regulate the enzyme responsible for hydrolysing $PtdIns(4,5)P_2$. Other models consider that $Ins(1,4,5)P_3$ is supplied at a constant level but that its action on the internal stores results in the periodic release of calcium. One of these models considers that Ins(1,4,5)P$_3$ releases calcium from one pool which then overloads an Ins(1,4,5)P$_3$-insensitive pool and induces it to release calcium through a process of calcium-induced calcium release (CICR). Ins(1,4,5)P$_3$ is not the only agent capable of initiating such oscillations based on CICR, any process which introduces a calcium load on the cell is likely to initiate oscillatory activity.

The fact that oscillator frequency has been found to vary with hormone concentration has led to the proposal that calcium transients might be part of a frequency-encoded signalling system. When a hormone arrives at a cell the information is translated into a train of calcium spikes – in effect the signal has been digitized. It seems that certain cells may convey information by varying the frequency of this digital signal.

REFERENCES

Allen, D. G., Eisner, D. A. and Orchard, C. H. (1984) Characterization of oscillations of intracellular calcium concentration in ferret ventricular muscle *J. Physiol.* **352**, 113–28.

Ambler, S. K., Poenie, M., Tsien, R. Y. and Taylor, P. (1988) Agonist-stimulated oscillations and cycling of intracellular free calcium in individual cultured muscle cells. *J. biol. Chem.* **263**, 1952–9.

Berridge, M. J. (1987) Inositol trisphosphate and diacylglycerol: two interacting second messengers *A. Rev. Biochem.* **56**, 159–93.

Berridge, M. J. (1988) Inositol trisphosphate-induced membrane potential oscillations in *Xenopus* oocytes. *J. Physiol.* **403**, 589–99.

Berridge, M. J. and Irvine, R. F. (1984) Inositol trisphosphate, a novel second messenger in cellular signal transduction. *Nature* **312**, 315–21.

Berridge, M. J., Cobbold, P. H. and Cuthbertson, K. S. R. (1988) Spatial and temporal aspects of cell signalling. *Phil. Trans. R. Soc. Lond. B.* **320**, 325–43.

Berridge, M. J. and Galione, A. (1988) Cytosolic calcium oscillators. *FASEB J.* **2**, 3074–82.

Busa, W. B., Ferguson, J. E., Joseph, S. K., Williamson, J. R. and Nuccitelli, R. (1985) Activation of frog (*Xenopus laevis*) eggs by inositol trisphosphate. I. Characterization of Ca^{2+} release from intracellular stores. *J. Cell Biol.* **101**, 677–682.

Cuthbertson, K. S. R., Whittingham, D. G. and Cobbold, P. H. (1981) Free Ca^{2+} increases in exponential phases during mouse oocyte activation. *Nature* **294**, 754–7.

Cuthbertson, K. S. R. and Cobbold, P. H. (1985) Phorbol ester and sperm activate mouse oocytes by inducing sustained oscillations in cell Ca^{2+}. *Nature* **316**, 541–2.

Endo, M., Tanaka, M. and Ogawa, Y. (1970) Calcium induced release of calcium from the sarcoplasmic reticulum of skinned skeletal muscle fibres. *Nature*, Lond. **228**, 34–36.

Evans, M. G. and Marty, A. (1986) Potentiation of muscarinic and α-adrenergic responses by an analogue of guanosine 5′-triphosphate. *Proc. natn. Acad. Sci. U.S.A.* **83**, 4099–4103.

Fabiato, A. and Fabiato, F. (1975) Contractions induced by a calcium triggered release of calcium from the sarcoplasmic reticulum of single skinned cardiac cells *J. Physiol.* **249**, 469–95.

Glitsch, H. G. and Pott, L. (1975) Spontaneous tension oscillations in guinea-pig atrial trabeculae. *Pflügers Arch.* **358**, 11–25.

Gray, P. T. A. (1989) Oscillations of free cytosolic calcium evoked by cholinergic and catecholaminergic agonists in rat parotid acinar cells. *J. Physiol.* (in press).

Harootunian, A. T., Kao, J. P. Y. and Tsien, R. Y. (1989) Agonist-induced calcium oscillations in depolarized fibroblasts and their manipulation by photoreleased Ins(1,4,5)P$_3$, Ca^{2+} and Ca^{2+} buffer. *Cold Spring Harbor Symp. on Quant. Biol.* **53**, 935–43.

Holl, R. W., Thorner, M. O., Mandell, G. L., Sullivan, J. A., Sinka, Y. N. and Leong, D. A. (1988) Spontaneous oscillations of intracellular calcium and growth hormone secretion. *J. biol. Chem.* **263**, 9682–5.

Iino, M., Kobayashi, T. and Endo, M. (1988) Use of ryanodine for functional removal of the calcium store in smooth muscle cells of the guinea-pig. *Biochem. biophys. Res. Commun.* **152**, 417–22.

Igusa, Y. and Miyazaki, S-I. (1983) Effects of altered extracellular and intracellular calcium concentration on hyperpolarizing responses of the hamster egg. *J. Physiol.* **340**, 611–32.

Igusa, Y. and Miyazaki, S-I. (1986) Periodic increase of cytoplasmic free calcium in fertilized hamster eggs measured with calcium-sensitive electrodes. *J. Physiol.* **377**, 193–205.

Jacob, R., Merritt, J. E., Hallam, T. J. and Rink, T. J. (1988) Repetitive spikes in cytoplasmic calcium evoked by histamine in human endothelial cells. *Nature* **335**, 40–5.

Kort, A. A., Capogrossi, M. C. and Lakatta, E. G. (1985) Frequency, amplitude and propagation velocity of spontaneous Ca^{2+}-dependent contractile waves in intact adult rat cardiac muscle and isolated myocytes. *Circulation Res.* **57**, 844–55.

Kruskal, B. A. and Maxfield, F. R. (1987) Cytosolic free calcium increases before and oscillates during frustrated phagocytosis in macrophages. *J. Cell Biol.* **105**, 2685–93.

Kuba, K. (1980) Release of calcium ions linked to the activation of potassium conductance in a caffeine-treated sympathetic neurone. *J. Physiol.* **298**, 251–69.

Kuba, K. and Nishi, S. (1976) Rhythmic hyperpolarizations and depolarization of sympathetic ganglion cells induced by caffeine. *J. Neurophysiol.* **39**, 547–63.

Lipscombe, D., Madison, D. V., Poenie, M., Reuter, H., Tsien, R. Y. and Tsien, R. W. (1988) Spatial distribution of calcium channels and cytosolic calcium transients in growth cones and cell bodies of sympathetic neurons. *Proc. natn. Acad. Sci. U.S.A.* **85**, 2398–402.

MacVicar, B. A., Crichton, S. A., Burnard, D. M. and Tse, F. W. Y. (1987) Membrane conductance oscillations in astrocytes induced by phorbol ester. *Nature.* **329**, 242–3.

Meyer, T. and Stryer, L. (1988) Molecular model for receptor-stimulated calcium spiking. *Proc. natn. Acad. Sci. U.S.A.* **85**, 5051–55.

Meyer, T., Holowka, D. and Stryer, L. (1988) Highly cooperative opening of calcium channels by inositol 1,4,5-trisphosphate. *Science, N.Y.* **240**, 653–6.

Miledi, R., Parker, I. and Sumikawa, K. (1982) Properties of acetylcholine receptors translated by cat muscle mRNA in *Xenopus* oocytes. *EMBO J.* **1**, 1307–12.

Miyazaki, S-I., Hashimoto, N., Yoshimoto, Y., Kishimoto, T., Igusa, Y. and Hiramoto, Y. (1986) Temporal and spatial dynamics of the periodic increase in intracellular free calcium at fertilization of golden hamster eggs. *Devl Biol.* **118**, 259–67.

Miyazaki, S-I. (1988) Inositol 1,4,5-trisphosphate-induced calcium release and guanine nucleotide-binding protein-mediated periodic calcium rise in golden hamster eggs. *J. Cell. Biol.* **106**, 345–53.

Nishizuka, Y. (1984) Turnover of inositol phospholipids and signal transduction. *Science, N.Y.* **225**, 1365–70.

Nishizuka, Y. (1988) The molecular heterogeneity of protein kinase C and its implication for cellular regulation. *Nature* **334**, 6651–65.

Payne, R., Walz, B., Levy, S. and Fein, A. (1988) The localization of calcium release by inositol trisphosphate in *Limulus* photoreceptor and its control by negative feedback. *Phil. Trans. R. Soc. B* **320**, 359–79.

Prentki, M., Glennon, M. C., Thomas, A. P., Morris, R. L., Matschinsky, F. M. and Corkey, B. E. (1988) Cell-specific patterns of oscillating free Ca^{2+} in carbamylcholine-stimulated insulinoma cells. *J. biol. Chem.* **263**, 11044–7.

Rapp, P. E. (1987) Why are so many biological systems periodic? *Prog. Neurobiol.* **29**, 261–73.

Rapp, P. E. and Berridge, M. J. (1981) The control of transepithelial potential oscillations in the salivary gland of *Calliphora erythrocephala. J. exp. Biol.* **93**, 119–32.

Sauve, R., Simoneau, C., Parent, L., Monette, R. and Roy, G. (1987) Oscillatory activation of calcium-dependent potassium channels in HeLa cells induced by histamine H_1 receptor stimulation: a single-channel study. *J. Membrane Biol.* **96**, 199–208.

Schlegel, W., Winiger, B. P., Mollard, P., Vacher, P., Wuarin, F., Zahnd, G. R., Wollheim, C. B. and Dufy, B. (1987) Oscillations of cytosolic Ca^{2+} in pituitary cells due to action potentials. *Nature* **329**, 719–21.

Shangold, G. A., Murphy, S. N. and Miller, R. J. (1988) Gonadotrophin-releasing hormone-induced Ca^{2+} transients in single identified gonadotropes require both intracellular Ca^{2+} mobilization and Ca^{2+} influx. *Proc. natn. Acad. Sci. U.S.A.* **85**, 6566–70.

Suematsu, E., Hirata, M., Hashimoto, T. and Kuriyama, H. (1984) Inositol 1,4,5-trisphosphate releases Ca^{2+} from intracellular store sites in skinned single cells of porcine coronary artery. *Biochem. biophys. Res. Commun.* **120**, 481–5.

Thierry, J. and Klee, C. B. (1986) Calcium modulation of inositol 1,4,5-trisphosphate-induced calcium release from neuroblastoma × glioma hybrid NG108-15 microsomes. *J. biol. Chem.* **261**, 16414–20.

Ueda, S., Oiki, S. and Okada, Y. (1986) Oscillations of cytoplasmic concentrations of Ca^{2+} and K^+ in fused L cells. *J. Membrane Biol.* **91**, 65–72.

Wilson, H. A., Greenblatt, D., Poenie, M., Finkelman, F. D. and Tsien, R. Y. (1987) Crosslinkage of a B lymphocyte surface immunoglobulin by anti-Ig or antigen induces prolonged oscillation of intracellular ionized calcium. *J. exp. Med.* **166**, 601–6.

Woods, N. M., Cuthbertson, K. S. R. and Cobbold, P. H. (1986) Repetitive transient rises in cytoplasmic free calcium in hormone-stimulated hepatocytes. *Nature* **319**, 600–2.

Woods, N. M., Cuthbertson, K. S. R. and Cobbold, P. H. (1987) Phorbol ester-induced alterations of free calcium ion transients in single rat hepatocytes. *Biochem. J.* **246**, 619–23.

Worley, P. F., Baraban, J. M., Supattapone, S., Wilson, V. S. and Snyder, S. H. (1987) Characterization of inositol trisphosphate receptor binding in brain: regulation of pH and calcium. *J. biol. Chem.* **262**, 12132–6.

Theoretical insights into the origin of signal-induced calcium oscillations

GENEVIÈVE DUPONT AND ALBERT GOLDBETER

Faculté des Sciences, Université Libre de Bruxelles, Campus Plaine, C.P. 231, B-1050 Brussels, Belgium

INTRODUCTION

In a variety of cells, stimulation by an external signal triggers a train of intracellular calcium spikes. Such oscillations have been observed upon hormonal or neurotransmitter stimulation in hepatocytes (Woods *et al.*, 1986, 1987), pituitary gonadotropes (Shangold *et al.*, 1988) and somatotropes (Holl *et al.*, 1988), muscle cells (Ambler *et al.*, 1988), fibroblasts (Harootunian *et al.*, 1989) and endothelial cells (Jacob *et al.*, 1988). The phenomenon is also observed in eggs upon fertilization (Cuthbertson and Cobbold, 1985; Miyazaki *et al.*, 1986; Miyazaki, 1988). Depending on the type of cell and on the nature of the signal, the oscillations of calcium have a variable period which generally ranges from 1 to 60 s (see the recent reviews by Berridge and Galione, 1988 and Berridge *et al.*, 1988, as well as the chapters by Cuthbertson and Berridge in this volume). The ubiquity of calcium oscillations indicates that they mediate the transduction of extracellular stimuli in a large variety of cells. Here we consider how theoretical insights may be gained into the origin of signal-induced calcium transients, and propose a minimal model for the generation of sustained calcium oscillations.

Broadly speaking, the oscillations may originate from one of the following mechanisms (Berridge *et al.*, 1988; Berridge and Galione, 1988; Jacob et al., 1988; Berridge and Cuthbertson in this volume): (i) A membrane oscillator, based on voltage-dependent conductances, as in nerve and muscle cells endowed with pacemaker properties; (ii) a cytosolic calcium oscillator; (iii) a biochemical oscillatory process driving the periodic evolution of cytosolic calcium. The second hypothesis appears to be the most likely (Berridge *et al.*,

1988; Berridge and Galione, 1988). Several processes might contribute to the onset of cytosolic calcium oscillations. First, the phenomenon could originate from an interaction between cytosolic calcium and inositol(1,4,5)-trisphosphate (Ins(1,4,5)P$_3$), given that the latter compound triggers the release of calcium from intracellular stores (Berridge and Irvine, 1984). A negative feedback of calcium on the synthesis of Ins(1,4,5)P$_3$, through protein kinase C, has been invoked as a source of oscillatory behaviour (Woods et al., 1987; Cuthbertson, in this volume). On the other hand, a model for calcium spiking has been proposed by Meyer and Stryer (1988), on the basis of an activation of Ins(1,4,5)P$_3$ synthesis by cytosolic calcium. In these models, Ins(1,4,5)P$_3$ oscillations necessarily accompany repetitive Ca^{2+} spikes.

An alternative view (Berridge et al., 1988; Berridge and Galione, 1988; Berridge, in this volume) holds that calcium oscillations might well occur in the absence of any oscillation of Ins(1,4,5)P$_3$. That such may be the case is suggested by the observation of calcium oscillations in hepatocytes (Capiod et al., 1987) and oocytes (Parker and Miledi 1986; Berridge, 1988) perfused with Ins(1,4,5)P$_3$. The mechanism responsible for the oscillations appears to be, in this case, the calcium-induced release of calcium from intracellular stores such as the endoplasmic reticulum (ER). The phenomenon of calcium-induced calcium release (CICR) was first described for skeletal muscle cells (Endo et al., 1970), and later reported for the sarcoplasmic reticulum (SR) in skinned cardiac cells (Fabiato and Fabiato, 1975; Fabiato, 1983). Calcium oscillations have been observed in the latter preparation (Fabiato and Fabiato, 1975). We analyse here the conditions for the occurrence of sustained oscillations based on the CICR mechanism.

The ubiquity of calcium oscillations raises the possibility that second messengers are frequency-encoded (Woods et al., 1987; Berridge et al., 1988; Berridge and Galione, 1988; Jacob et al., 1988; Berridge, in this volume). To investigate this question, we have used the model proposed for the generation of repetitive Ca^{2+} spikes to determine how signal-induced Ca^{2+} oscillations might be encoded in terms of their frequency through protein phosphorylation by a Ca^{2+}-activated protein kinase (Goldbeter et al., 1989).

A MINIMAL MODEL FOR CALCIUM OSCILLATIONS BASED ON THE SELF-AMPLIFIED RELEASE OF CALCIUM FROM INTRACELLULAR STORES

Can a model generate signal-induced calcium oscillations on the sole basis of the CICR mechanism? To investigate this question, we shall construct a minimal model which is schematized in Fig. 1. The model is based on the following assumptions, most of which are closely related to the mechanism proposed by Berridge (in this volume) for the oscillatory phenomenon:

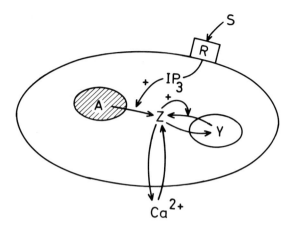

Fig. 1. Model for signal-induced Ca^{2+} oscillations based on the self-amplified release of calcium from intracellular stores. The external signal (S) binds to a membrane receptor (R) and thereby triggers the synthesis of $Ins(1,4,5)P_3$ (IP_3); the latter messenger elicits the release of Ca^{2+} from an $Ins(1,4,5)P_3$-sensitive store (hatched domain) whose Ca^{2+} content (A) is assumed to produce then a constant, net influx of cytosolic $Ca^{2+}(Z)$. The latter is pumped into an $Ins(1,4,5)P_3$-insensitive store; Ca^{2+} in this store (Y) is transported into the cytosol, in a process activated by cytosolic Ca^{2+}. Other arrows refer to the calcium influx into and extrusion from the cell.

(a) Cytosolic calcium (Z) is transported through the plasma membrane from the extracellular medium at a constant rate v_0, and extruded at a rate kZ.

(b) The external signal triggers the synthesis of $Ins(1,4,5)P_3$ which in turn induces the release of calcium from an intracellular store sensitive to $Ins(1,4,5)P_3$; this produces a net influx of cytosolic calcium, $v_1\beta$, which is proportional to the saturation function β of the receptor for $Ins(1,4,5)P_3$ on the ER or SR membrane. This source remains constant as long as the cell remains stimulated by the external signal.

(c) Cytosolic calcium is pumped into a second, $Ins(1,4,5)P_3$-insensitive store, at a rate v_2, and released from this store into the cytosol at a rate v_3; how these rates depend on the concentration of calcium in the cytosol and in the $Ins(1,4,5)P_3$-insensitive store will be specified later.

Assumption (a) merely allows one to set in a straightforward manner the basal level of cytosolic calcium as equal to the ratio v_0/k in the absence of external stimulation. Implicit in (b) is the assumption that the calcium concentration in the $Ins(1,4,5)P_3$-sensitive store remains constant; this implies that this store is replenished by extracellular calcium as quickly as it discharges into the cytosol. The term $v_1\beta$ might also be viewed as including an influx of extracellular calcium triggered by external stimulation or by a degradation product of Ins(1,4,5)P$_3$ (Putney, 1986; Hallam and Rink, 1989). We assume that the level of Ins(1,4,5)P$_3$ remains constant during stimulation, and proportional to the magnitude of the external signal.

The model is a minimal one as it contains only two variables, namely, the concentration of calcium in the cytosol (Z) and in the Ins(1,4,5)P$_3$-insensitive store (Y). The time evolution of these concentrations is governed by two kinetic equations:

$$\frac{dZ}{dt} = v_0 + v_1\beta - v_2 + v_3 - kZ$$

$$\frac{dY}{dt} = v_2 - v_3$$

(1)

In the above equations, concentrations Z and Y are both defined with respect to the total intracellular volume, as are the rates $v_i (i = 0,..,3)$ (if one writes the equation for Y in terms of the intravesicular volume, a factor, equal to the ratio of cellular to intravesicular volumes, multiplies the rates v_2 and v_3). The steady-state concentration of cytosolic calcium is given by:

$$Z_0 = (v_0 + v_1\beta)/k$$

(2)

This relation indicates that the external stimulation produces a rise in cytosolic calcium through the increased level of Ins(1,4,5)P$_3$.

Now, the question is how the external signal induces not just a rise in cytosolic calcium, but also the oscillations which are observed in a large variety of cells. The usefulness of two-variable models is that they are amenable to phase-plane analysis (Minorsky, 1967). In particular, there exists a powerful criterion due to Bendixson, which permits one to rule out the occurrence of sustained oscillations in two-variable systems. Although negative in nature, this criterion is particularly useful in cases like the present one, as it permits one to conclude, *a priori*, whether or not sustained oscillations are at all possible in a given two-variable system (Minorsky, 1967; Nicolis and Prigogine, 1977).

Applied to system (1), the Bendixson criterion states that sustained oscillations will never occur as long as the quantity B defined by eqn(3) cannot change sign:

$$B = \frac{\partial \dot{Z}}{\partial Z} + \frac{\partial \dot{Y}}{\partial Y}$$

(3)

In this equation $\dot{Z}$ and $\dot{Y}$ denote the time derivatives of Z and Y given by the kinetic eqns(1). Substituting the latter equations into (3), and taking into account the constancy of the terms v_0 and $v_1\beta$, leads to eqn(4):

$$B = -\frac{\partial v_2}{\partial Z} + \frac{\partial v_3}{\partial Z} - k + \frac{\partial v_2}{\partial Y} - \frac{\partial v_3}{\partial Y} \tag{4}$$

If the rate of release of calcium from the Y-containing store depends only on Y and not on cytosolic calcium, Z, the second term in (4) is nil, while the last one will be negative, given that the rate of release increases with Y. On the other hand, if one assumes that the rate of pumping of Z into Y only depends on the level of cytosolic calcium, the third derivative will vanish while the first will be negative. Then, quantity B will always be negative, thus ruling out sustained oscillations. One sees, therefore, that in the absence of additional time delays, the CICR from intracellular stores cannot induce oscillations in the case where this CICR process solely relies on the filling up of the store and on its discharge when a threshold level of intravesicular calcium is reached.

Given the negative nature of the first and last terms in (4), the quantity B will be capable of changing sign only if either one of the second and third derivatives is positive, i.e. if v_3 increases with Z and/or if v_2 increases with Y. The latter condition would correspond to the activation of Ca^{2+} pumping from the cytosol into the store by Ca^{2+} already sequestered in the latter compartment. Conversely, the former condition would imply the activation of the release process by calcium ions present in the cytosol. This activation is one possible form (Katz et al., 1977; Fabiato, 1985) of the phenomenon of CICR demonstrated in skeletal muscle and cardiac cells (Endo et al., 1970; Fabiato and Fabiato, 1975; Fabiato, 1983). While initially the accent was placed on the activation of the release by calcium present in the intracellular store, more recent studies appear to move the emphasis towards an activating role of cytosolic calcium in this process (Fabiato, 1985). As this form of the CICR mechanism, which is observed experimentally, provides a possible source of oscillatory behaviour according to the above analysis based on the Bendixson criterion, we shall investigate a model in which the rate v_3 of calcium release from the ER is activated by cytosolic calcium. The other possible source of oscillations, which relies on the activation of the pumping rate v_2 from the cytosol by calcium sequestered in the ER, has not been documented by experimental observations and will not be considered in the present, minimal model.

In these conditions, the rates v_2 and v_3 are given by eqns (5):

$$v_2 = V_{M2}\frac{Z^n}{(K_2{}^n + Z^n)}, \quad v_3 = V_{M3}\frac{Y^m}{(K_R{}^m + Y^m)} \cdot \frac{Z^p}{(K_A{}^p + Z^p)} \tag{5}$$

In the above equations, V_{M2} and V_{M3} denote the maximum rates of Ca^{2+}

pumping into the Ins(1,4,5)P$_3$-insensitive intracellular store and of Ca^{2+} release into the cytosol. The rates v_2 and v_3 have been written so as to allow for positive cooperativity in pumping and release, as well as in the activation of the latter process by cytosolic calcium; K_2, K_R and K_A denote the threshold constants for these processes, while n, m and p represent the Hill coefficients characterizing their degree of cooperativity (n, m, $p \geqslant 1$). The threshold constant K_R is defined as the concentration Y, with respect to the total intracellular volume; to obtain the actual value of Y and K_R in the intracellular store, one has to multiply these values by the ratio of the cellular to total storage volumes.

REPETITIVE SPIKES PRODUCED BY CICR

The steady-state concentration of sequestered calcium, Y_0, can readily be obtained by setting (dY/dt) equal to zero and solving the resulting algebraic equation, while taking into account eqns (2) and (5). A physically acceptable steady state will exist as long as the maximum release rate V_{M3} is sufficiently larger than the maximum pumping rate V_{M2}; otherwise, Ca^{2+} accumulation in the intracellular pool would occur (this situation, which is never reached when $V_{M3} \gg V_{M2}$, results from that the ATP-driven pump should revert its operation when sequestered calcium becomes too large as compared to cytosolic calcium; we shall consider, for simplicity, that the release takes place before this reversal becomes significant). On the other hand, a steady state always exists if a term $-k_f Y$ is added in the kinetic equation for Y in eqns (1), to describe a small, passive leak of Y into Z (the same term, with the opposite sign, would then appear in the equation for Z). Such a leak, whose existence is not required for oscillations, stabilizes the amplitude of the Ca^{2+} transients when the degree of stimulation increases (Goldbeter et al., 1989), a property which is observed in the experiments (Woods et al., 1987; Cuthbertson, 1989).

The stability properties of the unique steady state can be determined by linear stability analysis (Minorsky, 1967; Nicolis and Prigogine, 1977). Figure 2 shows some typical results of this analysis, as a function of parameters K_R and $(v_0 + v_1\beta)$. For low values of the threshold constant for release, K_R, the system only admits a stable steady state as the saturation function β rises upon increasing cellular stimulation. For sufficiently large values of K_R, however, the situation changes: the steady state is stable for $\beta = 0$ or at low, finite β values and corresponds then to a low Ca^{2+} level (Fig. 3, top panel); when β rises beyond a critical value, the steady state becomes unstable. Sustained oscillations in Z and Y occur in these conditions, with a period that decreases as the degree of stimulation – as reflected by β – rises (Fig. 3, middle panels). When the external signal is so strong that β exceeds a

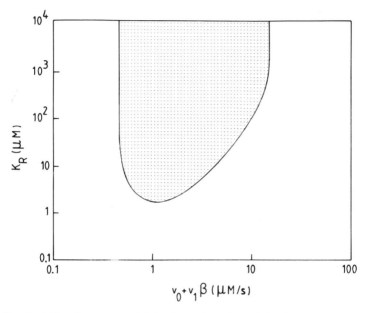

Fig. 2.   Stability diagram established as a function of the threshold constant for release, K_R, and of the total (basal plus signal-triggered) influx of cytosolic Ca^{2+}, $(v_0 + v_1\beta)$. The diagram is obtained by linear stability analysis of eqns (1) and (5) around the unique steady-state solution admitted by these equations. Parameter values are: $V_{M2} = 100\ \mu M\ s^{-1}$; $V_{M3} = 10 V_{M2}$; $m = n = p = 2$; $K_2 = 1\ \mu M$; $K_A = 2.5\ \mu M$; $k = 2\ s^{-1}$. The steady state is unstable in the dotted domain; sustained oscillations of Ca^{2+} occur in these conditions.

second, higher critical value, the steady state recovers its stability (Fig. 2) and the repetitive spikes of calcium disappear, to be replaced by a high, constant level of cytosolic calcium (Fig. 3, bottom panel).

The waveform of the oscillations predicted by the model for cytosolic calcium (Fig. 3) resembles that of the spikes observed for a number of cells stimulated by external signals. In particular, the rise in cytosolic calcium is preceded by a rapid acceleration which starts from the basal level; this pattern, which is reminiscent of the pacemaker potential that triggers autonomous spiking in nerve and muscle cells, has been observed in epithelial cells stimulated by histamine (Jacob *et al.*, 1988). As in the model of Meyer and Stryer (1988), the oscillations of calcium in the intracellular store have a sawtooth appearance (see the dashed curve in the second panel from top in Fig. 3). Here, however, the phenomenon does not require the periodic variation of $Ins(1,4,5)P_3$ whose constant level is reflected in parameter β.

The oscillations in Fig. 3 have a period of the order of a few seconds, as observed, for example, in cardiac cells (Fabiato and Fabiato, 1975) and pituitary gonadotropes (Shangold *et al.*, 1988). Such period results from the

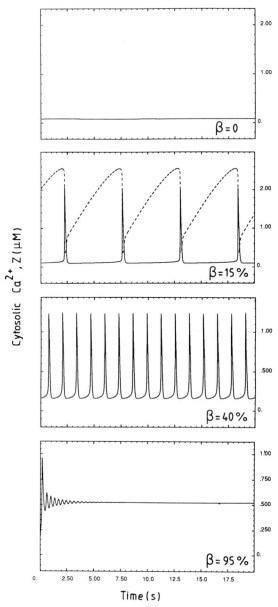

Fig. 3. Effect of an increase in stimulation on the time evolution of cytosolic calcium. The variation of cytosolic Ca^{2+} is shown for increasing values of the signal, measured by β. A stable steady state characterized by a low and a large value of cytosolic Ca^{2+} is established for $\beta=0$ and $\beta=95\%$, respectively. At intermediate values, oscillations occur whose frequency increases with the magnitude of the stimulation. The time evolution of Ca^{2+} in the intracellular store (Y) is shown (dotted line) for $\beta=15\%$. The curves are obtained by numerical integration of eqns (1) and (5). Parameter values are $k=8.7\,s^{-1}$; $m=n=2$; $p=4$; $v_0=0.74\,\mu M\,s^{-1}$; $v_1=4\,\mu M\,s^{-1}$; $K_2=1\,\mu M$; $K_R=6\,\mu M$; $K_A=0.9\,\mu M$; $V_{M2}=50\,\mu M\,s^{-1}$; $V_{M3}=10^4\,\mu M\,s^{-1}$.

choice of parameter values (see legend to Fig. 3). It is not uncommon to observe variations by more than two orders of magnitude between different cell types for parameters such as the pumping rate of calcium into the SR or ER (Carafoli and Crompton, 1978). Such variation would readily produce periods of the order of 1 min, as observed for Ca^{2+} oscillations in hepatocytes (Woods *et al.*, 1986, 1987), fibroblasts (Harootunian *et al.*, 1989), and epithelial cells (Jacob *et al.*, 1988).

In the phase plane formed by concentrations Y and Z, sustained oscillatory behaviour corresponds to the evolution towards a limit cycle (Fig. 4). Also indicated in this figure are the nullclines $(dZ/dt)=0$ and $(dY/dt)=0$, whose intersection defines the steady state. As in a model for glycolytic oscillations (Venieratos and Goldbeter, 1979), it is possible to show that the steady state is unstable when the two nullclines intersect in a region of sufficiently negative slope (dY/dZ) on the Z nullcline, as in the case of Fig. 4. As indicated by eqn (2), the abscissa of the steady state in the phase plane moves from left to right when β (i.e. the stimulation) increases. This explains why a sufficiently low value of β will correspond (as in Fig. 3, top panel) to a stable steady state, located in the region of positive slope on the Z nullcline, while an increase in β will induce oscillations as soon as the slope becomes sufficiently negative (see Fig. 3, middle panels). A further rise in stimulation

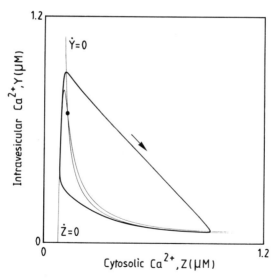

Fig. 4. Evolution towards a limit cycle in the phase plane. Sustained oscillations correspond to a rotation of the system on a closed curve in the phase plane formed by concentrations Y and Z. The nullclines $(dY/dt)=0$ and $(dZ/dt)=0$ are indicated; their intersection (dot) defines the (unstable) steady state. The curves are established for the following parameter values; $k=9.5\,s^{-1}$; $m=n=2$; $p=4$; $v_0=0.5\,\mu M\,s^{-1}$; $v_1\beta=0.6\,\mu M\,s^{-1}$; $V_{M2}=80\,\mu M\,s^{-1}$; $V_{M3}=5.10^3\,\mu M\,s^{-1}$; $K_2=1\,\mu M$, $K_R=0.5\,\mu M$; $K_A=0.9\,\mu M$.

will produce a stable steady state, corresponding to a high Z value (as in Fig. 3, bottom panel), if the abscissa given by eqn (2) is that of a point located on the Z nullcline in the region where the slope (dY/dZ) is either not sufficiently negative, or positive.

FREQUENCY ENCODING OF CALCIUM OSCILLATIONS

One likely way in which the oscillations of Ca^{2+} might produce their physiological effect is through protein phosphorylation by a Ca^{2+}-activated protein kinase. The simulations corresponding to such a situation show that the fraction of protein phosphorylated increases with Z – owing to activation of the kinase by cytosolic calcium – and decreases when Z goes to its minimum in the course of oscillations (Goldbeter et al., 1989). As predicted in the right panel of Fig. 3 in the chapter by Berridge in this volume, the average value of the fraction of phosphorylated protein increases with the frequency of Ca^{2+} oscillations because the phosphatase that reverts the effect of the kinase has less time to dephosphorylate the protein from one peak of cytosolic calcium to the next.

To allow for effective frequency-encoding, however, the phosphorylation system has to possess appropriate kinetic properties. Of particular importance in this respect are the absolute and relative rates of the kinase and phosphatase, the Michaelis constants of the two enzymes, and the value of the activation constant of the kinase by cytosolic calcium. When the kinase and phosphatase are saturated by their protein substrate, there exists a sharp threshold in the phosphorylation curve as a function of the ratio of kinase to phosphatase activity. In these conditions of 'zero-order ultrasensitivity' (Goldbeter and Koshland, 1981), the value of the mean fraction of phosphorylated protein W* varies over a much wider range (0.2–0.9) as a function of the frequency of Ca^{2+} oscillations than it does when the modifying enzymes are far from being saturated by their protein substrate (Goldbeter et al., 1989).

DISCUSSION

Owing to its relative simplicity and to the fact that it contains only two variables, the model for Ca^{2+} oscillations is certainly a crude representation of the dynamics of Ca^{2+} movements that follow stimulation of the cell by an external signal. The simplicity of the model permits us, however, to focus our attention on the manner in which repetitive Ca^{2+} spikes may arise, in the absence of $Ins(1,4,5)P_3$ oscillations, from the well-known phenomenon of calcium-induced calcium release from intracellular stores. Our analysis suggests that sustained oscillations may arise from this process if cytosolic calcium activates calcium release from the intracellular store.

The predictions of the model agree with the experimental observations (see Woods *et al.*, 1987; Berridge *et al.*, 1988; Berridge and Galione, 1988; Jacob *et al.*, 1988; Berridge, and Cuthbertson, in this volume) as to the effect of external stimulation: at low values of the signal, a stable steady state corresponding to low cytosolic Ca^{2+} is maintained; when the stimulus increases, oscillations of Ca^{2+} occur, whose frequency increases with the magnitude of the signal. At larger stimuli, a stable steady state is established, which corresponds to a high, constant level of cytosolic Ca^{2+}. On the other hand, the amplitude of the oscillations slightly decreases as the stimulation increases, while experimentally not much variation is seen (Woods *et al.*, 1987); as previously mentioned, the amplitude is further stabilized when a passive leak of Y into Z is considered in the model (Goldbeter *et al.*, 1989). The half-width of the Ca^{2+} spikes remains practically constant in the model as in the experiments.

A prediction of the model is that oscillations could spontaneously develop in the absence of stimulation ($\beta = 0$), for sufficiently large values of the basal calcium influx into the cell, v_0, or for sufficiently low values of the basal efflux measured by the rate constant k. This could explain how Ca^{2+} oscillations occur spontaneously in certain cells such as pituitary somatotropes (Holl *et al.*, 1988). The intriguing possibility arises that calcium oscillations might also develop spontaneously in other cells such as *Dictyostelium discoideum* amoebae. This would provide a plausible explanation for the existence of a second autonomous oscillator in these cells, that would be relatively independent of the mechanism generating periodic pulses of cyclic AMP (Nanjundiah and Wurster, in this volume). The fact that cyclic AMP signals induce the synthesis of $Ins(1,4,5)P_3$ (Europe-Finner and Newell, 1987) would explain how such a Ca^{2+} oscillator could be linked to cyclic AMP pulses in *Dictyostelium*. In support of these views, oscillations of extracellular Ca^{2+} have been observed in *D. discoideum* suspensions (Wurster, 1988), but intracellular Ca^{2+} oscillations have not yet been recorded.

The model presented here differs from that proposed by Meyer and Stryer (1988) in that it comprises only two variables instead of three, does not require oscillations in $Ins(1,4,5)P_3$ for the periodic generation of Ca^{2+} spikes, and takes explicitly into account the role of extracellular Ca^{2+}. On the other hand, the present model is closely related to that proposed by Kuba and Takeshita (1981) on the basis of the CICR mechanism for membrane potential oscillations associated with cytosolic calcium transients which occur in sympathetic neurones treated with caffeine. In this model, the release of calcium from intracellular stores was activated in a cooperative manner both by calcium in the ER and in the cytosol. In the present study we have extended the model of Kuba and Takeshita by taking into account explicitly the effect of the external stimulation, and specified, by means of phase-plane analysis, the minimal conditions for the occurrence of signal-induced calcium oscillations based on the calcium-induced calcium release mechanism.

The fact that complex Ca^{2+} waveforms are sometimes observed upon stimulation (Woods et al., 1987; Berridge et al., 1988, Berridge and Galione, 1988; Berridge, 1988) suggests that more than one instability-generating mechanism may be active at a given time within the cell. The interplay between two endogenous oscillatory mechanisms is indeed known to produce bursting or chaos (Decroly and Goldbeter, 1982; Martiel and Goldbeter, 1985). In these conditions, the present mechanism could operate in conjunction with one that would involve the oscillatory production of $Ins(1,4,5)P_3$ (Woods et al., 1987; Meyer and Stryer, 1988).

Finally, the analysis of a phosphorylation system coupled to Ca^{2+} oscillations via a protein kinase activated by cytosolic calcium indicates that such a reaction provides an efficient mode of frequency-encoding of the oscillatory process, provided that stringent conditions are met by the enzymes that catalyse covalent modification. Controlling the level of one or several phosphorylated proteins by modulating the frequency of calcium oscillations can be as effective, easier and more precise than controlling it through the modulation of a constant level of intracellular calcium, particularly in cases where a sharp threshold exists in the phosphorylation of the target proteins as a function of the cytosolic Ca^{2+} concentration.

Acknowledgements: We are grateful to Dr M. J. Berridge for fruitful discussions. This work was supported, under Convention BIO/08, by the Belgian National Incentive Program for Fundamental Research in the Life Sciences, launched by the Science Policy Programming Services of the Prime Minister's Office (SPPS).

REFERENCES

Ambler, S. K., Poenie, M., Tsien, R. Y. and Taylor, P. (1988) Agonist-stimulated oscillations and cycling of intracellular free calcium in individual cultured muscle cells. *J. Biol. Chem.* **263**, 1952–9.

Berridge, M. J. (1988) Inositol trisphosphate-induced membrane potential oscillations in *Xenopus* oocytes. *J. Physiol., Lond.* **403**, 589–99.

Berridge, M. J. and Galione, A. (1988) Cytosolic calcium oscillators. *FASEB J.* **2**, 3074–82.

Berridge, M. J. and Irvine, R. F. (1984) Inositol trisphosphate, a novel second messenger in cellular transduction. *Nature* **312**, 315–21.

Berridge, M. J., Cobbold, P. H. and Cuthbertson, K. S. R. (1988) Spatial and temporal aspects of cell signalling. *Phil. Trans. R. Soc. Lond. B* **320**, 325–43.

Busa, W. B., Ferguson, J. E., Joseph, S. K., Williamson, J. R. and Nuccitelli, R. (1985) Activation of frog (*Xenopus laevis*) eggs by inositol trisphosphate. I. Characterization of Ca^{2+} release from intracellular stores. *J. Cell Biol.* **101**, 677–82.

Capiod, T., Field, A. C., Ogden, D. C. and Sanford, C. A. (1987) Internal perfusion of guinea-pig hepatocytes with buffered Ca^{2+} or inositol 1,4,5-trisphosphate mimics noradrenaline activation of K^+ and Cl^- conductances. *FEBS Lett.* **217**, 247–52.

Carafoli, E. and Crompton, M. (1978) The regulation of intracellular calcium. *Curr. Topics Membr. Transp.* **10**, 151–216.

Cuthbertson, K. S. R. and Cobbold, P. H. (1985) Phorbol ester and sperm activate mouse oocytes by inducing sustained oscillations in cell Ca^{2+}. *Nature* **316**, 541–2.

Decroly, O. and Goldbeter, A. (1982) Birythmicity, chaos and other patterns of temporal self-organization in a multiply regulated biochemical system. *Proc. Natl. Acad. Sci. U.S.A.* **79**, 6917–21.

Endo, M., Tanaka, M. and Ogawa, Y. (1970) Calcium induced release of calcium from the sarcoplasmic reticulum of skinned skeletal muscle fibers. *Nature* **228**, 34–6.

Europe-Finner, G. N. and Newell, P. (1987) Cyclic AMP stimulates accumulation of inositol trisphosphate in *Dictyostelium*. *J. Cell Sci.* **87**, 221–9.

Fabiato, A. (1983) Calcium-induced release of calcium from the cardiac sarcoplasmic reticulum. *Am. J. Physiol.* **245**, C1–C14.

Fabiato, A. (1985) Simulated calcium current can both cause calcium loading in and trigger calcium release from the sarcoplasmic reticulum of a skinned canine cardiac Purkinje cell. *J. Gen. Physiol.* **85**, 291–320.

Fabiato, A. and Fabiato, F. (1975) Contractions induced by a calcium-triggered release of calcium from the sarcoplasmic reticulum of single skinned cardiac cells. *J. Physiol., Lond.* **249**, 469–95.

Goldbeter, A. and Koshland, D. E. Jr. (1981) An amplified sensitivity arising from covalent modification in biological systems. *Proc. Natl. Acad. Sci. U.S.A.* **78**, 6840–4.

Goldbeter, A., Dupont, G. and Berridge, M. J. (1989) A model for signal-induced Ca^{2+} oscillations and for their frequency encoding through protein phosphorylation. (Submitted for publication).

Hallam, T. R. and Rink, T. J. (1989) Receptor-mediated Ca^{2+} entry: diversity of function and mechanism. *Trends Biochem. Sci.* **10**, 8–10.

Harootunian, A. T., Kao, J. P. Y. and Tsien, R. Y. (1989) Agonist-induced oscillations in depolarized fibroblasts and their manipulation by photoreleased Ins(1,4,5)P_3, Ca^{2+}, and Ca^{2+} buffer. In: *Molecular Biology of Signal Transduction. Cold Spring Harbor Symp. Quant. Biol.* LIII, 935–43.

Holl, R. W., Thorner, M. O., Mandell, G. L., Sullivan, J. A., Sinha, Y. N. and Leong, D. A. (1988) Spontaneous oscillations of intracellular calcium and growth hormone secretion. *J. Biol. Chem.* **263**, 9682–5.

Jacob, R., Merritt, J. E., Hallam, T. J. and Rink, T. J. (1988) Repetitive spikes in cytoplasmic calcium evoked by histamine in human endothelial cells. *Nature* **335**, 40–5.

Katz, A. M., Repke, D. I., Dunnett, J. and Hasselbach, W. (1977) Dependence of calcium permeability of sarcoplasmic reticulum vesicles on external and internal calcium ion concentrations. *J. Biol. Chem.* **252**, 1950–6.

Kuba, K. and Takeshita, S. (1981) Simulation of intracellular Ca^{2+} oscillations in a sympathetic neurone. *J. Theor. Biol.* **93**, 1009–31.

Martiel, J. L. and Goldbeter, A. (1985) Autonomous chaotic behaviour of the slime mould *Dictyostelium discoideum* predicted by a model for cyclic AMP signalling. *Nature* **313**, 590–2.

Meyer, T. and Stryer, L. (1988) Molecular model for receptor-stimulated calcium spiking. *Proc. Natl. Acad. Sci. U.S.A.* **85**, 5051–5.

Minorsky, N. (1967) *Nonlinear Oscillations*. Van Nostrand, Princeton, NJ.

Miyazaki, S.-I. (1988) Inositol 1,4,5-trisphosphate-induced calcium release and guanine nucleotide-binding protein-mediated periodic calcium rises in golden hamster eggs. *J. Cell Biol.* **106,** 345–53.

Miyazaki, S.-I., Hashimoto, N., Yoshimoto, Y., Kishimoto, T., Igusa, Y., and Hiramoto, Y. (1986) Temporal and spatial dynamics of the periodic increase in intracellular free calcium at fertilization of golden hamster eggs. *Devl Biol* **118,** 259–67.

Nicolis, G. and Prigogine, I. (1977) *Self-Organization in Nonequilibrium Systems.* Wiley, New York.

Orchard, C. H., Eisner, D. A. and Allen, D. G. (1983) Oscillations in intracellular Ca^{2+} in mammalian cardiac muscle. *Nature* **304,** 735–8.

Parker, I. and Miledi, R. (1986) Changes in intracellular calcium and in membrane currents evoked by injection of inositol trisphosphate into *Xenopus* oocytes. *Proc. R. Soc. Lond. B* **228,** 307–15.

Putney, J. W., Jr. (1986) A model for receptor-regulated calcium entry. *Cell Calcium* **7,** 1–12.

Shangold, G. A., Murphy, S. N. and Miller, R. J. (1988) Gonadotropin-releasing hormone-induced Ca^{2+} transients in single identified gonadotropes requires both intracellular Ca^{2+} mobilization and Ca^{2+} influx. *Proc. Natl. Acad. Sci. U.S.A.* **85,** 6566–70.

Venieratos, D. and Goldbeter, A. (1979) Allosteric oscillatory enzymes: influence of the number of protomers on metabolic periodicities. *Biochimie* **61,** 1247–56.

Woods, N. M., Cuthbertson, S. K. R. and Cobbold, P. H. (1986) Repetitive transient rises in cytoplasmic free calcium in hormone-stimulated hepatocytes. *Nature* **319,** 600–2.

Woods, N. M., Cuthbertson, S. K. R. and Cobbold, P. H. (1987) Agonist-induced oscillations in cytoplasmic free calcium concentration in single rat hepatocytes. *Cell Calcium* **8,** 79–100.

Wurster, B. (1988) Periodic cell communication in *Dictyostelium discoideum.* In: *From Chemical to Biological Organization* (M. Markus, S. Müller and G. Nicolis, eds.) pp. 255–60. Springer, Berlin.

Part 6
Intercellular communication in *Dictyostelium*

Reversible phosphorylation of G-protein-coupled receptors controls cAMP oscillations in *Dictyostelium*

ROBERT E. GUNDERSEN, RON JOHNSON, PAMELA
LILLY, GEOFF PITT, MAUREEN PUPILLO, TZELI SUN,
ROXANNE VAUGHAN AND PETER N. DEVREOTES

*Department of Biological Chemistry, The Johns Hopkins
University School of Medicine, Baltimore, Maryland 21205, USA*

INTRODUCTION

Dictyostelium discoideum live as single amoebae which seek out and phagocytize bacteria. The bacteria are located by chemotaxis to secreted metabolites, such as folic acid (Pan *et al.*, 1972). Upon exhaustion of the food supply, growth ceases and a developmental programme is initiated. The developmental programme can be divided into several steps: (1) Aggregation: up to a million cells spontaneously migrate to an aggregation centre. (2) Pattern formation: the cells amass to form the pseudoplasmodium where development and segregation of the two cell types becomes apparent. (3) Morphogenesis: the multicellular structure undergoes a series of shape changes resulting in the formation of a fruiting body, composed of stalk cells and spore cells. Under the appropriate conditions the spores will germinate into single amoebae and the cycle will repeat itself (Bonner, 1982).

The spontaneous aggregation depends critically on a chemotactic response of the cells. For the species *D. lacteum* and *D. minutum*, small numbers of cells aggregate via chemotaxis up a stable gradient of attractants secreted by cells within the aggregation centre. A more intricate process has evolved in *D. discoideum* which results in the attraction of a larger number of cells. Cyclic-AMP, the chemoattractant during the aggregation phase, is first secreted by cells within the future aggregation centre. Instead of a simple linear gradient of attractant, however, concentric or spiral waves of cAMP are formed which move outward from a central point (Fig. 1A; Tomchik and Devreotes, 1981).

Cell to Cell Signalling: From
Experiments to Theoretical Models
ISBN 0–12–287960–0

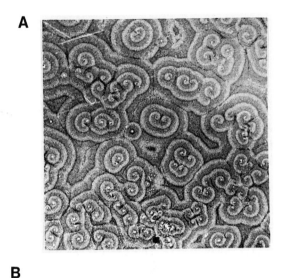

Fig. 1. cAMP-mediated aggregation in *Dictyostelium*. (A) Aggregating cells at 5 h in development. Dark-field photography reveals 1–2 cm diameter territories each containing ∼1 million cells. The spiral or circular pattern of cells results from the waves of cAMP which move out from the territory centre. (B) A signal transduction model illustrating the process of cAMP signalling. Binding of cAMP to the receptor (R) which is coupled to a G-protein (G) activates the G-protein and leads to stimulation of adenylate cyclase (AC). Prolonged exposure of the receptor to cAMP leads to adaptation through receptor phosphorylation (D). Extracellular cAMP-phosphodiesterases (PDE) break down the cAMP signal allowing dephosphorylation of the receptor and return to a cAMP-sensitive state.

Behaviourally, the cells respond to the approaching cAMP waves by chemotaxis (Konijn, 1970) and by producing and secreting additional cAMP (Roos *et al.*, 1975; Shaffer, 1975). It is this relay of the cAMP signal that produces the waves of cAMP. The waves are unidirectional, moving outward from the centre, due to an adaptive behaviour of the cells. When persistently stimulated by cAMP, the cells adapt within minutes and become non-responsive to additional cAMP; thus cells move toward the approaching cAMP but do not follow the wave as it passes. The cells de-adapt within a few minutes after the cAMP wave has passed. This adaptation/de-adaptation cycle allows the cells to create waves of cAMP which reach cells several centimetres away and induce their migration into the aggregation centre.

The biochemistry of the signal transduction system involved in the cAMP signalling response in *D. discoideum* is remarkably similar to that involved in the β-adrenergic stimulation of adenylate cyclase in higher organisms. Receptors in the plasma membrane bind extracellular ligands and activate receptor-coupled G-proteins which in turn leads to stimulation of adenylate cyclase (Fig. 1B). In *D. discoideum*, the increased cytoplasmic cAMP is secreted and can feed back on the cell for additional stimulation or diffuse to adjacent cells. Prolonged exposure of the receptors to the ligand causes adaptation, a process which appears to occur through phosphorylation of the receptor. This results in the uncoupling of the pathway and a return to the prestimulus level of adenylate cyclase activity. In *D. discoideum*, extracellular phosphodiesterase activity also contributes to signal termination by degrading excess cAMP. Once the signal has been removed, the receptors are dephosphorylated and return to their active state, ready to respond to the next stimulus. In *D. discoideum*, this process occurs spontaneously, triggering an oscillation with a periodicity of 6–7 min.

The oscillatory nature of cAMP signalling in *Dictyostelium* has allowed formulation of a model for oscillatory behaviours in nature (Martiel and Goldbeter, 1987). The model is based on the positive feedback of cAMP and receptor desensitization (see Fig. 1B). The model has recently been expanded upon to describe the formation of the cAMP wave pattern seen during cell aggregation shown in Fig. 1A (Tyson *et al.*, 1989). Pulsatile stimulation is an important factor in regulating the expression of several developmental genes in *D. discoideum* (Chisholm *et al.*, 1984), apparently involving a second signal transduction pathway mediated by the cAMP receptor. In an effort to understand the mechanisms of intercellular communication at work in *D. discoideum* during cell aggregation, we are systematically identifying and characterizing the components of the signal transduction pathways.

THE cAMP RECEPTOR

The first component in the signalling pathway is the surface cAMP receptor. The pharmacological properties of the cAMP receptor of *D. discoideum* have been extensively characterized through binding assays using radiolabelled cAMP and cAMP analogues. It has been shown to be distinct from other cAMP-binding proteins such as the regulatory subunit of cAMP-dependent protein kinase and cAMP phosphodiesterase (Van Haastert and Klein,

1983). The pharmacological data also suggest that the measured binding sites mediate the physiological responses to cAMP in *D. discoideum*.

The cAMP receptor was first visualized by photoaffinity labelling with 8-N_3-[^{32}P]cAMP (Theibert *et al.*, 1984). It appeared as a doublet on SDS–PAGE with an apparent molecular weight of 40–43 kDa. The labelled doublet displayed the same developmental regulation as the surface receptor and labelling was inhibited by cAMP and cAMP analogues in the correct order of potency.

The two receptor forms, designated R (MW = 40 kDa) and D (MW = 43 kDa) are known to be interconvertible (Klein *et al.*, 1985a). If cells undergoing cAMP oscillations are sampled at 1 min intervals and then photoaffinity labelled, the receptor is seen to oscillate between the R and D forms (Fig. 2). Cells isolated just prior to active cAMP synthesis contain predominantly the R form, but during the period of cAMP synthesis and secretion the R form decreases and there is a concomitant increase in the D form. Experiments in which the cAMP levels are controlled show that the

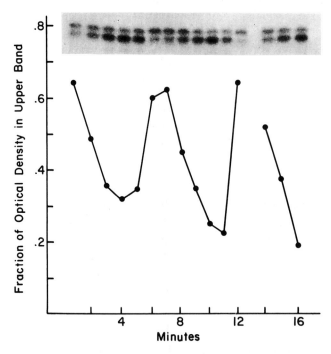

Fig. 2. Spontaneous cAMP receptor oscillations. Cells developed for 3.5 h were sampled at 1 min intervals (sample 13 was lost) during spontaneous oscillations and were photolabelled with 8-N_3-[^{32}P]cAMP. Membrane samples were separated by SDS–PAGE and autoradiographed. The labelled bands were scanned and the optical density of the upper band was plotted as a fraction of the total optical density in the doublet.

shift in electrophoretic mobility is dependent on extracellular cAMP. When the spontaneous oscillations are blocked, the R form predominates. Upon the addition of 1 μM cAMP, approximately 80% of the receptor shifts to the D form within 15 min. Removal of the stimulus allows for a transition back to the R form. Shifts between R and D forms are also observed at subsaturating levels of cAMP and the steady-state distribution of R and D forms is dependent on ambient concentration of extracellular cAMP (Devreotes and Sherring, 1985).

Current evidence suggests that the shift in electrophoretic mobility is due to phosphorylation of the receptor on serine and threonine residues (Klein *et al.*, 1985b). The R form is phosphorylated independently of cAMP, while the cAMP-induced conversion of the R form to the D form results in at least a four-fold increase in the level of phosphorylation. The increased level of receptor phosphorylation is dependent on receptor occupation by cAMP and not on the intracellular level of cAMP. Inhibition of the increase in intracellular cAMP has no effect on the kinetics or level of receptor phosphorylation, suggesting that cAMP-dependent protein kinase is not involved. The amount of steady-state receptor phosphorylation is also dependent on the concentration of extracellular cAMP. Both phosphorylation and the fraction of R form converted to D form are half maximal at 5 nM cAMP and saturating at 100 nM cAMP (Vaughan and Devreotes, 1988).

Ligand-induced phosphorylation of the receptor shows a biphasic response (Fig. 3A). At first, the receptor is rapidly phosphorylated. Incorporated $^{32}P_i$ can be detected within 5 s of addition of the stimulus. The fast response accounts for the phosphorylation of 60–70% of the total available sites. Maximum levels of phosphorylation are obtained at a slower rate requiring approximately 5 min for completion. The calculated half-time for complete receptor phosphorylation is about 45 s (Vaughan and Devreotes, 1988).

When the cAMP stimulus is removed, the receptor is dephosphorylated at a rate which is independent of the level of phosphorylation. Dephosphorylation is also a biphasic event (Fig. 3B). A rapid decrease in radioactivity occurs over 3 min, initially being detected by 30 s, and the remainder gradually declines over 10 min. The half-time for maximal dephosphorylation is 2 min. The rates of dephosphorylation were corrected for small losses in the specific activity of the $AT^{32}P$ pool, caused by the washing steps used to remove the cAMP stimulus. When the receptor is held in the D form by cAMP, phosphate turnover on the receptor occurs at about one-fifth the initial stimulus-induced rate. A similar rate of phosphate turnover occurs on the stimulus-independent phosphorylation site(s) of the R form (Vaughan and Devreotes, 1988).

The kinetics and dose response of receptor shift were analysed by immunoblotting and were found to be similar, but not identical, to those of phosphorylation. The similarity in kinetics between receptor phosphoryla-

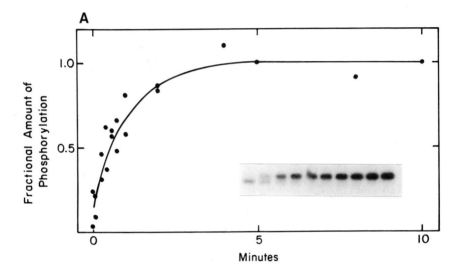

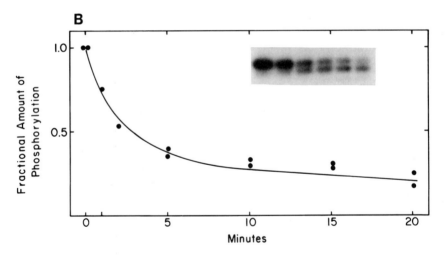

Fig. 3. Cyclic-AMP receptor phosphorylation and dephosphorylation. (A) The kinetics of receptor phosphorylation were followed by labelling cells with $^{32}P_i$ for 45 min and then stimulating with 10 μM cAMP. Labelled receptor was immuno-precipitated from samples and subjected to SDS–PAGE. The resulting autoradio-graph (inset) was scanned and time points were plotted as the fraction of phosphorylated D form. (B) The kinetics of receptor dephosphorylation were followed as described in A except that the cAMP stimulus was removed by washing prior to sampling. The level of dephosphorylation is plotted as loss of D form phosphorylation divided by loss of phosphorylated receptor.

tion and mobility shift strongly suggest a causal relationship; however, this remains to be confirmed. Estimates of the specific activity of the D form of the receptor indicate that not all sites are phosphorylated during the R to D conversion. When the shift occurs one-third to one-half of the available sites are phosphorylated. After the shift is completed the D form continues to be phosphorylated. Since the stimulated receptor contains several phosphates per molecule (4 ± 1), this may indicate that different sites are phosphorylated during and after the shift (Vaughan and Devreotes, 1988).

The kinetics of receptor phosphorylation and dephosphorylation also correlate with the kinetics of adaptation and de-adaptation in *Dictyostelium*. Both the stimulation of adenylate cyclase and the cAMP-induced cell shape change undergo adaptation and de-adaptation. Adaptation of these responses is first detected within 20 s of stimulation by cAMP and is completed in approximately 12 min (Dinauer *et al.*, 1980b). De-adaptation occurs with a half-time of about 4 min regardless of the level at which the receptors were stimulated (Dinauer *et al.*, 1980a). The similar kinetics strongly suggests a correlation between phosphorylation/dephosphorylation of the receptor and the adaptation/de-adaptation of receptor-mediated responses. Receptor phosphorylation has been correlated with adaptation of the β-adrenergic receptor and rhodopsin in mammals (Sibley and Lefkowitz, 1985). Definitive proof for the role for cAMP-receptor phosphorylation in adaptation is not yet available; however, recent strides in the approaches of molecular biology in *Dictyostelium* may soon allow this question to be answered.

The cAMP receptor has recently been cloned and sequenced using λgt11 expression libraries (Klein *et al.*, 1988). The deduced amino acid sequence yields a protein with a calculated molecular weight of 44,243. Analysis of the amino acid sequence for hydropathy (Kyte and Doolittle, 1982) revealed a protein with seven hydrophobic domains and a long hydrophilic C-terminus. This pattern is similar to bovine rhodopsin and the mammalian β-adrenergic receptor as well as several other recently reported G-protein-linked receptors (see Dohlman *et al.*, 1987). The proposed structure has an extracellular N-terminus followed by seven transmembrane segments which are connected by six short hydrophilic loops. The C-terminus is cytoplasmic and contains several serines and threonines as potential sites of phosphorylation.

In order to prove that the cloned cDNA encodes a cAMP-binding protein, it was expressed in *D. discoideum*. Cells were transformed with a vector which overexpresses the cDNA insert during the growth stage of development, a time when the receptor is not normally expressed. The number of cAMP binding sites was determined using standard [^{3}H]cAMP binding assays. The transformants expressed up to 35 times more binding sites than growing wild-type cells and a six-fold greater level than maximal expression of endogenous receptors in wild-type cells, which occurs after 5 h of development. Wild-type cells transformed with the parent vector showed no changes in the number or time course of expression of binding sites. A more detailed time course of receptor protein expression was also followed on immuno-blots. Wild-type and control-transformed cells show the typical pattern of expression of the receptor doublet, with the protein being absent in growing cells and peaking between 6 and 8 h of development. The overexpression transformants display large quantities of the same receptor doublet bands during growth and throughout development (Johnson and Devreotes, in preparation).

Overexpression of the receptor alone appears to be insufficient to induce development but does alter the time course of the developmental pattern by inducing a delay of 1–7 h in the onset of aggregation. The length of the delay appears to correspond with the expression level of receptor. The transfor-mants, however, do develop normally once aggregation begins (Johnson and Devreotes, in preparation). The reason for the delay is being investigated.

The most convincing evidence for confirming the primary role of the cAMP receptor in initiating the developmental cycle of *D. discoideum* comes from cells transformed with receptor antisense sequence. *In vitro* the anti-sense cDNA was shown to block translation of endogenous receptor mRNA (Klein *et al.*, 1988). A full-length receptor cDNA clone was inserted into a plasmid vector so that the non-coding strand would be transcribed. When plated on starvation plates, cells transformed with the antisense cDNA remain as flat cell monolayers, failing to aggregate for up to 24 h; whereas, wild-type cells normally complete aggregation and the formation of a fruiting body in that period of time. The absence of cAMP receptors was confirmed with [³H]cAMP binding assays and immunoblots of the transformants.

G-PROTEINS

The receptor-mediated events of certain signal transduction pathways have been shown to involve G-proteins in numerous mammalian systems. The term G-protein refers to a heterotrimer composed of an α subunit (39–45 kDa), a β subunit (35–36 kDa) and a γ subunit (8 kDa). G-proteins act to couple plasma membrane receptors to effectors. In the inactive state a molecule of GDP is bound to the α subunit, but activation of a G-protein by a ligand-occupied receptor allows a molecule of GTP to replace the GDP. The activated α subunit then dissociates from the βγ complex and is free to interact with an effector. The α subunit is a GTPase and hydrolysis of the bound GTP results in its inactivation and subsequent rebinding to a βγ complex.

The involvement of G-proteins in the pathway of cAMP-mediated stimu-lation of adenylate cyclase in *Dictyostelium* has been implied by a number of observations. First, adenylate cyclase is stimulated by the addition of GTP or GTPγS, a non-hydrolysable GTP analogue (Theibert and Devreotes, 1986). In addition, cAMP increases the level of GTP hydrolysis, and the addition of

GTP or GDP to receptor binding assays modulates cAMP binding affinity. This information is analogous to other mammalian receptors coupled to G-proteins.

The existence of G-proteins in *Dictyostelium* has recently been confirmed by obtaining cDNA sequences for three G-protein subunits. Two α subunits, $G_\alpha 1$ and $G_\alpha 2$ have been sequenced and their deduced amino acid sequences are 45% identical to one another. In comparison to other G_α subunits, they are 45% identical to mammalian $G_{\alpha i}$, $G_{\alpha o}$ and transducin but are only 35% identical to $G_{\alpha s}$ and yeast GPA1 (Pupillo *et al.*, 1989). The most striking homology present in all the G-protein α subunits, including those from *Dictyostelium* is believed to be the GTP binding and hydrolysis sites. $G_\alpha 1$ is composed of 356 amino acids with a calculated molecular weight of 40,621. $G_\alpha 2$ is made up of 357 amino acids and has a molecular weight of 41,323 (Pupillo *et al.*, 1989). The third G-protein cDNA obtained from *Dictyostelium* is a G_β sequence, which is 63% identical to the human β subunit. *Dictyostelium* G_β has 347 amino acids with a predicted molecular weight of 38,579 (Lilly and Devreotes, 1989).

To better understand the function of the three G-protein subunits, their expression was followed through development. For G_β, expression of a single 1.9 kb mRNA was observed on Northern blots; it is present during growth and throughout development. The G_α subunits, on the other hand, have varied expression patterns. $G_\alpha 1$ cDNA hybridizes to three mRNA species on Northern blots. The major band is 1.7 kb and is present in vegetative and developing cells, with a slight increase during aggregation. The $G_\alpha 2$ subunit has a complex pattern of mRNA expression which is developmentally regulated. $G_\alpha 2$ mRNA does not appear in growing cells, while in developing cells a variety of messages are present. The major mRNA is 2.7 kb and peaks during aggregation between 2 and 10 h. A 2.9 kb message is seen briefly early in development and two additional mRNAs of 1.9 and 2.3 kb appear sequentially late in development. The significance of the various mRNAs for the two G_α subunits is unknown at this time.

$G_\alpha 1$ and $G_\alpha 2$ proteins were followed on immunoblots using peptide antibodies generated against each G_α subunit (Fig. 4; Kumagai *et al.*, 1989). $G_\alpha 1$ appears as a 38 kDa protein which is present in growing and developing cells. The abundance of $G_\alpha 1$, however, appears to decrease over 6 h of development in contrast to the Northern blot data. The $G_\alpha 2$ protein has an apparent MW of 40 kDa on immunoblots and increases in amount over 2–8 h after which it declines. The difference in apparent MW of $G_\alpha 1$ and $G_\alpha 2$ is evident when the two antibodies are mixed (Fig. 4, lanes 9 and 10).

Having established the existence of G-proteins in *D. discoideum*, the next step is to identify the pathways in which each of the G_α subunits functions. Determining the function of $G_\alpha 2$ was aided by a developmental mutation, designated *Frigid A*. *Frigid A* mutants fail to initiate the developmental pathway upon starvation (Coukell *et al.*, 1983). Biochemical analysis of

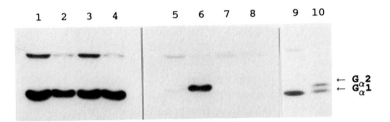

Fig. 4. G-Proteins of *Dictyostelium*, $G_\alpha 1$ and $G_\alpha 2$, as seen by immunoblotting. All samples are cell lysis pellets corresponding to $\sim 3 \times 10^6$ cells. Lanes 1–4 were stained for $G_\alpha 1$ and lanes 5–8 were stained for $G_\alpha 2$. Lanes 9 and 10 contained a mixture of the two antibodies to emphasize the difference in apparent MW. Wild type cells, vegetative (lanes 1, 5 and 9) and 6 h developed (lanes 2, 6 and 10) are compared with *frigid A* (allele HC85) cells, vegetative (lanes 3 and 7) and developed with exogenous cAMP pulsing (lanes 4 and 8). The band above $G_\alpha 1$ and $G_\alpha 2$ of varying intensity is the result of non-specific staining.

Frigid A (allele HC85) showed that cAMP failed to activate either adenylate or guanylate cyclases *in vivo*. *In vitro*, GTP inhibition of cAMP binding is greatly reduced yet GTP activation of adenylate cyclase is normal (Kesbeke *et al.*, 1989). Southern blots of the strain HC85 showed a large deletion in the $G_\alpha 2$ gene. Likewise, immunoblots on HC85 failed to detect any $G_\alpha 2$ protein, while $G_\alpha 1$ appears normal (Fig. 3). These results strongly suggest that the *Frigid A* alleles are $G_\alpha 2$ mutants and that $G_\alpha 2$ is linked to cAMP receptor-mediated events in development. At present no role has been assigned to the $G_\alpha 1$ subunit.

Additional components in the pathway of cAMP-mediated stimulation of adenylate cyclase are being revealed with *synag* mutants. *Synag* mutants fail to aggregate alone but will proceed through development if synergized with wild-type cells. These mutants behave as wild-type cells in the expression of the cAMP receptor and cAMP stimulation of guanylate cyclase if pulsed with cAMP. However, there is no GTP activation of adenylate cyclase and thus, without cAMP signalling, the cells fail to aggregate. In one of the *synag* mutants, *synag 7*, GTP activation of adenylate cyclase can be completely restored by the addition of a high-speed supernatant fraction from wild-type cells. The reconstituting activity in wild-type supernatants is a protein that has been named GRP (GTP-activating reconstitution protein) and appears to be required for GTP activation of adenylate cyclase in *Dictyostelium* (Theibert and Devreotes, 1986). *Synag 7* appears to lack GRP and efforts are currently underway to isolate GRP from wild-type supernatants using an adenylate cyclase assay with *synag 7* membranes.

CONCLUSION

The cAMP-mediated developmental cycle of *Dictyostelium discoideum* has proven to be an excellent system for studying signal transduction mechanisms. The oscillatory nature of the cAMP-signalling pathway arises from adaptation/de-adaptation of the cAMP receptor, where the underlying mechanism appears to be ligand-regulated receptor phosphorylation/dephosphorylation. Cloning of the cAMP receptor should now allow definitive confirmation of this relationship. By constructing receptors with deletions or substitutions, receptor phosphorylation sites can be removed or modified. Cells transformed to express the altered receptors should then allow these questions to be answered.

The importance of G-proteins and an additional component in the pathway have been demonstrated in *Dictyostelium* using the *Frigid A* and *synag* mutants. A variety of approaches are being employed to obtain a better understanding of the physiological functions of these proteins. For example, the cDNA sequences of the G-protein subunits are being used for overexpression or to 'knock out' the gene through either homologous recombination or antisense transformation. Following selection on neomycin, the resulting transformants can be quickly screened with antibodies directed against the protein of interest, $G_\alpha 1$, $G_\alpha 2$ or G_β. The transformants can then be observed for any changes in development and can be analysed for specific biochemical changes.

The developmental cycle of *Dictyostelium* is a finely tuned mechanism with precise dependence among the component parts. A greater insight into this signal transduction pathway should evolve from studies of the cAMP receptor, G-proteins and additional members of the pathway.

REFERENCES

Bonner, J. (1982) Comparative biology of cellular slime molds. In *The Development of* Dictyostelium discoideum (ed. W. F. Loomis), pp. 1–33. Academic Press, New York.

Chisholm, R., Fontana, D., Theibert, A., Lodish, H. F. and Devreotes, P. N. (1984) Development of *Dictyostelium discoideum*: chemotaxis, cell–cell adhesion and gene expression, In *Microbial Development* (ed. R. Losick and L. Shapiro), pp. 219–54. Cold Spring Harbor Laboratory, Cold Spring Harbor, New York.

Coukell, M. B., Lappano, S. and Cameron, A. M. (1983) Isolation and characterization of cAMP unresponsive (frigid) aggregation-deficient mutant of *Dictyostelium discoideum*. *Devl. Genet.* **3**, 283–97.

Devreotes, P. N. and Sherring, J. (1985) Kinetics and concentration dependence of reversible cAMP-induced modification of the surface cAMP receptor in *Dictyostelium*. *J. biol. Chem.* **260**, 6378–84.

Dinauer, M. C., Steck, T. L. and Devreotes, P. N. (1980a) Cyclic 3′,5′-AMP relay in *Dictyostelium discoideum*. IV. Recovery of the cAMP signaling response after adaptation to cAMP. *J. Cell Biol.* **86**, 545–53.

Dinauer, M. C., Steck, T. L. and Devreotes, P. N. (1980b) Cyclic 3′,5′-AMP relay in *Dictyostelium discoideum*. V. Adaptation of cAMP signaling response during cAMP stimulation. *J. Cell Biol.* **86**, 554–61.

Dohlman, H. G., Caron, M. G. and Lefkowitz, R. J. (1987) A family of receptors coupled to guanine nucleotide regulatory proteins. *Biochemistry* **26**, 2657–64.

Kesbeke, F., Snaar-Jagalska, B. E. and Van Haastert, P. J. M. (1989) Signal transduction in *Dictyostelium* Frd A mutants with a defective interaction between surface cAMP receptor and a GTP-binding regulatory protein. *J. Cell Biol.* **107**, 521–8.

Klein, P., Theibert, A., Fontana, D. and Devreotes, P. N. (1985a) Identification and cyclic AMP-induced modification of the cyclic AMP receptor in *Dictyostelium discoideum. J. biol. Chem.* **260**, 1757–63.

Klein, P., Fontana, D., Knox, B., Theibert, A. and Devreotes, P. N. (1985b) cAMP receptors controlling cell–cell interaction in the development of *Dictyostelium.* In *Cold Spring Harbor Symposia on Quantitative Biology*, Vol. 50, pp. 787–99.

Klein, P., Vaughan, R., Borleis, J. and Devreotes, P. N. (1987) The surface cyclic-AMP receptor in *Dictyostelium. J. biol. Chem.* **262**, 358–64.

Klein, P., Sun, T. J., Saxe, C. L., III, Kimmel, A. R., Johnson, R. L. and Devreotes, P. N. (1988) A chemoattractant receptor controls development in *Dictyostelium discoideum. Science, N.Y.* **241**, 1467–72.

Konijn, T. (1970) Microbial assay of cyclic $3',5'$-AMP. *Experientia* **26**, 367–9.

Kumagai, A., Pupillo, M., Gundersen, R., Miake-Lye, R., Devreotes, P. N. and Firtel, R. A. (1989) Regulation and function of G_α protein subunits in *Dictyostelium. Cell* **57**, 265–75.

Kyte, J. and Doolittle, R. F. (1982) A simple method for displaying the hydropathic character of a protein. *J. molec. Biol.* **157**, 105–32.

Martiel, J. -L. and Goldbeter, A. (1987) A model based on receptor desensitization for cyclic AMP signaling in *Dictyostelium* cells. *Biophys. J.* **52**, 807–28.

Pan, P., Hall, E. M. and Bonner, J. T. (1972) Folic acid as second chemotactic substance in the cellular slime moulds. *Nature (New Biol.)* **237**, 181–2.

Pupillo, M., Kumagai, A., Pitt, G., Firtel, R. and Devreotes, P. N. (1989) Multiple α subunits of guanine nucleotide-binding proteins in *Dictyostelium. Proc. Natl. Acad. Sci. U.S.A.* (in press).

Roos, W., Nanjundiah, V., Malchow, D. and Gerisch, G. (1975) Amplification of cAMP signals in aggregating cells of *Dictyostelium discoideum. FEBS Lett.* **53**, 139–42.

Shaffer, B. M. (1975) Secretion of cAMP induced by cAMP in the cellular slime mould *Dictyostelium discoideum. Nature* **255**, 549–52.

Sibley, D. and Lefkowitz, R. (1985) Molecular mechanisms of receptor desensitization using the β-adrenergic receptor-coupled adenylate cyclase system as a model. *Nature* **317**, 124–9.

Theibert, A. and Devreotes, P. N. (1986) Surface receptor mediated activation of adenylate cyclase in *Dictyostelium*: Regulation by guanine nucleotide in wild type cells and aggregation deficient mutants. *J. biol. Chem.* **261**, 15121–5.

Theibert, A., Klein, P. and Devreotes, P. N. (1984) Specific photoaffinity labeling of the cAMP surface receptor in *Dictyostelium discoideum. J. biol. Chem.* **259**, 12318–21.

Tomchik, K. J. and Devreotes, P. N. (1981) cAMP waves in *Dictyostelium discoideum* by a novel isotope dilution fluorography technique. *Science, N.Y.* **212**, 443–6.

Tyson, J. L., Alexander, K. A., Manoranjan, V. S. and Murray, J. D. (1989) Cyclic-AMP waves during aggregation of *Dictyostelium* amoebae. *Development* (in press).

Van Haastert, P. J. M. and Klein, E. (1983) Binding of cAMP derivatives to *Dictyostelium discoideum* cells. *J. biol. Chem.* **258**, 9636–42.

Vaughan, R. and Devreotes, P. N. (1988) Ligand-induced phosphorylation of the cAMP receptor from *Dictyostelium discoideum. J. biol. Chem.* **263**, 14538–43.

Is there a cyclic-AMP-independent oscillator in *Dictyostelium discoideum*?

VIDYANAND NANJUNDIAH[a] AND BERND WURSTER

Fakultät für Biologie, Universität Konstanz, 7750 Konstanz, FRG
[a]Permanent address: Departments of Microbiology and Cell Biology and Centre for Theoretical Studies, Indian Institute of Science, Bangalore 560012, India.

INTRODUCTION

Cyclic-AMP oscillations

Anywhere from 10^3 to 10^5 starved amoebae of *Dictyostelium discoideum* can gather together at common collecting points by performing a series of coordinated oscillatory movement steps (Bonner, 1967; Gerisch, 1968). These oscillations are believed to be due to the periodic synthesis and release of a chemoattractant, cyclic-AMP (cAMP), by one or a few amoebae. Other amoebae respond to the cAMP by both moving towards local sources and amplifying and relaying the cAMP signal (Konijn *et al.*, 1967; Roos *et al.*, 1975; Shaffer, 1975). In accordance with this, waves of cAMP can be demonstrated in aggregating fields of *D. discoideum* (Tomchik and Devreotes, 1981). Apart from being of interest for the light they throw on regulatory mechanisms, cAMP oscillations in *D. discoideum* are of functional importance. Calculations show that for a given average rate of release of cAMP, pulsed emission results in a longer signal range than steady release (Nanjundiah, 1973). Further, periodic – but not continuous – application of cAMP can induce differentiation (Darmon *et al.*, 1975; Gerisch *et al.*, 1975); in this respect periodic stimuli are also more effective than aperiodic stimuli (Nanjundiah, 1989). In short, cAMP oscillations both exist and serve a purpose; that of inducing morphogenesis and differentiation.

Cell to Cell Signalling: From
Experiments to Theoretical Models
ISBN 0–12–287960–0

Other periodic activities

A number of variables have been observed to oscillate in well-stirred suspensions of starved *D. discoideum* cells. Roughly in the order these variables were identified, they are: intensity of light scattering (Gerisch and Hess, 1974), redox state of cytochrome-b (Gerisch and Hess, 1974), intra- and extracellular cAMP (Gerisch and Wick, 1975), intracellular cGMP (Wurster *et al.*, 1977), extracellular pH (Malchow *et al.*, 1978a), extracellular CO_2 (Gerisch *et al.*, 1977), extracellular Ca^{2+} (Bumann *et al.*, 1986), extracellular K^+ (Aeckerle *et al.*, 1985), and cellular agglomeration (B. Wurster, in preparation).

Are all oscillations based on a cAMP oscillator?

What is striking is that in the case of most of the oscillatory variables listed above (the test remains to be done in the case of cytochrome-b) an external application of cAMP can induce changes consistent with the measured oscillations being a passive consequence of a basic cAMP oscillator: for example, cAMP stimulation can induce transient decreases in pH (Malchow *et al.*, 1978b) or a transient uptake of Ca^{2+} (Wick *et al.*, 1978; Bumann *et al.*, 1984).

Can one then conclude that there is a single, cAMP-dependent oscillator in *D. discoideum*? The purpose of this chapter is to highlight a number of results which do not fit this hypothesis. Specifically, the phenomena of sinusoidal oscillations, oscillations at low cell densities, and oscillatory activities in a mutant, Agip43, all indicate that *D. discoideum* cells can exhibit cAMP-independent periodic activity.

SINUSOIDAL OSCILLATIONS

Description

After being starved for about 3.5 h, the intensity of scattered light from a vigorously agitated suspension of wild-type (Ax2) *D. discoideum* cells starts to display spontaneous periodic variations presumably reflecting changes in the shape or size of single cells or cell groups. These variations have a relative amplitude of 1–2% and initially show up in tracings as sharply peaked spikes of optical density (OD) minima recurring every 8–9 min. In the course of time the period of the spikes gradually decreases and concomitantly, one sees a transformation of the waveform into a smoother, sinusoidal profile; the OD minima are now about 180° out of phase with respect to those of the spikes (Figs 1 and 2). Typically, spikes are found from about 3.5 to 5 h following

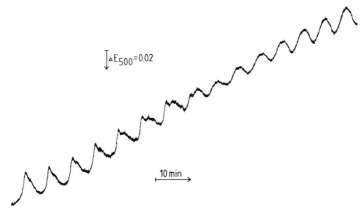

Fig. 1.   Spike-shaped and sinusoidal oscillations in light-scattering properties of a cell suspension. After starvation for 3.5 h in phosphate buffer a suspension of *D. discoideum* strain Ax2 (2 ml, 2×10^7 cells/ml) was transferred into an optical cuvette and agitated by bubbling oxygen through the suspension (Gerisch and Hess, 1974). The optical density was recorded at 500 nm. Only a section of the oscillation sequence is shown, the early spikes and the late sinusoidal oscillations were omitted.

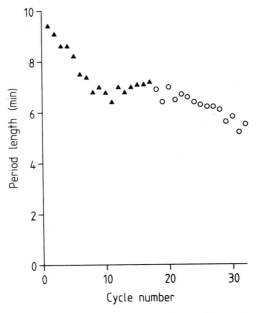

Fig. 2.   Dependence of the period length on the cycle number. Triangles and circles represent spike-shaped and sinusoidal oscillations, respectively.

starvation and sinusoids thereafter, with a transitional period intervening in between. An important point is that even though normal morphogenesis is blocked in suspension cultures the time course of differentiation to aggregation-competence is about the same as on a two-dimensional substratum (Gerisch, 1968 and unpublished observations). Since cell aggregation begins about 7 h after starvation, even after allowing for some lack of exactness in these times it is certain that when cells start to aggregate on plates the light-scattering oscillation profile in suspensions is sinusoidal and not spike-shaped (Fig. 3).

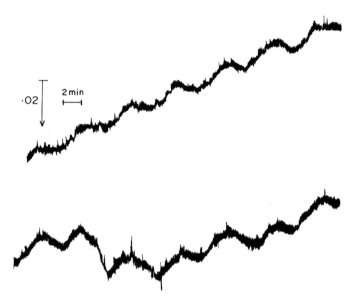

Fig. 3. A continuous tracing of oscillations monitored with cells allowed to develop on an agar plate till $t=5.8$ h, a stage when no signs of aggregation were visible, and then transferred into the spectrophotometer cuvette. The oscillations started after 1 h; aggregation had just begun on a companion plate. Judging by the absorbance, the density was about 3.5×10^7 cells/ml. Both records were made with *D. discoideum* Ax2 (clone B2). The optical density, recorded at 500 nm, decreases towards the top.

Absence of measurable cAMP changes

Spike-shaped oscillations are accompanied by intra- and extracellular cAMP oscillations of fairly large amplitude: up to a 100-fold increase in extracellular cAMP has been reported, with peak concentrations ranging from 0.1 μM to 1 μM (Gerisch and Wick, 1975). The most plausible explanation of the spikes is that they are a consequence of periodic cAMP synthesis and release.

For instance, the released cAMP could elicit light-scattering changes just as externally added cAMP does (Gerisch and Hess, 1974; Gerisch and Wick, 1975).

In contrast, cAMP levels in sinusoidally oscillating populations remain constant (Gerisch *et al.*, 1977); a typical figure for the extracellular cAMP concentration, measured with 2×10^7 cells/ml, is 2 nM. Therefore it appears that cAMP cannot be a participant in these oscillations. Extracellular Ca^{2+}

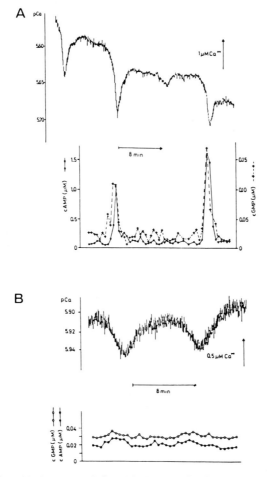

Fig. 4. Relationship between Ca^{2+}, cAMP, and cGMP oscillations. Extracellular Ca^{2+} concentrations were recorded by means of a calcium-sensitive electrode. During Ca^{2+} oscillations, 30 μl samples were withdrawn from the cell suspensions at intervals of 30 s. In these samples the total (extracellular and intracellular) cAMP content and cGMP content were determined by radioimmunoassay. The data are expressed as concentrations (μmol/l of cell suspension). (A) Spike-shaped oscillations in 1×10^8 cell/ml. (B) Sinusoidal oscillations in 2×10^8 cells/ml. (Taken from Bumann *et al.*, 1986.)

oscillations also exhibit a spike-shaped to sinusoidal transition in the course of development; sinusoidal Ca^{2+} oscillations occur at the same time as sinusoidal light-scattering oscillations. In one experiment (Bumann *et al.*, 1986) the following striking observation was made: a series of Ca^{2+} spikes was interrupted by a single Ca^{2+} oscillation of a smaller amplitude, similar to that seen during the sinusoidal cycles. Simultaneous measurements of cAMP (and cGMP) showed that cyclic nucleotide oscillations accompanied the spikes but were absent during the low-amplitude calcium oscillation (Fig. 4).

When cells are starved as agitated suspensions they tend to form clumps, and the point might be made that appreciable cAMP oscillations could exist locally, i.e. within the clumps, with average cAMP levels in the suspension as a whole always remaining very low. This is, however, highly implausible. First of all, it is difficult to imagine how a small molecular species like cAMP could be trapped between cells. Secondly, it is known that an external application of cAMP to cells of this stage can induce the synthesis of intracellular cGMP (Mato *et al.*, 1977; Wurster *et al.*, 1977; Mato and Malchow, 1978), so that cAMP oscillations would have to be accompanied by those in cGMP (as indeed they are during spikes). What one sees on the other hand is that cGMP levels too remain constant to within the accuracy of the assay. To sum up, sinusoidal oscillations seem to be based on oscillatory variables other than cAMP. A more cautious statement would be to say that if total cAMP levels change at all during sinusoidal light-scattering oscillations, the amplitude of the change must be in the range of 1–10 nM; this contrasts with the 10- to 100-fold higher changes seen during spikes.

Can cAMP influence sinusoidal oscillations?

Implications for synchronization

Undamped sinusoidal oscillations can continue with a relatively steady period for many cycles; the absence of any discernible damping implies that the oscillatory activity of different cells must be synchronized (Nanjundiah, 1986). Attempts to isolate a soluble synchronizer have failed so far; cAMP itself cannot synchronize the cells, because – in contrast to its effect on spikes – an external application of a cAMP pulse leaves sinusoidal oscillations in the same phase as before the application.

Phosphodiesterase addition

The implication that sinusoidal oscillations are not dependent on variations in extracellular cAMP is strengthened by experiments involving the addition of phosphodiesterase (PDE) to oscillating cell suspensions. Whereas up to 300 mU/ml of purified *D. discoideum* PDE (calculations based on the measured rate of hydrolysis of 20 nM [^{3}H]cAMP show that the total PDE activity translates into a cAMP half-life $t_{0.5}$ of 0.3 s in the suspension of 2×10^7 cells/ml) could be added to a sinusoidally oscillating population without stopping the oscillations, an addition of 50 mU/ml PDE ($t_{0.5} = 1.7$ s) to a population displaying spikes was sufficient to quench oscillatory activity.

Continuous cAMP addition or constant receptor occupancy

Even though cAMP does not participate in sinusoidal oscillations, it can interact with the system responsible for oscillatory activity. Continuous addition of cAMP at a rate of 0.05 nM/s can quench sinusoidal oscillations reversibly (Gerisch and Hess, 1974 and unpublished observations). In one experiment the addition resulted in a steady-state external cAMP level (5 nM) indistinguishable from the basal level prevailing before addition; in the case of spikes the same rate of addition caused a cAMP increase from 2 nM to 2.8 nM and resulted in irreversible quenching.

A possible explanatory framework

While we are yet to understand what causes sinusoidal oscillations, a few possibilities can be listed. To begin with, since the oscillations in light scattering are accompanied by those in (extracellular) H^+, Ca^{2+} and K^+, one or more of these ions could be participating in a cAMP-independent oscillatory loop. At the same time (or alternatively), given the failure of attempts to isolate a soluble synchronizer, and given that the size of cell agglomerates oscillates (B. Wurster, in preparation), one can envisage the following novel possibility. Namely, cell–cell contacts could be responsible for activating the synthesis of an intracellular intermediate which in turn would promote contact formation: a situation predisposing the system to oscillatory behaviour via positive feedback (Higgins, 1967).

Irrespective of the molecular basis of sinusoidal oscillations, the effects of continuous cAMP addition make it clear that cAMP can perturb them, presumably via the cAMP receptor. One way to account for this perturbation would be to say that spikes (which depend on cAMP) and sinusoids share one or more component(s) in common. The level of one such component – 'X' – could be linked to the state of the receptor with the linkage functioning as follows. A steady increase in extracellular cAMP, leading to a concentration plateau, would lead to an increase and eventual constancy in X; this forced constancy of X would phase-lock sinusoidal oscillations and stop all overt oscillatory behaviour. Once the addition of external cAMP was stopped, X would once again be free to vary and the oscillations would resume. On the other hand when the extent and/or duration of cAMP binding to the receptor is *decreased* (e.g. by adding PDE) there could be no constraint on the variation of X, permitting oscillations to continue. If continuous cAMP addition locks the oscillator at the phase at which addition begins the above explanation is consistent with the observation that a brief application of cAMP does not cause phase shifts. This assumption can be experimentally tested by applying cAMP continuously but for varying durations and at different oscillatory phases.

OSCILLATIONS AT LOW DENSITIES

Description

Sustained light-scattering oscillations can be observed over nearly a 1000-fold range in cell density, from about 3×10^5 to 3×10^8 cells/ml (V. Nanjundiah, in preparation). Observational problems at both the high-density end (increasing noise in the OD trace) and the low-density end (barely detectable signal amplitude) make it difficult to say whether the oscillations continue beyond this range, but the simplest assumption is that they do. The oscillation morphology can be clearly identified as either spike-shaped or sinusoidal at intermediate and high cell densities, but below about 5×10^6 cells/ml the distinction between the two forms is not immediately apparent. Figure 5, showing oscillation traces after two rapid and successive five-fold dilutions starting from 2×10^7 cells/ml, illustrates the problem. It seems likely – given that only three cycles intervene and that no transitory cycles (as in

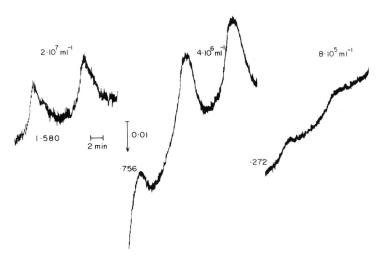

Fig. 5. Actual tracings of optical density during an experiment. The record begins soon after $t=3.5$ h and is to be 'read' continuously along the time (horizontal) axis. On the absorbance (vertical) axis, however, there is a decrease (i.e. upward shift) at each five-fold dilution which has been compensated in the tracing. The numbers by the tracings refer to actual optical densities at 500 nm.

Fig. 1) are seen – that the oscillations remain 'spikes' down to 8×10^5 cells/ml, meaning that their underlying basis remains the same as that at 2×10^7 cells/ml (Fig. 5). However, the small amplitude makes such an identification difficult in every case. One way around the problem is to assign all early oscillations, say those within the first ten cycles from the beginning to the spike category. If such an assignment is valid, spikes can be said to occur over the entire range of cell densities. If not, one can make the more conservative statement – based simply on oscillation morphology – that spikes occur above a density of 8×10^6 cells/ml. Sinusoidal oscillations, whether identified as such on the basis of morphology or late time of occurrence, are found over essentially the entire range of cell densities monitored.

The basic feature regarding these low-density oscillations is that the frequency does not show any systematic dependence on cell density. Rather, allowing for batch-to-batch variability in the cells, the frequency has an average value of 0.17 ± 0.03 cycles/min (the corresponding mean values for spikes, measured at 8×10^6 or more cells/ml, and sinusoids are 0.16 ± 0.03 cycles/min and 0.18 ± 0.05 cycles/min, respectively).

Comparison with models

There are at present two classes of detailed theoretical models seeking to explain oscillatory activity in *D. discoideum*. Both models have as their central feature the same positive feedback step, intracellular synthesis and subsequent release of cAMP activated by extracellular cAMP binding to receptors. The distinction between the models is that whereas Goldbeter and colleagues (Goldbeter, 1975; Goldbeter and Segel, 1977; Martiel and Goldbeter, 1987) assume that the intracellular compartment is thermodynamically open to ATP and (in the latest version) that cAMP binding leads to receptor desensitization, Monk and Othmer (1989) make the entirely different assumption that cAMP binding to receptors elevates intracellular Ca^{2+} and that *this*, rather than ATP depletion or receptor desensitization, is the essential reason for adenylate cyclase activity reverting to its basal level in each cycle.

Upon comparing either of these models with the data just described, two discrepancies stand out. The first is that the models predict a systematic dependence of oscillation frequency on cell density (admittedly a weak dependence in the case of Monk and Othmer), and this is not observed. The second discrepancy has its origin in a common, qualitative, feature of both classes of models. The point is that the models rely on a feedback-activation step based on cAMP binding, and on the face of it this suggests that at very low cell densities oscillations might not exist (the increase in cAMP binding caused by cAMP release at any one step becoming too weak to 'close' the feedback loop). This critical density below which oscillations fail is about

1×10^8 cells/ml* in the model of Martiel and Goldbeter (1987) and about 1×10^7 cells/ml in that of Monk and Othmer (private communication).

To repeat, the experimental observation is that morphological spikes occur at cell densities above 8×10^6 per ml and oscillations with the same underlying basis probably exist over the entire density range examined.

Implications

Thus sustained light-scattering oscillations take place at cell densities 30- to 300-times smaller than the minimum densities predicted by the models. It is possible that the discrepancy is not very serious and that minor tinkering with the models will lead to vastly better agreement. Also, the experimental situation might not be all that clear: the formation of cell clumps in suspensions, already alluded to, may mean that local cell densities are significantly higher than the average cell density over the entire volume. However, this can at best provide a partial explanation. The average particle size in clumps (as seen in an oscillating population whose mean cell density was 2×10^7 per ml) does not exceed 2–4 cells (B. Wurster, in preparation) and so this effect accounts for a factor of 4 at most in the density. Besides, at very low cell densities the early light-scattering oscillations which can be observed in freshly washed cells occur in the absence of any visible clumping.

The alternative is that at low densities there is a qualitative failure of the models; the reasons for the failure can be various. (i) The failure may only have to do with the specific issue of modelling a cAMP-based oscillator. In that case, and to the extent that the failure pertains to sinusoidal oscillations, it simply strengthens the case in favour of their being cAMP-independent. (ii) As already mentioned, there are sound reasons for believing that 'spikes' exist at low cell densities too. If these 'spikes' continue to be cAMP-based, the breakdown of existing models suggests that intracellular cAMP synthesis can be activated by some means *other* than cAMP binding to the receptor. In fact, the problem goes beyond the role played by cAMP as the activator, and therefore as the coupling agent, between cells. Whatever the agent, the intensity of cell to cell coupling must decrease at low cell densities if the coupling is by means of a diffusible metabolite. It is in order to get around this fact that one needs to invoke a feedback loop whereby cAMP synthesis is activated in some other manner, say by means of intercellular contact

*Own calculations. All the published models for cAMP oscillations by Goldbeter and colleagues (Goldbeter, 1975; Goldbeter and Segel, 1977; Martiel and Goldbeter, 1987) assume a constant, meaning cell density-independent, rate of extracellular cAMP hydrolysis. Unpublished experiments show on the other hand that (as might be expected) the rate is proportional to cell density. The calculations were therefore made by starting with the values for phosphodiesterase activity and cell density as in Martiel and Goldbeter (1987) and then scaling appropriately at other cell densities.

formation. (iii) The final, and in a sense the most interesting, possibility is that sustained oscillatory activity at very low cell densities implies the existence of a wholly intracellular periodic process which may or may not be linked (and if so, facultatively) to a cAMP oscillator. The experiments of Satoh *et al.* (1985) on periodic changes of shape and area in single *D. discoideum* cells lend support to an intracellular oscillatory mechanism. Also, their finding that cyanide leads to an increase in oscillation frequency fits our preliminary observations (V. Nanjundiah, unpublished) correlating frequency increase with a fall in oxygen tension in suspensions; this then suggests the presence of an oscillator controlled by respiration.

AGIP43

The oscillatory phenotype

Agip43 is a non-aggregating mutant of *D. discoideum*, strain Ax2. When monitored under vigorously agitated cell suspension conditions it exhibits spontaneous oscillations in light scattering (Wurster and Mohn, 1988). These oscillations resemble the early oscillations in the parent strain in that they are spike-shaped and appear fairly soon (within 3 h) after starvation. There are, however, differences between periodic light-scattering changes in Agip43 and spikes in Ax2: in Agip43 the period is about 1.5-times as long and crucially, simultaneous measurements of cAMP levels indicate no temporal variations within the accuracy of the experimental assays (Wurster and Mohn, 1988). In spite of the absence of cAMP variations during both, the Agip43 oscillations also differ from sinusoidal oscillations in Ax2 in an important respect. Whereas in Ax2 brief external addition of cAMP cannot change the phase of the sinusoids, in Agip43 such addition causes permanent phase-shifts. The phase–response curve in Agip43 is qualitatively different from that seen with Ax2 (with respect to spikes); its form suggests that whatever the initial phase cAMP can shift the oscillator to a unique final phase (B. Wurster, unpublished).

Implications

Once again we have an indication of a cAMP-independent periodicity in *D. discoideum*. This periodicity displays characteristics superficially resembling both the wild-type spikes (time of appearance, oscillation morphology, cAMP-induced phase-shifts) and the wild-type sinusoids (absence of concomitant cAMP changes). What causes oscillatory behaviour in Agip43 is at present unknown, but two general possibilities can be envisaged. It may be that Agip43 points to the existence of an entirely new oscillatory pathway.

On the other hand, in the light of the remarks made earlier that (in the wild-type) sinusoidal oscillations appeared to share some common component(s) with spikes, Agip43 may point to a hybrid oscillatory state in which cAMP levels remain approximately constant but cAMP pulses can still phase-shift the oscillator.

CONCLUDING REMARKS

Even at this preliminary stage a study of light-scattering oscillations in *Dictyostelium discoideum* has shown us qualitatively novel features. There appear to be at least two manifestations of oscillatory behaviour (sinusoids in the wild-type and spikes in Agip43) in which direct measurements show that cAMP is not an oscillatory component; in the case of low-density oscillations the lack of involvement of cAMP is at present only an inference. An extreme likelihood is that each of these three situations demands a *different* cAMP-independent oscillator. At the other extreme, there could be a single, multicomponent master oscillator capable of existing in different dynamical states. The possibility of an entirely intracellular oscillation deserves further attention, as does a scenario in which cells synchronize their phases by means of direct interactions like contact formation rather than via a diffusible molecule.

The most important questions still remain. If cAMP-independent oscillations occur what are their components? Do they have a function, or are they automatic consequences of complicated regulatory loops? These questions assume even greater importance in the light of the finding that it is the sinusoidal oscillations which are temporally correlated with cell aggregation.

Acknowledgements: V. Nanjundiah wishes to thank the Alexander von Humboldt Stiftung for the award of a research fellowship and D. Malchow for being of immense help as a host at the University of Konstanz and for his advice during the performance of some of the experimental work described here. Experimental work was supported by the Deutsche Forschungsgemeinschaft, Sonderforschungsbereich 156.

REFERENCES

Aeckerle, S., Wurster, B. and Malchow, D. (1985) Oscillations and cyclic AMP induced changes of the K^+ concentration in *Dictyostelium discoideum*. *EMBO J.* **4**, 39–43.

Bonner, J. T. (1967) *The Cellular Slime Molds* (2nd edn). Princeton University Press, Princeton, NJ.

Bumann, J., Wurster, B. and Malchow, D. (1984) Attractant-induced changes and oscillations of the extracellular Ca^{++} concentration in suspension of differentiating *Dictyostelium* cells. *J. Cell Biol.* **98**, 173–8.

Bumann, J., Malchow, D. and Wurster, B. (1986) Oscillation of Ca^{++} concentration during the cell differentiation of *Dictyostelium discoideum*. *Differentiation* **31**, 85–91.

Darmon, M., Brachet, P. and Pereira Da Silva, L. H. (1975) Chemotactic signals induce cell differentiation in *Dictyostelium discoideum*. *Proc. natn. Acad. Sci. U.S.A.* **72**, 3163–6.

Gerisch, G. (1968) Cell aggregation and differentiation in *Dictyostelium discoideum*. *Curr. Topics Devl Biol.* **3**, 157–97.

Gerisch, G. and Hess, B. (1974) Cyclic AMP-controlled oscillations in suspended *Dictyostelium* cells: their relation to morphogenetic cell interactions. *Proc. natn. Acad. Sci. U.S.A.* **71**, 2118–22.

Gerisch, G. and Wick, U. (1975) Intracellular oscillations and release of cyclic AMP from *Dictyostelium* cells. *Biochem. biophys. Res. Commun.* **65**, 364–70.

Gerisch, G., Fromm, H., Huesgen, A. and Wick, U. (1975) Control of cell contact sites by cAMP pulses in differentiating *Dictyostelium discoideum* cells. *Nature* **255**, 547–9.

Gerisch, G., Maeda, Y., Malchow, G., Roos, W., Wick, U. and Wurster, B. (1977) Cyclic AMP signals and the control of cell aggregation in *Dictyostelium discoideum*. In *Development and Differentiation in the Cellular Slime Molds* (eds P. Cappuccinelli and J. M. Ashworth), pp. 105–24. Elsevier/North Holland Biomedical Press, Amsterdam.

Goldbeter, A. (1975) Mechanism for oscillatory synthesis of cyclic AMP in *Dictyostelium discoideum*. *Nature* **253**, 540–2.

Goldbeter, A. and Segel, L. A. (1977) Unified mechanism for relay and oscillations of cAMP in *Dictyostelium discoideum*. *Proc. natn. Acad. Sci. U.S.A.* **70**, 1543–7.

Higgins, J. (1967) The theory of oscillating reactions. *Ind. Engng Chem.* **59**, 18–62.

Konijn, T. M., Van De Meene, J. G. C., Bonner, J. T. and Barkley, D. S. (1967) The acrasin activity of adenosine -3′,5′-cyclic phosphate. *Proc. natn. Acad. Sci. U.S.A.* **58**, 1152–4.

Malchow, D., Nanjundiah, V. and Gerisch, G. (1978a) pH oscillations in cell suspensions of *Dictyostelium discoideum*: their relation to cyclic AMP signals. *J. Cell Sci.* **30**, 319–30.

Malchow, D., Nanjundiah, V., Wurster, B., Eckstein, F. and Gerisch, G. (1978b) Cyclic AMP induced pH changes in *Dictyostelium discoideum* and their control by calcium. *Biochem. Biophys. Acta.* **538**, 473–80.

Martiel, J-L. and Goldbeter, A. (1987) A model based on receptor desensitization for cyclic AMP signalling in *Dictyostelium* cells. *Biophys. J.* **52**, 807–28.

Mato, J. M. and Malchow, D. (1978) Guanylate cyclase activation in response to chemotactic stimulation in *Dictyostelium discoideum*. *FEBS Lett.* **90**, 119–22.

Mato, J. M., Krens, F. A., Van Haastert, P. J. M. and Konijn, T. M. (1977) 3′,5′-Cyclic AMP-dependent 3′,5′-cyclic GMP accumulation in *Dictyostelium discoideum*. *Proc. natn. Acad. Sci. U.S.A.* **74**, 2348–51.

Monk and Othmer (1989) *Phil. Trans. R. Soc. Lond.* (in press).

Nanjundiah, V. (1973) Chemotaxis, signal relaying and aggregation morphology. *J. theor. Biol.* **42**, 63–105.

Nanjundiah, V. (1986) How rapidly do uncoupled oscillators desynchronize? *J. theor. Biol.* **121**, 375–9.

Nanjundiah, V. (1989) Periodic stimuli are more successful than randomly spaced ones for inducing development in *Dictyostelium discoideum*. *Biosci. Rep.* (in press).

Roos, W., Nanjundiah, V., Malchow, D. and Gerisch, G. (1975) Amplification of cAMP signals in aggregating cells of *Dictyostelium discoideum*. *FEBS Lett.* **53**, 139–42.

Satoh, H., Ueda, T. and Kobatake, Y. (1985) Oscillations in cell shape and size during locomotion and in contractile activities of *Physarum polycephalum*, *Dictyostelium discoideum*, *Amoeba proteus* and macrophages. *Expl Cell Res.* **156**, 79–90.

Shaffer, B. M. (1975) Secretion of cAMP induced by cAMP in the cellular slime mold *Dictyostelium discoideum*. *Nature* **255**, 549–52.

Tomchik, K. J. and Devreotes, P. N. (1981) Cyclic AMP waves in *Dictyostelium discoideum*: a demonstration by isotope dilution fluorography. *Science,* **212**, 443–6.

Wick, U., Malchow, D. and Gerisch, G. (1978) Cyclic AMP stimulated calcium influx into aggregating cells of *Dictyostelium discoideum*. *Cell Biol. Int. Rep.* **2**, 71–9.

Wurster, B. and Mohn, R. (1988) Cyclic AMP-independent cell communication in a mutant of *Dictyostelium discoideum*. *J. Cell Sci.* **87**, 723–30.

Wurster, B., Schubiger, K., Wick, U. and Gerisch, G. (1977) Cyclic GMP in *Dictyostelium discoideum*: oscillations and pulses in response to folic acid and cyclic AMP signals. *FEBS Lett.* **76**, 141–144.

Spiral order in chemical reactions

STEFAN C. MÜLLER AND BENNO HESS

Max-Planck-Institut für Ernährungsphysiologie, Rheinlanddamm 201, D-4600 Dortmund 1, FRG

INTRODUCTION

The spiral stands out as a widespread structural element in the large variety of shapes created by nature. It appears in the form of snail shells, the order of sunflower kernels, the structure of the DNA molecule, the clouds merging into the eye of hurricanes and the distant galaxies. This simple geometric form belongs to the archetypes of spatial organization.

Dynamic evolution of spiral-shaped structures occurs frequently in excitable systems capable of transmitting a local pulse of excitation into its resting, non-excited neighbourhood. Such pulses are known to spread out over large distances in a wave-like fashion. They leave behind a refractory area which, after a given period of time of recovery, becomes ready again to participate in a subsequent transmission process, triggered by any pulse approaching from adjacent regions. Wave-like transmission processes of this kind involve the excitation kinetics of chemical substances and form a basic mechanism in cell to cell signalling, as exemplified in the slime mould *Dictyostelium discoideum* (Gerisch, 1982; Martiel and Goldbeter, 1987; Klein *et al.*, 1988). Similar spatially extended waves have been studied in the development of embryos (Gierer and Meinhardt, 1972); in phage-infected bacteria (Ivanitsky *et al.*, 1984); in chicken retina (Gorelova and Bures, 1983); in spreading depressions in the cerebral cortex (Bures *et al.*, 1984); in the physiology of heart tissue (Allessie *et al.*, 1977; Winfree, 1983); in pheronome emission (odour song) (Bossert, 1968), among others. In a number of these biological systems spiral-shaped excitation waves have been observed.

A chemical model system displaying such spatio-temporal self-organization and especially spiral propagation of waves of excitation is the Belousov–

Cell to Cell Signalling: From
Experiments to Theoretical Models
ISBN 0–12–287960–0

Zhabotinskii (BZ) reaction (Busse, 1969; Zaikin and Zhabotinskii, 1970; Winfree, 1972; Busse and Hess, 1973; Field and Burger, 1985; Müller *et al.*, 1985a; Ross *et al.*, 1988). Many of the structural features displayed in this reactive solution can be taken as easily reproducible model cases from which we may learn how macroscopic dynamical behaviour is connected with non-linear mechanisms at the molecular level. The properties of these spatial patterns are closely related to the underlying temporal oscillations. It is most interesting to study how complexity in time is reflected in spatial phenomena via a coupling to physical transport processes such as molecular diffusion or hydrodynamic convection.

In the following we present results on chemical spiral waves as recently obtained in our laboratory. After a short description of chemical wave propagation and modern methods for their quantitative analysis we focus on the properties of spiral structures and summarize recent experiments on the verification of curvature effects on the propagation velocity, yielding useful parameters for the modelling of spirals. Hydrodynamic flow is frequently involved in the pattern dynamics and some effort has been devoted to quantify such influences. Finally, some analogies between chemical spirals and spirals in the development of *Dictyostelium discoideum* patterns are discussed.

CHEMICAL WAVES AND THEIR QUANTITATIVE ANALYSIS

Chemical fronts and waves are variations in concentrations of chemical species which travel in space and occur in non-linear systems far from equilibrium (Glansdorff and Prigogine, 1971). Different types of waves may occur, but here we focus on the so-called trigger waves produced in quiescent, excitable reaction mixtures. They are described by solutions of reaction–diffusion equations of the form

$$\frac{\partial \psi}{\partial t} = D\nabla^2 \psi + F(\psi)$$

where ψ is a vector of time- and space-dependent state variables, such as the concentrations of chemical species, D is a matrix of diffusion coefficients, and $F(\psi)$ represents variations in time that arise from the chemical reactions. For a given system both terms can be determined experimentally.

Quite a few examples of chemical systems exhibiting wave propagation have been found during the last years (DeKepper *et al.*, 1982), but the BZ reaction is investigated to the greatest detail in its mechanism. Here, the overall reaction is the oxidative bromination of malonic acid by bromine in acidic solution catalysed by cerium, ferroin, or other redox compounds. A detailed mechanism in homogeneous solution describing chemical oscillations has been given by Field *et al.*, (1972) and Field and Försterling, (1986).

The fronts of waves of chemical activity may assume various shapes depending on the geometry and volume of the container. All the characteristic features of these shapes are best realized in quiescent solutions in which trigger waves can be excited. In a thin layer (about 1 mm) of reactive solution in a Petri dish there occur several types of two-dimensional forms as shown in Fig. 1, namely circular concentric waves, parallel wave trains, or spiral-shaped waves. (We are not concerned here with three-dimensional geometries of higher complexity (Welsh *et al.*, 1983; Winfree and Strogatz, 1984).)

Fig. 1. Complex pattern in a thin layer of the Belousov–Zhabotinskii reaction consisting of circular waves and spiral waves. Dish diameter: 7 cm.

During the early phase of research on this phenomenon, many reports were based on photographic observations, which have now been replaced by space-resolved photometric methods and digital evaluation of local concentrations. Experimental studies on global properties of the waves, such as the dispersion relation (Pagola *et al.*, 1988), the characterization of the geometric shape of fronts (Müller *et al.*, 1987), the relation between curvature and wave velocity (Foerster *et al.*, 1988) have appeared only in the last 3 years. The development of computerized one- (Wood and Ross, 1985) or two-dimensional (Müller *et al.*, 1986) techniques for the quantification of light transmission patterns taken at specific wavelengths of absorption allow a detailed study of reaction kinetics with high space resolution. Indeed, video cameras with 512×512 pixel raster resolution connected to large memory computers yield precise non-invasive measurements of the spatial distribution of chemical species without disturbing the structures (which would be the case, for instance, by immersing ion-sensitive electrodes).

The outcome of a two-dimensional measurement of the concentration of the oxidized catalyst (ferriin) in the BZ reaction, applied to a section of a single circular wave, is presented in Fig. 2A. In this digital image two narrow concentration levels are marked by assigning the 'colours' black and white to the picture elements (pixels) having the corresponding grey levels. In a three-dimensional perspective plot (Fig. 2B) the local concentrations of one-half of a circular wave are plotted along a third coordinate perpendicular to the plane of observation in which the pattern evolves. It clearly shows the steep concentration fronts propagating outward and the shallow trough forming in the inner part of the pattern. The applied technique of space-resolved light

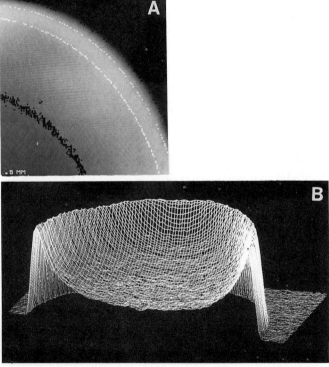

Fig. 2. (A) Digital image of a quarter section of a circular wave in the BZ reaction obtained with a computerized two-dimensional spectrophotometer. (B) Three-dimensional perspective representation of one-half of the same circular wave.

detection proves to be a very helpful tool for the quantification of such patterns in terms of one chemical variable, in this case the concentration of the catalyst.

SPIRAL-SHAPED WAVE FRONTS

Starting from circular waves evolving in a thin solution layer of the BZ reaction, spirals can be created by disrupting a circular front with, for instance, a gentle air blast from a pipette. The disturbance results in open ends of the circular wave fronts which then curl up to form rotating spirals, in which the spiral tip turns inwards while the chemical fronts move outwards. Such spirals are often arranged in symmetric pairs. Upon collision on the symmetry line the waves annihilate each other, a property to be further discussed in the next section.

The details of the evolving spiral geometry depend on the choice of initial concentrations. Chemical variables such as the acid and the bromate concentration determine whether the spirals have a shape close to that of an Archimedian spiral of exhibit characteristic assymmetric features. In Fig. 3 we present five examples of spirals observed with different concentrations of H^+ ions. Variation of this parameter leads to remarkable changes both in size and shape of the rotating wave patterns. With all other concentrations kept constant, the spirals are almost Archimedian for $[H^+] > 0.26$ N. From $[H^+] = 0.7$ N down to 0.23 N (Fig. 3A–C) the spiral pitch increases from 0.77 mm to 2.3 mm, the rotation period increases from 8.5 s to 50 s. At still lower $[H^+]$ not only pitch and period further decrease, but the spiral shape becomes increasingly asymmetric (Fig. 3D). In fact, at the lowest concentration $[H^+] = 0.15$ N (Fig. 3E) the open end of the previously disrupted wave front hardly curls up to a spiral but remains a rather isolated, slowly propagating wave tip surrounded by a large area of unexcited solution.

We have analysed in detail the properties of a chemical spiral in the ferroin-catalysed BZ reaction, corresponding to the 'standard' recipe of Fig. 3B, by evaluating data from two-dimensional spectrophotometric measurements (Müller et al., 1987). The shape of iso-concentration lines extracted from the video images turns out to be well represented by an Archimedian spiral, but within the experimental error a fit to the measured data by the involute of a circle fits equally well. This second approach is more appropriate, because it takes into account that there exists a small area around the centre of spiral rotation, the core of the spiral, in which the chemical processes are different from those occurring outside this area. Theoretical modelling requires the existence of such an area, as well (Keener and Tyson, 1986). So far, its spatial extent can only be determined by experiment.

In fact, by digitally overlaying a sequence of spiral images obtained with this recipe and taken during one revolution of the spiral around its centre, we

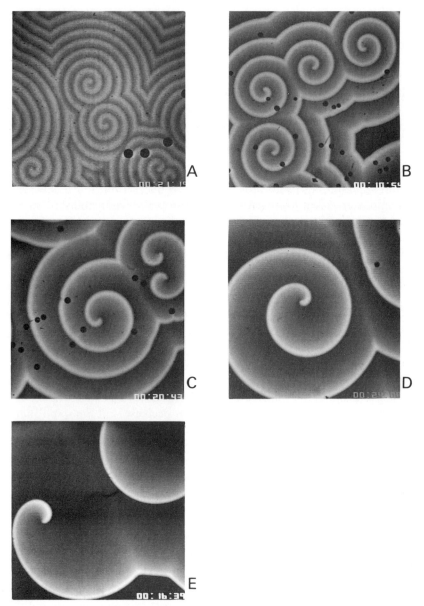

Fig. 3.   Five images of spirals in the BZ reaction as a function of acid concentration. Initial concentrations 0.12 M $CH_2(COOH)_2$, 0.33 M $NaBrO_3$, 0.06 M NaBr, 0.003 M ferroin for all systems; and 0.7 M (A), 0.37 M (B), 0.23 M (C), 0.19 M (D); 0.15 M H_2SO_4 (E). Area of images: 15×15 mm².

obtained the following: there is a small circular area (diameter ≈ 0.7 cm) in which the maximum degree of excitation, which is characteristic for the propagating wave crests, is never reached (Fig. 4A) (Müller *et al.*, 1985a). The spiral tip rotates on a stable circle around the centre of rotation, which is located right in the middle of the black dot of the superposed image. At this point of rotation there is a singular site at which the chemical state remains quasi-stationary, while at all points outside this cone-shaped area the chemistry varies in time between the two extremes of the chemical oscillation induced by wave propagation. Thus, in such regular chemical spirals we are confronted with the remarkable phenomenon that in a liquid layer there are at least two dynamical states: that of periodic oscillations (in the points outside the core) and that of stationary behaviour (in the centre of the core) (Müller *et al.*, 1985a). The theoretical descriptions by reaction–diffusion equations yield so far spiral geometry outside the core region (Keener and Tyson, 1986), but the modelling of the structure of the core itself remains a challenge for the future.

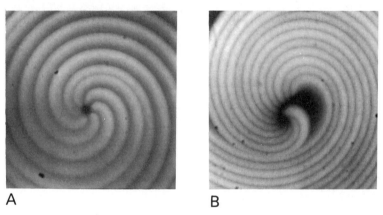

A   **B**

Fig. 4.   (A) Overlay of six images of a spiral produced in a mixture as in Fig. 3B, taken at 3 s intervals. (B) Overlay of 12 images of a spiral produced in a mixture as in Fig. 3D, taken at 6 s intervals.

In a three-dimensional perspective representation, a composite spiral picture, obtained from several revolutions of the spiral, results in a funnel-like structure (Fig. 5); it clearly shows that within the dark spot shown in Fig. 4A the maximum of excitation decreases towards the singular site at the centre of rotation, depicting the core in a way that resembles much the shape of a chemical 'tornado'.

A similar analysis has been recently applied to spiral patterns exhibiting asymmetric shapes and less stability with respect to the location of spiral rotation (cf. Agladze *et al.*, 1988; Winfree and Jahnke, 1989; and Plesser, Müller and Hess, in preparation). For example, for the pattern of Fig. 3D a

Fig. 5.   The core structure of an overlayed spiral system similar to that in Fig. 4A. The upper envelope of the concentration variation is shown in three-dimensional perspective.

digital overlay was calculated for 12 subsequent images, as shown in Fig. 4B. Here the [H$^+$] was decreased by about a factor of two as compared with Fig. 4A. The spiral tip now follows a path which is quite different from that observed in the case of the regular Archimedian spiral. For a short while this path is almost straight before it starts to bend rapidly into a small circle, similar to the regular circular rotation in the standard case (see Fig. 4B). Thus, if one follows the spiral tip in time, a 'looping' motion is detected in which almost rectilinear propagation alternates with fast turns, such that the spiral tip traces a curve similar to a cycloid. Our preliminary analysis indicates that, with the given initial concentrations, the spiral tip first moves in a weakly curved fashion into an unexcited area, but then makes a quite rapid turn in order to lead that area, which was previously left unexcited, to excitation. The experiments demonstrate that dynamics and shape of the waves are governed by a specific part of the reaction mechanism and that a macroscopic pattern might well be the result of the time-dependent state of one elementary reaction step of the BZ system. Such an understanding of the detailed chemical dynamics underlying the spiral motion calls for further work in theory as well as in experiments.

FRONT CURVATURE OF A CHEMICAL WAVE

Models of spiral wave propagation suggest that besides the dispersion relationship (Pagola *et al.*, 1988), the relation between propagation velocity and curvature of wave fronts plays a predominant role in pattern formation (Zykov and Morozova, 1979; Kuramoto, 1980; Keener and Tyson, 1986). There are positive and negative curvatures in such a pattern. We found that the formation of the sharp cusps upon wave collision present structures with negative curvature, which are useful for an experimental verification of its relationship with the propagation velocity (Foerster *et al.*, 1988).

Note that in chemical waves the phenomenon of wave collision is quite different from that commonly known in elastic waves in physics: there are no interference phenomena but waves running toward each other interact chemically when they collide, leaving typical cusp-like structures of the components analysed spectrophotometrically behind. An example is given in Fig. 6, which displays different iso-concentration levels in a pair of colliding circular waves being visualized by a 'pseudo-colouring' in black and white.

A detailed evaluation of the front curvature K of such iso-concentration levels at the vertices of the cusps and their instantaneous normal propagation velocity N yielded, indeed, a simple relationship

Fig. 6. Colliding circular waves presented in a perspective surface plot. Several iso-intensity levels are marked by different shadings.

$$N = c - D\,K$$

where c = velocity of plane waves and D = diffusion coefficient of the autocatalytic species. For a wave structure evolving in the standard recipe (cf. Fig. 3B) we found $c = 100\,\mu m/s$ and $D = 2.0 \times 10^5\,cm^2/s$ in accordance with previous assumptions. Using these values numerical simulations were carried out with the model proposed by Keener and Tyson (1986) and a remarkably good fit between predicted and measured iso-concentration lines could be obtained (Fig. 7).

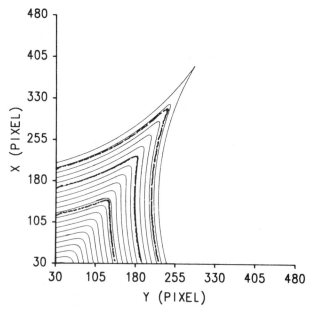

Fig. 7.   The temporal evolution of a specific intensity level during wave collision (dotted lines) is compared to numerical simulations (full curves) based on a model given by Keener and Tyson (1986).

Very recently this relationship was also obtained for positive curvatures as produced in small circular waves triggered by a silver-coated microelectrode in the same chemical medium. From these results a critical radius of about $20\,\mu m$ could be calculated, below which a circular wave will not propagate into its surrounding medium (Foerster et al., 1989). This result is particularly important for a clarification of open problems in theory concerning the role of this critical quantity for spiral core and circular wave nucleation (Keener and Tyson, 1986; Davydov et al., 1989).

HYDRODYNAMIC FLOW

In order to explain chemical wave propagation one considers the diffusion of chemical compounds generated by an autocatalytic process. However, other transport phenomena may also play a significant role. In particular, in a liquid reaction layer, the occurrence of hydrodynamic instabilities has to be taken into account. For instance, in an experimental set-up, the reaction dish containing a thin reaction-solution layer is frequently covered with a glass plate in order to minimize evaporation (leaving a small air gap between layer surface and cover), but if it is left uncovered, a variety of structural phenomena are observed, e.g. stationary 'mosaic' patterns, transient structures, or distortion and decomposition of waves (Boiteux and Hess, 1980; Orban, 1980; Agladze et al., 1984; Müller et al., 1985b, 1986). In such a case, the influence of convective flows is of major importance. These arise from temperature gradients caused by evaporative cooling and from inhomogeneities in surface tension as a result of differences in temperature and/or chemical composition (Rayleigh–Bénard and Marangoni-type instabilities) (Normand et al., 1977; Linde, 1982; Müller et al., 1985c).

An example of wave distortion and decomposition is shown in Fig. 8. Starting from a spiral wave in the BZ reaction (similar to the spirals in Fig. 3B) that has evolved in a covered dish, the glass cover is removed. After about 3 min, the onset of convective flow leads to a strong distortion of the spiral geometry (Fig. 8A) and subsequently to a pronounced wave decomposition (Fig. 8B) – a transition from periodicity to aperiodicity has taken place. On the other hand, after covering the dish again about 5 min later, partial reorganization into a number of regular spiral structures sets in (Fig. 8C), as soon as the hydrodynamic flow due to evaporative cooling has been 'switched off'. In this system we find that high spatial complexity in spiral order is induced by an order–disorder transition due to mutual interaction of diffusive and convective transport. Image sequences of this kind were treated by image analysis techniques in order to find numerical criteria for order and disorder in such patterns (Markus et al., 1987). The question whether spatial chaos occurs remains to be investigated.

While in such observations the effects of convection are only recorded qualitatively, the correlation between chemical pattern dynamics and hydrodynamic flow velocities was recently investigated quantitatively by applying two-dimensional velocimetry and two-dimensional spectrophotometry based on microscope video imaging techniques. For flow measurements small polystyrene particles (diameter 0.48 μm) serving as scattering centres were mixed into the BZ solution and illuminated by a narrow beam of He–Ne laser light (632.8 nm, slightly tilted with respect to the layer normal). Together with a homogeneous transmitted light beam (490 nm, perpendicular to the layer) we could observe hydrodynamic flow and propagation of chemical activity simultaneously (Miike et al., 1988a).

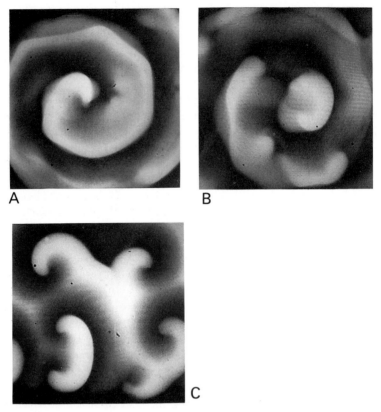

Fig. 8.   Decomposition of a single spiral wave due to convection in the liquid layer (A,B) and reorganization into several spirals after suppression of convection (C).

It turns out, in fact, that the hydrodynamic stability of the layer is almost never guaranteed. Independent of the effects of evaporative cooling any single wave may induce a flow that travels along with the front and initiates liquid motion in a large portion of the dish. If a spiral wave is produced, which, during its rotation, creates an almost periodic wave train, this induced flow may become oscillatory in that at a given point in space it alternates its direction with each passing or with each second or third passing wave, depending sensitively on the sample preparation and the detailed location and geometry of the pattern. A measurement of flow velocities with very pronounced oscillation amplitude is shown in Fig. 9. Such oscillations result in periodic deformations of wave fronts (Miike *et al.*, 1988b). They can cause wave disruption if their amplitudes reach values substantially higher than the wave propagation speed, as indeed observed in the experiment from which the data of Fig. 9 were obtained.

These findings prove that in chemical patterns there exists a direct

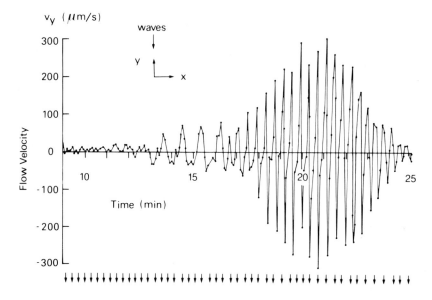

Fig. 9. Oscillatory flow velocity induced by periodic passage of a chemical wave train emerging from a rotating chemical spiral. The arrows indicate the passage of waves through the point of observation.

interaction between reaction–diffusion processes and hydrodynamic motion, a most interesting but complicated coupling that has to be elucidated by further applications of space-resolved velocimetry (Miike *et al.*, 1988b).

CELL BIOLOGY

Since its discovery the mechanism of the diffusion-coupled BZ reaction was considered to be functionally analogous to biological systems such as the slime mould *Dictyostelium discoideum* or even the more complex cellular networks of neural functions (Busse and Hess, 1973). In all these systems autocatalytic reaction mechanisms lead to diffusion and transport of chemicals, serving as chemical coupling transmitters between reactive volume elements in a spatial territory. Whereas in chemical systems the critical size of a volume element has been found experimentally and theoretically to be in the order of a few microlitres, the size of the analogous volume elements in biological systems is still not clear. In addition, it must be realized that in biological systems such volume elements are structurally bounded by cellular membranes and the critical size of an elementary excitable volume unit might be just one single cell or a few cells forming a cooperative functional unit. Furthermore, the mechanism of signal generation and intercellular transmission of chemical signals in a biological system, exemplified in the case of a slime mould, is rigidly controlled by multiple enzyme and receptor functions which are genetically determined and adaptable, whereas in the chemical BZ system the sum of all interacting chemical species is strictly conserved and the advance of the elementary reactions is determined by the initial nature of the complex chemical medium.

In spite of these differences, the macroscopic dynamic patterns observed in the BZ reaction and its biological slime mould analogue are almost identical. The slime mould system readily forms target patterns, spirals or disordered structures (Gerisch, 1982) just as in case of the BZ reaction. This illustrates that the macroscopic structures in chemical and topobiological mapping processes are generated by the same general mechanism of diffusion-coupled oscillating chemical or enzymic reaction systems as suggested by Turing as well as Glansdorff and Prigogine (1971) long ago.

In the slime mould, the occurrence of waves of cyclic-AMP, generated by oscillating dynamics of receptors, enzyme functions as well as ion channels have been identified (Gerisch and Hess, 1974; Gerisch, 1982; Klein et al., 1988). In order to understand the mechanism of wave collision, we are currently investigating the propagation of wave phenomena in the system with the help of two-dimensional transmission video techniques. Figure 10A shows a typical light-scattering picture obtained by dark-field illumination of high spatial resolution representing the spiral aggregation motion of many thousands of slime mould cells. In Fig. 10B the collision process at the border between two territories of such cells is displayed in an overview. These studies extend earlier observations (Gerisch, 1982; Klein et al., 1988) and allow one to look deeply into the core and the border of the process where essential

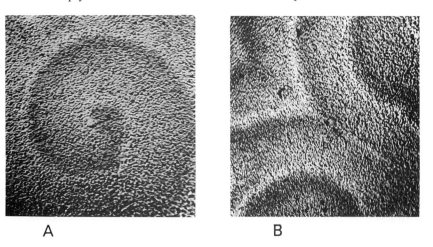

A   B

Fig. 10. (A) Spiral-shaped aggregation pattern in *Dictyostelium discoideum*. (B) Cusp-like structures due to wave collision in the same system (Foerster, Müller and Hess, unpublished).

nucleation events are expected to occur. A detailed analysis of the diffusion process of the intercellular chemical signals as well as the establishment of intercellular boundaries between territories and the curvature-dependent dynamics of their evolution is under way. In these experiments of high magnification, the general dynamical analogy between the slime mould and the chemical BZ reaction system is well documented.

OUTLOOK

The elucidation of overall mechanisms of spatial pattern formation in chemical and biological systems has to deal with three different subprocesses: (1) the elementary and global chemical reaction mechanisms, (2) molecular diffusion and active transport phenomena, and (3) convective flows and hydrodynamic phenomena interacting with local processes of accumulation and degradation of materials.

While the experimental study of complex reaction kinetics is well under-way in many laboratories using classical as well as open reactor instrumentation, suitable means for the study of dynamic pattern formation focusing on diffusion processes and hydrodynamic perturbation have only become available recently. Our studies revealed chemical events of macroscopic wave collision processes and allowed us to derive experimentally the wave front geometry in more detail. The critical collision distance and local diffusion coefficient could be measured. In addition, experimental evidence is at hand which illustrates the significance of hydrodynamic flow as an essential local element in chemical waves.

With respect to the elucidation of the molecular mechanism of pattern formation in the BZ reaction, it is obvious that the more components we detect, the more details we obtain on the intrinsic chemical sources. Indeed, it is necessary to vary the chemistry in order to identify the critical reaction steps which define a macroscopic pattern, whether a target pattern, spiral or cycloid.

In general, all studies show that each pattern might well be uniquely determined by its underlying physical chemistry and/or its cellular physiology, the latter genetically fixed. Thus, any change of a given reaction mechanism should lead to a change of a macroscopic pattern.

One of the prerequisites for the generation of target waves or spiral rotation is the autocatalytic character of the underlying chemical or biochemical processes which allow in proper regimes limit cycle dynamics in macroscopic volumes and highly cooperative distribution phenomena. Whenever such a reaction system operates periodically and couples to transport, macroscopic patterns are formed. In any of these systems the problems of pattern initiation, of its nucleation as well as pattern switching mechanisms are fully open and a challenge for future research. In biology, the control of pattern formation is exerted by dynamic functions within the genome of a cellular system. Here, we are dealing with extremely complex interactions and a new field of molecular topobiology is currently developing. No doubt, molecular biology will borrow new concepts and ideas from physical chemistry as well as mathematical phenomenology and vice versa.

REFERENCES

Agladze, K. I., Krinsky, V. I., Pertsov, A. M. (1984) Chaos in non-stirred Belousov–Zhabotinskii reaction is induced by interaction of waves and stationary dissipative structures. *Nature* **308**, 834–5.

Agladze, K. I., Panfilov, A. V. and Rudenko, A. N. (1988) Nonstationary rotation of spiral waves: three-dimensional effect. *Physica* **29D**, 409–15.

Allessie, M. A., Bonke, F. I. M. and Schopman, F. J. G. (1977) Circus movement in rabbit atrical muscle as a mechanism of tachycardia. *Circulation Res.* **41**, 9–18.

Boiteux, A. and Hess, B. (1980) Spatial dissipative structures in yeast extracts. *Ber. Bunsenges. Phys. Chem.* **84**, 393–398.

Bossert, W. H. (1968) Temporal patterning in olfactory communication. *J. theor. Biol.* **18**, 157–70.

Bures, J., Koroleva, V. I. and Gorelova, N. A. (1984) Leão's spreading depression, an example of diffusion-mediated propagation of excitation in the central nervous system. In *Self-Organization – Autowaves and Structures Far from Equilibrium* (ed. V. I. Krinsky), pp. 180–3. Springer-Verlag, Berlin.

Busse, H. G. (1969) A spatial periodic homogeneous chemical reaction. *J. Phys. Chem.* **73**, 750.

Busse, H. G. and Hess, B. (1973) Information transmission in a diffusion-coupled oscillatory chemical system. *Nature* **244**, 203–5.

Davydov, V. A., Mikhailov, A. S. and Zykov, V. S. (1989) Kinematical theory of autowave patterns in excitable media. In *Nonlinear Waves in Active Media* (eds D. G. Crighton and U. Engelbrecht). Springer-Verlag, Berlin (to be published).

DeKepper, P., Epstein, R., Kustin, K. and Orban, M. (1982) Batch oscillations and spatial wave patterns in chlorite oscillating systems. *J. Phys. Chem.* **86**, 170–1.

Field, R. J. and Burger, M. (eds) (1985) *Oscillations and Travelling Waves in Chemical Systems.* John Wiley, New York.

Field, R. J. and Försterling, H. -D. (1986) On the oxybromine chemistry rate constants with cerium ions in the Field–Körös–Noyes mechanism on the Belousov–Zhabotinskii reaction: The equilibrium $HBrO_2 + BrO_3^- + H^+ \rightleftarrows 2\,BrO_2 + H_2O$. *J. Phys. Chem.* **90**, 5400–7.

Field, R. J., Körös, E. and Noyes, R. M. (1972) Oscillations in chemical systems, Part 2. Thorough analysis of temporal oscillations in the $Ce–BrO_3^-$-malonic acid system. *J. Am. Chem. Soc.* **94**, 8649–64.

Foerster, P., Müller, S. C. and Hess, B. (1988) Curvature and propagation velocity of chemical waves. *Science, N.Y.* **241**, 685–7.

Foerster, P., Müller, S. C. and Hess, B. (1989) Critical size and curvature of wave formation in an excitable chemical medium. *Proc. Natl. Acad. Sci. U.S.A.* (in press).

Gerisch, G. (1982) Chemotaxis in *Dictyostelium*. *A. Rev. Physiol.* **44**, 535–52.

Gerisch, G. and Hess, B. (1974) Cyclic-AMP-controlled oscillations in suspended *Dictyostelium* cells: Their relation to morphogenetic cell interactions. *Proc. natn. Acad. Sci. U.S.A.* **71**, 2118–22.

Gierer, A. and Meinhardt, H. (1972) A theory of biological pattern formation. *Kybernetik* **12**, 30–9.

Glansdorff, P. and Prigogine (1971) *Thermodynamic Theory of Structure, Stability and Fluctuations.* Wiley-Interscience, New York.

Gorelova, N. A. and Bures, J. (1983) Spiral waves of spreading depression in the isolated chicken retina. *J. Neurobiol.* **14**, 353–63.

Ivanitsky, G. R., Kunisky, A. S. and Tzyganov, M. A. (1984) Study of "target patterns" in a phage-bacterium system. In *Self-Organization – Autowaves and Structures Far from Equilibrium* (ed. V. I. Krinsky), pp. 214–7. Springer-Verlag, Berlin.

Keener, J. P. and Tyson, J. J. (1986) Spiral waves in the Belousov–Zhabotinskii reaction. *Physica* **21D**, 307–24.

Klein, P. S., Sun, T. J., Saxe, III, Ch. L., Kimmel, A. R., Johnson, R. L. and Devreotes, P. N. (1988) A chemoattractant receptor controls development in *Dictyostelium discoideum. Science, N.Y.* **241**, 1467–72.

Kuramoto, Y. (1980) Instability and turbulence of wave fronts in reaction–diffusion systems. *Prog. Theor. Phys.* **63**, 1885–903.

Linde, H. (1982) Marangoni-instabilities. In *Convective Transport and Instability Phenomena* (eds J. Zierep and H. Oertel), pp. 265–96. Braun, Karlsruhe.

Markus, M., Müller, S. C., Plesser, Th. and Hess, B. (1987) On the recognition of order and disorder. *Biol. Cybern.* **57**, 187–95.

Martiel, J. -L. and Goldbeter, A. (1987) A model based on receptor desensitization for cyclic AMP signaling in *Dictyostelium* cells. *Biophys. J.* **52**, 807–27.

Miike, H., Müller, S. C. and Hess, B. (1988a) Oscillatory hydrodynamic flow induced by chemical waves. *Chem. Phys. Lett.* **144**, 515–20.

Miike, H., Müller, S. C. and Hess, B. (1988b) Oscillatory deformation of chemical waves induced by surface flow. *Phys. Rev. Lett.* **61**, 2109–12.

Müller, S. C., Plesser, Th. and Hess, B. (1985a) The structure of the core of the spiral wave in the Belousov–Zhabotinskii reaction. *Science, N.Y.* **230**, 661–3.

Müller, S. C., Plesser, Th. and Hess, B. (1985b) Surface tension driven convection in chemical and biochemical solution layers. *Ber. Bunsenges. Phys. Chem.* **89**, 654–8.

Müller, S. C., Plesser, Th., Boiteux, A. and Hess, B. (1985c) Pattern formation and Marangoni convection during oscillating glycolysis. *Z. Naturforsch.* **40c**, 588–91.

Müller, S. C., Plesser, Th. and Hess, B. (1986) Two-dimensional spectrophotometry and pseudo-color representation of chemical reaction patterns. *Naturwissenschaft.* **73**, 165–79.

Müller, S. C., Plesser, Th. and Hess, B. (1987) Two-dimensional spectrophotometry of spiral wave propagation in the Belousov–Zhabotinskii reaction. 2. Geometric and Kinematic parameters. *Physica* **24D**, 87–96.

Normand, C., Pomeau, Y. and Velarde, M. G. (1977) Convective instability: A physicist's approach. *Rev. Mod. Phys.* **49**, 581–624.

Orban, M. (1980) Stationary and moving structures in uncatalyzed oscillatory chemical reactions. *J. Am. Chem. Soc.* **102**, 4311–4.

Pagola, A., Vidal, C. and Ross, J. (1988) Measurement of dispersion relation of chemical waves in an oscillatory reacting medium. *J. Phys. Chem.* **92**, 163–6.

Ross, J., Müller, S. C. and Vidal, C. (1988) Chemical waves. *Science, N.Y.* **240**, 460–5.

Welsh, B. J., Gomatam, J. and Burgess, A. E. (1983) Three-dimensional chemical waves in the Belousov–Zhabotinskii reaction. *Nature* **304**, 611–4.

Winfree, A. T. (1972) Spiral waves of chemical activity. *Science, N.Y.* **175**, 634–6.

Winfree, A. T. (1983) Sudden cardiac death. A problem of topology. *Scient. Am.* **248**(5), 144–61.

Winfree, A. T. and Jahnke, W. (1989) *J. Phys. Chem.* (in press).

Winfree, A. T. and Strogatz, S. H. (1984) Singular filaments organize chemical waves in three dimensions. *Physica* **13D,** 221–33.

Wood, P. M. and Ross, J. (1985) A quantitative study of chemical waves in the Belousov–Zhabotinskii reaction. *J. Chem. Phys.* **82,** 1924–36.

Zaikin, N. and Zhabotinskii, A. N. (1970) Concentration wave propagation in a two-dimensional, liquid-phase self-oscillating system. *Nature* **225,** 535–7.

Zykov, V. S. and Morozova, G. L. (1979) Speed of spread of excitation in a two-dimensional excitable medium. *Biophysics* **24,** 739–44.

Cyclic-AMP waves in *Dictyostelium*: Specific models and general theories

JOHN J. TYSON

Department of Biology, Virginia Polytechnic Institute and State University, Blacksburg, VA 24061, USA

COMMUNICATION BY TRAVELLING WAVES IN EXCITABLE MEDIA

When amoebae of the cellular slime mould *Dictyostelium discoideum* exhaust their food supply, they begin signalling each other with a chemical messenger, cyclic adenosine 3',5'-monophosphate (cAMP). cAMP diffuses through the extracellular milieu and binds to receptor proteins on the plasma membrane of amoebae. Upon binding cAMP the membrane receptor initiates a sequence of intracellular responses, including activation of adenylate cyclase, the enzyme that synthesizes cAMP from ATP. Newly synthesized cAMP is excreted by stimulated cells, thus passing on the signal to their neighbours. Amplification of the signal is limited by desensitization of the membrane receptor: on prolonged exposure to cAMP the receptor system loses its ability to stimulate adenylate cyclase activity. In the absence of cAMP synthesis, cAMP concentration decreases by the action of intracellular and extracellular phosphodiesterases, which hydrolyse cAMP to 5'AMP.

In a field of signalling amoebae spread over an agar surface, pulses of extracellular cAMP travel across the domain as expanding concentric circular waves or as rotating spiral waves. As a wave of cAMP passes an amoeba, it stimulates cAMP synthesis by the cell, thereby regenerating the signal, and it stimulates chemotactic movement of the cell in the direction from which the wave came, thereby causing the amoebae to aggregate at the centre of the pattern. The multicellular slug that is formed in this manner develops further into a fruiting body, from which spores are disseminated to start new colonies in (hopefully) more favourable environments.

The aggregation phase of the *Dictyostelium* life cycle is a paradigm of cell

Cell to Cell Signalling: From
Experiments to Theoretical Models
ISBN 0–12–287960–0

to cell signalling. Like hormones and pheromones, a chemical messenger is excreted by a source (a gland or an ovum), the messenger passes by diffusion and/or convection through the separating medium (the circulatory system or external milieu), the messenger is received by binding to membrane proteins on target cells (e.g. another organ within the body, or sperm cells in the vicinity), and these receptors translate the message into appropriate action (changes in intracellular biochemistry and/or chemotactic movement). Unlike hormonal systems, however, the cells of *Dictyostelium* function simultaneously as source and target cells, which causes the signal to be amplified locally and to propagate actively across the aggregation field. In this respect, communication among *Dictyostelium* amoebae resembles propagation of action potentials in neural and muscular tissues. Indeed, the expanding target patterns and rotating spiral waves of cAMP observed during amoebal aggregation (Alcantara and Monk, 1974; Gross *et al.*, 1976; Tomchik and Devreotes, 1981; Devreotes *et al.*, 1983) are remarkably similar to waves of electrical excitation observed in mammalian heart preparations (Allessie *et al.*, 1973, 1984; Winfree, 1987). Similar travelling waves of chemical, electrical or mechanical activity may also mediate intercellular communication during certain developmental processes (Cooke and Zeeman, 1976; Odell *et al.*, 1981; Oster and Odell, 1984). Even in non-living systems, actively propagating waves of chemical reaction are observed (Zaikin and Zhabotinskii, 1970; Winfree, 1972, 1973, 1987; Field and Burger, 1985). In all of these cases the medium through which the signal passes is excitable: rather than passively transmitting a signal which steadily dissipates, the medium actively regenerates the signal, causing it to propagate indefinitely with constant speed and amplitude. Because of these analogies, a thorough understanding of travelling waves of cAMP in *Dictyostelium* aggregation fields should provide valuable insight into cell communication by wave propagation through excitable media.

SIGNAL RECEPTION AND AMPLIFICATION IN *DICTYOSTELIUM*

To construct a realistic and accurate model of cAMP wave propagation during aggregation of *Dictyostelium* amoebae, we must gather the best current biochemical information about the membrane receptor, adenylate cyclase and phosphodiesterase. This synthesis of experimental data into a coherent mathematical model of the cAMP signalling system in *Dictyostelium* has been carried out recently by Martiel and Goldbeter (1987). Their model relies heavily on the 'receptor box' picture of the receptor–cAMP interaction, which was outlined theoretically by Seger *et al.*, (1986) and confirmed experimentally by Devreotes and Sherring (1985).

In this picture, the receptor exists in four forms:

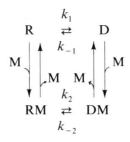

The two forms of the receptor (R and D) are interconverted by relatively slow phosphorylation and dephosphorylation reactions (described by rate constants k_1, k_{-1}, k_2, k_{-2}), and they can sense messenger (M) by fast, reversible binding steps (dissociation constants K_R and K_D). According to the measurements of Devreotes and Sherring, in the absence of cAMP, most receptor molecules are in form R (R:D = 8:1). When cAMP first appears to unstimulated cells, most of it binds rapidly to the R form, but this shifts the equilibrium in favour of form D (RM:DM = 1:4). Assuming, as do Martiel and Goldbeter, that only form RM is active in stimulating adenylate cyclase activity, we see that, on exposure to cAMP, adenylate cyclase activity first increases in response to RM formation, but then decreases as RM is converted to the inactive form DM ('receptor desensitization'). As messenger is removed by phosphodiesterase, form D reverts slowly to form R and the cell recovers its ability to respond to extracellular cAMP.

Stimulation of adenylate cyclase by receptor–cAMP complexes (RM) is a complicated process, mediated by 'protein G', and is currently under vigorous study (Klein *et al.*, 1988). Martiel and Goldbeter (1987) assume a simple cooperative activation step, $2RP + C \rightleftharpoons E$, where E is the 'active' form of adenylate cyclase (100-fold greater affinity for ATP, and 100-fold greater turnover number). To the kinetics of the receptor box and adenylate cyclase reaction, Martiel and Goldbeter add steps for the passive transport of cAMP across the plasma membrane and for the action of intracellular and extracellular phosphodiesterases. Altogether their model can be expressed as three kinetic equations

$$d\beta/dt = \underset{\text{(synthesis)}}{\sigma\phi(\rho,\gamma)} - \underset{\text{(secretion)}}{k_t\beta} - \underset{\text{(hydrolysis)}}{k_i\beta} \tag{1}$$

$$d\gamma/dt = \underset{\text{(secretion)}}{(k_t/h)\beta} - \underset{\text{(hydrolysis)}}{k_e\gamma} \tag{2}$$

$$d\rho/dt = \underset{\text{(desensitization)}}{-f_1(\gamma)\rho} + \underset{\text{(resensitization)}}{f_2(\gamma)(1-\rho)} \tag{3}$$

where β and γ are, respectively, the intracellular and extracellular cAMP

Table 1. Kinetic parameters

Name	Description	Experimental range*	Values used in calculations[†]			
			Set A	Set B	Set C	Set D
k_1	Rate constant	0.012 per min	0.036	0.036	0.12	0.036
k_{-1}	Rate constant	0.104 per min	0.36	0.36	1.2	0.36
k_2	Rate constant	0.22 per min	0.67	0.67	2.2	0.67
k_{-2}	Rate constant	0.055 per min	0.0033	0.0033	0.011	0.0033
K_R	Dissociation constant	10^{-7}–10^{-9} M	10^{-7}	—	—	—
K_D	Dissociation constant	10^{-8} M	10^{-8}	—	—	—
c	K_R/K_D	0.1–10	10	—	—	—
λ_1	Dimensionless constant	~10^{-4}	10^{-4}	10^{-3}	10^{-3}	10^{-3}
λ_2	Dimensionless constant	~1	0.26	2.4	2.4	2.4
σ	Rate constant	10^2–10^4 per min	1.8×10^3	1.8×10^3	6×10^3	2.6×10^3
k_t	Transport coefficient	0.3–0.9 per min	0.9	0.9	3.0	5.5
k_i	Rate constant	1.7 per min	1.7	1.0	3.3	1.7
k_e	Rate constant	2–12 per min	5.4	3.6	12	3.6
h	Volume ratio	5–100	5	—	—	—
D	Diffusion coefficient	0.024 mm²/min[‡]	0.024	—	—	—

*From Martiel and Goldbeter (1987).

[†]Units are the same as in column giving experimental range. The symbol — denotes that all four parameter sets assume the same value.

[‡]Dworkin and Keller (1977).

concentrations relative to K_R, the dissociation constant for the receptor–cAMP complex ($RM \rightleftharpoons R + M$), and ρ is the fraction of receptor in the active form:

$$\beta = \frac{[cAMP]_{intra}}{K_R}, \quad \gamma = \frac{[cAMP]_{extra}}{K_R}, \quad \rho = \frac{[R]+[RM]}{[R]+[RM]+[D]+[DM]}$$

The function

$$\varphi(\rho,\gamma) = \frac{\lambda_1 + Y^2}{\lambda_2 + Y^2}, \quad Y = \frac{\rho\gamma}{1+\gamma} \qquad (4)$$

describes the relative activity of adenylate cyclase (as a function of receptor state and extracellular cAMP concentration), and the functions

$$f_1(\gamma) = \frac{k_1 + k_2\gamma}{1+\gamma}, \quad f_2(\gamma) = \frac{k_{-1} + k_{-2}c\gamma}{1+c\gamma} \qquad (5)$$

describe the specific rates of desensitization and resensitization of the membrane receptor. The kinetic parameters in the model equations (1)–(5) are specified in Table 1.

Using eqns (1)–(5) Martiel and Goldbeter (1987) calculated cAMP oscillations expected in well-stirred suspensions of *Dictyostelium* cells (see Fig. 1). For parameter set A in Table 1, the period of oscillation (10 min) and the amplitude of oscillation (intracellular cAMP varies between 1 and 20 µM) agree well with experimental observations (Gerisch and Wick, 1975). For slightly different values of the parameters (set B in Table 1), the model equations no longer admit sustained oscillations, but the locally stable steady state is excitable. When perturbed by a sufficiently large pulse of extracellular cAMP, the model equations respond by a burst of synthesis of intracellular cAMP: the relay response (Fig. 2). The timing and magnitude of the relay

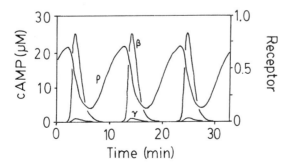

Fig. 1. cAMP oscillations in the MG model, eqns (1)–(3) for parameter set A in Table 1. (From Martiel and Goldbeter, 1987, used by permission.)

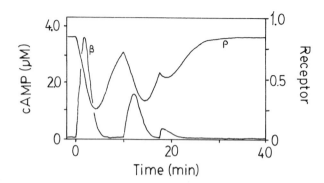

Fig. 2. Relay response to cAMP pulses, calculated from eqns (1)–(3) for parameter set B in Table 1. (From Martiel and Goldbeter, 1987, used by permission.) For $t<0$, the system is at a stable steady state. At $t=0$, the system is perturbed by a pulse of 3×10^{-8} M extracellular cAMP, which elicits the relay response: transient synthesis of cAMP intracellularly. Ten minutes after the first pulse, the system is perturbed by a second pulse of the same magnitude, which elicits a smaller relay response because the membrane receptor has not recovered full sensitivity by the time of the second stimulation. A third stimulus, 7.5 min after the second, elicits a still smaller response.

response and the adaptation of the system to repeated stimulation with extracellular cAMP agree well with experiments, as discussed in detail by Martiel and Goldbeter (1987).

In the calculations reported in Figs 1 and 2, Martiel and Goldbeter used experimentally available values for all parameters except the rate constants for the $R \rightleftarrows D$ and $RM \rightleftarrows DM$ transitions in the receptor box. To get the experimentally observed time scale for oscillations and relay response, Martiel and Goldbeter had to take values of these rate constants three times larger than the values determined experimentally by Devreotes and Sherring (1985). These rate constants have recently been redetermined by a more sensitive experimental method, and the revised values are two times larger than originally reported (Devreotes, private communication).

SPIRAL WAVES OF cAMP IN *DICTYOSTELIUM* AGGREGATION FIELDS

To model cAMP wave propagation in fields of signalling amoebae, we must modify eqns (1)–(3) to introduce spatial coupling by diffusion of cAMP through the extracellular milieu. Thus, eqn (2) becomes

$$\partial\gamma/\partial t = (k_t/h)\beta - k_e\gamma + D\nabla^2\gamma \tag{2a}$$

where $\nabla^2 = \partial^2/\partial x^2 + \partial^2/\partial y^2$ is the Laplacian operator in two spatial dimensions, and D is the diffusion coefficient of cAMP in dilute aqueous solution (see Table 1). For simplicity, we shall assume that the cells maintain a uniform distribution in space, so that terms for random cell movements and chemotactic movements are not necessary in eqns (1) and (3). This is a reasonable assumption early in the aggregation stage, when cell movement is negligible compared to cAMP diffusion. So, our model for cAMP wave propagation at the start of the signalling phase consists of eqns (1), (2a) and (3).

Tyson *et al.*, (1989) have computed spiral wave solutions of the reaction–diffusion system (1), (2a), (3) for parameter set B (Table 1), suggested originally by Martiel and Goldbeter (1987) to model the relay response of *Dictyostelium* amoebae in well-stirred cell suspensions. When this excitable system is distributed over a two-dimensional spatial domain and allowed to communicate only by diffusion of extracellular cAMP, then, given appropriate initial conditions, a rotating spiral wave of cAMP develops in the field of cells (Fig. 3). The period of rotation of the spiral wave is 48 min, and the speed of propagation of the wave is 0.15 mm/min. Devreotes and co-workers (Tomchik and Devreotes, 1981; Devreotes *et al.*, 1983) have made careful measurements of spiral cAMP waves observed by a clever competitive-binding assay. The spiral waves they measured had a common rotation period of 7 min and a wave speed of 0.3 mm/min. The spiral wave calculated by Tyson and co-workers for the MG model agrees reasonably well with these observations considering the fact that parameter set B was developed by Martiel and Goldbeter to account for signal relaying in well-stirred cell suspensions, a preparation quite different from the agar-surface cultures employed for wave propagation studies. By modifying some of the kinetic parameters in set B, we can change the time and space scales for the spiral wave illustrated in Fig. 3. For example, for parameter set C, the spiral has a rotation period of 14 min and a wave speed of 0.28 mm/min. A more thorough investigation of the parameter space might uncover a reasonable parameter set with even better agreement to experiment.

For a given parameter set, there is a unique spiral wave solution to the reaction–diffusion equations, but there is a family of target pattern solutions (i.e. expanding concentric circular waves). The target patterns within this family differ from each other according to their temporal period, but once the temporal period is specified the wave speed and the wavelength of the pattern is uniquely determined. For target patterns in *Dictyostelium* agar-surface cultures, the wave speed is roughly constant (0.26 mm/min) independent of temporal period in the range 4–10 min (Alcantara and Monk, 1974). The dependence of wave speed on period predicted by the MG model (Tyson *et al.*, 1989) is compared with the measurements of Alcantara and Monk (1974) in Fig. 4. The agreement is excellent.

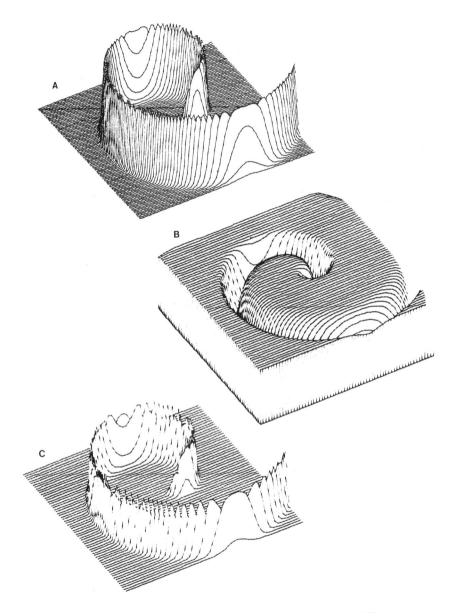

Fig. 3. Spiral wave solution of the three-component reaction–diffusion equations (1), (2a), (3). (A) Extracellular cAMP; (B) active receptor; (C) intracellular cAMP. The solution is identical for parameter sets B and C, except for a change in time and space scales: set B period 48 min, wavelength 7 mm; set C period 14 min, wavelength 4 mm. (From Tyson *et al.*, 1989, used by permission.)

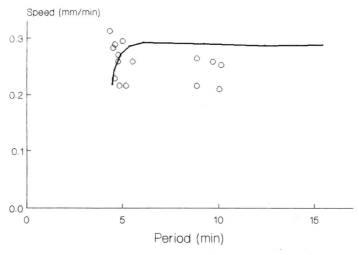

Fig. 4. Relation between wave speed and temporal period for target patterns, as predicted by the MG model, parameter set C (solid line), and as measured by Alcantara and Monk (open circles). (From Tyson *et al.*, 1989, used by permission.)

GENERAL THEORY OF TRAVELLING WAVES IN EXCITABLE MEDIA

A general theory of travelling waves in excitable media has developed around reaction–diffusion equations of the form

$$\frac{\partial u}{\partial t} = \varepsilon \nabla^2 u + \frac{1}{\varepsilon} f(u,v) \tag{6}$$

$$\frac{\partial v}{\partial t} = \varepsilon \delta \nabla^2 v + g(u,v) \tag{7}$$

where u,v are state variables (e.g. chemical concentrations, receptor activity, membrane potential, membrane conductance), ∇^2 is the Laplacian operator in one, two or three spatial dimensions, and $\delta = D_v/D_u$ is the ratio of diffusion coefficients of the two state variables ($\delta \geqslant 0$). The functions $f(u,v)$ and $g(u,v)$ describe the non-linear kinetic interactions between u and v, and ε is a small positive parameter ($0 < \varepsilon \ll 1$) representing a separation of time scales. Roughly speaking, u is a 'fast' variable and v a 'slow' variable. A typical phase plan for the reaction system $\varepsilon \dot{u} = f(u,v)$, $\dot{v} = g(u,v)$ (no diffusion) is illustrated in Fig. 5.

The MG model for cAMP waves, eqns (1), (2a), (3), is not in the general

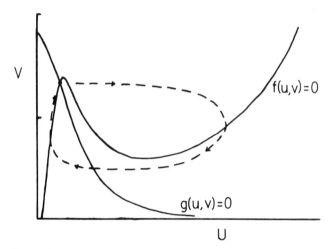

Fig. 5. Phase plane for the system $\varepsilon\dot{u}=f(u,v)$, $\dot{v}=g(u,v)$, with f and g given by eqns (8) and (9). The solid lines are the 'nullclines' $f(u,v)=0$ and $g(u,v)=0$. The dashed line is a typical trajectory illustrating excitability of the system. For ε small ($0<\varepsilon\ll1$), the trajectory moves faster in the u-direction (horizontally) than in the v-direction (vertically). (Adapted from Tyson *et al.*, 1989, used by permission.)

form of system (6), (7), but it can be cast into this form. For parameter set D in Table 1, Tyson *et al.* (1989) have shown that the activity of the membrane receptor is the 'slow' variable, extracellular cAMP concentration is the 'fast' variable, and intracellular cAMP concentration is the 'fastest' variable. Because $\beta=[cAMP]_{intra}/K_R$ changes on the fastest time scale, we can eliminate eqn (1) by expressing β as a function of γ and ρ:

$$\beta=[\sigma/(k_t+k_i)]\varphi(\rho,\gamma)$$

Equations (2a) and (3) become

$$\frac{\partial}{\partial t}=\frac{k_t\sigma}{h(k_t+k_i)}\varphi(\rho,\gamma)-k_e\gamma+D\nabla^2\gamma$$

$$\frac{\partial}{\partial t}=-f_1(\gamma)\rho+f_2(\gamma)(1-\rho)$$

which can be cast into the dimensionless form (6), (7) by introducing dimensionless time and space variables:

$$k_1t\rightarrow t,\quad \frac{k_1x}{\sqrt{(k_eD)}}\rightarrow x$$

In this case, $u = \gamma$ = extracellular cAMP, $v = \rho$ = membrane receptor in active form, $\varepsilon = k_1/k_e \cong 0.01$, $\delta = 0$, and

$$f(u,v) = s\varphi(v,u) - u \qquad (8)$$

$$g(u,v) = -\frac{1+\kappa u}{1+u}v + \frac{L_1+\kappa L_2 cu}{1+cu}(1-v) \qquad (9)$$

with $s = k_t \sigma/k_e h(k_t + k_i) \cong 750$, $\kappa = k_2/k_1 = 18.5$, $L_1 = k_{-1}/k_1 = 10$, and $L_2 = k_{-2}/k_2 = 0.005$.

For ε small, system (6), (7) can be analysed by singular perturbation theory (Ortoleva and Ross, 1975; Ostrovskii and Yakhno, 1975; Fife, 1976; Tyson and Fife, 1980; Keener, 1986; Tyson and Keener, 1988). In one spatial dimension, the basic result is the existence of propagation fronts that switch the system, at constant v, from the left hand branch of $f(u,v) = 0$ to the right hand branch (see Fig. 5), or vice versa. In the case of *Dictyostelium* such fronts correspond to waves of cAMP synthesis (left to right in Fig. 5) or cAMP degradation (right to left). From such fronts one can construct periodic travelling waves of excitation, and the speed of these waves depends on their period as in Fig. 6. This dependence of speed on period, $c = C(T)$, is called the dispersion relation of the excitable medium.

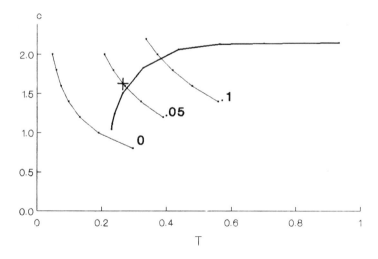

Fig. 6. Dispersion and curvature relations. The dispersion relation (solid line) was calculated from eqns (6)–(9), parameter set D. The curvature relations (dashed lines) were calculated from eqn (12) for $r_0 = 0$, and eqn (12a) for $r_0 = 0.05$ and 0.10. The + marks the location of the computed spiral wave. (From Tyson *et al.*, 1989, used by permission.)

The characteristic periodic patterns in two spatial dimensions, targets and spirals, must satisfy the dispersion relation because sufficiently far from the centre of either pattern in a radial direction both targets and spirals are identical to one-dimensional periodic travelling waves. Target patterns need only satisfy the dispersion relation. That is, given any temporal period (T) above the minimum period in Fig. 6, the asymptotic speed (c) of propagation is fixed by the dispersion relation and the wavelength is simply $\lambda = cT$. Spiral waves on the other hand seem to obey another constraint in addition to the dispersion relation because in a given medium the rotation frequencies of all spiral waves are the same.

The additional constraint on spiral waves arises from consideration of the effects of wavefront curvature on speed of propagation (Zykov, 1980a,b, 1988; Fife, 1984; Keener, 1986; Keener and Tyson, 1986; Tyson and Keener, 1988). These authors show that the normal velocity of a wavefront (N) is equal to the speed of plane-wave propagation (c) adjusted by an amount proportional to the curvature (K) of the front:

$$N = c + \varepsilon K \tag{10}$$

For positive curvature (wavefront curved in the direction of propagation) $N > c$, whereas for negative curvature (wavefront curved away from its direction of propagation) $N < c$.

To see how eqn (10) constrains spiral waves, consider the parametric equations for a rigidly rotating one-armed spiral

$$x = r \cos[\theta(r) - \omega t]$$
$$y = s \sin[\theta(r) - \omega t]$$

Here $\theta(r)$ determines the shape of the spiral (at fixed t) and ω is the angular frequency of rotation ($\omega = 2\pi/T$). Our problem is to determine both $\theta(r)$ and ω. Since N depends on $\theta'(r)$ and ω, and K depends on $\theta'(r)$ and $\theta''(r)$, eqn (10) is really an ordinary differential equation for the unknown function $\theta(r)$ in terms of two parameters c and ω:

$$\frac{\omega r}{(1 + r^2\theta'^2)^{1/2}} = c + \varepsilon \frac{(\theta' + r\theta'') + \theta'(1 + r^2\theta'^2)}{(1 + r^2\theta'^2)^{3/2}} \tag{10a}$$

Insisting that the solution of (10a) corresponds to an Archimedean spiral far from the origin, i.e. $\theta(r) = \omega r/c$ for r large, we find (Tyson and Keener, 1988) that the solution of (10a) is given approximately by

$$\theta(r) \cong \frac{\omega r}{c} + \frac{\omega \varepsilon}{c^2} \ln\left(\sqrt{(3)} + \frac{cr}{\varepsilon}\right) \tag{11}$$

provided that

$$\frac{\omega\varepsilon}{c^2} = \frac{3-\sqrt{3}}{4} \cong 0.32 \tag{12}$$

Equation (12) is a relation between wave speed (c) and period ($T = 2\pi/\omega$) which must be satisfied by spiral waves in addition to satisfying the dispersion relation $c = C(T)$.

The curvature relation (12) was derived assuming that the spiral wave can be continued all the way to the origin; see eqn (11). More realistically, the spiral wavefront ends at a finite radius, r_0, measuring the size of the 'core' of the spiral. Tyson and Keener (1988) have derived a more complicated expression for the curvature relation in this case:

$$\frac{4r_0\omega}{c} = \frac{1 + 4(cr_0/\varepsilon) - \sqrt{(1 + 8(cr_0/\varepsilon))}}{1 + (cr_0/\varepsilon)} \tag{12a}$$

Curvature relations (12) and (12a) are illustrated in Fig. 6 along with the dispersion relation. Spiral waves are expected at the intersection of the dispersion relation with the curvature relation of correct 'core' size r_0. In Fig. 6 the spiral wave is expected to have a core size $r_0 = 0.05$; the computed core size was $r_0 = 0.03$.

This view of curvature and spiral waves has been tested several ways. Direct experimental confirmation of eqn (10) has been obtained for oxidation waves in the Belousov–Zhabotinskii reaction (Foerster et al., 1988), and the theory has been compared in detail with experimental measurements of spiral waves in this reaction (Keener and Tyson, 1986). Also, the theory has been compared in detail with numerical solutions of PDE (6)–(7) for Oregonator kinetics (Tyson and Keener, 1988), FitzHugh–Nagumo kinetics (Tyson and Keener, 1987), and Martiel–Goldbeter kinetics (Tyson et al., 1989). In all cases there is good agreement among theory, numerics and experimental observations.

In three-dimensional space, spiral waves become scroll-shaped waves rotating around a one-dimensional filament which threads through the spatial domain, either meeting the boundary or closing on itself in a ring. During the course of many rotations of the scroll wave around the filament, the filament itself moves through space. The filament moves because it is pulled about by the rotating scroll wave which at any instant in any local region is attempting to move with normal velocity $N = c + \varepsilon(K_1 + K_2)$, where K_1 and K_2 are the principal curvatures of the wavefront surface. Keener (1988) has used this notion to derive a set of equations describing the motion of the filament:

alteration in rotation
rate of scroll wave $\quad = c_1\kappa - a_1 w^2 + b_1\dfrac{\partial w}{\partial s}$ $\qquad$ (13)
around filament

$$\begin{matrix} \text{normal component} \\ \text{of velocity of filament} \end{matrix} = b_2\kappa - a_2 w^2 + c_2\frac{\partial w}{\partial s} \tag{14}$$

$$\begin{matrix} \text{binormal component} \\ \text{of velocity of filament} \end{matrix} = c_3\kappa - a_3 w^2 + c_4\frac{\partial w}{\partial s} \tag{15}$$

where $s =$ arclength along filament, $\kappa(s,t) =$ curvature of filament, and $w(s,t) =$ twist rate of scroll wave around a filament as measured in the laboratory frame of reference. The coefficients a_i, b_i, c_i are constants which depend on the matrix of diffusion coefficients and the form of the spiral wave solution to the two-dimensional problem. In the simple case of equal diffusion coefficients ($\delta = 1$), $b_1 = b_2 = D$ (the diffusion coefficient) and $c_1 = c_2 = c_3 = c_4 = 0$. If, furthermore, the filament is untwisted and untorted ($w = 0$), then Keener's equations reduce to the simple relation $n = D\kappa$, where n is the normal velocity of the filament in its tangent plane.

The equation $n = D\kappa$ is the simplest equation of motion for a scroll-wave filament. It has been derived by many people in diverse ways (Panfilov and Pertsov, 1984; Panfilov et al., 1986; Keener and Tyson, 1988) and applied primarily to the case of scroll rings. If r is the radius of a circular filament, then $n = D\kappa$ implies that $dr/dt = -D/r$, or $r(t) = (r_0^2 - 2Dt)^{1/2}$. That is, scroll rings should shrink and vanish in finite time. Such behaviour is observed in numerical calculations on PDE (6)–(7) (Panfilov et al., 1986) and in experimental observations of BZ scroll rings (Welsh et al., 1983). Keener and Tyson (1988) have also applied $n = D\kappa$ to the case of elongated spiral waves and elongated target patterns observed in thick layers of BZ reagent (Winfree, 1973, 1974), and they found remarkable agreement between theory and experiment.

CONCLUSIONS

(1) cAMP signalling in *Dictyostelium* aggregation fields provides a stunning example of cell to cell communication by wave propagation in excitable media. Other examples of such communication include nerve impulse propagation, cardiac contraction waves, and morphogenetic movements.

(2) The cAMP signalling system is known in enough biochemical detail that a realistic mathematical model can be constructed. Numerical simulations of this model (the Martiel–Goldbeter model) agree in quantitative detail with experimental observations of cAMP oscillations, the cAMP relay-response, and cAMP wave propagation.

(3) The Martiel–Goldbeter model can be cast in a general form suited to

mathematical analysis of wave propagation in excitable media, thus strengthening the association of travelling cAMP waves with other more thoroughly studied systems (nerve, heart and BZ reagent).

(4) A general theory of wave propagation in one-, two- and three-dimensional excitable media, which has recently come to maturity, is quite successful in accounting for typical properties of such waves: dispersion, spiral rotation, and scroll-wave motion. This theory provides a general framework for understanding cell to cell communication by wave propagation in diverse excitable systems.

Acknowledgements: Lengthy discussions over many years with Jim Keener, Art Winfree and Paul Fife have shaped my ideas about travelling waves in excitable media. Advice and encouragement from Albert Goldbeter and Peter Devreotes was essential in modelling cAMP waves in *Dictyostelium*. This work was supported by NSF Grant DMS-8810456.

REFERENCES

Alcantara, F. and Monk, M. (1974) Signal propagation during aggregation in the slime mould *Dictyostelium discoideum. J. gen. Microbiol.* **85**, 321–34.

Allessie, M. A., Bonke, F. I. M. and Schopman, F. J. G. (1973) Circus movement in rabbit atrial muscle as a mechanism of tachycardia. *Circulation Res.* **33**, 54–62.

Allessie, M. A., Lammers, W. J. E. P., Bonke, F. I. M. and Hollen, J. (1984) Intraatrial reentry as a mechanism for atrial flutter induced by acetylcholine and rapid pacing in the dog. *Circulation* **70**, 123–35.

Cooke, J. and Zeeman, E. C. (1976) A clock and wavefront model for control of the number of repeated structures during animal morphogenesis. *J. theor. Biol.* **58**, 455–76.

Devreotes, P. N., Potel, M. J. and MacKay, S. A. (1983) Quantitative analysis of cAMP waves mediating aggregation in *Dictyostelium discoideum. Devl Biol.* **96**, 405–15.

Devreotes, P. N. and Sherring, J. A. (1985) Kinetics and concentration dependence of reversible cAMP-induced modification of the surface cAMP receptor in *Dictyostelium. J. biol. Chem.* **260**, 6378–84.

Dworkin, M. and Keller, K. H. (1977) Solubility and diffusion coefficient of adenosine 3':5' monophosphate. *J. biol. Chem* **252**, 864–5.

Field, R. J. and Burger, M. (1985) *Oscillations and Traveling Waves in Chemical Systems.* John Wiley, New York.

Fife, P. C. (1976) Pattern formation in reacting and diffusing systems, *J. Chem. Phys.* **64**, 554–64.

Fife, P. C. (1984) Propagator-controller systems and chemical patterns. In *Non-Equilibrium Dynamics in Chemical Systems* (eds C. Vidal and A. Pacault), pp. 76–88. Springer-Verlag, Berlin.

Foerster, P., Müller, S. C. and Hess, B. (1988) Curvature and propagation velocity of chemical waves, *Science, N.Y.* **241**, 685–7.

Gerisch, G. and Wick, U. (1975) Intracellular oscillations and release of cyclic AMP from *Dictyostelium* cells. *Biochem. biophys. Res. Commun.* **65**, 364–70.

Gross, J. D., Peacey, M. J. and Trevan, D. J. (1976) Signal emission and signal propagation during early aggregation in *Dictyostelium discoideum*. *J. Cell Sci.* **22**, 645–56.

Keener, J. P. (1986) A geometrical theory for spiral waves in excitable media, *SIAM J. appl. Math.* **46**, 1039–56.

Keener, J. P. (1988) The dynamics of three dimensional scroll waves in excitable media. *Physica* **31D**, 269–76.

Keener, J. P. and Tyson, J. J. (1986) Spiral waves in the Belousov–Zhabotinskii reaction. *Physica* **21D**, 307–24.

Keener, J. P. and Tyson, J. J. (1988) The motion of untwisted untorted scroll waves in Belousov–Zhabotinsky reagent. *Science, N.Y.* **239**, 1284–6.

Klein, P. S., Sun, T. J., Saxe, C. L., Kimmel, A. R., Johnson, R. L. and Devreotes, P. N. (1988) A chemoattractant receptor controls development in *Dictyostelium discoideum*. *Science, N.Y.* **241**, 1467–72.

Martiel, J. L. and Goldbeter, A. (1987) A model based on receptor desensitization for cyclic-AMP signalling in *Dictyostelium* cells. *Biophys. J* **52**, 807–28.

Odell, G. M., Oster, G., Alberch, P. and Burnside, B. (1981) The mechanical basis of morphogenesis. I. Epithelial folding and invagination. *Devl Biol.* **85**, 446–62.

Ortoleva, P. and Ross, J. (1975) Theory of propagation of discontinuities in kinetic systems with multiple time scales: fronts, front multiplicity, and pulses. *J. Chem. Phys.* **63**, 3398–408.

Oster, G. F. and Odell, G. M. (1984) The mechanochemistry of cytogels. *Physica* **12D**, 333–50.

Ostrovskii, L. A. and Yakhno, V. G. (1975) Formation of pulses in an excitable medium. *Biophysics* **20**, 498–503.

Panfilov, A. V. and Pertsov, A. M. (1984) Vortex ring in a three-dimensional active medium described by reaction–diffusion equations. *Dokl. Biophys.* **274**, 58–60.

Panfilov, A. V., Rudenko, A. N and Krinskii, V. I. (1986) Vortex rings in 3-dimensional active media with diffusion in two components. *Biofizika* **31**, 850–4.

Segel, L. A., Goldbeter, A., Devreotes, P. N. and Knox, B. E. (1986) A mechanism for exact sensory adaptation based on receptor modification. *J. theor. Biol.* **120**, 151–79.

Tomchik, K. J. and Devreotes, P. N. (1981) Adenosine 3′,5′-monophosphate waves in *Dictyostelium discoideum*: a demonstration by isotope dilution-fluorography. *Science, N.Y.* **212**, 443–6.

Tyson, J. J. and Fife, P. C. (1980) Target patterns in realistic model of the Belousov–Zhabotinskii reaction, *J. Chem. Phys.* **73**, 2224–37.

Tyson, J. J. and Keener, J. P. (1987) Spiral waves in a model of myocardium. *Physica* **21D**, 215–22.

Tyson, J. J. and Keener, J. P. (1988) Singular perturbation theory of traveling waves in excitable media (a review). *Physica* **32D**, 327–61.

Tyson, J. J., Alexander, K. A., Manoranjan, V. S. and Murray, J. D. (1989) Cyclic-AMP waves during aggregation of *Dictyostelium* amoebae, *Physica* **34D**, 193–207.

Welsh, B. J., Gomatam, J. and Burgess, A. E. (1983) Three-dimensional chemical waves in the Belousov–Zhabotinskii reaction. *Nature* **304**, 611–4.

Winfree, A. T. (1972) Spiral waves of chemical activity. *Science, N.Y.* **175**, 634–6.

Winfree, A. T. (1973) Scroll-shaped waves of chemical activity in three dimensions. *Science, N.Y.* **181**, 937–9.

Winfree, A. T. (1974) Rotating chemical reactions. *Sci. Am.* **230**(6), 82–95.

Winfree, A. T. (1987) *When Time Breaks Down*. Princeton University Press, Princeton, NJ.

Zaikin, A. N. and Zhabotinskii, A. M. (1970) Concentration wave propagation in two-dimensional liquid-phase self-oscillating system. *Nature* **225**, 535–7.

Zykov, V. S. (1980a) Kinematics of the steady circulation in an excitable medium. *Biophysics* **25**, 329–33.

Zykov, V. S. (1980b) Analytical evaluation of the dependence of the speed of an excitation wave in a two-dimensional excitable medium on the curvature of its front. *Biophysics* **25,** 906–11.
Zykov, V. S. (1988) *Modelling of Wave Processes in Excitable Media.* Manchester University Press, Manchester, UK.

Part 7
Signal propagation in the heart

On the non-linear motions of the heart: Fractals, chaos and cardiac dynamics

ARY L. GOLDBERGER AND DAVID R. RIGNEY

Department of Medicine, Beth Israel Hospital and Harvard Medical School, 330 Brookline Avenue, Boston, MA 02215, USA

INTRODUCTION

This chapter describes some applications of non-linear dynamics to cardiac physiology and to cell to cell signalling in circulatory control. Our goal is to introduce the general reader to concepts that offer new perspectives on problems of historic interest, including the anatomy of the cardiovascular system, the mechanism of physiologic heart rate variability, and the dynamics of sudden cardiac death.

Although physiology and clinical medicine are among the richest sources of non-linear behaviour, concepts from non-linear science are still alien to most investigators in these disciplines (Goldberger *et al.*, 1985a; Gleick, 1987; Goldberger and West, 1987; Koslow *et al.*, 1987; West and Goldberger, 1987; Glass and Mackey, 1988). As implied by the name, the term non-linear applies to systems whose output is not proportional to the strength of some stimulus (input). For example, the hysteresis loop describing the relationship between pressure and volume in the ventricles through the course of an entire cardiac cycle is non-linear, as is the voltage–current relationship in cardiac pacemaker cells. The classical Hodgkin–Huxley equation describing voltage changes in nerve cells as a function of ionic fluxes is also non-linear (Hodgkin and Huxley, 1952).

NON-LINEAR DIAGNOSTICS

Unfortunately, the precise laws of motion governing the dynamics of complex physiological systems are ordinarily unknown. How can an investigator in the laboratory or a clinician at the bedside infer the presence of a non-linear mechanism simply by examining the behaviour of the system? Are there any 'diagnostic' markers of non-linear dynamics?

Several dynamic patterns of behaviour do in fact provide evidence for underlying non-linear mechanisms including (1) sustained autonomous oscillations, (2) abrupt changes (bifurcations), and (3) chaos (fractal dynamics). Linear systems may also oscillate but the oscillations in such cases will damp-out over time unless they are driven by an external source. In non-linear systems, in contrast, autonomous oscillations may continue for minutes, hours, days or longer. Furthermore, such oscillations may start and stop abruptly. In the parlance of non-linear dynamics, sudden changes of this kind are referred to as bifurcations.

Finally, the concept of chaos or fractal dynamics refers to the apparently erratic, aperiodic fluctuations in the behaviour of a complex system. When the variables of such a system are plotted in a phase space representation, nearby trajectories diverge exponentially. This sensitivity to initial conditions is a characteristic feature of chaotic dynamics.

The frequency spectrum is also helpful in deciding whether a process is periodic or chaotic. The spectral representation of a periodic process is narrowband, that is, the spectrum shows one or a few closely spaced frequency peaks with low background. In contrast, the spectral respresentation of chaos is broadband, sometimes with superimposed frequency peaks (noisy periodicity) (Lorenz, 1980). Furthermore, the broadband noise of a chaotic or fractal process shows a $1/f$-like distribution, so named because of the inverse relationship between frequency and power.

The concept of chaos is virtually synonymous with that of fractal time (Shlesinger, 1987). The notion of a fractal is used in two related contexts in non-linear dynamics: spatial and temporal. From a spatial or geometric viewpoint, a fractal object is one with a fractional (non-integer) dimension. Such objects show a scaling invariance referred to as self-similarity (Mandelbrot, 1982). Magnification of a fractal object reveals multiple layers of small-scale detail that resemble the larger scale form. Such fractal architecture is exemplified by tree-like objects in nature that have multiple generations of self-similar branchings. These kinds of fractal structures, in contrast to classical Euclidean forms, lack a characteristic scale of length.

The fractal concept also extends to the temporal domain where it applies to complex processes that do not have a characteristic scale of time (i.e. they are not periodic). Instead, these chaotic processes, as just described, appear to be noisy and generate fluctuations over many scales of time. In a fractal or chaotic process, the fluctuations over different time scales are self-similar, resulting in the $1/f$-like distribution of the frequency spectrum.

FRACTAL SELF-ORGANIZATION OF THE CIRCULATORY SYSTEM

The branching circulatory system is a fractal-like system that provides a mechanism for distributing the blood volume (dimension = 3) in the large arteries to the capillary surface area (dimension = 2). The tracheobronchial tree serves a similar function for gas transport and distribution in the lungs. We have previously identified examples of fractal-like architecture in the heart including the branching coronary arteries, the chordae tendineae that support the mitral and tricuspid valves, the branching of certain muscle bundles inside the heart, and the ramifying conduction fibres that convey the electrical impulse to the ventricles (His–Purkinje system) (Fig. 1). The fractal model of the ventricular conduction system developed by Goldberger *et al.* (1985b) was recently shown by Eylon *et al.* (1989) to account not only for the high-frequency content of the electrocardiographic signal (QRS complex) but also for the appearance of pathologic late potentials associated with damage to the conduction system.

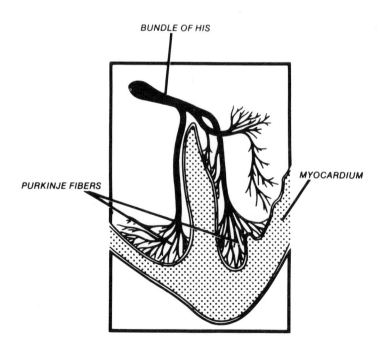

Fig. 1. The specialized conduction fibres (His–Purkinje network) that transmit electrical signals from the atria to ventricles are part of a fractal-like branching system. (Adapted from Goldberger *et al.*, 1985b.)

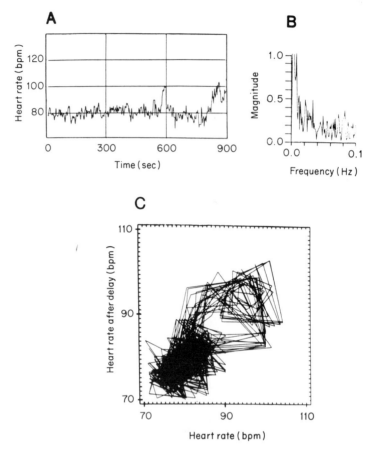

Fig. 2. Heart rate time series, frequency spectrum and two-dimensional phase space representation of 900 s of data from a healthy subject. (A) Shows the instantaneous heart rate which exhibits an erratic, aperiodic pattern. The spectrum is broadband, with a l/f-like distribution (B). (C) Shows the phase space trajectory of the heart rate vector where the first variable is the current heart rate, and the second variable is the heart rate after a fixed delay of 12 s. The strange-like attractor reveals heart rate trajectories converging on a dense central region from which they then appear to diverge. Compare this physiologic pattern with the abnormal pattern in Fig. 3. (Adapted from Goldberger and Rigney, 1988.)

CHAOS AND HEART RATE CONTROL

The fractal concept may also be applied to understanding the mechanism of physiologic heart rate variability, viz. fluctuations in the beat to beat sinus node rate (Kobayashi and Musha, 1982; Goldberger *et al.*, 1984; West and Goldberger, 1987). Under physiologic conditions, the heart rate is regulated by the firing of the pacemaker cells in the sinus node. This sinus rhythm mechanism, in turn, is controlled by signals from the involuntary (autonomic) branch of the nervous system. Contrary to conventional belief, the heart rate in healthy subjects at rest is not strictly regular or periodic. Instead, the heart rate in sinus rhythm fluctuates is an apparently erratic fashion with a time series representation consistent with chaos (Fig. 2). The spectrum of this healthy variability is broadband, with a $1/f$-like distribution (Kobayashi and Musha, 1982; Goldberger et al., 1984; West and Goldberger, 1987). The heart rate fluctuations are the result of a complex neurohumoral feedback system, involving the divergent influences of the sympathetic (accelerative) and parasympathetic (decelerative) branches of the autonomic nervous system. Currently, efforts are underway in our laboratory to model realistically the non-linear neurohumoral interactions that give rise to chaotic (fractal) variability in heart rate. In an important sense, these beat to beat fluctuations provide a sensitive and highly accessible way of assaying the dynamics of neural control. Thus, the non-linear dynamics of the heart rate can be viewed as epiphenomenal since they are under the influence of neural control, i.e. chaos of the heart beat reflects neurohumoral chaos.

That cell to cell signalling in neural control produces healthy chaos would seem an apparent violation of W. B. Cannon's principle of homeostasis (Cannon, 1929). In Cannon's formulation, physiological systems seek a constant state. In contrast, the example of heart rate variability (Fig. 2) indicates that chaos, not constancy, may be the 'wisdom of the body'. If heart rate is controlled by a non-linear feedback system and if the parameters regulating these variations are in a region where chaotic fluctuations occur, then the heart rate will not tend to settle down to a constant value (Goldberger et al., 1985; West and Goldberger, 1987; Glass and Mackey, 1988). It should be noted, however, that the presence of chaos is *not* inconsistent with certain constraints on the range of the system. In the case of heart rate control, these contraints will limit the maximum and minimum sinus heart rates in adults (e.g. usually between $200\,\text{min}^{-1}$ and $40\,\text{min}^{-1}$).

THE DYNAMICS OF SUDDEN CARDIAC DEATH

It has been postulated that cardiac chaos describes the dynamics of sudden cardiac death (Smith and Cohen, 1984). For reasons discussed below, this view is controversial (Goldberger and Rigney, 1988). Sudden cardiac death results most commonly from a rapid abnormal activation of the heart from a site or sites outside the sinus node (ectopic rhythms). The most lethal rhythms are generated by rapid electrical activity in the ventricles including the various types of ventricular tachycardia and ventricular fibrillation. During these ventricular tachyarrhythmias, the normal activation of the ventricles via the fractal-like His–Purkinje network no longer takes place. Instead, the activation site is peripheral, either in the myocardium or in the more distal part of the Purkinje network.

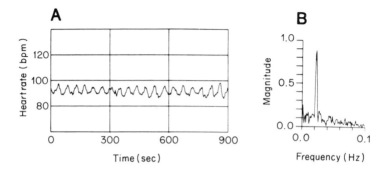

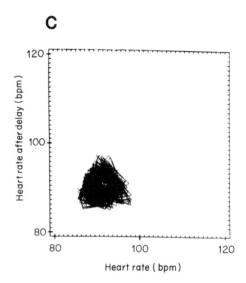

Fig. 3. Heart rate time series, frequency spectrum and two-dimensional phase space representation of 900 s of heart rate data from a patient who died suddenly 8 days after this recording. Note the highly periodic oscillations of heart rate (A) associated with a sharp spectral peak at about 0.02 Hz (B). The phase space plot (C) shows periodic orbits. These patterns contrast with the physiological chaos seen in the dynamics of a healthy heart beat (Fig. 2). (Adapted from Goldberger and Rigney, 1988.)

Consider now the dynamics that occur *during* these lethal arrhythmias, as well as alterations in the dynamics of the 'normal' heart rhythm (sinus rhythm) *prior to* the abrupt appearance of one of these electrical disturbances. Tachyarrhythmias such as conventional ventricular tachycardia, torsades de pointes, and ventricular flutter are usually highly periodic. On the other hand, ventricular fibrillation has been assumed to be chaotic because of its apparently irregular appearance on the surface electrocardiogram (Smith and Cohen, 1984). However, spectral analyses (Battersby, 1965; Herbschleb *et al.*, 1978; Goldberger *et al.*, 1986) and electrophysiological mapping studies (Ideker *et al.*, 1981; Worley *et al.*, 1985) indicate that ventricular fibrillation also has a periodic type of dynamics. The electrophysiological mechanisms for these tachycardias remain uncertain. Traditional explanations have focused on altered conduction (re-entry) or increased automaticity. Non-linear dynamics, however, suggests alternative possibilities that may complement and extend these classical models. For example, the types of non-linear scroll waves described by Winfree (1987) have features reminiscent of both a periodically discharging automatic focus and a re-entrant loop.

Work in our laboratory most recently has focused on the dynamics of the heartbeat preceding the onset of these malignant ventricular tachyarrhythmias. Contrary once again to the conventional notion of sudden death as cardiac chaos, we have found that a loss of fractal variability in heart rate dynamics typically antedates cardiac arrest. In particular, patients at high risk of sudden death commonly show one of the following two dynamical patterns in the organization of their sinus rhythm heartbeats: (1) a marked overall loss of heart rate variability associated with a relatively flat spectral pattern particularly for frequencies (<0.1 Hz), or (2) prominent low frequency (0.01–0.06 Hz) heart rate oscillations associated with one or more sharp spectral peaks (Goldberger *et al.*, 1988). Of interest, similar dynamics have been described in the foetal distress syndrome (Modanlou and Freeman, 1982) and in space sickness syndrome (Goldberger, *et al.*, 1987), conditions associated with perturbed neuroautonomic control. We have also

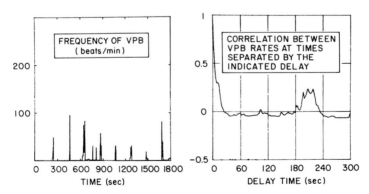

Fig. 4. Patients at high risk of sudden cardiac death may show complex periodic patterns of ventricular ectopic beat activity. For this case, periodic bursts of ventricular premature beats (VPB), occurred shortly before ventricular fibrillation. The left panel shows the times of VPB bursts in a patient who went into fibrillation at 1855 s. The regularity of interburst intervals is demonstrated in the right panel. It is seen that a VPB burst is likely to be followed by another one after about 3.5 min.

observed complex periodic patterns of ventricular ectopic beats in patients at high risk of sudden death (Fig. 4).

The mechanisms underlying these pathological dynamical patterns and the loss of fractal heart rate variability are not yet understood. Clearly, this loss of heart rate variability and the emergence of low frequency oscillations reflect autonomic instability. Although these abnormal heart rate dynamics do not themselves cause sudden death, they do appear to be direct or indirect markers of other factors that may destabilize the cardiac electrophysiological system. Such factors could include decreased parasympathetic tone, increased sympathetic tone, low cardiac output, hypoxemia, electrolyte fluctuations or drug toxicity (Goldberger et al., 1988).

SUMMARY AND PROSPECTS

Non-linear dynamics offers new ways of analysing the structure and function of the cardiovascular system. The fractal-like geometry of the vascular system and heart has important implications for the morphogenesis of these complex organs. The $1/f$-like spectrum of healthy heart rate variability suggests a fractal (chaotic) neurohumoral control mechanism governing physiologic fluctuations of the sinus node pacemaker rate. Neuroautonomic perturbations associated with sudden cardiac death syndrome commonly cause a loss of this fractal variability, sometimes with low frequency heart rate oscillations. These dynamic patterns may provide new ways of monitoring patients at high risk of sudden death, as well as of assessing the efficacy and toxicity of pharmacologic agents. The precise mechanisms of the arrhythmias actually causing sudden death, such as ventricular tachycardia and fibrillation, are not yet established. However, evidence now suggests that the route to sudden death is not a bifurcation to physiological chaos but instead is characterized by a loss of the fractal (chaotic) dynamics of the healthy heartbeat (Goldberger and Rigney, 1988).

Acknowledgements. This work was partly supported by grants from the National Aeronautics and Space Administration (NAG 2-514) and the G. Harold and Leila Y. Mathers Charitable Foundation. The authors thank J. Mietus for excellent technical assistance and A. J. Mandell for many helpful discussions.

REFERENCES

Battersby, E. J. (1965) Pacemaker periodicity in atrial fibrillation. *Circulation Res.* **17**, 296–302.
Cannon, W. B. (1929) Organization for physiological homeostasis. *Physiol. Rev.* **9**, 399–431.

Eylon, D., Sadeh, D., Kantor, Y. and Abboud, S. (1989) A model of the heart's conduction system using a self-similar (fractal) structure. In *Computers in Cardiology*. Computer Society Press of the IEEE, Washington, DC.

Glass, L. and Mackey, M. C. (1988) *From Clocks to Chaos: The Rhythms of Life*. Princeton University Press, Princeton, NJ.

Gleick, J. (1987) *Chaos: Making a New Science*. Viking Press, New York.

Goldberger, A. L. and Rigney, D. R. (1988) Sudden death is not chaos. In *Dynamic Patterns in Complex Systems* (eds J. A. S. Kelso, A. J. Mandell and M. F. Shlesinger), pp. 248–64. World Scientific Publishers, Singapore.

Goldberger, A. L. and West, B. J. (1987) Fractals in physiology and medicine. *Yale J. biol. Med.* **60**, 421–35.

Goldberger, A. L., Findley, L. J., Blackburn, M. R. and Mandell, A. J. (1984) Nonlinear dynamics in heart failure: Implications of long-wavelength cardiopulmonary oscillations. *Am. Heart J.* **107**, 612–5.

Goldberger, A. L., West, B. J. and Bhargava, V. (1985a) Nonlinear mechanisms in physiology and pathophysiology. Toward a dynamical theory of health and disease. *Proceedings of the 11th International Association for Mathematics and Computers in Simulation World Congress*, Oslo, Norway. (eds B. Wahlstrom, R. Henrickson and N. P. Sundby), Vol. 2, pp. 239–42. North Holland Publishing, Amsterdam.

Goldberger, A. L., Bhargava, V., West, B. J. and Mandell, A. J. (1985b) On a mechanism of cardiac electrical stability: the fractal hypothesis. *Biophys. J.* **48**, 525–8.

Goldberger, A. L., Bhargava, V., West, B. J. and Mandell, A. J. (1986) Some observations on the question: Is ventricular fibrillation 'chaos'? *Physica* **19D**, 282–9.

Goldberger, A. L., Thornton, W., Jarisch, W. R. *et al.* (1987) Low frequency heart rate oscillations in shuttle astronauts: a potential new marker of susceptibility to space motion sickness. *Space Life Science Symposium: Three Decades of Life Science Research in Space*. Washington, D.C., pp. 78–80.

Goldberger, A. L., Rigney, D. R., Mietus, J. *et al.* (1988) Nonlinear dynamics in sudden cardiac death syndrome: heartrate oscillations and bifurcations. *Experientia* **44**, 983–7.

Herbschleb, J. N., Heethaar, R. M., van der Tweel, I. *et al.* (1978) Signal analysis of ventricular fibrillation. In *Computers in Cardiology*, pp. 245–8. IEEE Computer Society, Long Beach, CA.

Hodgkin, A. L. and Huxley, A. F. (1952) A quantitative description of membrane current and its application to conduction and excitation in nerve. *J. Physiol., Lond.* **117**, 500–44.

Ideker, R. E., Klein, G. J., Harrison, L. *et al.* (1981) The transition to ventricular fibrillation induced by reperfusion after acute ischemia in the dog: a period of organized epicardial activation. *Circulation* **63**, 1371–9.

Kobayashi, M. and Musha, T. (1982) 1/f fluctuation of heartbeat period. *IEEE Trans. Biomed. Eng.* **29**, 456–7.

Koslow, S. H., Mandell, A. J. and Shlesinger, M. F. (eds) (1987) *Perspectives in Biological Dynamics and Theoretical Medicine. Ann. N.Y. Acad. Sci.* **504**, 1–313.

Lorenz, E. N. (1980) Noisy periodicity and reverse bifurcation. *Ann. N.Y. Acad. Sci.* **357**, 282–91.

Mandelbrot, B. B. (1982) *The Fractal Geometry of Nature*. W. H. Freeman, New York.

Modanlou, H. D. and Freeman, R. K. (1982) Sinusoidal fetal heart rate pattern: its definition and significance. *Am. J. Obstet. Gynecol.* **142**, 1033–8.

Shlesinger, M. F. (1987) Fractal time and 1/f noise in complex systems. *Ann. N.Y. Acad. Sci.* **504**, 214–28.

Smith, J. M. and Cohen, R. J. (1984) Simple finite-element model accounts for wide range of cardiac dysrhythmias. *Proc. Natl. Acad. Sci. U.S.A.* **81,** 233–7.

West, B. J. and Goldberger, A. L. (1987) Physiology in fractal dimensions. *Am. Scient.* **75,** 354–65.

Winfree, A. T. (1987) *When Time Breaks Down. The Three-Dimensional Dynamics of Electrochemical Waves and Cardiac Arrhythmias.* Princeton University Press, Princeton, NJ.

Worley, S. J., Swain, J. L., Colavita, P. G. *et al.* (1985) Development of an endocardial-epicardial gradient of activation rate during electrically-induced, sustained ventricular fibrillation. *Am. J. Cardiol.* **55,** 813–20.

Alternans in periodically stimulated isolated ventricular myocytes: Experiment and model

MICHAEL R. GUEVARA,[a] FRANCISCO ALONSO,[a]
DOMINIQUE JEANDUPEUX[a] AND ANTONI C. G. VAN
GINNEKEN[b]

[a]Department of Physiology and Centre for Nonlinear Dynamics
in Physiology and Medicine, McGill University, 3655 Drummond
Street, Montreal, Quebec, Canada H3G 1Y6
[b]Fysiologisch Laboratorium, Universiteit van Amsterdam,
Academisch Medisch Centrum, Meibergdreef 15, 1105 AZ
Amsterdam, The Netherlands

> The alternating contraction of the ventricle is not a simple pathological phenome-
> non but rather a general physiological function, that makes its appearance in many
> circumstances, ... We have, in fact, to deal with a capacity of the cardiac muscle
> which enables the ventricle to go on with rhythmical contractions even under
> abnormal conditions.
>
> Muskens (1907–08)

INTRODUCTION

Beat to beat alternation in the intensity of the peripheral arterial pulse was
first described using graphical recording techniques more than a century ago.
This alternation is usually attributed to an alternation in the force of
contraction of the left ventricle of the heart. Perhaps the simplest way in
which alternation could occur is if there would be a simultaneous concordant
alternation of the force of contraction in all cells of the left ventricle. In 1882
Gaskell put forth an alternative hypothesis stating that a beat to beat
alternation in the intensity of the heart beat could be due to spatial
inhomogeneities in contractile activity in different areas of the ventricle: 'The
explanation, therefore, of this alternation in the force of the contractions
must be sought for in the muscular tissue itself, and it seems to me that the

Cell to Cell Signalling: From
Experiments to Theoretical Models
ISBN 0–12–287960–0

most probable explanation is that a larger amount of tissue contracts when the beats are large than when they are small, and that therefore, in all probability, certain portions of the ventricle respond only to every second impulse, while other portions respond to every impulse'. A third possibility, that there could be a 2:1 response in two different subpopulations, with the conducted beat occurring on alternate beats in the two subpopulations, was suggested by Mines (1913).

With the advent of electrical recording techniques, electrical alternans was observed in whole hearts using extracellular (Hellerstein and Liebow, 1950) and intracellular (Downar *et al.*, 1977) recording techniques. In the latter case, even though one records the transmembrane potential of a single cell, it is impossible to say whether or not the beat to beat alternation in the morphology of the action potential seen in that cell is intrinsic to the cell or is due to an electrotonic injection of current from neighbouring cells that are coupled to the cell in question by low-resistance gap-junctional pathways. In the latter case, the alternans can be at least partly due to the fact that the stimulus current delivered to the cell is itself alternating. For example, in virtually all types of cardiac tissue, alternation of action potential morphology can be seen in the region of block when 2:1 block occurs (see Guevara, 1984, for references). Modelling work indicates that in one such instance alternation is indeed secondary to the fact that the flow of stimulus current into the region of block from regions distal to the site of block is itself alternating: the alternation is thus not intrinsic to the cells in question (Guevara, 1988, figure 3D).

In light of the above considerations, it seemed timely for us to investigate: (i) whether or not a single, isolated, quiescent ventricular cell could be made to display alternans; (ii) whether or not alternans could be seen in a mathematical model of a patch of isopotential membrane.

METHODS

Experimental

Single ventricular cells are isolated from rabbit hearts using standard techniques (Giles and van Ginneken, 1985; Giles and Imaizumi, 1988). The heart is removed from the animal, and the coronary arteries perfused with a heated (37°C), oxygenated, buffered salt solution flowing through the aorta in the retrograde direction. The heart is first perfused using a solution with a physiological Ca^{2+} concentration (2.2 mM) to wash out the blood from the coronary bed, then with a low-Ca^{2+} solution, and finally with a low-Ca^{2+} solution containing a mixture of the enzymes collagenase (Sigma, type 1A: 500 U/ml) and trypsin (Sigma, type 11: 249 u/ml). The low-Ca^{2+} solution is made by adding 15 μm Ca^{2+} to a nominally Ca^{2+}-free solution. The composition (mM) of the nominally zero $[Ca^{2+}]$ solution is: NaCl 121, KCl 5.0, $MgCl_2$ 1.0, $NaHCO_3$ 15, Na_2HPO_4 1.0, $Na-CH_2COOH$ 2.8, glucose 5.5. All solutions are gassed with carbogen (95% O_2–5% CO_2). Strips of tissue are then cut out of the right ventricle. These strips are cut into chunks, which are placed into a heated (37°C) bath containing low-Ca^{2+} solution sitting on a gyrating table rotating at 70 r.p.m.. This combined mechanical and chemical digestion breaks the chunks up, yielding a suspension of single cells.

The cells are then placed into a chamber sitting at room temperature (24–27°C) on the stage of an inverted microscope. The above isolation procedure produces quiescent, rod-shaped cells which are 'calcium-tolerant', i.e. which do not die when the experimental chamber is perfused with solution at a physiological Ca^{2+} concentration of 2.2 mM. A glass suction microelectrode is then positioned using a micromanipulator to come into contact with the membrane of a cell, whereupon an electrode–membrane seal of resistance 2–15 GΩ is formed upon application of slight suction. A suction pulse then disrupts the patch of membrane sealed off by the tip of the microelectrode. Since the electrolytic solution within the pipette (composition (mM): KCl 140, EGTA 5, $CaCl_2$ 1.54, $MgCl_2$ 1, Na_2ATP 5, HEPES 5, buffered to pH = 7.0 with KOH) is then contiguous with the intracellular fluid, the potential difference between the microelectrode and a second electrode placed in the bathing fluid surrounding the cell can then be measured. This potential difference is the transmembrane potential. A periodic train of current pulses of constant amplitude can then be delivered to the cell through the same microelectrode to stimulate the cell.

Numerical

We know of no published ionic models of the rabbit ventricular heart cell. For this reason, we investigate the model of Beeler and Reuter (1977) which describes ungulate ventricular fibres. There are five ionic currents in this model: the fast inward sodium current (I_{Na}), the slow inward calcium current (I_s), the time-dependent outward potassium plateau current (I_{x_1}), and the time-independent potassium and sodium leakage currents (I_{K_1}, $I_{Na,c}$). Numerical integration is carried out in single precision (approximately seven significant decimal digits) on a Hewlett-Packard minicomputer (model 1000F) using an efficient variable time-step Euler algorithm (Victorri et al., 1985) previously used by us in a modelling study of phase-resetting in Purkinje fibre (Guevara and Shrier, 1987). The maximum change in the transmembrane potential ΔV allowed in iterating from time t to time $t + \Delta t$ is 0.4 mV. When a value of ΔV larger than this upper limit results, the integration time step Δt is successively halved and the calculations redone until ΔV is less than 0.4 mV. When ΔV is less than 0.2 mV, Δt is doubled for the following iteration. Under these conditions, the value of Δt ranges between 1 μs and 8.192 ms, and the voltage waveform is very close to that appearing in the original description of the model (Beeler and Reuter, 1977). In advancing from time t to time $t + \Delta t$, the contribution of the transmembrane current to ΔV is calculated using the formula appearing in footnote 2 of Victorri *et al.* (1985). Since the time step is variable, it must be adjusted when current pulses are delivered so that the current starts and stops at exactly the right times. The time constants and asymptotic values of the various activation and inactivation variables (m, h, j, d, f, x_1) are stored in a table at steps of 0.2 mV and linear interpolation is used to extract out their values at any given voltage V. L'Hôpital's rule must be applied when necessary in calculating the rate constant α_m and the current I_{K_1}. Initial conditions on activation and inactivation variables were those appropriate to the membrane resting at $V = -84$ mV for an infinitely long time, and the initial condition on internal $[Ca^{2+}]$ was $[Ca^{2+}] = 0.1$ μM. This voltage of -84 mV is close to the resting potential in the model.

RESULTS

Experimental

For the pulse amplitude (A) sufficiently low, one obtains only a subthreshold response to each stimulus of the pulse train. This 1:0 response can be seen at any value of t_s, the time between stimuli. For A sufficiently high at a given t_s, the stimulus is suprathreshold, and so one can obtain action potentials. In that case, for t_s sufficiently large, one sees a 1:1 rhythm, in which each stimulus produces an action potential (Fig. 1A). (An $N{:}M$ rhythm is a periodic pattern consisting of repeating $N{:}M$ cycles each containing N stimuli and M action potentials. The period of repetition of the waveform is thus Nt_s. The M action potentials will in general have M different morphologies.)

For A chosen correctly, as t_s is decreased, the 1:1 rhythm will be converted directly into a 2:2 rhythm, without any other pattern being seen. Figure 1B shows an example of a 2:2 rhythm, showing an alternans in the morphology of the action potential. While there is an alternation in many of the parameters that can be used to describe an action potential, the alternation is probably most striking in action potential amplitude and duration. As t_s is decreased further, the alternans becomes more marked, with the amplitude and duration of the smaller action potential decreasing. Eventually, the 2:2 rhythm converts into a 2:1 rhythm, with every second stimulus no longer producing an action potential (Fig. 1C).

There is both hysteresis and bistability present in the behaviour shown in Fig. 1. If t_s is first decreased and then increased so that the border between two patterns is traversed twice, the value of t_s at which the transition occurs is different, depending upon the direction of change of t_s. This 'frequency hysteresis' has a counterpart in 'amplitude hysteresis', in which the pattern

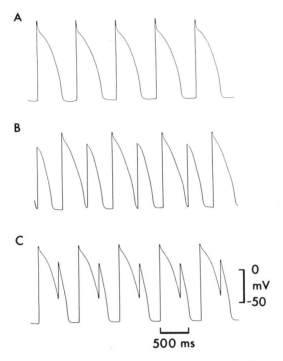

Fig. 1. Alternans in an isolated rabbit ventricular cell. The transmembrane potential is shown as a function of time. (A) 1:1 rhythm, t_s=700 ms. (B) 2:2 rhythm, t_s=430 ms. (C) 2:1 rhythm, t_s=350 ms. Pulse amplitude=1.0 nA, pulse duration=10 ms. Temperature=26°C. Sufficient time has been allowed to let transients pass. Stimulus artefact retouched.

seen at a particular amplitude at fixed t_s depends on whether A was increased or decreased to arrive at that value. If a well-timed extra stimulus pulse is added to the basic current pulse train, it is possible to convert a 2:1 rhythm into a 2:2 or a 1:1 rhythm. Injecting an extra stimulus or dropping one stimulus of the periodic drive train can also convert a 1:1 or 2:2 rhythm into a 2:1 rhythm. This illustrates that bistability is present, with the cell being capable of supporting two different periodic rhythms at a fixed combination of A and t_s.

Modelling

Figure 2A shows that, for t_s and A sufficiently large, 1:1 synchronization results. As t_s is decreased, the point is arrived at where one begins to see an alternation of action potential morphology. Figure 2B shows an example of such a 2:2 rhythm. As in the case of the experimental work, the alternation is perhaps most obvious in action potential amplitude and duration, and becomes more marked as t_s is decreased. Eventually, as in the experimental work, a 2:1 rhythm results (Fig. 2C).

Assuming a typical apparent membrane surface area of $5200\ \mu m^2$ and an apparent specific capacitance of $1.4\ \mu F/cm^2$ for a single rabbit ventricular cell (Giles and Imaizumi, 1988), one can calculate that the current of 1 nA used in Fig. 1 corresponds to a true current density of about $14\ \mu A/cm^2$, provided that the true specific capacitance is $1\ \mu F/cm^2$ in the absence of infoldings in the cell membrane. For a 20 ms duration pulse, an equivalent effect would be produced at a current density of about $7\ \mu A/cm^2$, which is quite close to the $3\ \mu A/cm^2$ actually used in the model in Fig. 2. The current density employed in the model also compares well with experimental results in isopotential spontaneously beating (Guevara, 1984) and quiescent (Guevara *et al.*, 1984)

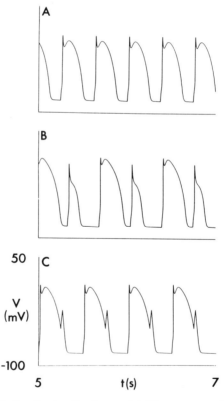

Fig. 2. Alternans in the Beeler–Reuter model. The transmembrane potential (V) is shown as a function of time (t). (A) 1:1 rhythm, $t_s=375$ ms. (B) 2:2 rhythm, $t_s=355$ ms. (C) 2:1 rhythm, $t_s=250$ ms. Pulse amplitude$=3\ \mu A/cm^2$, pulse duration$=20$ ms. The first stimulus pulse is injected at $t=0$, and 5 s are allowed to let transients pass.

aggregates of embryonic chick ventricular cells, where a current density of about $6\,\mu A/cm^2$ is needed to produce alternans when current pulses of duration $20\,ms$ are used (aggregate diameter $\simeq 100\,\mu m$, total surface area $\simeq 2.5 \times 10^{-3}\,cm^2$, pulse amplitude $\simeq 16\,nA$).

DISCUSSION

Alternans in ventricular tissue

The trace shown in Fig. 1B answers more than a century later a question raised in the work of Gaskell (1882). It demonstrates that it is indeed possible to obtain intrinsic alternation in a single, quiescent, ventricular myocyte. However, it does not answer the question of what is happening in the left ventricle during electrical alternans. There are relatively few recordings of the transmembrane potential in that circumstance, either in the intact ventricle or in isolated pieces of ventricular muscle. In most such experiments, only one impalement is made, and so one does not know if all cells are alternating in a 2:2 rhythm. One experiment in which four simultaneous intracellular impalements were made during ischaemia indicates that a 1:1 pattern was present in the non-ischaemic myocardium, while 1:0, 2:1 and 2:2 patterns were present in the ischaemic area (Downar et al., 1977, figure 10). In that case, the pattern of alternation that would be ascribed to the ventricle as a whole is due to the existence of an alternating rhythm in one subpopulation of cells as well as a 2:1 rhythm of block in yet another subpopulation. The existence of either one of these supopulations alone would theoretically be sufficient to generate an overall alternation in the ventricle. When both 2:2 and 2:1 rhythms occur together, it is impossible to decide which of these two rhythms would be the more important in generating alternation of the QRST complex of the surface electrocardiogram or mechanical alternans, since this would depend both on the relative sizes of the two subpopulations and also on the degree of alternation present in the 2:2 response. Since mechanical alternation is sometimes discordant with electrical alternation, with the larger action potential producing the weaker contraction (Spear and Moore, 1971), one might even imagine a situation in which an overall mechanical alternans might not be seen, provided that the balance between the 2:2 and 2:1 subpopulations was just right.

In the case when 1:1, 2:2 and 2:1 rhythms are seen in different areas of the ventricle, it is possible that the 2:1 pattern is due to block of propagation of the cardiac impulse, with the 2:2 pattern being seen in the region of block. The 2:2 pattern seen in this circumstance is due to decremental conduction and can be seen when 2:1 block occurs in tissue taken from virtually all areas of the heart (see Guevara, 1984, for references). The possibility then exists that the 2:2 pattern in the region of block is partly due to electrotonic injection of current from the distal 2:1 region, in a manner recently demonstrated in modelling work on a one-dimensional strand of Purkinje fibre (Guevara, 1988). The question of whether or not it is possible to obtain a ventricle with a 2:2 pattern in all cells, or a 2:2 pattern in one circumscribed area and a 1:1 pattern everywhere else, remains open. This question can be answered only when techniques that allow recording of the transmembrane potential from many sites simultaneously (e.g. using potential-sensitive optical dyes) would be applied.

Ionic basis of alternans

The ionic basis of the response shown in Fig. 1B remains undetermined. However, the electrophysiological origin of the 2:2 rhythm is essentially due to the fact that, at the stimulation rates employed in Fig. 1, the duration of an action potential decreases monotonically with a decrease in recovery time since the immediately preceding action potential (Hauswirth *et al.*, 1972; Boyett and Jewell, 1978; Guevara *et al.*, 1984). Since in modelling work one can plot out not only the transmembrane potential, but also the various currents, activation variables and inactivation variables, it will be possible in future work to determine the ionic mechanisms producing the 2:2 response shown in Fig. 2.

Loss of 1:1 synchronization

We have shown above that the sequence of rhythms $\{1:1\rightarrow2:2\rightarrow2:1\}$ can be seen as t_s is decreased (Fig. 1). However, this occurs only at a sufficiently high stimulus amplitude. We have also observed on several occasions in isolated cells a direct transition from a 1:1 to a 2:1 rhythm, with no evidence of the 2:2 rhythm, when t_s is changed in steps of 10 ms. (Perhaps the earliest definitive example of a direct $\{1:1\rightarrow2:1\}$ transition is to be found in the work of Mines (1913), who studied the atropinized ventricle of the frog.) In addition to this direct $\{1:1\rightarrow2:1\}$ sequence, one can also see Wenckebach-like rhythms between the 1:1 and 2:1 rhythms at t_s is decreased, provided that the stimulus amplitude is just suprathreshold and t_s is quite large (Guevara *et al.*, 1989a). The existence of these three qualitatively different ways in which 1:1 synchronization can be lost has been previously described in periodically stimulated, spontaneously beating, embryonic chick ventricular heart-cell aggregates with the $\{1:1\rightarrow2:2\}$ transition occurring at high stimulus amplitude, the $\{1:1\rightarrow2:1\}$ transition at intermediate amplitude, and Wenckebach rhythms at low amplitude (Guevara, 1984; Guevara *et al.*, 1989b).

Reduction to a one-dimensional map

We have shown previously that dynamics similar to that shown in Fig. 1 seen in periodically stimulated, quiescent chick ventricular aggregates can be reduced to the study of the iteration of a one-dimensional finite-difference equation or map (Guevara *et al.*, 1984). This map gives the ith action potential duration as a function of the immediately preceding one and is obtained from consideration of the curve describing how action potential duration is restored as a function of the recovery time since the immediately preceding action potential. In that case, it can be shown that alternans occurs when the slope of the map at the fixed point becomes more negative than -1, leading to a period-doubling bifurcation (Guevara *et al.*, 1984, figure 4). The sequence of rhythms predicted from the map is then $\{1:1 \to 2:2 \to 2:1\}$, which agrees with what is seen experimentally. A similar analysis should be possible in the case of Figs 1 and 2. The first analysis of electrical alternans formulated using an iterative technique seems to have been that of Nolasco and Dahlen (1968), while mechanical alternans has been similarly studied (Mahler and Rogel, 1970).

When the recovery of action potential duration is sufficiently rapid, all points on the map have a slope more positive than -1, and so a period-doubling bifurcation cannot occur. In that circumstance, one obtains a direct $\{1:1 \to 2:1\}$ transition. Note that in this instance, the transition is not due to a period-doubling bifurcation (cf. Chialvo and Jalife, 1987). The 2:1 rhythm corresponds to a period-1 orbit on the map, while the 2:2 rhythm seen if the map is sufficiently steep corresponds to a period-doubled period-2 orbit (Guevara *et al.*, 1984, figure 4). The transition $\{2:2 \to 2:1\}$ seen in a simple model of a limit-cycle oscillator (Guevara and Glass, 1982) and in experimental work on spontaneously beating heart-cell aggregates (Guevara, 1984, figure 5-16) is due to a change in rotation number.

As mentioned previously, when stimulating with large t_s (~ 1 s) and small A, Wenckebach rhythms can be observed at values of t_s intermediate to those at which 1:1 and 2:1 rhythms occur. At such low stimulation frequencies in the rabbit ventricle, the action potential duration shortens with increasing recovery time (Saxon and Safronova, 1982; Giles and Imaizumi, 1988) – it does not prolong as at the higher stimulation frequencies employed in Fig. 1. One thus obtains a different class of maps which have two branches, each of which has positive slope everywhere (Guevara *et al.*, 1989a). Thus, a period-doubling bifurcation cannot occur. These maps predict the existence of Wenckebach rhythms, and are qualitatively similar to a class of maps derived in recent work on Wenckebach rhythms in the human atrioventricular node (Shrier *et al.*, 1987). Wenckebach rhythms can also be seen in periodically stimulated, spontaneously beating, embryonic chick ventricular aggregates (Guevara *et al.*, 1988); in that case, one can sometimes reduce the dynamics to iteration of a circle map whose slope is everywhere positive (e.g. Guevara, 1984, figure 5-4).

Alternans and ventricular arrhythmias

Electrical alternans has been described in many circumstances that are associated with ventricular fibrillation. These include coronary occlusion (Downar *et al.*, 1977; Adam *et al.*, 1984; see Guevara, 1984, for further references), abnormality of the sympathetic nervous system (Crampton, 1978), hypothermia (Adam *et al.*, 1984), and ventricular tachycardia (Sano *et al.*, 1958; Hogancamp *et al.*, 1959). In several of the above reports, intracellular impalements were made, and beat to beat alternation of the action potential morphology was recorded.

The close temporal connection between the phase of alternans and the phase of induction of malignant ventricular tachyarrhythmias might be entirely circumstantial. We think not. It has been suggested that the alternans seen in this instance and in other circumstances is due to a period-doubling bifurcation (Guevara *et al.*, 1981, 1984, 1989b; Guevara and Glass, 1982; Goldberger *et al.*, 1984; Guevara, 1984; Ritzenberg *et al.*, 1984; Adam *et al.*, 1984), and the implication has been made that this (first) bifurcation might be the initial part of a cascade of such bifurcations that would eventually result in chaos, i.e. fibrillation (Adam *et al.*, 1984).

However, at the present time, there is no experimental evidence for the existence of higher-order period doublings in the ischaemic heart. While period-4 rhythms have been identified on many occasions in experimental and modelling work on cardiac muscle, one must exercise caution in claiming that they result from two successive period-doubling bifurcations. There are at least four instances on record where the period-4 rhythm is clearly not the result of two such bifurcations (Guevara, 1984, figure 4-8C, figure 4-26C; Chialvo and Jalife, 1987, figure 2B (third trace); Guevara, 1988, figure 3B).

In our work with isolated rabbit ventricular cells, the rhythms 1:1, 2:2, 2:1, 4:2, 3:1 and 6:2, as well as apparently non-periodic rhythms, are often seen as t_s is decreased in steps of 10 ms. Note that while three period-doubling bifurcations occur $\{1:1 \rightarrow 2:2\}$, $\{2:1 \rightarrow 4:2\}$ and $\{3:1 \rightarrow 6:2\}$, they do not form part of a period-doubling cascade. For example, we have not seen the 2:2 rhythm period-double to a period-quadrupled 4:4 rhythm, as occurs with stimulation at low frequencies in spontaneously beating aggregates of embryonic chick ventricular cells (Guevara *et al.*, 1981).

In the case of induction of ventricular arrhythmias, we feel that one should leave open the possibility that there might be a direct transition from a period-doubled 2:2 or 4:2 pattern to a state of ventricular tachycardia or fibrillation. With regard to the latter eventuality, there are apparently mathematical schemes in which it is possible to generate chaotic behaviour following a single bifurcation (Tresser *et al.*, 1980). Should the first qualitative change occurring in the ischaemic ventricle be a transition from a 1:1 response in all cells to a 2:2 response in some subpopulation within the ischaemic zone, there would be still one action potential for each supraventricular stimulus in all cells of the ventricle, and the degree of spatient inhomogeneity would not be too pronounced. However, should some of the cells showing a 2:2 response proceed to the stage of showing a 2:1 response as the ischaemia becomes more profound, there would then be a greater degree of spatial inhomogeneity, with some areas demonstrating action potentials only on alternate beats. This might predispose the ventricle to formation of re-entrant rhythms.

In summary, while we have shown above that alternans can be generated by a single cell, it will take much more work to put this finding into perspective once one realizes that cells in the heart talk to one another.

Acknowledgements. We thank Tim Lewis for help with computer programming, Christine Pamplin for typing the manuscript, and Robert Lamarche for photographing the figures. Supported by grants to M. G. from the Quebec Heart Foundation and the Medical Research Council of Canada.

REFERENCES

Adam, D. R., Smith, J. M., Akselrod, S., Nyberg, S., Powell, A. O. and Cohen, R. J. (1984) Fluctuations in T-wave morphology and susceptibility to ventricular fibrillation. *J. Electrocardiol.* **17,** 209–18.

Beeler, G. W. and Reuter, H. (1977) Reconstruction of the action potential of ventricular myocardial fibres. *J. Physiol., Lond.* **268,** 177–210.

Boyett, M. R. and Jewell, B. R. (1978) A study of the factors responsible for rate-dependent shortening of the action potential in mammalian ventricular muscle. *J. Physiol., Lond.* **285,** 359–80.

Chialvo, D. R. and Jalife, J. (1987) Non-linear dynamics of cardiac excitation and impulse propagation. *Nature* **330,** 749–52.

Crampton, R. S. (1978) Another link between the left stellate ganglion and the long Q-T syndrome. *Am. Heart J.* **96,** 130–2.

Downar, E., Janse, M. J. and Durrer, D. (1977) The effect of acute coronary occlusion on subepicardial transmembrane potentials in the intact porcine heart. *Circulation* **56,** 217–24.

Gaskell, W. H. (1882) On the rhythm of the heart of the frog, and on the nature of the action of the vagus nerve. *Phil. Trans. R. Soc. Lond.* **173,** 993–1033.

Giles, W. R. and Imaizumi, Y. (1988) Comparison of potassium currents in rabbit atrial and ventricular cells. *J. Physiol., Lond.* **405,** 123–45.

Giles, W. R. and van Ginneken, A. C. G. (1985) A transient outward current in isolated cells from the crista terminalis of rabbit heart. *J. Physiol., Lond.* **368,** 243–64.

Goldberger, A. L., Shabetai, R., Bhargava, V., West, B. J. and Mandell, A. J. (1984) Nonlinear dynamics, electrical alternans, and pericardial tamponade. *Am. Heart J.* **107,** 1297–9.

Guevara. M. R. (1984) Chaotic cardiac dynamics. PhD thesis, McGill University, Montreal.

Guevara, M. R. (1988) Spatiotemporal patterns of block in an ionic model of cardiac Purkinje fibre. In *From Chemical to Biological Organization* (eds M. Markus, S. C. Müller and G. Nicolis), pp. 273–81. Springer, Berlin.

Guevara, M. R. and Glass, L. (1982) Phase locking, period doubling bifurcations and chaos in a mathematical model of a periodically driven oscillator: a theory for the entrainment of biological oscillators and the generation of cardiac dysrhythmias. *J. Math. Biol.* **14,** 1–23.

Guevara, M. R. and Shrier, A. (1987) Phase resetting in a model of cardiac Purkinje fiber. *Biophys. J.* **52,** 165–75.

Guevara, M. R., Glass, L. and Shrier, A. (1981) Phase locking, period-doubling bifurcations, and irregular dynamics in periodically stimulated cardiac cells. *Science* **214,** 1350–3.

Guevara, M. R., Ward, G., Shrier, A. and Glass, L. (1984) Electrical alternans and period-doubling bifurcations. In *Computers in Cardiology,* pp. 167–70. IEEE Computer Society, Silver Spring, MD.

Guevara, M. R., Shrier, A. and Glass, L. (1988) Phase-locked rhythms in periodically stimulated heart cell aggregates. *Am. J. Physiol.* **254,** H1–H10.

Guevara, M. R., Jeandupeux, D., Alonso, F. and Morissette, N. (1989a) Wenckebach rhythms in isolated ventricular heart cells. In *Singular Behaviour and Nonlinear Dynamics* (eds St. Pnevmatikos, T. Bountis and Sp. Pnevmatikos). World Scientific, Singapore (in press).

Guevara, M. R., Shrier, A. and Glass, L. (1989b) Chaotic and complex cardiac rhythms. In *Cardiac Electrophysiology: From Cell to Bedside* (eds D. P. Zipes and J. Jalife). Saunders, Philadelphia (in press).

Hauswirth, O., Noble, D. and Tsien, R. W. (1972) The dependence of plateau currents in cardiac Purkinje fibres on the interval between action potentials. *J. Physiol., Lond.* **222,** 27–49.

Hellerstein, H. K. and Liebow, I. M. (1950) Electrical alternation in experimental coronary artery occlusion. *Am. J. Physiol.* **160,** 366–74.

Hogancamp, C. E., Kardesch, M., Danforth, W. H. and Bing, R. J. (1959) Transmembrane electrical potentials in ventricular tachycardia and fibrillation. *Am. Heart J.* **57,** 214–22.

Mahler, Y. and Rogel, S. (1970) Interrelation between restitution time-constant and alternating myocardial contractility in dogs. *Clin. Sci.* **39,** 625–39.

Mines. G. R. (1913) On dynamic equilibrium in the heart. *J. Physiol., Lond.* **46,** 349–83.

Muskens, L. J. J. (1907–08) Genesis of the alternating pulse. *J. Physiol., Lond.* **36,** 104–12.

Nolasco, J. B. and Dahlen, R. B. (1968) A graphic method for the study of alternation in cardiac action potentials. *J. Appl. Physiol.* **25,** 191–6.

Ritzenberg, A. L., Adam, D. R. and Cohen, R. J. (1984) Period multiplying-evidence for nonlinear behaviour of the canine heart. *Nature* **307,** 159–61.

Sano, T., Tsuchihashi, H. and Shimamoto, T. (1958) Ventricular fibrillation studies by the microelectrode method. *Circulation Res.* **6,** 41–6.

Saxon, M. E. and Safronova, V. G. (1982) The rest-dependent depression of action potential duration in rabbit myocardium and the possible role of the transient outward current. A pharmacological analysis. *J. Physiol., Paris* **78,** 461–6.

Shrier, A., Dubarsky, M., Rosengarten, M., Guevara, M. R., Nattel, S. and Glass, L. (1987) Prediction of complex atrioventricular conduction rhythms in humans with use of the atrioventricular nodal recovery curve. *Circulation* **76,** 1196–205.

Spear, J. F. and Moore, E. N. (1971) A comparison of alternation in myocardial action potentials and contractility. *Am. J. Physiol.* **220,** 1708–16.

Tresser, C., Coullet, P. and Arneodo, A. (1980) On the existence of hysteresis in a transition to chaos after a single bifurcation. *J. Phys. Lett.* (Paris) **41,** L243–L246.

Victorri, B., Vinet, A., Roberge, F. A. and Drouhard, J. -P. (1985) Numerical integration in the reconstruction of cardiac action potentials using Hodgkin–Huxley-type models. *Comp. Biomed. Res.* **18,** 10–23.

The role of anisotropic impulse propagation in ventricular tachycardia

M. A. ALLESSIE, M. J. SCHALIJ, C. J. H. J. KIRCHHOF,
L. BOERSMA, M. HUYBERS AND J. HOLLEN

Department of Physiology, Biomedical Center, University of
Limburg, Maastricht, The Netherlands

INTRODUCTION

Cardiac arrhythmias are caused either by the generation of new impulses (abnormal automaticity, triggered activity), or by failure of the propagation of the depolarization wave to extinguish after having excited all cardiac fibres. This last mechanism, called re-entrant excitation, is responsible for the majority of clinical tachyarrhythmias. A variety of excellent articles and books are available reviewing both the experimental and clinical literature on re-entry (Cranefield et al., 1973; Moe, 1975; Wit and Cranefield, 1978; Pick and Langendorf, 1979; Hoffman and Rosen, 1981; Speare and Moore, 1982; Janse, 1986).

Experimental and clinical studies have revealed three basically different kinds of re-entry: (1) circuits which are based on macroanatomic pathways; (2) functionally determined circuits in the syncitium of myocardial cells without the involvement of a gross anatomical obstacle (leading circle re-entry); and (3) re-entry in uniform or non-uniform anisotropic tissue.

Anatomically defined re-entry was first studied by Mines (1913) in rings of excitable tissue. In the in situ heart, anatomical circuits may be formed by the bundle branches of the specialized ventricular conducting system (Moe et al., 1965), by tissue around the orifices of the venae cavae (Lewis, 1920, 1925) or the AV ring (Frame et al., 1983, 1986, 1987), or in patients with WPW syndrome by the atria, AV conducting system, ventricles and the accessory pathway (Gallagher et al., 1975). In all these examples, during re-entrant tachycardia, the impulse continuously circulates in one direction around a central barrier.

Cell to Cell Signalling: From
Experiments to Theoretical Models
ISBN 0–12–287960–0

The second type of re-entry is not dependent on a non-conductile central obstacle, but the pathway through which the impulse circulates is defined by the functional electrophysiological properties of the fibres composing the circuit (Allessie *et al.*, 1973, 1977, 1984, 1985; Allessie and Bonke, 1984). Thus re-entry can occur when at a critical site, a properly timed premature impulse blocks transiently in one direction where the refractory period is somewhat prolonged, while it conducts through other regions where refractoriness is shorter. When the time that the impulse turns around the area of block is long enough, the fibres proximal to the line of block will be re-excited by the turning impulse. A functional re-entrant circuit is thus established and re-entry may continue by the leading circle mechanism (Allessie *et al.*, 1973, 1977, 1984, 1985).

The properties of re-entry resulting from these two different mechanisms and the arrythmias they cause, are different. These differences are mostly a result of the relationship between the wavelength of the conducting impulse and the length of the path comprising the re-entrant circuit. The wavelength of the cardiac impulse is defined as the distance travelled by the depolarization wave during the refractory period (wavelength of refractoriness) (Smeets *et al.*, 1986; Rensma *et al.*, 1988). For re-entry to continue, the wavelength of the circulating impulse cannot be longer than the path length, or the depolarization wave would run into its own refractory tail and the circulating impulse would be blocked. In a re-entrant circuit around an anatomical obstacle the wavelength can be significantly shorter than the path length of the circuit, leaving a segment of the re-entrant circuit fully or partially excitable. The presence of such an excitable gap has several significant influences on the properties of the re-entrant arrhythmia. First, the revolution time of the impulse around the circuit is determined primarily by the conduction velocity of the impulse and will not be significantly influenced by (moderate) changes in the refractory period. Second, it is unlikely that the circulating impulse will block somewhere around the re-entrant loop and the chances for spontaneous termination of such an arrhythmia are low. Third, it is possible to perturb the re-entrant impulse by applying appropriately timed stimuli and the re-entrant rhythm can be either reset or terminated by programmed electrical stimulation (Karagueuzian *et al.*, 1979).

On the other hand, in a functional re-entrant circuit caused by the leading circle mechanism, the length of the circuit is not fixed by an anatomical barrier, but instead is determined by the electrophysiological properties of the tissue. In this type of re-entry the impulse circulates in the smallest possible pathway in which the stimulating efficacy is just enough to excite the tissue ahead which is in its relative refractory phase – 'in this smallest circuit possible which is designated as the leading circle, the head of the circulating wave front is continuously biting in its own tail of refractoriness' (Allessie *et al.*, 1977). Because of this tight fit, the length of the circulating pathway is equal to the wavelength of the circulating impulse. Functional circuits of this kind may therefore be expected to be small, extremely rapid and unstable. Because there is a tight fit between the crest and the tail of the impulse, changes in refractory period will have significant influences both on the revolution time and the size of the circuit. The absence of an excitable gap also implies that a wave front (or stimulus) of greater efficacy than the circulating impulse is required to interfere with the leading circle (Allessie *et al.*, 1977). This property protects leading circle re-entry from being terminated by stimuli applied to the heart during the arrhythmia and may explain why some re-entrant arrhythmias cannot be terminated by pacemakers (Waldo *et al.*, 1977).

Recently a third type of re-entry has been described in anisotropic myocardium (anisotropic re-entry). This type of re-entry is also functional – there is usually no gross anatomical obstacle and the re-entrant circuits are determined by the functional properties of the cardiac tissue (Wit *et al.*, 1982; Dillon *et al.*, 1988; Schalij, 1988). However, unlike leading circle re-entry a clear excitable gap is present, allowing capture of the myocardium outside the circuit by programmed electrical stimulation. An important feature of the geometry of the ventricles is that the fibres are orientated parallel to each other. Conduction of the cardiac impulse (including circulating excitation) is strongly influenced by the micro-architecture of the myocardium. One of the most striking characteristics is that the impulse propagates about 3 times faster in a direction parallel to the long axis of the myocardial fibres compared to conduction in a transverse direction (Clerc, 1976). This strong anisotropy in conduction has been explained by differences in cell coupling in different directions (Clerc, 1976; Spach *et al.*, 1982a,b, 1981). Anisotropy has been further subdivided by Spach *et al.* (1981, 1982b). According to their definition uniform anisotropy exists in cardiac muscle in which the fibres are all arranged parallel to each other. Non-uniform anisotropy is present when in addition, microscopic non-conductile barriers are formed by connective tissue.

Anisotropic properties of cardiac muscle may play an important role in the genesis of re-entrant arrhythmias. Spach *et al.* (1981) proposed that it may cause conduction block of premature impulses because of a lower safety factor for conduction in the longitudinal direction. On the other hand there are a number of studies showing preferential block in a transverse direction (Schmitt and Erlanger, 1928; Myerburg *et al.*, 1973; Tsuboi *et al.*, 1985). In a computer simulation of impulse conduction Van Capelle (1983, and in this volume) was unable to produce longitudinal block in a two-dimensional anisotropic sheet of cardiac cells. However, they found that a critically timed premature impulse could block preferentially in a transverse direction.

Anisotropy may also modify the properties and behaviour of sustained re-entrant circuits. Because of the differences in conduction velocity based on the anisotropy in geometry, the conduction velocity of the impulse around a re-entrant circuit is not uniform. It is more rapid in the limbs of the circuit which run parallel to the fibre orientation and more slowly in the segments where the impulse propagates transverse to the fibre axis (Dillon *et al.*, 1988). It has been proposed by Frame and Hoffman (1984) that such differences in properties in different parts of the circuit might lead to the appearance of an excitable gap.

A NEW EXPERIMENTAL MODEL OF VENTRICULAR TACHYCARDIA IN ANISOTROPIC MYOCARDIUM

To study the role of anisotropy in re-entrant arrhythmias, a two-dimensional preparation of ventricular myocardium was used. In a Langendorff-perfused rabbit heart the endocardial and intramural layers of the left ventricle were destroyed by freezing with liquid nitrogen ($-192°C$). The perfused heart was immersed in a tissue bath, containing perfusion fluid of $30°C$. A cryoprobe was installed in the left ventricular cavity and the coronary circulation was temporarily interrupted. The cryoprobe was then filled with liquid nitrogen and the heart was frozen for 7 min. During this period an intramural gradient of temperatures developed between $-192°C$ at the endocardium and $30°C$ at the epicardium, the freezing point being about 1 mm from the epicardial surface. When after 7 min the cryoprobe was removed and coronary circulation restored (temperature of coronary perfusion $37°C$), the endocardial and intramural layers of the left ventricle were completely destroyed. At the free wall of the left ventricle, however, a thin epicardial layer of about 1 mm thick had survived this procedure. The coronary circulation was also left intact and after freezing, coronary flow was less than 10% different compared to control (perfusion pressure 50 mm Hg). To evaluate the cryoprocedure, at the end of the experiment the heart was perfused with TTC (a buffered 2,3,5-triphenyl tetrazolium chloride solution, Merck). This substance is a specific indicator of dehydrogenase enzymes, yielding a bright red formazan pigment. No staining occurs at sites where the myocardium has lost its dehydrogenase activity. This macroscopic staining method has been shown to correlate well with the ultrastructural signs of myocardial necrosis as demonstrated with standard histological or electron microscopical techniques (Nachlas and Shnitka, 1963; Fishbein *et al.*, 1981; Vivaldi *et al.*, 1985).

After staining, the ventricles were cut in 2 mm thick sections parallel to the AV ring from base to apex. The transition between surviving and dead tissue was sudden. Histological examination showed no islands of viable tissue in the destroyed parts of the myocardium and no dead fibres in the surviving epicardium. The thickness of the surviving epicardium was 1.0 ± 0.4 mm ($n = 99$). The variation in local thickness of the surviving muscle layer was caused by the presence of epicardial fat or major blood vessels. At places where epicardial fat or a coronary vessel was present, the surviving rim of the underlying myocardium was thinner. Obviously, the interposition of these structures between the epicardium and the tissue bath decreased the protecting effect of the warm surrounding perfusion fluid during freezing.

For epicardial recording and stimulation a spoon-shaped electrode was used which fitted to the total free wall of the left ventricle. This 'spoon' electrode contained 384 individual electrodes (silver wires diameter 0.3 mm) at regular distances of 2 mm. Unipolar electrograms were recorded using the stainless steel aorta cannula as indifferent electrode. Details of the mapping system enabling simultaneous recording of 256 electrograms have been given elsewhere (Allessie *et al.*, 1984; Hoeks *et al.*, 1988). Programmed electrical stimulation was applied with a programmable constant-current stimulator delivering square pulses of 2 ms duration and 2–4 times diastolic threshold to any selected pair of electrodes of the mapping device. The heart was paced at various rates, and conduction velocity and refractory period were measured at multiple epicardial locations.

Both local refractory periods and conduction velocity parallel to the epicardial fibre orientation were the same as before the freezing procedure. However, the amplitude of the recorded epicardial electrograms was clearly diminished, caused by the decreased thickness of myocardium under the electrodes. Another effect of removing the inner $\frac{4}{3}$th part of the ventricular wall was that slow conduction perpendicular to the epicardial fibre orientation proceeded over a longer distance than in the intact heart. Obviously in the intact heart the deeper layers provide alternative routes for impulse conduction resulting in epicardial breakthrough ahead and before the slowly propagating transverse activation wave. Absence of these deeper layers unmasks the true anisotropic properties of the ventricular myocardium.

However, there was an important difference in the induction of ventricular arrhythmias in the isolated rabbit heart before and after the cryo procedure. During control, programmed electrical stimulation (burst pacing) only resulted in the induction of ventricular fibrillation. In contrast, after endocardial freezing ventricular fibrillation could not be induced anymore. Instead, rapid pacing frequently resulted in the induction of sustained ventricular tachycardia. Epicardial mapping of the sequence of excitation during sustained ventricular tachycardia showed that the arrhythmia was caused by continuous circus movement of the impulse in the thin surviving epicardial layer. The cycle length of these tachycardias was constant and ranged from 110 to 165 ms in different experiments. The tachycardias were stable; if not disturbed by electrical stimulation or the administration of cardiac drugs they could last for several hours without any change in cycle length or electrogram morphology. Due to the anisotropy of the epicardium the conduction velocity of the circulating impulse around the circuit varied from about 60 cm/s during longitudinal conduction to less than 20 cm/s during slow transverse conduction. At the line of block which was oriented parallel to the fibre direction, no insulating anatomical barriers could be found. On the basis of extracellular recordings, the length of the functional conduction block in the centre of the circuit was estimated to be about 2 cm.

THE EXCITABLE GAP IN ANISOTROPIC RE-ENTRY

There are several indications that this type of re-entrant tachycardia in an anisotropic sheet of ventricular myocardium has an excitable gap: (1) Measurement of refractory periods during the same pacing rate as the rate of the tachycardia revealed that the refractory period of the epicardium was about 30 ms shorter than the cycle length of the tachycardia. (2) During the tachycardias the ventricles could be captured by single premature stimuli up to cycle lengths 20–40 ms shorter than the cycle length of the tachycardia. (3) The re-entrant circuits possessed a much higher degree of stability compared to functional re-entry in the atria.

The presence of an excitable gap during anisotropic re-entry is the most important difference compared to leading circle re-entry in the atria (Allessie *et al.*, 1977). The existence of an excitable gap of about 20–40 ms in a functionally determined circuit in uniform anisotropic myocardium might be explained by three possible mechanisms.

(1) Microanatomical barriers at the pivoting points

Although no gross anatomical barriers were involved, the presence of microanatomical obstacles at the pivoting points of the circuit could not be ruled out. If present, they may enlarge the central, functionally determined line of block of the circuit. It will also stabilize the position of the re-entrant loop at a fixed location in the myocardium. Small conduction barriers may exist when, due to the interposition of collageneous septa, adjacent myocardial fibres become electrically separated.

(2) Block at the pivoting points because of high electronic load

The stimulating efficacy of a propagating action potential is influenced by sudden changes in the axial current load (Spach *et al.*, 1981, 1982b, 1987; Spach and Dolber, 1986). Such a sudden increase of the current load occurs at branching sites of the myocardial fibres or when an abrupt change in direction of impulse propagation occurs. A depression of the stimulating efficacy may lead to decremental conduction or conduction block despite the fact that the cells have already recovered their excitability.

At the pivoting points of anisotropic re-entry, the slowly conducting transverse wavefront enounters a sudden increase of axial current load when it tries to excite the longitudinal limb of the circuit. It is feasible that at the transition from transverse to longitudinal conduction, this sudden increase of electronic load leads to a temporary local halt of propagation and the returning longitudinal limb of the circuit will not be activated until a larger part of the wavefront has rotated around the pivoting point. This mismatch between the excitatory current generated by the cells at the pivoting point and the axial current load imposed on these cells by the fibres beyond the pivoting point will create an excitable gap because it causes a conduction delay of about 30 ms after the cells have recovered their excitability. This will lead to a functional lengthening of the central line of conduction block and consequently in a re-entrant cycle length about 30 ms longer than the refractory period of the ventricular myocardium.

(3) Electrotonic prolongation of the action potential at the pivoting points

Spatial differences of the action potential duration may also contribute to the creation of an excitable gap. It is known that during regular pacing the action potential near the site of stimulation is prolonged by electrotonic interaction with more remote cells which are activated later in time. Recently, Osaka *et al.* (1987) demonstrated that the amount of prolongation of the action potential depends on the conduction time between the cells and the axial resistance. During centrifugal impulse spread in anisotropic myocardium the degree of prolongation of the action potential was most marked. This prolongation of the action potential lengthened the local refractory period. During anisotropic re-entry considerable differences in activation time are present at both sides of the pivoting points. Because electrical coupling is intact, these differences in activation time over only a few millimetres may result in action potential prolongation of the cells at the pivoting points and consequently in a local lengthening of the refractory period. Due to this local prolongation of refractoriness there is a tight fit between the crest of the depolarization wave and its tail of refractoriness at the pivoting points of the anisotropic circuit. However, in all other parts of the myocardium an excitable gap will exist.

MICROELECTRODE RECORDINGS FROM THE ANISOTROPIC CIRCUIT

In a preliminary study in which we used microelectrodes to record the cellular characteristics of anisotropic re-entry, we found clear electrotonic interaction across the central line of conduction block. During sustained ventricular tachycardia a regular 1 to 1 activation and normal action potentials were recorded from cells lying outside the line of block. During the tachycardia cycle these cells showed full repolarization and a phase 4 of about 30 ms in which the cells were at their normal resting potential. Within the central area of the circuit local electrotonic responses were recorded of twice the frequency of the encircling tachycardia. These electrotonic humps, which were due to electrotonic interaction between the two longitudinal limbs of the circuit, led to prolongation of the plateau of the action potential. As a result of this, the cells in the centre of the circuit showed various degrees of local conduction block, leading to regular 2:1 or 3:2 patterns.

At the pivoting points evidence was found for local conduction block caused by a mismatch between the excitatory current generated by the cells at the pivoting points and the large electrotonic load imposed by the cells distal to the U-turns of the circuit. At those parts of the circuit the action potentials were prolonged not by an electrotonic prolongation of the plateau phase, but by a prolongation of the depolarization phase. These cells showed a step-like depolarization which resulted in local conduction delay of about 30 ms. As a result of the step-like depolarization of the cells at the pivoting points, there was not enough time for these cells during the tachycardia cycle to fully repolarize. Consequently at the U-turns of the circuit the fibres did not reach their resting potential and the next depolarization occurred during the last portion of phase 3 of the action potential. Thus, unlike the other fibres of the ventricles, at the pivoting points of the anisotropic circuit, the action potentials did not exhibit a phase 4 and because of the absence of a clear diastolic interval did not possess an excitable gap. The results of these microelectrode mapping studies indicate that the excitable gap during functional anisotropic re-entry is caused by a lengthening of the action potentials at the pivoting points of the anisotropic circuit related to the impulse delay caused by the sudden increase of electrotonic load as the impulse changes direction from transverse to longitudinal.

There are indications that anisotropic conduction may play an important role in the pathogenesis of clinical cardiac arrhythmias. The greater incidence of atrial fibrillation with age may be caused by redistribution of collagen and progressive electrical uncoupling of myocardial fibres. The resulting higher degree of non-uniform anistoropy may facilitate the occurrence of unidirectional conduction block and provide small areas of slow conduction for intra-myocardial micro-reentry (Spach and Dolber, 1986). Spear *et al.* (1983) showed that after myocardial infarction the degree of anisotropy is enhanced by collageneous septa dividing the myocardial fibres. In the studies of Wit *et al.* (1982) and El-Sherif *et al.* (1981) on ventricular tachycardia 3–4 days after myocardial infarction, re-entrant circuits were found in the thin epicardial layer overlying the infarcted area. The conduction properties of this substrate of ventricular tachycardia also exhibited greatly enhanced (non-uniform) anisotropy (Dillon *et al.*, 1988).

REFERENCES

Allessie, M. A. and Bonke, F. I. M. (1984) Atrial arrhythmias: Basic concepts. In *Cardiac Arrhythmias: Their Mechanisms, Diagnosis and Management* (ed. W. J. Mandel). J. B. Lippincott, Philadelphia.

Allessie, M. A., Bonke, F. I. M. and Schopman, F. J. G. (1973) Circus movement in rabbit atrial muscle as a mechanism of tachycardia. *Circulation Res.* 32, 54–62.

Allessie, M. A., Bonke, F. I. M. and Schopman, F. J. G. (1977) Circus movement in rabbit atrial muscle as a mechanism of tachycardia. III. The "leading circle" concept: a new model of circus movement in cardiac tissue without the involvement of an anatomic obstacle. *Circulation Res.* 41, 9–18.

Allessie, M. A., Lammers, W. J. E. P., Bonke, F. I. M. and Hollen, J. (1984) Intra-atrial reentry as a mechanism for atrial flutter by acetylcholine and rapid pacing in the dog. *Circulation* 70, 123–35.

Allessie, M. A., Lammers, W. J. E. P., Bonke, F. I. M. and Hollen, J. (1985) Experimental evaluation of Moe's multiple wavelet hypothesis of atrial fibrillation. In *Cardiac Arrhythmias* (eds D. P. Zipes and J. Jalife), pp. 265–76. Grune & Stratton, New York.

Clerc, L. (1976) Directional differences of impulse spread in trabecular muscle from mammalian heart. *J Physiol., Lond.* 255, 335–46.

Cranefield, P. F., Wit, A. L. and Hoffman, B. F. (1973) Genesis of cardiac arrhythmias. *Circulation* 47, 190–204.

Dillon, S., Allessie, M. A., Ursell, P. C. and Wit, A. L. (1988) Influences of anisotropic tissue structure on reentrant circuits in the epicardial border zone of subacute canine infarcts. *Circulation Res.* 63, 182–206.

El-Sherif, N., Smith, A. and Evans, K. (1981) Canine ventricular arrhythmias in the late myocardial infarction period. 8. Epicardial mapping of reentrant circuits. *Circulation Res.* 49, 255–65.

Fishbein, M. C., Meerbaum, S., Rit, J., Lando, U., Kanmatsuse, Mercier, J. C., Corday, E. and Ganz, W. (1981) Early phase acute myocardial infarctsize quantification: validation of the triphenyl tetrazolium chloride tissue enzyme staining technique. *Am. Heart J.* 101, 593–600.

Frame, L. H. and Hoffman, B. F. (1984) Mechanisms of tachycardia. In *Tachycardias* (eds B. Surawicz, C. Pratrap Reddy and E. N. Prystowsky). Martinus Nijhoff, the Hague.

Frame, L. H., Page, R. L. and Hoffman, B. F. (1986) Atrial reentry around an anatomic barrier with a partially refractory excitable gap. *Circulation Res.* 58, 495–511.

Frame, L., Page, R., Boyden, P., Fenoglio, J. J. Jr and Hoffman, B. F. (1987) Circus movement in the canine atrium around the tricuspid ring during experimental atrial flutter and during reentry *in vitro*. *Circulation Res.* 76, 1155–75.

Gallagher, J. J., Gilb, M., Sevenson, R. H., Sealy, W. C., Kasell, J. and Wallace, A. G. (1975) Wolff–Parkinson–White syndrome: the problem, evaluation and surgical correction. *Circulation* 51, 767–85.

Hoeks, A. P. G., Schmitz, G. M. L., Allessie, M. A., Jas, H., Hollen, S. J. and Reneman, R. S. (1988) Multichannel storage and display system to record the electrical activity of the heart. *Med. Biol. Eng. Comp.* 26, 434–8.

Hoffman, B. F. and Rosen, M. R. (1981) Cellular mechanisms for cardiac arrhythmias. *Circulation Res.* 49, 1–15.

Janse, M. J. (1986) Reentrant rhythms. In *The Heart and Cardiovascular System* (eds H. A. Fozzard, E. Haber, R. B. Jennings, A. M. Katz and H. E. Morgan). Raven Press, New York.

Karagueuzian, H. S., Fenoglio, J. J. Jr, Weiss, M. B. and Wit, A. L. (1979) Protracted ventricular tachycardia induced by premature stimulation of the canine heart after coronary artery occlusion and reperfusion. *Circulation Res.* **44,** 833–48.

Lewis, T. (1920) Observations upon flutter and fibrillation. Part IV. Impure flutter; theory of circus movement. *Heart* **7,** 293–331.

Lewis, T. (1925) *The Mechanism and Graphic Registration of the Heart Beat.* Shaw & Sons, London.

Mines, G. R. (1913) On dynamic equilibrium in the heart. *J. Physiol.* **46,** 349–83.

Moe, G. K. (1975) Evidence for reentry as a mechanism for cardiac arrhythmias. *Rev. Physiol. Biochem. Pharm.* **72,** 56–66.

Moe, G. K., Mendez, C. and Han, J. (1965) Aberrant AV impulse propagation in the dog heart: a study of functional bundle branch block. *Circulation Res.* **16,** 261–86.

Myerburg, R. J., Nilsson, K., Befeler, B., Castellanos, A. and Gelband, H. (1973) Transverse spread and longitudinal dissociation in the distal A–V conducting system. *J. clin. Invest.* **52,** 885–95.

Nachlas, M. M. and Shnitka, T. K. (1963) Macroscopical identification of early myocardial infarcts by alteration in dehydrogenase activity. *Am. J. Pathol.* **4,** 379–405.

Osaka, T., Kodama, I., Tsuboi, N., Toyama, J. and Yamada, K. (1987) Effects of activation sequence and anisotropic geometry on the repolarization phase of action potential of dog ventricular muscles. *Circulation* **76,** 226–36.

Pick, A. and Langendorf, R. (1979) *Interpretation of Complex Arrhythmias.* Lea & Febiger, Philadelphia.

Rensma, P. L., Allessie, M. A., Lammers, W. J. E. P., Bonke, F. I. M. and Schalij, M. J. (1988) The length of the excitation wave as an index for the susceptibility to reentrant atrial arrhythmias. *Circulation Res.* **62,** 395–410.

Schalij, M. J. (1988) Anisotropic conduction and ventricular tachycardia. Thesis, University of Limburg.

Schmitt, F. O. and Erlanger, J. (1928) Directional differences in the conduction of the impulse through heart muscle and their possible relation to extrasystolic and fibrillatory contractions. *Am. J. Physiol.* **87,** 326–47.

Smeets, J. L. R. M., Allessie, M. A., Lammers, W. J. E. P., Bonke, F. I. M. and Hollen, J. (1986) The wavelength of the cardiac impulse and reentrant arrhythmias in isolated rabbit atrium. *Circulation Res.* **58,** 96–108.

Spach, M. S. and Dolber, P. C. (1986) Relating extracellular potentials and their derivatives to anisotropic propagation at a microscopic level in human cardiac muscle. *Circulation Res.* **58,** 356–71.

Spach, M. S., Miller, W. T., Geselowitz, D. B., Barr, R. C., Kootsey, J. M. and Johnson, E. A. (1981) The discontinuous nature of propagation in normal canine cardiac muscle. *Circulation Res.* **48,** 39–54.

Spach, M. S., Kootsey, J. M. and Sloan, J. D. (1982a) Active modulation of electrical coupling between cardiac cells of the dog: A mechanism for transient and steady state variations in conduction velocity. *Circulation Res.* **51,** 347–62.

Spach, M. S., Miller, W. T., Dolber, P. C., Kootsey, M., Sommer, J. R. and Mosher (1982b) The functional role of structural complexities in the propagation of depolarization in the atrium of the dog. Cardiac conduction disturbances due to discontinuities of effective axial resistivity. *Circulation Res.* **50,** 175–91.

Spach, M. S., Dolber, P. C., Heidlage, J. F., Kootsey, J. M. and Johnson, E. A. (1987) Propagating depolarization in anisotropic human and canine cardiac muscle: apparent directional differences in membrane capacitance. A simplified model for selective directional effects of modifying the sodium conductance on V_{max}, t_{foot}, and the propagation safety factor. *Circulation Res.* **60,** 206–19.

Spear, J. F. and Moore, E. N. (1982) Mechanisms of cardiac arrhythmias. *A. Rev. Physiol.* **44**, 485–97.

Spear, J. F., Michelson, E. L. and Moore, N. E. (1983) Cellular electrophysiologic characteristics of chronically infarcted myocardium in dogs susceptible to sustained ventricular tachyarrhythmias. *J. Am. Coll. Cardiol.* **1**, 1099–110.

Tsuboi, N., Kodama, I., Toyama, J. and Yamada, K. (1985) Anisotropic conduction of properties of canine ventricular muscles. Influences of high intracellular K concentration and stimulation frequency. *Jap. Circ. J.* **49**, 487–98.

Van Capelle, F. J. L. (1983) Slow conduction and cardiac arrhythmias. Thesis, University of Amsterdam.

Vivaldi, M. T., Kloner, R. A. and Schoen, F. J. (1985) Triphenyltetrazolium staining of irreversible ischemic injury following coronary artery occlusion in rats. *Am. J. Pathol.* **121**, 522–30.

Waldo, A. L., MacLean, W. A. H., Karp, R. B., Kouchoukos, N. T. and James, T. N. (1977) Entrainment and interruption of atrial flutter with atrial pacing: studies in man following open heart surgery. *Circulation* **56**, 737–45.

Wit, A. L. and Cranefield, P. F. (1978) Re-entrant excitation as a cause of cardiac arrhythmias. *Am. J. Physiol.* **235**, H1–H17.

Wit, A. L., Allessie, M. A., Bonke, F. I. M., Lammers, W. J. E. P., Smeets, J. and Fenoglio, J. J. Jr (1982) Electrophysiologic mapping to determine the mechanism of experimental ventricular tachycardia initiated by premature impulses. *Am. J. Cardiol.* **49**, 166–85.

Computer simulation of anisotropic impulse propagation: Characteristics of action potentials during re-entrant arrhythmias

F. J. L. VAN CAPELLE[a] AND M. A. ALLESSIE[b]

[a]Department of Experimental Cardiology, University of
Amsterdam, Academic Medical Centre, Meibergdreef 9, 1105 AZ
Amsterdam, The Netherlands
[b]Department of Physiology, Biomedical Center, University of
Limburg, PO Box 616, 6200 MD Maastricht, The Netherlands

RE-ENTRY OF THE CARDIAC ACTIVATION WAVE AS A MECHANISM OF TACHYCARDIA

Circus movement of the cardiac activation wave was demonstrated experimentally as early as 1913 by G. R. Mines (Mines, 1913). Making an incision in the auricle of the isolated tortoise heart he obtained a ring-shaped preparation in which by mechanical stimulation he could set up an excitation wave that would circulate indefinitely. On the basis of these experiments he speculated that in the human heart the mechanism of circus movement of the excitation wave might be responsible for certain clinical forms of tachycardia.

This early prediction has since been confirmed. Yet a direct demonstration of the re-entrant nature of a particular tachycardia has remained extremely difficult: it requires recording, preferably simultaneously, of electrograms at many places along the circuit and, ideally, the demonstration that making a cut through the circuit stops the tachycardia. The practical difficulties of mapping the circuit of a clinical circus movement tachycardia have been recently illustrated by de Bakker et al. (1988) who have performed extensive endocardial mapping of life-threatening tachycardias in infarct survivors who underwent anti-arrhythmic surgery. Although the mechanism of that particular type of tachycardia turned out to be circus movement of the activation, the principal aspect of the maps obtained was that of a concentric spread of activation from an apparent rapidly firing 'focus'. The part of the circuit that carried back the activation to the apparent focus frequently escaped detection by the multichannel mapping probe because it consisted of isolated surviving fibres buried in the healed infarct.

It is no wonder then that the clinical criteria for determining whether a tachycardia is re-entrant in nature are largely derived from indirect evidence (Brugada and Wellens, 1984). The distinction between focal and re-entrant tachycardias is of clinical importance, however. Focal tachycardias are caused by electrical instability of the cell membranes in the focal region, whereas re-entrant tachycardias depend on the electrophysiological properties of the circuit that carries the circulating activation wave. Therefore, different drugs are required for treatment of the two types of tachycardia

Most of the current knowledge about circus movement tachycardias has been derived from animal experiments. Since the fundamental experiments of Mines a substantial body of evidence regarding the mechanisms and characteristics of these tachycardias has become available. In particular it has been experimentally demonstrated by Allessie et al. (1973, 1976, 1977) that it is not necessary for a circus movement of the activation to run around a 'hole' or anatomically fixed obstacle. They started a tachycardia in the isolated left atrium of the rabbit heart, using programmed stimulation. By careful mapping with extra- and intracellular recordings they located the central area of the vortex-like activation sequence and demonstrated that this region consisted of normal excitable cells that showed normal action potentials during a paced basic rhythm and nevertheless showed inactivity or conduction block during the circus movement tachycardia. They coined the term 'leading-circle tachycardia' for this type of circus movement tachycardia, to distinguish it from the more classical circus movement tachycardias around an anatomical obstacle.

It is noteworthy that Gul'ko and Petrov (1972) in an elegant hybrid computer study, had demonstrated that this type of circus movement activation could be evoked in homogeneous excitable sheets. Subsequent theoretical work, especially by Russian groups, has now provided a solid theoretical foundation for the concept of circus movement of activity in two- and three-dimensional media (see Winfree, 1987).

An important difference between fixed-obstacle and leading-circle tachycardias is the way in which their cycle length is determined. In a fixed-length circuit the cycle length is the product of the conduction velocity and the length of the circuit. For the tachycardia to exist at all, the circuit must be large enough for the cycle length to exceed the longest refractory period occurring in the circuit. The cycle length of a stable fixed-obstacle circus movement tachycardia will always exceed the refractory period by a certain amount, and therefore a temporal 'excitable gap', i.e. an interval during which the excitable element can be successfully stimulated must always be present in the cycle. In contrast, a leading circle tachycardia adjusts and minimizes its circuit length, reducing its cycle length to the refractory period of the tissue. This should make it much more difficult to influence a leading circle tachycardia by interposed electrical stimuli than a fixed-obstacle tachycardia.

In reality the situation is even more complicated. Anisotropy of the tissue has a profound influence on the geometry of the leading circle. By freezing the interior part of the heart Allessie (Allessie, in this volume) developed a two-dimensional preparation consisting of a surviving subepicardial layer of ventricular muscle. The functional area of conduction block during leading circle tachycardias in this preparation was usually a line of 2–3 cm oriented parallel to the fibre direction. Within the area of conduction block local responses were recorded, often at twice the frequency of the encircling tachycardia. The activation wave spent a large part of its cycle in the neighbourhood of the two pivoting points at either side of the line of functional block. Possibly as a result of this local delay at the pivoting points an 'excitable gap' of 30–40 ms appeared to be present in this type of circus movement tachycardia. As explained elsewhere (Allessie, in this volume) there are basically three possible mechanisms that might explain the presence of an excitable gap during anisotropic re-entry. The delay at the pivoting points could be explained by the presence of microanatomical barriers or alternatively by the sudden increase of the axial current load at the transition of transverse to longitudinal propagation when the impulse finishes its U-turn around the pivoting point. Furthermore, electrotonic lengthening of the action potential proximal to the pivoting points might play an independent role.

Yet it is not intuitively clear why anisotropy should result in the appearance of an excitable gap. In continuous cable theory conduction velocity scales with the square root of axial conductivity, leaving the time course of the propagating action potential unaltered. One would therefore anticipate merely a linear scaling of the activation maps transforming the leading circle into a leading ellipse with its long axis parallel to the fibre direction, but there is certainly no à priori reason for an excitable gap to appear. However, it is well-known that cable theory, while being a reasonably good approximation for longitudinal conduction, does not describe transverse propagation equally well (Spach et al., 1981). We wondered whether a qualitative difference between longitudinal and transverse propagation, the first behaving like a continuous cable and the second like a lumped cable with its resistance concentrated at discrete barriers, could account for the presence of an excitable gap during anisotropic re-entry. We explored this possibility using a simple interactive computer model.

THE MODEL

The computer model has been described in more detail elsewhere (van Capelle and Durrer, 1980). Essentially it consists of an arbitrary number of excitable elements which can be connected in various ways in a two-dimensional matrix. The behaviour of the individual elements is determined by a set of functions and parameters that can be adjusted to trim their electrophysiological properties (e.g. the duration of the refractory period). For a simulation run, up to 650 elements can be connected in various ways. The geometrical arrangement of the elements, together with the values of the coupling resistances is specified interactively using a graphic screen. Since the differential equations controlling the behaviour of the elements must be solved for many elements simultaneously, making variable time-step techniques impractical, we used a primitive model for the underlying excitatory mechanism. Two voltage–current relations $i_0(V)$ and $i_1(V)$ describe the behaviour of the membrane. $i_0(V)$ has a region of negative chord-conductance and can therefore generate an action potential upstroke. The other current, $i_1(V)$, is used to carry the membrane back to its resting potential. An excitability parameter, Y, which assumes values between 0 (maximum excitability) and 1 (complete inexcitability) determines how much of each current is switched on at a particular time. The instantaneous voltage–current relation of the membrane is given by $i(V) = Yi_1(V) + (1 - Y)i_0(V)$. Y itself moves toward a steady-state value $Y_{inf}(V)$, an S-shaped curve which corresponds with excitability at the resting potential and with inexcitability for depolarized elements. The membrane is thus characterized by the two equations:

$$C\,dV/dt = -Yi_1(V) - (1 - Y)i_0(V) + i_{ex}$$

$$T\,dY/dt = Y_{inf}(V) - Y$$

where C is the membrane capacity, T is the time constant of the activation/inactivation process, and i_{ex} is the current entering the element through its connections with other elements. The elements used in the two-dimensional sheets in the present study (coupling resistances were 0.3 and 1.2, in longitudinal and transverse directions) were defined by the continuous functions $Y_{inf}(V)$, $i_0(V)$ and $i_1(V)$ as follows: $Y_{inf}(V)$ was piecewise linear, zero for $V < -80$ and unity for $V > -60$. $i_1(V)$ was also piecewise linear, passing through $(-70, 5)$ and $(0, 6)$; its slope was 0.5 for $V < -70$ and 0.425 for $V > 0$. $i_0(V)$ was equal to the sum of $i_1(V)$ and an auxiliary function $f(V)$. $f(V)$ increased linearly with a slope of 2.0 until the point $(-74.3, 7.84)$ was reached; it then proceeded as a cubic, with a maximum in $(-70, 12)$ and a minimum in $(-30, -100)$ until the point $(-27.8, -98.84)$, where it became linear once more with a slope of 1.71. The cubic was fitted in such a way that f and its derivative were continuous everywhere. C was 0.1 and T was 0.5.

It must be stressed that the ionic mechanism employed in this model is very much oversimplified. The model should therefore be used with caution; ideally the results ought to be verified using more sophisticated membrane models such as the Beeler and Reuter model (Beeler and Reuter, 1977). However, solving the full set of differential equations for all elements in the sheet requires extensive programming and the use of very large computers, whereas the present model can easily be implemented in a small laboratory computer.

SIMULATIONS

In Fig. 1 the influence of increasing coupling resistance between the elements of a cable on the shape of the action potential and on its conduction velocity is illustrated. Forty elements were coupled by identical coupling resistances R

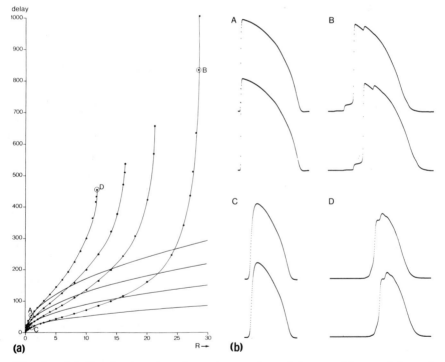

(a)   (b)

Fig. 1. (a) Conduction time through a section of linear cable consisting of identical excitable elements as a function of the coupling resistance in arbitrary values. Thick lines are theoretical parabolic relations as predicted by continuous cable theory for four levels of membrane excitability. Thin lines connect simulation results. Points A–D correspond to the action potential configurations of (b). (b) Shape of the propagating action potentials in two contiguous elements in the simulations corresponding to points A–D in (a). Low resistance coupling results in smooth upstrokes (A, C) and high-resistance coupling results in notched upstrokes (B, D). Depressed elements (C, D) are characterized by a low upstroke velocity and a lower amplitude.

to form a one-dimensional cable. In order to eliminate end-effects of the cable, we measured the conduction time between elements 10 and 30 when element 1 was being stimulated. The conduction velocity in the cable was varied in two ways: (1) by changing the coupling resistance between successive elements and (2) by changing the maximal value of the regenerative inward current of the elements. In Fig. 1a the conduction time is plotted against the coupling resistance R (normalized by division by the resting membrane resistance R_m) for four levels of membrane excitability (relative values of the maximum regenerative inward current were 1.0, 0.3, 0.2, 0.15, respectively). With increasing R, the observed values of the conduction time in the model (thin lines) soon begin to exceed the values predicted by the continuous cable theory (thick lines). The increase of the conduction delay then becomes increasingly steep until conduction block occurs.

Comparing the decrease of the conduction velocity effected by depressing the excitability of the cells or by increasing the coupling resistance, we note that non-depressed elements tolerate a much larger increase of the coupling resistance than depressed ones before conduction block occurs. As a consequence the longest conduction times in Fig. 1 were obtained using non-depressed elements and high coupling resistances. Action potentials of adjoining cells are shown in Fig. 1b for four combinations of excitability and coupling resistance (A–D). (A) Represents normal excitability and low coupling resistance; conduction is fast and the action potential configuration is smooth. (B) Represents the same cells when the coupling resistance is so large that conduction through the cable almost fails. Conduction is saltatory, prominent 'foot' and 'hump' configurations in the action potentials reflecting the moment of activation of the neighbour elements. (C) and (D) are analogous to (A) and (B), but for depressed elements; again we see a smooth upstroke when the coupling resistance is low enough for the cable approximation to hold, and a foot-and-hump configuration when conduction nearly fails. Yet the quality of the action potentials is much poorer, as can be seen from the reduction of the upstroke velocity and of the action potential amplitude.

We simulated anisotropic conduction in a rectangular sheet by connecting the elements with different resistances in the horizontal and vertical directions. Figure 2 shows a circus movement of the activation wave in a sheet of 41×10 identical excitable elements. The elements (indicated as small '+' signs in the sheet) were coupled to their immediate neighbours in a square lattice. The vertical coupling resistances were four times as large as the horizontal ones, resulting in a marked difference of the propagation velocity of the impulse in the two directions. A horizontal slit was made in the middle of the sheet by deleting the transverse coupling resistances between opposite elements for 20 element pairs (horizontal thick line). The circulating wave was started by temporarily extending the slit at one end to the left edge of the sheet and them stimulating an element in the upper left corner. In this way a

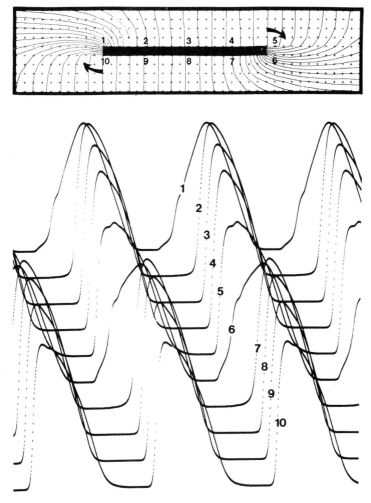

Fig. 2. Action potentials of selected elements during circus movement of the activation around a slit in an anisotropic rectangular sheet of excitable elements (above). Isochronal lines (10 ms apart) indicate activation times of the elements. One complete cycle (540 ms) is depicted in the upper panel. Conduction delay at the pivoting points, as apparent by the crowding of isochronal lines, corresponds to broadened action potentials of the elements at the distal side of the pivot.

wavefront was set up that travelled through the upper half of the sheet to the right, turned around the slit and travelled back to the left through the lower part of the sheet. Just before the returning wavefront reached the left pivoting point of the original slit, the extended line of conduction block was removed by connecting the elements again, and as a result the impulse re-entered the upper region and started circulating around the slit.

The thin lines drawn in the sheet are isochronal lines for the activation moments, interpolated between the elements. After scaling the refractory period of the elements to about 330 ms, the isochronal lines are 10 ms apart. One complete cycle (540 ms) of the 'tachycardia' is depicted. The isochronal lines are much farther apart when the impulse travels horizontally than when it propagates in the transverse direction. Around both pivoting points there is a distinct crowding of the isochronal lines; the circulating impulse spends about half of the cycle in the immediate neighbourhood of the pivoting points.

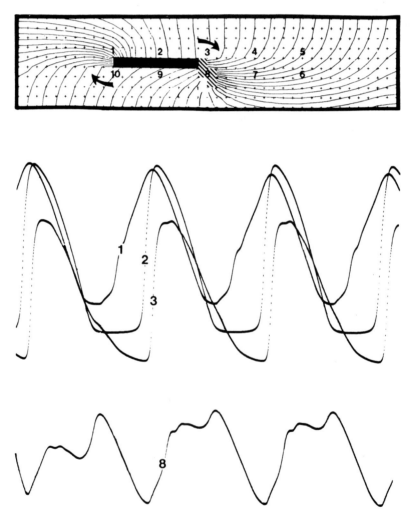

Fig. 3. Action potentials and isochronal lines after sudden shortening of the slit in Fig. 2. The cycle length has reduced to 480 ms. The left pivoting point retains its character of an anatomical barrier, while the right pivoting point behaves as a functional obstacle.

In the lower part of Fig. 2 the action potentials of 10 cells adjacent to the slit are shown. When the impulse travels horizontally along the slit, smooth action potentials are registered (elements 2–4, 7–9) but elements 1 and 6 at the pivoting points are clearly different. They have broad action potentials characterized by a very slow notched upstroke. This is caused by the sudden increase of axial current load when the impulse changes direction from vertical to horizontal and suddenly 'sees' the tightly coupled not-yet-excited cells downstream. As a consequence, the excitable gap, which roughly equals the time the elements spend at their resting potential, is different in various parts of the circuit; it is longest where the wavefront moves longitudinally and shortest at the pivoting points.

Starting from the simulation presented in Fig. 2 we tried to introduce functional block by suddenly reducing the length of the slit. We therefore reconnected 10 element pairs at the right side of the slit (Fig. 3) at the time that the impulse was rotating around the left pivoting point. The tachycardia continued (with a cycle length of 480 ms) and the character of the action potentials at the pre-existing left pivoting point did not change during the first cycles. In contrast, the newly introduced functional pivoting point at the

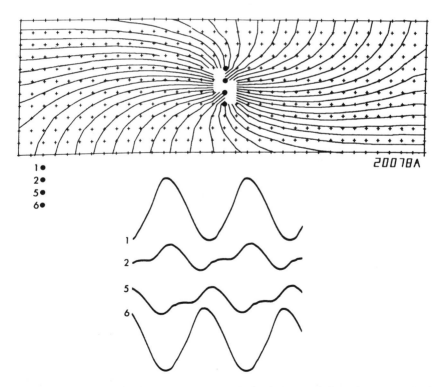

Fig. 4. Action potentials and isochronal lines as they stabilize after removal of the slit. The cycle length is now 420 ms. The zone of functional block is concentrated in a relatively small central area.

right behaved quite differently. It was not possible for the impulse to immediately cross the previous line of anatomical conduction block, and a marked delay occurred while the wavefront moved farther to the right, and then swept down to turn to the left only when it approached the bottom edge of the sheet. The block zone thus became more or less L-shaped, the anatomical pivoting point positioned at the end of a long leg of the L, and the functional one in the short leg. The element in the centre of the pivot, element 8, showed slow, distorted complexes of a long duration. At the pivoting points there was no excitable gap left, but elsewhere in the circuit (element 2) the elements showed a clear diastolic interval.

In Fig. 4 the activation pattern is shown when the tachycardia has stabilized after reconnecting the slit over its complete length. The interval (420 ms) is still shorter than in Fig. 3, and no excitable gap is present in any of the tracings. The wacefront encircles a completely functional zone of block where the two pivoting points coincide. The elements in the centre are never fully activated but show local (block) responses originating from the elements above and below the centre. Although the local responses are not of the same size, it can be seen that their frequency in the centre is double that of the encircling wavefront.

DISCUSSION

The results reported by Allessie (in this volume) show that during circus movement tachycardia in a thin layer of subepicardial ventricular tissue the impulse circulates around a 2–3 cm long line of functional conduction block, and that an excitable gap is present under these circumstances. It can be seen from our simulations (Figs 2 and 3) that the presence of an excitable gap in the elements in the longitudinal stretch of the circuit may well be caused by lengthening of the action potentials at the pivoting points of the anisotropic circuit related to the impulse delay caused by the sudden increase of electrotonic load as the impulse changes direction from transverse to longitudinal. However, we needed an anatomical barrier in at least part of the circuit for an excitable gap to appear at all. This leads us to suspect that the functional line of block observed in Allessie's experiments may, in fact, be a preferential site of block, appearing only during the stressed situation of a very fast rhythm but not when relatively slow, stimulated, basic rhythm impulses with a good margin of safety invade the area. This idea is also in agreement with observations during the experiment of local responses at twice the frequency of the tachycardia in the central area. Naturally, such local responses were not seen in the simulations around a barrier, but they did appear during the anisotropic leading circle tachycardia of Fig. 4. Finally, the concept of preferential sites of block could also explain the stability of the position of the block in Allessie's experiments. In the simulations the block position was also very stable, but in the absence of an anatomical barrier this was almost certainly caused by restriction of the centre of the circuit by the close neighbourhood of the boundaries because of the small size of the sheet. Indeed, it is well-known (Yermakova and Pertsov, 1986) that the centre of a leading circle tachycardia in homogeneous excitable sheets tends to move through the sheet in a way that is controlled by its relative position with respect to the sheet boundaries.

The simulations presented above suggest that the appearance of an excitable gap during anisotropic re-entry in Allessie's experiments may be due to local delay at the pivoting points of the linear block zone, which in turn is caused by the sudden increase of electrotonic load as the impulse finishes its U-turn. It should be stressed, however, that (1) the elements used are primitive caricatures of cardiac excitable tissue, (2) the dimensions of the sheets in the simulation are rather small and the spatial discretization is coarse, and (3) the perfect homogeneity of the sheets, all elements being rigorously identical, may make them behave differently in comparison with real tissue.

REFERENCES

Allessie, M. A., Bonke, F. I. M. and Schopman, F. J. G. (1973) Circus movement in rabbit atrial muscle as a mechanism of tachycardia. *Circulation Res.* **32,** 54–62.

Allessie, M. A., Bonke, F. I. M. and Schopman F. J. G. (1976) Circus movement in rabbit atrial muscle as a mechanism of tachycardia. II. The role of nonuniform recovery of excitability in the occurrence of unidirectional block, as studied with multiple microelectrodes. *Circulation Res.* **39,** 168–77.

Allessie, M. A., Bonke, F. I. M. and Schopman, F. J. G. (1977) Circus movement in rabbit atrial muscle as a mechanism of tachycardia. III. The 'leading circle' concept: A new model of circus movement in cardiac tissue without the involvement of an anatomic obstacle. *Circulation Res.* **41,** 9–18.

Beeler, G. W. and Reuter, H. (1977) Reconstruction of the action potential of ventricular myocardial fibres. *J. Physiol.* **268,** 177–210.

Brugada, P. and Wellens, H. J. J. (1984) Programmed electrical stimulation of the human heart. In *Tachycardias: Mechanisms, Diagnosis, Treatment* (eds M. E. Josephson and H. J. J. Wellens) Lea and Febiger, Philadelphia.

de Bakker, J. M. T., van Capelle, F. J. L., Janse, M. J., Wilde, A. A. M., Coronel, R., Becker, A. E., Dingemans, K. P., van Hemel, N. M. and Hauer, R. N. W. (1988) Reentry as a cause of ventricular tachycardia in patients with chronic ischemic heart disease: electrophysiologic and anatomic correlation. *Circulation* **77,** 589–606.

Guk'ko, F. B. and Petrov, A. A. (1972) Mechanisms of the formation of closed pathways of conduction in excitable media. *Biofizika* **17,** 261–70.

Mines, G. R. (1913) On dynamic equilibrium of the heart. *J. Physiol., Lond.* **46,** 349–83.

Spach, M. S., Miller, W. T. III, Geselowitz, D. B., Barr, R. C., Kootsey, J. M. and Johnson, E. A. (1981) The discontinuous propagation in normal canine cardiac muscle: Evidence for recurrent discontinuities of intracellular resistance that affect the membrane currents. *Circulation Res.* **48,** 39–54.

van Capelle, F. J. L. and Durrer, D. (1980) Simulation of arrhythmias in a network of coupled excitable elements. *Cirulation Res.* **47,** 454–66.

Winfree, A. T. (1987) *When Time Breaks Down. The Three-Dimensional Dynamics of Electrochemical Waves and Cardiac Arrhythmias.* Princeton University Press, Princeton, NJ.

Yermakova, Y. A. and Pertsov, A. M. (1986) Interaction of rotating spiral waves with a boundary. *Biofizika* **31,** 855–61.

A mathematical model for the initiation of ventricular tachycardia in myocardium

J. P. KEENER

Department of Mathematics, University of Utah, Salt Lake City, Utah 84112, USA

INTRODUCTION

In 1981, Spach and co-workers (Spach *et al.*, 1981) published the results of an experiment that has since provided the impetus for much research and controversy. In their experiment, a stimulating electrode was positioned on a section of crista terminalis of an adult dog. Four recording electrodes were placed at various distances from the stimulus site along the fibre axis, and two recording electrodes were placed transversely to the fibre axis from the stimulus site. The experiment consisted of providing a series of evenly timed stimuli followed by one premature stimulus, and then observing the electrical activity at the recording sites.

The results of the experiment were surprising and remain unexplained in the cardiology literature. The experiment showed, as expected, that when the premature stimulus was very early (during the absolute refractory period at the stimulus site), propagation was blocked in all directions. When the premature stimulus was not so early, the action potential propagation was successful in all directions. But there was a small window of stimulus times during which a premature stimulus elicited a response that propagated in only one direction and was blocked in the opposite (orthogonal) direction. Furthermore, the action potential propagated successfully away from the stimulus site for some distance and them appeared to turn around and propagate in the retrograde direction on parallel pathways, eventually stimulating the cells that were at first blocked from activity.

The fact that propagation block was directionally dependent is a surprise, at least if one is accustomed to thinking of the myocardium as a continuum (a syncytium). It is easy to show that the success or failure of action potential propagation cannot be directionally dependent in a continuum.

Cell to Cell Signalling: From
Experiments to Theoretical Models
ISBN 0–12–287960–0

It is also a surprise that re-entry can occur when a spatially homogeneous medium is stimulated at only one site. All previous theories of initiation of re-entry require spatial inhomogeneity or loss of symmetry of some sort, either through inhomogeneity of the medium or from asymmetry of multiple stimuli locations, or both.

Probably the biggest surprise of this experiment was that the direction in which the propagation first failed was the longitudinal direction (the long cell axis) of fibre orientation. This was completely counterintuitive because the longitudinal direction is the direction of fastest propagation, and therefore, it was believed, the direction of greatest safety factor. According to this line of reasoning, the direction of first failure should have been the lateral direction, and since it was not, some investigators suggested that the results of the experiments were incorrect,

Two additional observations pertaining to the shape of the action potentials were made that went against common understanding. For a continuous medium, the maximal rate of rise of the action potential $(dv/dt)_{max}$, and the time constant of the exponential foot of the rising action potential, τ_{foot}, are independent of the coupling strength. However, it was observed that they were not the same for the two directions of spread for a successful stimulus, but that the maximal rate of rise appeared to decrease and τ_{foot} appeared to increase as the velocity of propagation changed from slow to fast.

All of these observations suggest that the common understanding of propagation of action potentials in myocardium is flawed, and it is reasonable to point to continuous models of myocardium as the culprit. Indeed, it has been argued that our understanding of the myocardium is incorrect since it is based on continuum models (mostly cable theory), and that there is a need for investigation of the implications of discrete models (Spach and Kootsey, 1983; Spach and Dolber, 1985).

The purpose of this paper is to show that the principal dilemmas of the Spach experiment can be adequately explained using discrete models for the myocardium. We will demonstrate that a discrete anisotropic cellular medium can have propagation failure in only one direction, and, more significantly, that this can lead to re-entry and permanently rotating action potentials, similar to a state of ventricular tachycardia. We will also show, using data from trabecular bundle of calf, that the direction of propagation failure is the long direction of cells, as suggested by the experiment of Spach. Indeed, we calculate that lateral coupling of cardiac cells is stronger than longitudinal coupling, in spite of the fact that lateral propagation is slower than longitudinal propagation. This conclusion is contrary to current dogma, and since it is difficult for many to believe and crucial for the validity of this model, we give a detailed argument showing why it is reasonable. Other investigators have concluded that discrete models are inadequate to explain all the features of the Spach experiment for the principal reason that in all previous models, longitudinal coupling is assumed to be stronger than lateral coupling. Therefore, although there is nothing faulty with our theory mathematically, as an explanation of the Spach experiment it succeeds if the coupling of cells is as we calculate here, and not as is usually assumed.

PROPAGATION AND RE-ENTRY IN A DISCRETE EXCITABLE MEDIUM

Propagation of action potentials in an excitable medium is usually described by differential equations of the form

$$\frac{\partial v}{\partial t} = Lv + f(v,w_i), \quad \frac{\partial w_i}{\partial t} = g(v,w_i), \quad i = 1, 2, \ldots, n \tag{1}$$

where v represents the voltage potential and w_i represent various other variables such as gating variables, chemical concentrations, etc. For example, the Hodgkin–Huxley (1952) (nerve), McAllister–Nobel–Tsien (1975) (Purkinje fibre), Yanagihara–Noma–Irisawa (1980) (sinoatrial node), and Beeler–Reuter (1977) (myocardium) models as well as the FitzHugh–Nagumo caricature (FitzHugh, 1961) are in this form. The term Lv is a linear operator representing that cells are coupled spatially through the voltage. The exact form of the operator Lv depends on what is being modelled. For example, for a long continuous cable, or single cell, it is usual to use second partial derivatives in space,

$$Lv = Dv_{xx} \tag{2}$$

The equations (1) with spatial coupling (2) is generally referred to as a cable model. In a two- or three-dimensional collection of cells it is common to take

$$Lv = D_x v_{xx} + D_y v_{yy} + D_z v_{zz} \tag{3}$$

where D_x, D_y, D_z are diffusion coefficients in the x, y, z coordinate directions, and may be different if the medium is anisotropic (as is the myocardium, see Clerc, 1976).

For a discrete cellular medium, such as myocardium, continuous coupling may not be appropriate and may give incorrect results. To model a collection of isopotential cells, we assume that the coupling is purely resistive, so that the current flow between cells depends only on the potential difference between the cells divided by the resistance (Ohm's law). Thus, we take for our coupling operator the second-order difference operator,

$$Lv = d(v_{n-1} - 2v_n + v_{n+1}) \tag{4}$$

for a one-dimensional collection of cells, or

$$Lv = d_1(v_{n-1,k} - 2v_{n,k} + v_{n+1,k}) + d_2(v_{n,k-1} - 2v_{n,k} + v_{n,k+1}) \tag{5}$$

for a two-dimensional collection of cells on a rectangular grid. The coupling coefficients d_1 and d_2 are the inverse of the product of the resistances in the respective directions with the membrane capacitance. Of course, since we must keep track of the voltage potential at individual cells, the variables v and w_i must be subscripted to denote the location of the cell. If we use a non-rectangular grid, then a different discrete operator (discussed in the next section) is necessary.

The assumption that cells are isopotential is an approximation that simplifies the analysis of these systems enormously. It is shown to be warranted by Kawato et al. (1986) using formal perturbation arguments. In a number of recent numerical investigations, the grid on which numerical simulations were done included some large number of grid points for each cell (Joyner, 1982; Diaz et al., 1983; Spach and Dolber, 1985; Cole et al., 1988). However, since the intercellular resistance is more important than the intracellular cytoplasmic resistance, the numerical simulations show that propagation within the cell is fast compared with propagation across cell boundaries. That is, the propagation along connected strands appears to move discontinuously. This saltatory type of propagation is more pronounced as the rate of firing is increased or as the coupling between cells is weakened (Cole et al., 1988) and this is precisely the parameter range with which we are concerned.

The next simplification we make is to assume that the propagation of action potentials is determined by a simplified set of equations in which recovery variables vary slowly in the leading edge of the action potential. For example, in myocardium, the sodium current is fast compared to potassium and calcium currents. Thus, much of our discussion will focus on a single equation, the Nagumo equation, given by

$$\frac{\partial v}{\partial t} = Lv + f(v,w) \tag{6}$$

where L is the appropriate spatial coupling operator, and w is held fixed. To be specific, in this paper we use

$$f(v,w) = 12\sqrt{3}v(1 - v)(v - 0.5) - w \tag{7}$$

with w fixed. For this equation an excitation front is defined as a solution of (6) that takes v at any point in space from the smallest zero of the cubic (7) to the largest zero of (7) as time progresses from minus infinity to plus infinity. With this particular scaling, the value $w = -1$ corresponds to a fully recovered medium while for $w > 0$ the medium is absolutely refractory, and does not support excitation fronts.

The usual first step in the analysis of eqn (6) is to look for travelling solutions of the form $v(x,t) = V(x - ct)$. If the coupling is continuous as in (2), then the existence of travelling wave solutions does not depend on the size of coupling D, since the solution can be expressed as

$$v(x,t) = V_0 \left(\frac{x - c_0\sqrt{Dt}}{\sqrt{D}} \right) \tag{8}$$

where the function V_0 satisfies the differential equation

$$V_0'' + c_0 V_0' + f(V_0, w) = 0 \tag{9}$$

independent of D. The speed of propagation of such a travelling solution is $c = \sqrt{D}\, c_0$ where c_0 depends only on the recovery variable w. It is apparent from (8) that the time-dependent behaviour of the solution is independent of D. Thus the properties of propagating action potentials measured as a function of time at a point in a continuous medium are independent of D.

In contrast, if fronts are governed by the eqn (6) with discrete coupling (4), propagating solutions need not exist. It is easy to see that for d small enough, there are no travelling wave solutions for the eqns (4) and (6). It is also true, but much more difficult to verify (Chi et al., 1986; Keener, 1987; Zinner, 1988) that for d large enough, eqn (4) with coupling (6) has travelling wave solutions of the form $v_n(t) = V(t - n\tau)$ with asymptotic speed of propagation given approximately by $1/\tau = c = c_0\sqrt{d}$.

The failure of propagation for small d is easily seen to be true. If the possible range of the dependent variables during an action potential is bounded, and if the resting state is stable (which it is for all of the non-oscillatory excitable media models), then when d is very small the influence on a cell by its neighbours is also small, and the dominant influence on the cell is the local dynamics of the cell, which returns it to rest. Thus, for small coupling, the stimulating currents from neighbouring cells are small and a cell cannot be raised by the action of its neighbours to a potential that exceeds its threshold, so action potentials cannot propagate.

In Fig. 1 is shown a sketch of the speed of propagation as a function of coupling strength for the eqns (6) with discrete coupling (4) (solid line) and for continuous coupling (2) (dashed line) with $w = -0.5$. The plot shows the nearly square root dependence of speed on coupling for large coupling in the discrete model, while for d below a critical value d^*, travelling wave solutions cease to exist. It is important to note that the speed of propagation for the discrete model is always measured as the number of cells per unit time that are stimulated by the travelling solution.

For the discrete problem, a travelling wave solution exists only if the coupling d is above some critical value d^*. Necessarily d^* depends on the value of recovery w, and it is reasonable to believe that a well-recovered medium has a lower critical value of coupling d^* than a relatively unreco-

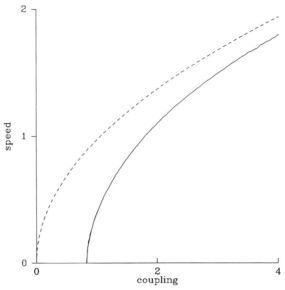

Fig. 1.   Speed as a function of coupling strength for the Nagumo equation with $w=-0.5$, shown for discrete coupling (solid curve) and continuous coupling (dashed curve).

vered medium. This follows logically from the observation that in a well-recovered medium, it is easier to stimulate a cell, so that the minimal required current flow between neighbouring cells is less for recovered cells than for unrecovered cells. In other words, coupling need not be as strong in a well-recovered medium as in an unrecovered medium to sustain a travelling front of excitation.

Two properties of action potentials which are significant to physiologists are the maximum velocity of the upstroke of the action potential, $(dv/dt)_{max}$, and the time constant of the rising action potential τ_{foot}. These quantities are seen from (8) to be independent of the coupling strength for a continuous model. When the maximum derivative of v is attained, the second derivative of v is zero so that a simplified relationship between $(dv/dt)_{max}$ and currents can be obtained from (6). The quantity $(dv/dt)_{max}$ is taken to be a diagnostic of the fast sodium current, since the only current that participates significantly in the upstroke, and hence the principal determinant of $(dv/dt)_{max}$, is the sodium current. In discrete media, $(dv/dt)_{max}$ is not independent of coupling, since, when $(dv/dt)_{max}$ is attained, although the second time derivative of v is zero, the spatial coupling operator is not zero. Diaz et al. (1983) and Cole et al. (1988) calculated the relationship between $(dv/dt)_{max}$ and the coupling coefficient d for the Beeler–Reuter model and showed it to be biphasic. For large d, $(dv/dt)_{max}$ approaches the correct value for continuous coupling. The biphasic nature of $(dv/dt)_{max}$ is disturbing, since a biphasic curve does not have obvious diagnostic capability.

The time constant τ_{foot} can be calculated from the linearization of (6) about the resting steady state. The time constant is defined so that the rise of the action potential from the resting potential as a function of time is an exponential of the form $\exp(t/\tau_{foot})$. If $f(v_0,w)=0$, and $f'(v_0,w)=\rho_0$, then $\tau_{foot}=1/\lambda$, where λ satisfies the characteristic equation (obtained by linearization)

$$\lambda^2 - c_0^2\lambda + c_0^2\rho = 0 \tag{10}$$

in the continuous case (2), and

$$4d \sinh^2\left(\frac{\lambda}{2c}\right) - \lambda + \rho = 0 \tag{11}$$

in the discrete case (4). Using what is known about the speed of propagation from Fig. 1, it can be shown that the time constant τ_{foot} is a monotone decreasing function of coupling coefficient d which approaches the time constant for the continuous problem in the limit of large d, but which is infinitely large in the neighbourhood of the critical coupling d^*. The experiments of Spach et al. (1981) show that the time constant τ_{foot} for propagation transverse to the cell longitudinal axis is smaller than the time constant τ_{foot} for propagation along the cell longitudinal axis. In the usual interpretation that cells are coupled more strongly along their longitudinal direction than in the transverse direction, this experimental finding is completely opposite from the theoretical result. A number of investigators have used this observation to conclude that discrete models do not agree with experiments. But if cells are more stongly coupled in their transverse direction than in their longitudinal direction (as we claim), then the theoretical result is in agreement with the experimental result.

In a two-dimensional anisotropic continuous medium, propagation success does not depend on direction. In a continuous medium, the evolution of fronts of excitation is governed by (6) with coupling (3), and plane wave solutions are found in the form

$$v(x,y,t) = V_0\left(\frac{x \sin\theta}{\sqrt{D_x}} + \frac{y \cos\theta}{\sqrt{D_y}} - c_0 t\right) \tag{12}$$

where V_0 satisfies the differential equation (9). Thus, while the speed of propagation is given by

$$c(\theta) = \frac{\sqrt{(D_x D_y)}c_0}{\sqrt{(D_x \sin^2\theta + D_y \cos^2\theta)}} \tag{13}$$

and depends in a non-trivial way on θ, the existence of a travelling solution is independent of θ and the coupling coefficients D_x and D_y. If propagation is successful at all, it will spread in all directions, although with different speeds.

The same is not true in a discrete medium. For a two-dimensional grid of identical, isopotential cells, coupled discretely over a rectangular grid, the dynamics of spread of excitation is governed by eqn (6) with coupling (5). If one of the coupling coefficients is larger and the other is smaller than d^*, then there can be propagation success in one direction, but not the other. As a result, a two-dimensional discrete medium can support isolated one-dimensional paths of propagation, and waves of excitation travelling along these one-dimensional paths may fail to spread laterally to excite the adjacent medium.

The primary consequence of discrete anisotropic coupling is in the response to a current stimulus. Depending on the condition of the medium at the time of the stimulus, one of three things can happen. If the medium is well recovered so that d^* is small, propagation in both directions will be successful. If the medium is refractory (d^* large), no propagation is possible. However, for intermediate ranges of recovery, a third possibility exists, and propagation in the direction of strongest coupling succeeds while propagation fails in the direction of weakest coupling. The directional differences of propagation in a discrete medium are explained by the fact that since the resistances are different, the amount of stimulating current provided to a cell by an approaching action potential may be insufficient to reach threshold if resistance is high, but adequate if resistance is low.

In a discrete medium, if the weak coupling is below the critical coupling, then propagation in the weak direction will fail. (For the remainder of this section, we shall refer to the strong direction of coupling as the axial direction, and the weak direction as the lateral direction. Since we have not yet calculated the sizes of the couplings for myocardium, this nomenclature must not be construed to correspond to any particular orientation for myocardial cells.) Furthermore, as long as lateral coupling fails, the cells adjacent to those which are being excited act as a current sink to the travelling front, and therefore depress the speed with which the front travels. The size of this decrement depends on a number of factors, such as the recovery state of the medium and the strength of the strong coupling. It is possible for lateral current losses to completely kill propagation in the axial direction. In other words, the presence of weak lateral coupling changes the critical size of axial coupling required to sustain single fibre propagation. In Fig. 2 are shown the regions of coupling parameter space in which the different possibilities occur. The solid curve corresponds to the critical value of axial coupling as a function of lateral coupling necessary to sustain single fibre propagation. For a relatively unrecovered medium, axial propagation can be eliminated by the presence of lateral propagation failure. For a well-recovered medium (not shown), lateral current losses increase the critical

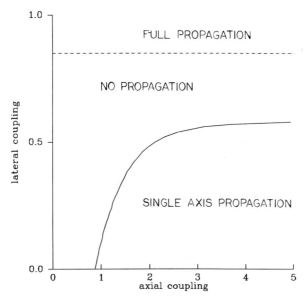

Fig. 2.   Plot of regions in the axial coupling–lateral coupling plane with recovery fixed $w=-0.5$, in which (i) no propagation is possible, (ii) propagation in both coordinate directions is successful, and (iii) propagation succeeds only along a single fibre in the strong coupling direction.

coupling strength, but axial propagation is not eliminated before lateral propagation becomes successful.

It is illustrative to see how these three possibilities depend on the recovery variable w. In Fig. 3 these three regions are replotted, but now with the recovery variable and lateral coupling as variables and with axial coupling fixed. (The two solid lines are for two different fixed values of axial coupling.) We see that in an unrecovered medium no propagation is possible, and in a well-recovered medium, propagation in both axial and lateral directions is possible. Additionally, if lateral coupling is not too strong, there is an intermediate range of recovery for which single fibre propagation is possible. We designate this region as the vulnerable region for the medium.

Until now we have nominally viewed propagation as if the medium into which an action potential propagates is at rest. This is an approximation to the truth, although the time and space scales on which an action potential front propagates are generally very much shorter than those of other processes in the medium. For example, in myocardium, the upstroke of the action potential is due primarily to the fast sodium ion current. Other ionic currents such as potassium or calcium, act on a much slower time scale. The main point is that a depolarizing wavefront is actually propagating into a medium with properties that are changing only slowly in both space and

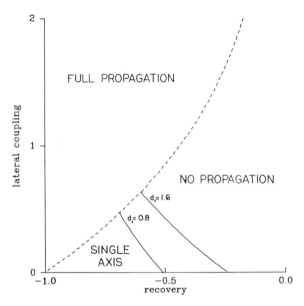

Fig. 3. Plot of regions in the recovery–lateral coupling plane with axial coupling d_1 fixed ($d_1=0.8$, 1.6), in which (i) no propagation is possible, (ii) propagation in both coordinate directions is successful, and (iii) propagation succeeds only along a single fibre in the strong coupling d_1 direction (vulnerable period).

time, but at any point in space and time it is reasonable to assume that the wavefront properties are determined by the local properties of the medium.

Suppose a cell on a fully recovered patch of medium is stimulated with a superthreshold current stimulus. If the medium is sufficiently excitable at rest (equivalently, if the weak coupling is not too small), then propagation will be successful in all directions, and the entire medium will be excited by the spreading stimulus. Some time after the initial stimulus, the medium returns to the resting state during which the medium recovers its excitability. The consequences of a successive stimulus depend very much on the elapsed time since the first stimulus was applied. We can determine from Fig. 3 what happens. If the stimulus is applied very early while the medium is not recovered, propagation fails completely. If the stimulus is applied after a long recovery period so that the medium has nearly returned to its resting state, then the action potential propagates in both directions from the stimulus site. For a continuous medium these are the only two choices. However, if the medium is discrete and the coupling in the weak coupling direction is small, then a stimulus applied during the vulnerable phase will result in a wave of excitation that propagates only along a single fibre (or group of fibres) in the direction of strongest coupling.

If the second stimulus occurs during the phase of vulnerability of the medium, then an action potential propagates in the strong direction, but not in the weak direction. Since the action potential propagates along a single fibre into relatively unrecovered medium, the speed of propagation of this wavefront is slower than its predecessor for two reasons. First, the speed of propagation in unrecovered medium is slower than in recovered media, and second, when laterally neighbouring cells are not stimulated by the propagating wavefront, they act as current sinks, further depressing the speed of wavefront propagation. The net result is that as the wavefront moves along a single fibre, the medium into which it is moving is gradually becoming less refractory, since the new action potential is travelling slower than the recovery that precedes it. Eventually, if the medium is long enough, the patch of medium that is being stimulated by the wavefront is no longer in the vulnerable phase and lateral propagation once again becomes possible. At this point propagation becomes successful in the lateral direction and then propagation becomes retrograde along parallel well-recovered fibres. The excitation front moves along the neighbouring well-recovered pathways back toward the original stimulus site and the wave of excitation eventually returns to its source. If the timing of this retrograde step is correct, the excitation will once again excite the stimulus site and the wave will continue to circulate forever, without further stimuli.

It is not guaranteed that a circulating wavefront will be formed. However, a premature stimulus timed earlier in the vulnerable period rather than later has a better chance of becoming a self-sustained rotating pattern. A stimulus late in the vulnerable phase will propagate only a short distance into the medium along a one-dimensional pathway before lateral propagation becomes successful and the retrograde excitation occurs. Such a returning stimulus will return too early to initiate a permanently rotating pattern. To create re-entry, one should stimulate earlier rather than later in the vulnerable period.

In Fig. 4 we show the results of numerical simulation for identical anisotropically coupled FitzHugh–Nagumo excitable cells that demonstrate precisely the scenario we have just described. A grid of discretely coupled identical excitable cells, with no-flux boundary conditions at the edges, was stimulated at one point with two successive current stimuli. The tracings are those of the potential at numbered sites as a function of time. In the tracing for the stimulus site one can see the artifacts of the initial stimulus and the subsequent premature stimulus. Later action potential deflections are the consequence of a rotating pattern of excitation, not from externally applied stimuli.

CALCULATION OF COUPLING STRENGTHS

The scenario for the initiation of re-entry described here is known to occur in myocardium. Spach *et al.* (1981) and Spach and Dolber (1985) show tracings

(a)

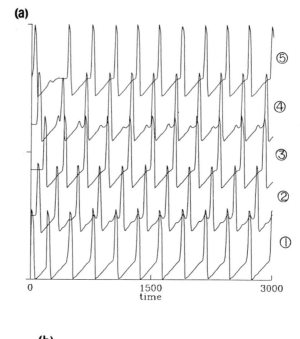

(b)

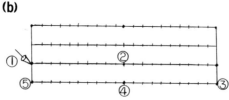

Fig. 4.   Trace of the potential of five cells from a 50×4 grid of anisotropically coupled cells with coupling coefficients $d_1=0.3$ and $d_2=0.011$. The model equations were the FitzHugh–Nagumo equations with $f(v,w)=v(v-0.1)$ $(1-v)-w, g(v,w)=0.005\,v.$

of the electrical activity of a collection of cells with precisely the qualitative behaviour as shown in Fig. 4. In this experiment, a premature stimulus failed to propagate in one direction, but succeeded in the orthogonal direction, and eventually a retrograde stimulus was observed to attempt re-entry at the stimulus site.

To see in which direction the block should occur, we must calculate the relative strengths of the coupling coefficients for myocardium. The assumption behind the discrete models is that there is a collection of isopotential cells coupled by intercellular currents and the coupling coefficient is proportional to the inverse of the electrical resistance between the cells. The electrical resistance per unit length is therefore the coupling resistance per cell times the number of cells per unit length. In a similar way, the velocity of propagation is the number of cells activated per unit time divided by the number of cells per unit length. Thus, to compare velocity of propagation with coupling strength, it is necessary to know the cell dimensions.

For the models discussed here, the velocity of propagation, when measured as number of cells excited per unit time, is an increasing function of coupling strength. However, if the number of cells per unit length is high, this can translate into a small apparent velocity, measured as length per unit time. Expressed in a formula, if c_i, S_i, and l_i are the velocity in cells per unit time, the velocity in length per unit time, and the length of the cell, respectively, in the ith direction, then

$$\frac{c_1}{c_2} = \left(\frac{S_1}{l_1}\right)\left(\frac{l_2}{S_2}\right) = \left(\frac{S_1}{S_2}\right)\left(\frac{l_2}{l_1}\right) \tag{14}$$

and for well-recovered media it is reasonable (see, for example, Fig. 1) to assume that

$$\frac{d_1}{d_2} = \left(\frac{c_1}{c_2}\right)^2 \tag{15}$$

so that

$$\frac{d_1}{d_2} = \left(\frac{S_1}{S_2}\right)^2\left(\frac{l_2}{l_1}\right)^2 \tag{16}$$

A more precise calculation of the coupling coefficients uses Kirkhoff's laws. We suppose that a small subthreshold constant stimulus is applied to some cell. If the cells are isopotential, the voltage along a one-dimensional strand of cells must satisfy the linear difference equation

$$d(v_{n+1} - 2v_n + v_{n-1}) - \frac{v_n}{r_m} = 0 \tag{17}$$

where d is inversely proportional to the resistance between the cells, and $r_m = R_m/S_m$, where R_m is the resistivity of the cell membrane at rest, and S_m is the total cell membrane surface area. Solutions of (17) are of the form $v_n = v_0\mu^n$, where μ satisfies the characteristic equation

$$d(\mu - 1)^2 - \frac{\mu}{r_m} = 0 \tag{18}$$

Of course, μ can be measured experimentally, and is related to the space constant Σ by

$$\mu = e^{-(l/\Sigma)} \tag{19}$$

where l is the length of cells on the direction of the measurement. It follows that the two coupling coefficients are related by

$$\frac{d_1}{d_2} = \frac{\mu_1}{\mu_2}\frac{(\mu_2-1)^2}{(\mu_1-1)^2}, \quad \mu_i = e^{-(l_i/\Sigma_i)} \tag{20}$$

Of course, space constants and speeds scale together so that

$$\frac{S_1}{S_2} = \frac{\Sigma_1}{\Sigma_2} \tag{21}$$

In the special case that l_i/Σ_i is small, (20) is well approximated by (16).

Cells in the trabecular bundle of the right ventricle of a calf are about 120 μm long and 20 μm wide, and the velocity of propagation is about three times faster in the longitudinal direction than in the transverse direction (absolute numbers are 0.49 m/s in the longitudinal direction versus 0.17 m/s in the transverse direction) (Wiedmann, 1970; Clerc, 1976). To calculate the coupling strength between cells and estimate the anisotropic vulnerability, we must take into account the cell dimensions. Using (16), we determine that for cells in the trabecular bundle of a calf, $d_1/d_2 = 1/4$, where d_1 is the coupling strength in the longitudinal direction of cells, and d_2 is the coupling strength in the direction transverse to cell longitudinal axis.

According to this calculation, the longitudinal coupling between cells in the trabecular bundle of calf is about four times *weaker* than the lateral coupling. Since this result is viewed with scepticism by cardiologists, it is worthwhile to give further evidence of why it is valid. It is important to keep in mind that any calculation of coefficients from data depends strongly on the model that is assumed. The model that was assumed for the above calculation was of isopotential cells on a rectangular grid, and it may be that the calculation gave this result because of the assumption of a rectangular grid, which may be wrong for myocardium.

Suppose we assume instead that the cells are arranged in a non-rectangular grid, with cells of length l_1 and width l_2, aligned horizontally and displaced horizontally by an amount pl_1 between rows, $0 \leqslant p \leqslant 1$. We take the coupling between cells to be d_1 in the horizontal direction and d_2 times the proportion of contact between cells in the vertical direction, either p or $q = 1 - p$. We make no assumption about the relative sizes of l_1 and l_2. To describe the coupling operator for this arrangement of cells, we suppose that cells are isopotential, and represent the voltage of a cell whose centre is at coordinates x and y by $v(x,y)$. Using Kirkhoff's laws we find that the coupling operator is

$$\begin{aligned} Lv = &\ d_1(v(x+l_1,y) - 2v(x,y) + v(x-l_1,y)) \\ &+ qd_2(v(x-pl_1,y+l_2) - 2(x,y) + v(x+pl_1,y-l_2)) \\ &+ pd_2(v(x+ql_1,y+l_2) - 2v(x,y) + v(x-ql_1,y-l_2)) \end{aligned} \tag{22}$$

If we suppose that the potential does not change abruptly from cell to cell, we can approximate the difference operator (22) with the differential operator

$$Lv = (d_1 + d_2 p(1-p))l_1^2 v_{xx} + d_2 l_2^2 v_{yy} \tag{23}$$

Thus, if S_1 and S_2 are the speeds of propagation in the horizontal and vertical directions, respectively, we see that

$$\frac{d_1 + d_2 p(1-p)}{d_2} = \left(\frac{S_1}{S_2}\right)^2 \left(\frac{l_2}{l_1}\right)^2 \tag{24}$$

This formula reduces exactly to (16) when $p=0$ for a rectangular grid, while $p=1/2$ corresponds to a hexagonal grid.

If we take the horizontal direction to be the longitudinal axis of cells, then we find that

$$d_1 = d_2(\tfrac{1}{4} - p(1-p)) \tag{25}$$

whereas if we take the vertical direction to be the longitudinal axis of the cells we find that

$$d_2 = \frac{d_1}{4 - p(1-p)} \tag{26}$$

In eqn (25) the longitudinal coupling of cells is given by d_1 while in (26) the longitudinal coupling is given by d_2. In both cases the longitudinal coupling between cells is significantly smaller than the transverse coupling between cells, with a ratio that ranges from zero to 4/15. We conclude that different cell arrangements do not significantly change the observation that transverse coupling is stronger than longitudinal coupling.

Perhaps the explanation of this result is that we assumed cells are isopotential when in fact cells are not isopotential. To explore this possibility, we consider a one-dimensional strand of rectangular cells of length l_1, width l_2 and depth l_3, which are not isopotential. We suppose that a constant subthreshold stimulus is applied at some point along the strand and we want to calculate how the voltage decays as a function of distance away from the stimulus site. To do so, we assume that the voltage does not vary in the lateral direction. (To achieve this in a laboratory setting, the stimulus must be applied from a row of electrodes rather than from a single site.) In the longitudinal direction a core conductor model is valid, so that the voltage satisfies the differential equation

$$\frac{A}{R_c} v_{xx} - \frac{P}{R_m} v = 0 \tag{27}$$

in the interior of cells. Here A is the cross sectional area of cells, P is the perimeter, R_c is the cytoplasmic resistivity (in ohm-cm) and R_m is the membrane resistivity (in ohm-cm^2) of cells. At the ends of cells, two boundary conditions must be applied to account for the current flow between cells. These are

$$\left(-\frac{A}{R_c}v_x - \frac{A}{R_m}v \right)_{x=\text{left}} = \frac{v(x=\text{left}) - v(x=\text{right})}{r_g} = \left(-\frac{A}{R_c}v_x + \frac{A}{R_m}v \right)_{x=\text{right}} \qquad (28)$$

where $x=\text{left}$ refers to the left side of an intercellular junction and $x=\text{right}$ refers to the right side. The constant r_g is the total intercellular resistance across the end of the cells.

The solution $v(x)$ of the system of eqns (27) and (28) is smooth and gradually decreasing inside the cells with jumps across the boundaries between cells. The relative sizes of these two types of decrements depend on the values of resistivity chosen. On average, the voltage decreases geometrically with a rate μ that is related to the space constant of the medium through (19). Since space constants can be measured, we can use eqn (27) with jump conditions (28) to determine the gap resistance. We find that

$$r_g\frac{\mu}{2\rho}\left(E(\beta+\rho)^2 - \frac{1}{E}(\beta-\rho)^2 \right) = \mu^2 - \mu\left(E(1+\frac{\beta}{\rho}) + \frac{1}{E}(1-\frac{\beta}{\rho}) \right) + 1 \qquad (29)$$

where

$$\beta = \frac{A}{R_m}, \ \rho = \frac{A\Lambda}{R_c}, \ E = e^{\Lambda l}, \ \Lambda = \sqrt{\left(\frac{R_c}{\alpha R_m} \right)} \qquad (30)$$

and α is the aspect ratio for the cell $\alpha = A/P$.

We can apply the formula (29) in the two coordinate directions to calculate the ratio of coupling coefficients for myocardium. We suppose that cells are 120 µm long and 20 µm wide and deep, and that the space constant in the transverse direction is one-third the space constant in the longitudinal direction. In Fig. 5 are plotted two solid curves showing the ratio of longitudinal coupling to lateral coupling as a function of cytoplasmic resistance for two different values of longitudinal space constant. For the larger of the two curves, the space constant is 1200 µm, and for the smaller, the space constant is 900 µm, both reasonable values. For both curves, the membrane resistivity was taken to be 7000 ohm-cm^2. Reasonable values of membrane resistivity are between 7000 and 9100 ohm-cm^2. The dashed curves have membrane resistivity of 9100 ohm-cm^2.

From Fig. 5 it is apparent that for a cytoplasmic resistivity anywhere below 300 ohm-cm if the space constant is 900 µm or below 180 ohm-cm if the space constant is 1200 µm, the ratio of longitudinal to lateral coupling is

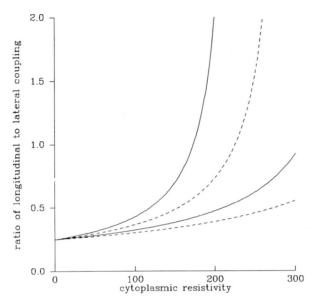

Fig. 5. Ratio of longitudinal to lateral coupling as a function of cytoplasmic resistance for myocardial cells of length 120 µm and width 20 µm assuming a space constant of 900 or 1200 µm and membrane resistance of 7000 ohm-cm² (solid curves) and 9100 ohm-cm² (dashed curves).

below 1. For zero cytoplasmic resistivity we recover our earlier result (21) for isopotential cells. A reasonable value of cytoplasmic resistivity is about 120 ohm-cm (Hume and Giles, 1981). This value is reliable because it comes from measurements on single cells and does not include the effects of resistances from other sources. Measurements that use only average properties of fibres with many cells usually suggest higher values of cytoplasmic resistivity, because it is not known exactly how to eliminate the influence of the gap junctions. Nonetheless, these less reliable measurements still suggest that cytoplasmic resistivity is below 180 ohm-cm.

Regardless of which coupling is stronger, eqn (29) shows how to determine the coupling resistance for cells in a rectangular grid with known physical properties. The formula for a non-rectangular grid is more complicated, but gives a similar result. This formula is substantially different than the formula used in most of the cardiology literature to calculate average resistance of a fibre, because the usual formula is derived by lumping the cell into one resistor in series with the gap junctions, rather than taking into account the spatial distribution of cells. Notice that the curves in Fig. 5 blow up at that value of cytoplasmic resistivity for which gap junction resistance can be taken to be zero and a homogeneous core conductor cable gives the correct average resistance per unit length. Larger values of cytoplasmic resistance are inconsistent with the assumed space constant.

We see that for all of the models considered here, the longitudinal coupling of cells is considerably weaker than the lateral coupling. The main consequence of this calculation is that the direction of highest safety factor in all cases is the transverse direction, and not the longitudinal direction, as is commonly assumed. That is, the direction in which propagation is most likely to fail under early stimulus is the longitudinal direction. This agrees exactly with the experiments shown in Spach *et al.* (1981). The important point is that an estimate of the size of the period of vulnerability is determined by the difference between coupling strengths, and not the speed or resistances per unit length.

SUMMARY

We have shown that in a discrete anisotropic excitable medium there is a period of vulnerability during which a premature stimulus will initiate a re-entrant pattern of excitation. This period of vulnerability can be explicitly calculated for specific models of excitable tissue via numerical computation, and mathematically rigorous estimates of this vulnerable phase can be given for the FitzHugh–Nagumo model (Keener, 1987, 1988). The mechanism proposed here cannot work in a continuous medium. In a continuous medium, stimuli applied at one point cannot break the circular symmetry to form re-entrant waves. Premature stimuli in continuous media must be applied at different points if a re-entrant wave is to be formed (Winfree, 1982; Chen *et al.*, 1988).

We have further shown that the relative strength of coupling between cells depends on the ratio of the speed of propagation to the length of the cell in that direction, so it may be that, if cells are long and narrow, as they are in trabecular bundle, the coupling is smallest in the direction of fastest normal propagation.

These results show complete agreement with the experiments of Spach *et al.* (1981). Since coupling in the lateral direction is stronger than in the longitudinal direction, propagation is first observed to fail in the longitudinal direction. As a consequence, and as observed, the time constant of the rising action potential τ_{foot} is smaller in the lateral direction than in the longitudinal direction. There is no predicted relationship for $(dv/dt)_{max}$, since this is a biphasic function of coupling. Presently, there is no other adequate explanation of these experimental results.

Acknowledgements. This research was supported in part by NSF grant DMS-8801446.

REFERENCES

Beeler, G. W. and Reuter, H. (1977) Reconstruction of the action potential of ventricular myocardial fibres. *J. Physiol.* **268**, 177–210.

Chen, P. -S., Wolf, P. D., Dixon, E. G., Daniely, N. D., Frazier, D. W., Smith, W. M. and Ideker R. E. (1988) Mechanism of ventricular vulnerability to single premature stimuli in open-chest dogs. *Circulation Res.* **62**, 1191–209.

Chi, H., Bell, J. and Hassard, B. (1986) Numerical solution of a nonlinear advance–delay-differential equation from nerve conduction theory. *J. Math. Biol.* **24**, 583–601.

Clerc, L. (1976) Directional differences of impulse spread in trabecular muscle from mammalian heart. *J. Physiol.* **225**, 335–46.

Cole, W. C., Picone, J. B. and Sperelakis, N. (1988) Gap junction uncoupling and discontinuous propagation in the heart. *Biophys. J.* **53**, 809–18.

Diaz, P. J., Rudy, Y. and Plonsey, R. (1983) Intercalated discs as a cause for discontinuous propagation in cardiac muscle: A theoretical simulation. *Ann. Biomed Eng.* **11**, 177–89.

FitzHugh, R. (1961) Impulse and physiological states in models of nerve membrane. *Biophys J.* **1**, 445–66.

Hodgkin, A. L. and Huxley, A. F. (1952) A quantitative description of membrane current and its application to conduction and excitation in nerve. *J. Physiol.* **177**, 500–44.

Hume, J. R. and Giles, W. (1981) Active and passive electrical properties of single atrial cells. *J. gen. Physiol.* **78**, 19–42.

Joyner, R. W. (1982) Effects of the discrete pattern of electrical coupling on propagation through an electrical syncytium. *Circulation Res.* **50**, 192–200.

Kawato, M., Yamanaka, A., Urushiba, S., Nagata, O., Irisawa, H. and Suzuki, R. (1986) Simulation analysis of excitation conduction in the heart: Propagation of excitation in different tissues. *J. theor, Biol.* **120**, 389–409.

Keener, J. P. (1987) Propagation and its failure in coupled systems of discrete excitable cells. *SIAM J. appl. Math.* **47**, 556–72.

Keener, J. P. (1988) On the formation of circulating patterns of excitation in anisotropic excitable media. *J. Math. Biol.* **26**, 41–56.

McAllister, R. E., Noble, D. and Tsien, R. W. (1975) Reconstruction of the electrical activity of cardiac Purkinje fibres. *J. Physiol.* **251**, 1–59.

Rudy, Y. and Quan, W. -L (1987) A model study of the effects on the discrete cellular structure on electrical propagation on cardiac tissue. *Circulation Res.* **61**, 815–23.

Spach, M. S. and Dolber, P. C. (1985) The relation between discontinuous propagation in anisotropic cardiac muscle and the "vulnerable period" of reentry. In *Cardiac Electrophysiology and Arrhythmias* (eds D. P. Zipes and J. Jalife), pp. 241–52. Grune and Stratton, New York.

Spach, M. S. and Kootsey, J. M. (1983) The nature of electrical propagation in cardiac muscle. *Am. J. Physiol.* **244**, H3–H22.

Spach, M. S., Miller, W. T., Geselowitz, D. B., Barr, R. C., Kootsey, J. M. and Johnson, E. A. (1981) The discontinuous nature of propagation in normal canine cardiac muscle. *Circulation Res.* **48**, 39–54.

Wiedmann, S. (1970) Electrical constants of trabecular muscle on mammalian heart. *J. Physiol., Lond.* **118**, 348–60.

Winfree, A. T. (1982) The rotor as a phase singularity of reaction–diffusion problems and its possible role in sudden cardiac death. In *Nonlinear Phenomena in Chemical Dynamics*, (eds C. Vidal and A. Pacault), pp. 156–9. Springer Verlag, Berlin.

Yanagihara, K., Noma, A., and Irisawa, H. (1980) Reconstruction of sino-atrial node pacemaker potential based on the voltage clamp experiments. *Jap. J. Physiol.* **30,** 841–57.

Zinner, B. (1988) Traveling wavefront solutions for the discrete Nagumo equation. PhD thesis, University of Utah.

Vortex re-entry in healthy myocardium

A. T. WINFREE

*Department of Ecology and Evolutionary Biology, University of
Arizona, Tucson, AZ 85721, USA*

> Ventricular tachyarrhythmias probably cause most episodes of sudden cardiac
> death that account for over 1000 fatalities per day in the United States.
>
> (Prystowski *et al.*, 1983)

Because re-entrant tachycardias can lead directly to fibrillation and sudden
death it is of interest to understand them more thoroughly. The way
suggested here is to regard myocardium as an instance of an excitable
medium. 'Tachyarrhythmias' in other excitable media have been investigated
both in the laboratory and in the computer, and are better understood than
those in cardiac muscle, which are notoriously difficult to examine. Some
testable predictions of new electrophysiological phenomena were made on
the basis of the analogy of myocardium to generic excitable media. Their
confirmation in canine myocardium (Winfree, 1989a,b) is summarized in the
first section. Understanding of these phenomena might be either facilitated
or obstructed by their emergence from several seemingly distinct theoretical
schemes. The next section addresses the relationship between those schemes.
In particular, a question arises about possible reinterpretation of the theory
of 'non-uniform dispersions of refractoriness', as the conceptual scheme I
present here does not involve it. This question is considered separately in the
last section.

MYOCARDIUM AS A GENERIC EXCITABLE MEDIUM

The commonest mechanism of ventricular tachyarrhythmia is re-entry, the
circulation of an action potential around a long enough chain of myocardial
fibres to allow repolarization and restoration of excitability before the pulse
returns. This is the textbook understanding of re-entry. Textbooks are
usually a little bit out of date, in this case because it turns out that re-entry in

Cell to Cell Signalling: From
Experiments to Theoretical Models
ISBN 0–12–287960–0

the two-dimensional or three-dimensional context of human ventricular myocardium has at least one qualitatively different mechanism, with correspondingly different understanding of its onset and of implications for its management. This mechanism involves the vortex-like activity which is created not in a ring of fibres but simultaneously throughout a substantial area or volume of myocardium. It has been familiar to theorists since the mid-1960s and to experimental physiologists since Allessie and co-workers confirmed it experimentally in the mid-1970s (Allessie *et al.*, 1973, 1976, 1977). By then theorists understood why these vortices commonly arise in mirror-image pairs, and in the early 1980s El-Sherif *et al.* (1985) and El-Sherif (1985) named such pairs 'figure-eight reentry'. The existence of a coherent theory regarding the origin and dynamics of these 'rotors', as they are called individually, has recently contributed to the design and interpretation of experiments in various excitable media, including myocardium. One avenue of laboratory work leads to continuous, isotropic excitable media such as the Belousov–Zhabotinsky chemical reagent (Winfree, 1985; Jahnke *et al.*, 1988, 1989; Winfree and Jahnke, 1989). Another leads to cellular excitable media such as fields of social amoebae (Othmer, 1989; Tyson *et al.*, 1989; Tyson and Murray, 1989), and intact atrial (two-dimensional) and ventricular (three-dimensional) myocardium. These diverse systems differ in many idiosyncratic details affecting their re-entrant vortices, but they also share several universal features.

Such features depend not on the medium being quiescent unless provoked, nor on its being spontaneously oscillatory, nor on its being architecturally cellular, but on three central features:

(1) that in an appropriate state space the individual cell or volume element follows a closed or nearly closed loop of excitation, depolarization, refractoriness, repolarization and recovery of excitability leading back to near-quiescence;
(2) that adjacent volumes are coupled by diffusion of molecular species and/or the mathematically indistinguishable spread of electric potential by ionic currents, so that neighbours maintain similar states from moment to moment; and
(3) that the excitability is abrupt enough to sustain the steepness of a front propagating by diffusion from neighbour to neighbour.

On the basis of these features and the analogy between living ventricular myocardium and generic excitable media it is possible to make specific predictions of new electrophysiological phenomena (Winfree, 1982, 1983, 1986, 1987). These are associated with the onset of re-entrant tachycardia in myocardium, particularly with its instigation by a single stimulus during the so-called 'vulnerable period' (the existence of which was also independently 'predicted' by this conceptual scheme, but a half century too late). It was not long before these predicitions were tested in canine ventricle using an electric stimulus. The following results were anticipated and confirmed:

(1) The existence of stable vortex-like action potentials in uniformly healthy two-dimensional and three–dimensional myocardium. As noted above they were found in the two-dimensional case, and have recently been sought experimentally (Medvinsky and Pertsov, 1982; Medvinsky et al., 1983, 1984) and convincingly shown (Frazier et al., 1989; Shibata et al., 1988) in the three-dimensional case.

(2) The creation of rotors in mirror-image pairs potentially only 1–2 cm apart, near the sites where two critical contours cross within the myocardium: where local stimulus magnitude ranges through a critical value S^* and simultaneously local phase in the cycle of excitation and recovery ranges through a critical value T^*. Both of these ranges depend on the nature of the stimulus. By varying the timing of an electrical shock in canine left ventricle (and as a mirror-image control experiment, in right ventricle) Shibata et al. (1988) and Frazier et al. (1989) confirmed this in detail, measuring the critical values and determining that they are nearly constant in diverse situations.

(3) The existence not just of a vulnerable period, as noted above, but in more detail, of both lower and upper limits of stimulus magnitude outside which fibrillation does *not* occur, thus changing the name from 'vulnerable period' to 'vulnerable domain'. This domain appears in the centre of a bull's-eye on a 'vulnerability diagram' on which stimuli are characterized by timing and size. It should be surrounded closely on all sides by concentric fringes in which the corresponding stimuli elicit only a few repetitive beats without starting a persistent tachycardia. The bull's-eye and its surround were mapped out in detail, with upper and lower limits of vulnerability assayed quantitatively (Shibata et al., 1988).

PRINCIPLES OF VULNERABILITY DESCRIBED IN VARIOUS PARADIGMS

Various representations of excitability are more or less successful quantitatively for application to myocardium in two or three dimensions. Even discrete-state, discrete-space cellular automata have their appeal (Auger et al., 1988), especially for conservation of computer time in three-dimensional calculations (Winfree et al., 1985; Aoki et al., 1986). But they also miss the mark in significant respects concerning re-entry (explicitly listed in Zykov, 1984, and in Winfree, 1987). This failing can be traced to their discontinuous handling of cell to cell coupling by electrotonic currents. While there are situations in which discontinuity deserves emphasis, macro-reentry in undamaged tissue is apparently not one of them.

V. S. Zykov has further developed this theme by mathematics and computations based on cardiologically-motivated cable equation models (Zykov, 1984; for earlier history, starting with German/American mathematician Wiener and Mexican cardiologist Rosenblueth in the late 1940s, and ending in the 1970s with Russian biophysicists Krinsky, Panfilov, Pertsov, etc., see Winfree, 1980). In my opinion the central, visual concept (not what we need for quantitative comparison with experiment, but useful for understanding) underlying all of it is roughly as follows (abstracted from my own papers of the 1970s as in Winfree, 1980).

At any instant each microscopic volume in any excitable medium (not necessarily electrophysiological) has a certain state defined by its excitedness and the intensity of some local recovery process (Fig. 1, right side). These abstractions in any specific excitable medium translate into particular physical quantities involving chemical concentrations, local potentials, channel conductivities, etc. (e.g. FitzHugh, 1960, 1961: Zykov, 1984). Each volume's quantities are near to those of its immediate neighbours due to the intercellular coupling that binds them into a medium (e.g. diffusion of molecules in a chemically excitable medium such as *Dictyostelium* monolayers (Othmer, 1989; Tyson and Murray, 1989; Tyson et al., 1989), or electrotonic currents, alias diffusion of electric potential, in an electrophysiologically excitable medium). So the image of the medium in this descriptive space is continuous (Fig. 2). Even quantities that *don't* diffuse (e.g. channel states) are indirectly kept in order by the ones that do during rapid rotation. Within this image, each excited volume if free of neighbours would follow a certain sequence or path of changes leading back to near-quiescence, mostly along the standard closed or nearly-closed loop of excitation, depolarization, refractoriness, repolarization and recovery of excitability. If not free of neighbours, each moves instead in a way compromising the tendencies of the neighbourhood. Since the overall pattern of paths is rotary, the whole image rotates persistently along the loop, with some tiny volume caught in a stagnant zone of restricted change somewhere inside the loop.

For decades biophysicists in the USSR, the Netherlands, Germany, the USA and Japan have explored such phenomena using diverse numerical models of membrane-like excitability (Scott, 1977; Zykov, 1984). The FitzHugh–Nagumo model and close relatives are commonly deployed as generic representatives. Winfree (1989b) exhumes an attempt to design vortex-inducing experiments in myocardium quantitatively using the Fitz-Hugh–Nagumo generic membrane model with fibres coupled by the electrophysiologist's cable equation, and with time and space scales fitted to the known conduction speed and rotation time of action potentials in canine ventricular myocardium. ('Exhumes' is the correct verb: this crude prototype was not intended for presentation — until laboratory experiments (Shibata *et al.*, 1988) verified the main results and electrophysiologically more realistic models substituting for FitzHugh–Nagumo turned out to do no better despite enormously greater computational effort).

A central feature in such kinetics is the 'separator' trajectory at the core of the bundle of trajectories passing diagonally into the loop in Fig. 1. This is a path of indecision followed by membrane perturbed from rest in such a delicate way that it never quite makes up its mind to execute a full action

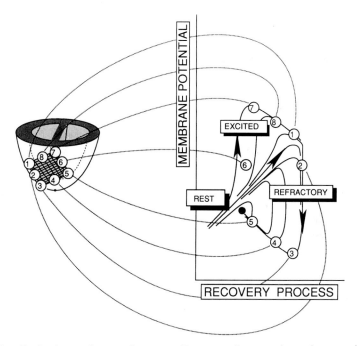

Fig. 1. Each tiny volume of myocardium, small enough to be considered electrically uniform, has some instantaneous membrane potential and intensity of recovery process (and many more specific measures of ionic channel activity, but these two suffice for present purposes). On a diagram with those coordinates, each volume can be mapped adjacent to its neighbours. An isolated small volume would spontaneously change membrane potential and recovery intensity in a direction determined by its instantaneous values of both. In a typical excitable medium these vectors describe trajectories approaching a loop that recedes from a state near rest, through excited states then refractory states, then back toward rest. Trajectories inside the excitation–recovery loop diverge, some rapidly depolarizing upward and some rapidly repolarizing downward, from a separator trajectory that passes near the rest state; it determines the strength-interval curve of classical physiology. During one-dimensional propagation all volumes follow the standard excitation–recovery loop from rest back to rest. Here a one-dimensional ring of tissue (just the border of the grid shown) is mapped into the (membrane potential, recovery process intensity) plane while an action potential circulates.

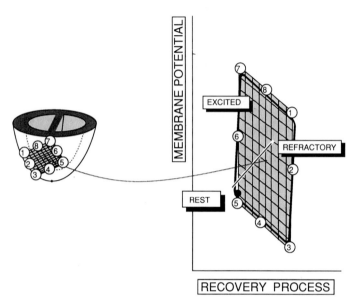

Fig. 2. The entire coordinate grid embossed on the tissue in Fig. 1 is mapped into the (membrane potential, recovery process intensity) plane (here just a cartoon of its representation in Fig. 1) in such a way that its perimeter maps along the excitation–recovery loop as in Fig. 1. Such mappings arise when local phase T and local stimulus intensity S are transversely graded through critical values T^* and S^*, spanning the range needed to cover the excitation–recovery loop in a piece of excitable medium large enough to support such diverse states within its confines.

potential, but neither does it give up and return directly to rest. Though the separator is the 'seat of the excitability' of any excitable medium, it has no clear mathematical definition, being only a segment of a trajectory from which others split off violently upward and downward to almost discontinuously change the timing of action potentials (Winfree, 1982). It ends vaguely inside the excitation–recovery loop where the splitting away is no longer violent, where all trajectories, having reached the extreme of the membrane's recovery intensity, turn downward, repolarizing. The separator plays the role of a membrane potential threshold for regenerative excitation: it is higher the stronger is the recovery process (to the right), or the briefer is the time or coupling interval since prior excitation (clockwise around the loop from full depolarization).

Now suppose the membrane potential and recovery intensity of every fibre within a rectangular piece of tissue could be adjusted just as we please. In order around the border of the piece we set fibres to those combinations of potential and intensity of recovery process that occur in succession during a normal cycle of excitation and recovery, just as though an action potential were circulating around the border. Each point on the border can then be mapped to a unique state on Fig. 1, along the conspicuous loop that departs from the neighbourhood of the rest state and eventually returns. If each fibre now behaves normally, then each with its neigbours follows nearly the same path toward the resting state as it would in isolation. Any cell caught up in normal propagation of an action potential follows this loop. Only fibres near the separator behave peculiarly. Those below it, being electrically coupled to fibres on the other side, cannot remain at rest while the latter depolarize. The potential difference creates compromising currents, and the fibres below threshold are pulled up over threshold, against their parochial tendencies. This is nothing other than a *visual* way of writing the cable equation for continuous electrotonic coupling. It shows the wavefront (where fibres cross threshold) moving (pulling into immediate excitation fibres nearby which, uncoupled, would instead directly approach rest). As a result the entire border persistently circulates on this diagram like the stream in M. C. Escher's 1961 'Waterfall': the action potential keeps moving around the border.

Circulation fails if the border is so short (its physical perimeter, not its loop image in this diagram) that it is stretched too taut around this loop, and the loop either breaks (if continuity fails on account of cellular structure) or pulls across the centre like a rubber band. 'Too taut' means either that geographically adjacent cells come to be on opposite sides of the excitation–recovery loop (continuity fails) or that cells are driven to short cut across the loop due to strong electrotonic currents from geographically near neighbours at quite different potentials (continuity prevails). Either way the 1:1 map of border to loop degenerates to a 2:1 map in which the border covers only part of the loop, i.e. it is no longer the case that at any moment *some* place on the border is crossing threshold. Re-entry has been broken. Homogeneity soon follows as all cells slide toward rest with no neighbour across threshold to pull them out again (see Winfree, 1978, for a series of diagrams).

But let us suppose that circulation persists. If there were *only* a border and tissue outside it through which action potentials propagate, then theory could stop here. But instead of emptiness inside the border, we have a continuous expanse of the same tissue. Instead of an anatomical hole or obstacle for a normal action potential to propagate around, we have a region of physiologically normal and normally coupled cells. Where does that tissue map to in Fig. 1? To the interior of the loop, as in Fig. 2. The whole grid rotates just as before, for the same reasons, but now some fibres in the interior of the rectangle are caught in the interior of the loop, serving at least momentarily as a pivot for the rotating image. They constitute the eye of a vortex of excitation, the tiny region around which an action potential circulates without penetrating. If this whole balancing act is unstable, and the inside of the grid sloshes back and forth irregularly while the perimeter sedately turns, then the eye is a region of occasional depolarization and graded action potentials, more or less as has been observed with intracellular electrodes (Allessie *et al.*, 1977).

The arrangement of Fig. 2 can come about quite easily. Suppose a region of the excitable medium had been activated by a passing wavefront some time ago, so that it is now on the excitation–recovery loop in the range of phase T indicated by 3–4–5 in Fig. 1. This gradient of phase spans the late stages of repolarization (centred near T^* in Fig. 3), when a depolarizing stimulus might re-excite region 5, but less-recovered regions 3–4 may be still too refractory. Each region homogeneously labelled 3, or 4, or 5 is a band of medium between isochronal contours. A geographically graded stimulus would now affect the various parts of each band with graded severity (unless the stimulus S happens to be exactly uniform along each band). In a region where the stimulus grades through intensities vertically spanning the excitation–recovery loop (centred near S^* in Fig. 3), it stretches each band vertically from state 3 to state 1, from 4 to 8, and from 5 to 7, thus opening the original, collapsed image of the region from 3–4–5 to the shaded

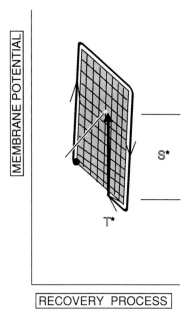

Fig. 3. The right side of Figs 1 and 2 is labelled to show the critical stimulus magnitude S^* needed to perturb membrane to mid-cycle from the critical phase T^* during normal excitation and recovery. Achievement of these critical values (and a range around them) is essential for induction of vortex re-entry. The loop is covered if and only if S straddles S^* widely enough and T straddles T^* widely enough. The resulting vortex is unstable if the pattern is compressed into too small an area, but is stable otherwise. The minimum area seems to be 1–2 cm² in mammalian ventricular myocardium.

rectangular image of Fig. 2. Thereupon, it tumbles clockwise, following the generally clockwise trajectories indicated. I interpret this mechanism as the induction of vortex re-entry by a premature stimulus given in the vulnerable phase (Winfree, 1989b).

It is interesting while viewing the moving image of the medium on this (membrane potential, recovery intensity) plane, to go back to geographical coordinates to view a locus of uniform membrane potential (the fibres found on any horizontal line on Fig. 2) or of uniform recovery intensity (the fibres on any vertical line on Fig. 2). Are they isochronal contours, the geographical loci along which the action potential arrives simultaneously? In other words are they the wavefront and its past and future positions? Not quite, but any isochronal contour in the tissue runs close to *half* of one such locus in state space: the half along which fibres are rising (or falling) through the chosen level of potential (or level of recovery intensity, or in fact any other line extending radially from the middle of the loop). The other half of any line through the middle of the loop corresponds to the *back* of the wave, another isochronal contour on the tissue (pictures not reproduced here: see Winfree, 1974a,b, 1978).

Physiologists implicitly, but almost never explicitly, think in this visual language when solving membrane kinetic equations. More commonly a computer is left to deal with equations in multiple interacting variables while the human investigator turns his own mind to a simplified model in which everything is reduced to the single variable that can be continuously monitored in the laboratory: membrane potential. This simplification is equivalent to projecting the right side of Fig. 1 onto the membrane potential axis. It is clear that much sense is lost in the process. For example the delicate separator locus is smeared along the axis. Its conceptual replacement by a 'threshold' point on the surviving membrane potential axis makes many phenomena incomprehensible until the reduced single-variable model is supplemented by special provisos (e.g. an absolute refractory period, strength-interval curves, anode break phenomena, zones of 'no response', graded action potentials, conduction block) that, in this framework, appear as idiosyncrasies of the experimental preparation. This piecemeal restoration of the necessary minimum complexity of the kinetics described by the membrane equations eventually becomes more awkward than simply retaining the required variables.

Nonetheless the language of single-variable models has much intuitive appeal: maybe projecting in a different way would be more appropriate, for example in a polar way rather than a Cartesian way, measuring circular phase on the excitation–recovery loop. This is really the *only* other choice, since phase (as local time since excitation) is as easily monitored as membrane potential, but it is hard to think of anything else so convenient. Adopting a phase measure is also quite natural in a preparation that is periodically driven (e.g. by the sinus rhythm or by a tachycardia). It is marginally legitimate even in a system that is excited but once, since it indicates position along the standard loop trajectory regardless of whether the system spontaneously or responsively (or neither) initiates the next cycle. Maps of local phase versus geographical position in the medium then correspond to conventional isochronal contours. Their occasional orderly convergence to a point in complete sets spanning one full cycle corresponds both to vortex-like re-entry in the laboratory and to well-developed ideas about phase singularities in the world of theory. This leads to a vision of tachycardia in terms of the creation and annihilation of phase singularities (Winfree, 1982, 1983, 1987). If the early stages of ventricular fibrillation (prior to ischemia, failure of gap junctions, etc.) can be thought of as vortex-like re-entry at multiple sites, then an understanding of phase singularities in excitable media might stimulate useful physiological inquiries into fibrillation and defibrillation.

It is also interesting to contemplate the geographical locus of the separator: it is the wave*front*, ending indistinctly where the re-entrant vortex pivots. (Extending the separator past the middle of the loop in state space to cross the loop again in the refractory zone, we have the wave*back*, as described above.) This correspondence between the separator, the wavefront, and the coupling-interval-dependent threshold constitutes the link between electrophysiological kinetics in the multidimensional language of ionic current dynamics, and the classical language of strength-interval curves and thresholds. In his doctoral thesis Foy (1974), following Brooks *et al.* (1955), derived much of the behaviour of cardiac rotors in terms of strength-interval curves, but it remained unpublished. Chen *et al.* (1988) and Shibata *et al.* (1988) also implicitly emphasize this locus in Figs 1 and 2, presenting the strength-interval interpretation as an alternative to existent theories of vortex-like re-entry. But reinterpretation of the state-space analysis of rotors in such terms, without their second dimension, requires us to construct the rotor from an appropriate selection of special behaviours, Rube–Goldberg-fashion, as though it were an idiosyncrasy of the particular medium (canine ventricular myocardium in this case), without seeing its geometrical simplicity in generic excitable media.

The oldest and most venerable physiological interpretation of re-entry and fibrillation provided almost the only paradigm for interpreting observations of vulnerability to re-entry and fibrillation for the past 20 years. This is the language of non-uniform dispersion of refractoriness or non-uniform recovery of excitability in *dis*continuous excitable media.

NON-UNIFORM DISPERSION OF REFRACTORINESS

The notion of non-uniform dispersion of refractoriness originated with Moe (1962), Han and Moe (1964), and Moe *et al.* (1964). Its clearest description took the form of a cellular automaton model implemented in a digital computer. In this model excited cells in a two-dimensional hexagonal array excite their neighbours unless the neighbour still remains too refractory after its prior excitation. Duration of refractoriness is an intrinsic property of each cell, and cells differ randomly according to a specified statistical distribution. A stimulus introduced inside the array near the time and place at which many cells are still refractory but many have recovered can elicit complex perpetual percolation of sparks of excitation, referred to as 'wandering wavelets'. Thus a depolarizing stimulus during the vulnerable period might elicit fibrillation in a medium characterized by non-uniform dispersion of refractoriness. The original phrasing was actually 'non-uniform recovery of excitability', which might correspond to the orderly phase gradients and recovery process gradients contemplated here. The notions involved became more vague and the experiments more exact during a 20-year exploitation of this paradigm. Thus it is now clear that real myocardium is more vulnerable to fibrillation if it is artificially made more heterogeneous, but it is not clear whether that heterogeneity must be intrinsic and irregular (as in patches of ischemia, or patches irregularly innervated by sympathetic or parasympathetic neuro-transmitters), or may also be ephemeral and orderly (as in a gradient of electric potential provided by an extracellular electrode). It is also not clear whether the vulnerability addressed is only vulnerability toward fibrillation, or also toward re-entry, and whether in real myocardium maintenance of re-entry (or fibrillation) or its instigation depends on non-uniform dispersion, or non-uniform recovery, or exactly what 'non-uniform' means.

Moe addressed some of these questions, as they pertain to the model, in a numerical experiment. After 'fibrillation' was established he made all cells identical. Re-entry persisted in the form of orderly vortices, simulating a tachycardia such as flutter or the earliest stage of fibrillation. (Presumably, restoring the variance of intrinsic refractory period durations would have restored fibrillation.) It seems to follow that fibrillation, at least in this model, results from the propagation of short-period waves (from re-entrant vortices or other sources of tachycardia) in a statistically irregular medium. Fibrillation is interpreted as the fragmentation of these waves.

But is non-uniform dispersion of refractoriness also needed for *initiation* of re-entry? Certainly not; this has been known, at least in models, since the time of Moe's computer simulations. Experiments also now show that in myocardium, as well, vortex re-entry can arise in the absence of much measured dispersion of intrinsic refractoriness (Allessie *et al.*, 1976; Boineau *et al.*, 1980; El-Sherif *et al.*, 1981; Mehra *et al.*, 1983; Dillon *et al.*, 1985; Gough *et al.*, 1985; Chen *et al.*, 1988; Shibata *et al.*, 1988; Frazier *et al.*, 1989; Ideker *et al.*, 1989, and theoretical discussion in Winfree, 1982, 1985, 1986, 1987). Several of these papers implicate orderly gradients of intrinsic refractoriness in the onset of re-entry, and others show a similar outcome even in the absence of discernible gradients of intrinsic refractoriness. (None, however, strictly exclude the possibility that unobserved non-uniform dispersion of refractoriness is an essential prerequisite for onset of re-entry.)

In idealized excitable media non-uniform dispersion of refractoriness is not necessary for maintenance of re-entry nor for its inital induction, nor has necessity been demonstrated in real heart muscle. But it is necessary for transition from mere re-entry to fully-developed fibrillation. There is no 'dispersion of refractoriness' in uniform continuous excitable media of chemical or numerical mechanism, and 'fibrillation' has not yet been demonstrated in such media. So long as fibrillation remains an idiosyncrasy of (discretely cellular) media, the idea of non-uniform dispersion of refractoriness remains an important candidate for the central insight into its mechanism; more recent notions involving phase singularities and rotors might turn out to constitute only a new description of macro-reentry. Even macro-reentry might even be encompassed within the domain of 'non-uniform dispersion of refractoriness' by stretching some definitions ('non-uniform' means 'graded', 'dispersion' means 'distribution', 'refractoriness' means 'momentary state of the membrane') or by emphasizing the parts of this literature that stress 'non-uniform recovery of excitability'. The orderly gradients of phase and of stimulus magnitude or effect necessary to create rotors could be viewed as well-structured non-uniformities in the recovery of excitability and as ephemeral smoothly graded dispersion of induced refractoriness.

The focus on dispersion can be traced back to practical cardiology. Most people with life-threatening dysrhythmias owe them to ischaemia and the resulting infarcts, which definitely make the myocardium heterogeneous. Cardiologists are necessarily pre-occupied with myocardium in which irregular regions have been damaged by failure of the blood supply in ways that do not lend themselves to quantitative description. In this context statistical measures of dispersion of refractoriness seem appropriate ingredients for any theory aspiring to predict degrees of vulnerability. But in all the theory and much of the experimentation mentioned above, the subject matter is healthy, undamaged myocardium.

In both healthy and infarcted preparations the stimuli contemplated mostly have a depolarizing effect. This restriction of purview may be responsible for the fixation on 'refractoriness' and 'recovery'. If the stimulus were hyperpolarizing rather than depolarizing then according to Figs 1–3 the critical phase T^* would not be in the late stages of repolarization and recovery, but rather during early plateau phase. Or if the stimulus were a neurotransmitter that instantly turns on the full recovery process without (yet) changing the membrane potential, then the arrow in Fig. 3 would point from left to right rather than low to high or high to low: the vulnerable period would be in the upstroke and the critical magnitude of stimulus would be about half the width (rather than half the height) of the excitation–recovery loop (Winfree, 1989b). Vulnerable phases for such stimuli would have no connection with the timing of repolarization or the moment of maximum dispersion of refractoriness (the T-wave of the EKG). If either of these experiments could be performed, and produced the results here suggested then we would have to regard the coincidence of the conventional vulnerable period with that moment as a mere accident dependent on the particular nature of the stimulus chosen.

Another test, also technically impossible at present, would subject the myocardium to a geographically uniform stimulus. If re-entry is necessary as preamble to fibrillation, and re-entry arises only by transverse crossing of a gradient of timing by a gradient of stimulus intensity centred at S^* and spanning the diameter of the excitation–recovery loop, then neither will arise in response to a uniform stimulus. (Care must be taken, of course, to ensure that the stimulus is physiologically uniform, not just physically uniform: fibre orientation and phase-dependent conductivity both play a role in determining local sensitivity to a physically uniform stimulus. It might be possible to ignore phase-dependent sensitivity in practice: if this theory is correct then only gradients of stimulation transverse to the phase gradient play a role in the induction of re-entry, but phase-dependent sensitivity only contributes a component of the stimulus gradient parallel to any pre-existing phase gradient.) This prediction contrasts sharply with expectation based on the idea of microscopic non-uniform dispersion of intrinsic refractoriness or of recovered excitability: the classical expectation is that re-entry and fibrillation should arise wherever such tissue is stimulated at several times diastolic threshold, uniformly or not, during the familiar vulnerable period.

CONCLUSION

Excitation propagates in cellular anisotropic tissue through gap junctions in ways that are dauntingly complex. Especially if the tissue is heterogeneously damaged, it is susceptible to diverse pathological modes of self-sustained activity. In cardiac muscle they are called tachycardias and fibrillation. The modern approach to understanding them physiologically has entailed increasingly sophisticated attention to anatomical complexity and to membrane channel complexity, with emphasis on discontinuities.

One can alternatively oversimplify the anatomy even to the extreme of an undamaged isotropic continuum with a minimum of ionic channel types (two), so as to harvest more insights from classical electrophysiology. From this different perspective certain aspects of cardiac dysrhythmias become more readily understandable. In particular, unforeseen phenomena associated with vulnerability to electric shock were predicted despite their counter-intuitive appearances, then sought and found in the laboratory. They can be understood in terms of visual images of the medium in its state space (Winfree, 1980), or in terms of geographical patterns of phase punctuated by phase singularities (Winfree, 1987), or in terms of classical concepts of excitation propagation, conduction block, and graded recovery of excitability (Chen *et al.*, 1988; Shibata *et al.*, 1988; Frazier *et al.*, 1989). Specific quantitative meaning is thus provided for classical intuitions about dispersion of recovery, now in terms of the measurable geographic patterns of stimulation and of isochronal contours.

Acknowledgements. I thank A. Goldbeter and NATO for the opportunity to attend this meeting. This material was also presented verbally at the Gordon Research Conference on Theoretical Biology, June 1988 and at the Canadian Society of Electrophysiologists meeting, June 1988. Research support came from the US National Science Foundation (Cellular Physiology), the John von Neumann Supercomputer Center, and the John D. and Catherine T. MacArthur Foundation. I thank James McClelland, M.D. (Oregon Health Sciences University), Bruce Steinhaus (Telectronics, Inc., Engelwood, CO), and Marc Courtemanche and Chris Henze (University of Arizona) for helpful corrections while trying to turn this text into readable English.

REFERENCES

Allessie, M. A., Bonke, F. I. M. and Schopman, J. G. (1973) Circus movement in rabbit atrial muscle as a mechanism of tachycardia. *Circulation Res.* **33,** 54–62.
Allessie, M. A., Bonke, F. I. M. and Schopman, F. J. G. (1976) Circus movement in rabbit atrial muscle as a mechanism of tachycardia: II. The role of nonuniform recovery of excitability. *Circulation Res.* **39,** 168–77.
Allessie, M. A., Bonke, F. I. M. and Schopman, F. J. G (1977) Circus movement in rabbit atrial muscle as a mechanism of tachycardia: III. The leading circle concept. *Circulation Res.* **41,** 9–18.
Aoki, M., Okamoto, Y., Musha, T. and Harumi, K. (1986) 3-Dimensional computer simulation of depolarization and repolarization processes in the myocardium. *Jap. Heart. J.* **27** Suppl, 225–34.
Auger, P., Bardou, A., Coulombe, A. and Degonde, J. (1988) Computer simulation of ventricular fibrillation. *Math. Comput. Modelling* **11,** 813–22.
Boineau, J., Schleusser, R., Moone, C., Miller, C., Wylds, A., Hudson, R., Borremans, J. and Brockus, C. (1980) Natural and evoked atrial flutter due to circus movement in dogs. *Am. J. Cardiol.* **45,** 1167–81.
Brooks, C., Hoffman, B., Suckling, E. and Orias, O. (1955) *The Excitability of the Heart.* Grune & Stratton, New York.
Chen, P. S., Wolf, P. D., Dixon, E. G., Danieley, N. D., Frazier, D. W., Smith, W. M. and Ideker, R. E. (1988) Mechanism of ventricular vulnerability to single premature stimuli in open chest dogs. *Circulation Res.* **62,** 1191–209.
Dillon, S., Ursell, P. C. and Wit, A. L. (1985) Pseudo-block caused by anisotropic conduction: A new mechanism for sustained reentry. *Circulation* **72** Suppl. III. 1116.
El-Sherif, N. (1985) The figure 8 model of reentrant excitation in the canine postinfarction Heart. In *Cardiac Electrophysiology and Arrhythmias* (eds D. P. Zipes and J. Jalife), pp. 363–78. Grune & Stratton, Orlando, FL.

El-Sherif, N., Smith, R. A. and Evans, K. (1981) Canine ventricular arrhythmias in the late myocardial infarction period. *Circulation Res.* **49**, 255–65.

El-Sherif, N., Gough, W. B., Zeiler, R. H. and Hariman, R. (1985) Reentrant ventricular arrhythmias in the late myocardial infarction period. 12. Spontaneous vs induced reentry and intramural vs epicardial circuits. *J. Am. Coll. Cardiol.* **6**(1), 124–32.

FitzHugh, R. (1960) Thresholds and plateaus in the Hodgkin–Huxley nerve equations. *J. gen. Physiol.* **43**, 867–96.

FitzHugh, R. (1961) Impulses and physiological states in theoretical models of nerve membrane. *Biophys. J.* **1**, 445–66.

Foy, J. L. (1974) A computer simulation of impulse conduction in cardiac muscle. Technical Report No.166, University of Michigan, Department of Computer Science.

Frazier, D. W., Wolf, P. D., Wharton, J. M., Tang, A. S. L., Smith, W. M. and Ideker, R. E. (1989) Stimulus-induced critical points: a mechanism for electrical induction of reentry in normal myocardium. *J. Clin. Invest.* **83**, 1039–1052.

Gough, W. B., Mehra, R., Restivo, M., Zeiler, R. H. and El-Sherif, N. (1985) Reentrant ventricular arrythmias in the late myocardial infarction period in the dog: 13. Correlation of activation and refractory maps. *Circulation Res.* **57**, 432–42.

Han, J. and Moe, G. K. (1964) Nonuniform recovery of excitability in ventricular muscle. *Circulation Res.* **14**, 44–60.

Ideker, R. E., Frazier, D. W., Krassowska, W., Chen, P. S. and Wharton, J. M. (1989) Physiologic effects of electrical stimulation in cardiac muscle. (Submitted).

Jahnke, W., Henze, C. and Winfree, A. T. (1988) Chemical vortex dynamics in three-dimensional excitable media. *Nature* **336**, 662–665.

Jahnke, W., Skaggs, W. E. and Winfree, A. T. (1989) Chemical vortex dynamics in the Belousov–Zhabotinsky reaction and in the 2-variable oregonator model. *J. Phys. Chem.* **93**, 740–749.

Medvinsky, A. B. and Pertsov, A. M. (1982) Initiation mechanism of the first extrasystole in a short-lived atrial arrhythmia (in Russian). *Biofizika* **27**, 895–9.

Medvinsky, A. B., Pertsov, A. M., Polishuk, G. A. and Fast, V. G. (1983) Mapping of vortices in myocardium. In *The Electrical Field of the Heart* (eds O. Baum, M. Roschevsky and L. Titomir), pp. 38–51. Nauka, Moscow (in Russian).

Medvinsky, A. B., Panfilov, A. V. and Pertsov, A. M. (1984) Properties of rotating waves in three dimensions. Scroll rings in myocardium. In *Self Organization: Autowaves and Structures far from Equilibrium* (ed. V. I. Krinsky). Springer-Verlag, Berlin.

Mehra, R., Zeiler, R. H., Gough, W. B. and El-Sherif, N. (1983) Reentrant ventricular arrhythmias in the late myocardial infarction period. 9. Electrophysiologic-anatomic correlation of reentrant circuits. *Circulation* **67**(1), 11–24.

Moe, G. K. (1962) On the multiple wavelet hypothesis of atrial fibrillation. *Arch. Int. Pharmacodyn.* **140**, 183–8.

Moe, G. K., Rheinboldt, W. C. and Abildskov, J. A. (1964) A computer model of atrial fibrillation. *Am. Heart J.* **67**, 200–20.

Monk, P. B. and Othmer, H. (1989) Relay, oscillations, and wave propagation in a model of *Dictyostelium discoideum. Lect. Math. Life Sci.* **21**, 87–122.

Prystowsky, E. N., Heger, J. J., Lloyd, E. A. and Zipes, D. P. (1983) Clinical electrophysiology of ventricular tachycardia. *Cardiol. Clinics* **1**(2), 253–70.

Scott, A. C. (1977) *Neurophysics.* John Wiley & Sons, New York.

Shibata, N., Chen, P. -S., Dixon, E. G., Wolf, P. D., Danieley, N. D., Smith, W. M. and Ideker, R. E. (1988) The influence of epicardial shock strength and timing on the induction of ventricular arrhythmias in dogs. *Am. J. Physiol.* **255**, H891–901.

Tyson, J. J. and Murray, J. D. (1989) Cyclic-AMP waves during aggregation of *Dictyostelium* amoebae. *Devl Biol.* July, (in press).

Tyson, J. J., Alexander, K. A., Manoranjan, V. S. and Murray, J. D. (1989) Spiral waves of cyclic AMP in a model of slime mold aggregation. *Physica D* **34**, 193–207.

Winfree, A. T. (1974a) Rotating solutions to reaction/diffusion equations. *SIAM/AMS Proc.*(ed. D. Cohen) **8**, 13–31.

Winfree, A. T. (1974b) Rotating chemical reactions. *Scient. Am.* **230**(6), 82–95.

Winfree, A. T. (1978) Stably rotating patterns of reaction and diffusion. *Prog. theor. Chem.* **4**, 1–51.

Winfree, A. T. (1980) *The Geometry of Biological Time.* Springer-Verlag, New York.

Winfree, A. T. (1982) Fibrillation as a consequence of pacemaker phase-resetting. In *Cardiac Rate and Rhythm* (eds L. N. Bouman and H. J. Jongsma), pp. 447–72. M. Nijhoff, The Hague.

Winfree, A. T. (1983) Sudden cardiac death — a problem in topology? *Scient. Am.* **248**(5), 144–61.

Winfree, A. T. (1985) Organizing centers for chemical waves in two and three dimensions. In *Oscillations and Traveling Waves in Chemical Systems* (eds R. Field and M. Burger), Chap.12, pp. 441–72. John Wiley, New York.

Winfree, A. T. (1986) The vulnerable phase and reentrant waves in 3D. *Proc. Int. Union Physiol. Sci.* **16**, 287.

Winfree, A. T. (1987) *When Time Breaks Down.* Princeton University Press, Princeton, NJ.

Winfree, A. T. (1989a) Ventricular reentry in three dimensions. In *Cardiac Electrophysiology, from Cell to Bedside* (eds D. P. Zipes and J. Jalife). W. B. Saunders, San Diego.

Winfree, A. T. (1989b) Electrical instability in cardiac muscle: Phase singularities and rotors. *J. theor. Biol.* **138**(3) (in press).

Winfree, A. T. and Jahnke, W. (1989) Three-dimensional scroll ring dynamics in the Belousov–Zhabotinsky reaction and in the 2-variable Oregonator model. *J. Phys. Chem.* **93**, 2823–2832.

Winfree, A. T., Winfree, E. M. and Seifert, H. (1985) Organizing centers in a cellular excitable medium. *Physica* **17D**, 109–15.

Zykov, V. S. (1984) *Simulation of Wave Processes in Excitable Media* (in Russian), Nauka, Moscow; also in 1988 translation by Manchester University Press (ed A. T. Winfree).

Index

Figures in *italic* type
Tables in **bold** type